A Critical Review of
SPACE NUCLEAR POWER AND PROPULSION
1984–1993

A Critical Review of SPACE NUCLEAR POWER AND PROPULSION 1984–1993

EDITOR

MOHAMED S. EL-GENK

UNIVERSITY OF NEW MEXICO
INSTITUTE FOR SPACE NUCLEAR
POWER STUDIES

American Institute of Physics **New York**

L.C. Catalog Card No. 94-70780
ISBN 1-56396-317-5

Printed in the United States of America.

Table of Contents

Introduction vii

A Tutorial Review of Radioisotope Power Systems 1
Robert G. Lange and Edward F. Mastal, U. S. Department of Energy, Germantown, MD

Summary of Space Nuclear Reactor Power Systems (1983-1992) 21
David Buden, Idaho National Engineering Laboratory, Albuquerque, NM

Silicon-Germanium: An Overview of Recent Developments 87
Cronin B. Vining and Jean-Pierre Fleurial, Jet Propulsion Laboratory/ California Institute of Technology, Pasadena, CA

Review of Thermionic Technology: 1983 to 1992 121
Richard C. Dahlberg, Lester L. Begg, Joe N. Smith, Jr., General Atomics, San Diego, CA; Gary O. Fitzpatrick and Daniel T. Allen, Advanced Energy Technologies, La Jolla, CA; G. Laurie Hatch, John B. McVey, and Ned Rasor, Rasor Associates, Inc., Sunnyvale, CA; and David P. Lieb and Gabor Miskolczy, ThermoTrex Corporation, Waltham, MA

Space Nuclear Power Heat Pipe Technology Issues Revisited 167
Michael A. Merrigan, Los Alamos National Laboratory, Los Alamos, NM

Fuels for Space Nuclear Power and Propulsion: 1983-1993 179
R. Bruce Matthews, Ralph E. Baars, H. Thomas Blair, Darryl P. Butt, Richard E. Mason, Walter A. Stark, Edmund K. Storms, and Terry C. Wallace, Los Alamos National Laboratory, Los Alamos, NM

Prelude to the Future: A Brief History of Nuclear Thermal Propulsion in the United States 221
Gary L. Bennett, NASA Headquarters, Washington, DC; Harold B. Finger, Consultant, Potomac, MD; Thomas J. Miller, NASA Lewis Research Center, Cleveland, OH; William H. Robbins, Analytical Engineering Corp., North Olmsted, OH; and Milton Klein, Consultant, Menlo Park, CA

U. S. Space Nuclear Safety: Past, Present, and Future 269
Joseph A. Sholtis, Jr. and Robert O. Winchester, Air Force Safety Agency, Kirtland AFB, NM; Neil W. Brown, General Electric, San Jose, CA; Andrew Potter, NASA Johnson Space Center, Houston, TX; Leonard W. Connell, Albert C. Marshall, and William H. McCulloch, Sandia National Laboratories, Albuquerque, NM; and James E. Mims, Advanced Sciences, Inc., Albuquerque, NM

Dynamic Power Conversion Systems for Space Nuclear Power 305
James E. Dudenhoefer, James E. Cairelli, Jeffrey G. Schreiber, Wayne A. Wong, Lanny G. Thieme, Roy C. Tew, Steven M. Geng, and Harvey S. Bloomfield, NASA Lewis Research Center, Cleveland, OH; Donald L. Alger and Jeffrey S. Rauch, Sverdrup Technology, Inc., Brook Park, OH; and David M. Overholt, Allied-Signal Aerospace Company, Tempe, AZ

Nuclear Electric Propulsion: Status and Future 381
James H. Gilland and Roger M. Myers, Sverdrup Technology, Inc., Brook Park, OH; James S. Sovey, NASA Lewis Research Center, Cleveland, OH; and John R. Brophy, Jet Propulsion Laboratory, Pasadena, CA

A Review of Advanced Radiator Technologies for Spacecraft Power Systems and Space Thermal Control 407
A. J. Juhasz, NASA Lewis Research Center, Cleveland, OH and
G. P. Peterson, Texas A&M University, College Station, TX

Advanced Static Energy Conversion for Space Nuclear Power Systems 443
C. P. Bankston, Jet Propulsion Laboratory, Pasadena, CA

List of Peer Reviewers 459

Author Index 461

Key Word Index 463

INTRODUCTION

In January 1993, we celebrated the tenth anniversary of the Symposia on Space Nuclear Power and Propulsion. The meeting has grown over the years from being a small specialist meeting to being the largest of its kind in the world. It has exceeded the expectation of everyone involved. It provides a unique, effective, professional, technical forum for the exchange of information between developers and potential users and for the interaction among program managers and technologists from industry, government, and universities. During the Tenth Anniversary Symposium, the number of technical papers presented on the different topics, as well as the number of attendees, achieved a record of more than 200 and 700, respectively. The number of exhibits on display at the conference by various DOE Laboratories, NASA, U.S. Air Force Centers, universities, and industry also reached a record level of 34. The tenth symposium, three-volume proceedings was the largest ever.

The annual symposia have motivated a younger generation to pursue their advanced studies in the different areas of space nuclear power and propulsion. These young professionals are benefiting from the technical talent presented by many senior members in the community. Without the presentations, publications, and other interactions of the symposia, a large body of experience and expertise would not have been passed on. The initiative has been further encouraged by the technical community at the annual symposia through the establishment of the Manual Lujan, Jr. Student Paper Award and the General Ernest C. Hardin, Jr. Scholarship Award. The Lujan Award recognizes outstanding contributions by students to the field of space nuclear power and propulsion through papers presented at the annual symposia. The Hardin Scholarship recognizes, annually, a number of outstanding freshmen, undergraduate, and graduate students who are pursuing their studies in the field of nuclear engineering with a space nuclear power option at the University of New Mexico (UNM).

In addition to its unique role in serving the technical community and inspiring college level students, the annual symposia include an effective and successful outreach component in the form of an annual High School Special Session and the Space Design Competition. These events, organized annually by the UNM-ISNPS Education Outreach Advisory Board and coordinated by Ms. Irene El-Genk, Teacher at West Mesa High School in Albuquerque, NM, and Professor David Kauffman, Associate Dean of Engineering at UNM, have attracted 150-200 high school students and teachers from throughout the State of New Mexico. The sponsoring organizations of these events are NASA Lewis Research Center, UNM-ISNPS, American Nuclear Society Trinity Section, and New Mexico Space Grant Consortium. The students, selected for the best space design project, and their sponsoring teachers are recognized annually at the symposium luncheon and presented with monetary awards and certificates of recognition.

To recognize distinguished contributions and excellence by those actively engaged in the field of space nuclear power and propulsion, the Schreiber-Spence Space Achievement Award was established. The nomination to the award is made by members of the community and the selection for this prestigious award is made by a special committee. The award honors Raemer E. Schreiber and Roderick W. Spence for their pioneering and technical contributions to concepts and designs for nuclear propulsion in space while both were members of the Los Alamos National Laboratory team. At the tenth anniversary conference this year, another prestigious award was approved to recognize outstanding paper(s) at the conference annually.

The annual symposia have successfully archived past and current technical development in the fields of space nuclear power and propulsion in books, proceedings, transactions, and this anniversary book. A demanding task that could not have been accomplished without the dedication and the effort of many individuals from the community, too many to mention by name, who served on the Technical and Advisory Committees over the years, those who contributed many hours of their busy schedule to review technical articles, and the administrative staff at the University of New Mexico's Institute for Space Nuclear Power Studies (UNM-ISNPS). Special thanks to Dr. Mark D. Hoover, who developed the style guide for publications and served as the Co-Editor of the Space Nuclear Power Systems Book Series and the Symposium Proceedings during his tenure as the symposia publication chairman (1984-1993). During the last ten years more than 2000 articles have been published in eleven volumes of the Space Nuclear Power and Propulsion book series, published by Orbit Book Company in Malabar, Florida, in nine volumes of the Proceedings, published by the American Institute of Physics, and six Transactions volumes, published by UNM-ISNPS.

On the national level, the space symposia have witnessed the conclusion of many successful missions to Jupiter, Mars, and Saturn, which have been enabled by Radioisotope Thermoelectric Generators (RTGs). Technological advancements in the SP-100 Program occurred in many areas including the fabrication and successful testing of UN fuel at Los Alamos National Laboratory, fabrication and manufacturing techniques of refractory structures, the development of advanced thermoelectric materials at Jet Propulsion Laboratory, development of fabrication techniques for multi-couple units, with significant increase in power density, development and testing of dual-loop electromagnetic pumps at General Electric (GE), successful demonstration of Stirling engines and advanced light weight heat pipe technology at NASA Lewis Research Center, and many others.

On the international level, the annual symposia have attracted contributions and attendees from many countries including the countries of the Commonwealth of Independent States (CIS), formerly the USSR, France, Germany, Italy, Japan, Norway, Netherlands, Spain, Republic of China, and the United Kingdom. The Sixth Symposium witnessed the first encounter in more than twenty-five years between the scientists of the former Soviet Union in the fields of space nuclear power and propulsion and those of the United States and other countries. At the Eighth Symposium, the TOPAZ-II system, representing the most advanced Russian technology of space nuclear power, was on display. The following year, this international interaction at the symposium resulted in the establishment of the Thermionic System Evaluation Test Facility in Albuquerque as a joint program between the U.S. and Russia. Since then, many joint ventures and agreements have been concluded between industry in the U.S. and various organizations of the CIS Republics for joint development of applications of advanced technologies in the fields of high temperature materials, electrical propulsion, and space nuclear power and propulsion. With the current participation of France and the United Kingdom, the TSET program has become a model for successful international programs in space nuclear power systems technology.

Since January of 1993, when the community celebrated the tenth anniversary of the Annual Space Nuclear Power and Propulsion Symposia, many events and developments have happened that impacted the technical community. The SP-100 program was terminated, the 40-kWe thermionic program is in question, the budget for the TSET program has been reduced, the Air Force Space Nuclear Thermal Propulsion Program, although not officially terminated, has been inactive, and NASA activities in space nuclear power and propulsion have been stopped.

On the optimistic side, the support for the RTG program has continued, yet with some changes in direction and focus in response to the changing need of the user organization, NASA, for future interplanetary missions. The TSET program continues with a wider international participation, new programs are being explored in the area of bi-modal space nuclear power and propulsion systems which could lead to the birth of a new national program, and a number of industry initiatives to form a consortia that would capitalize on the current developments in the fields of space nuclear power and propulsion technologies are being explored. Also, the French Government recently announced placing space nuclear propulsion on its top priority list for funding in the coming years.

In the midst of all these uncertainties and mix of optimism and pessimism, many prominent members of the community have accepted the invitation to undertake the most demanding task of writing review papers summarizing key technological advancements that occurred during the last ten years in their respective areas of expertise. It took more than eighteen months of continuous planning, writing, peer reviewing, revising, and reminding ourselves, the authors, and reviewers of the deadlines. And the publisher, American Institute of Physics, reminding us of the deadline for production in order to have the anniversary book available to the attendees of the eleventh symposium, to be held January 9-13, 1994.

This book is by no means a complete coverage of the many technical areas in the field because some authors had difficulties after January, 1993, committing the time and the resources to complete this difficult and demanding task. Nonetheless, the coverage of the many topics in the book will prove to be an invaluable reference source in the field for many years to come. I wish to thank the authors who have worked against the odds to complete their review and incorporated the comments of many reviewers.

In the preparation of this book for publication, I greatly appreciate the support provided to the authors and the reviewers by their organizations during this time of uncertainty and ever limited resources. Reliable scientific literature cannot be produced without a dedicated effort by everyone involved. A listing of the authors and of the reviewers and their respective organizations are included in the book. I am also grateful for the effort and input of many individuals, who are not mentioned by name, but who contributed to the successful compiling of this book. I am especially grateful for the help provided by Ms. Maureen Alaburda and Ms. Mary Bragg of UNM-ISNPS in coordinating the peer review, editing and, most of all, in the demanding task of reminding everyone involved of the deadlines.

On behalf of the space nuclear power and propulsion community, I wish to thank the many sponsoring and participating organizations from government, industry, national laboratories, NASA centers, the universities, and many countries for their input and contributions over the years. Without the involvement of these organizations and their employees, the annual symposia would not have been a reality nor would it have achieved its goals of serving the space nuclear power and propulsion community.

My final and heartfelt thanks go to my wife Irene El-Genk and my son Hussein El-Genk for their understanding, patience, and continued encouragement and support through the demanding task of completing this book.

Mohamed S. El-Genk
Editor and Symposia Technical Chairman

A TUTORIAL REVIEW OF RADIOISOTOPE POWER SYSTEMS

Robert G. Lange and Edward F. Mastal
U.S. Department of Energy
19901 Germantown Road
Germantown, Maryland 20874
(301) 903-4362

INTRODUCTION

Radioisotope power systems are nuclear power systems which derive their energy from the spontaneous decay of radionuclides as distinguished from nuclear fission energy created in reactor power systems. Radioisotope power systems have two major components: an isotope heat source and an energy conversion system. Heat is produced during the decay process within the radioisotope fuel and its encapsulation (the isotope heat source). This heat is partially transformed into electricity in the energy conversion part of the power system (the converter). The unconverted heat must be removed from the power system at all times, but can be used to heat payload components.

Radioisotope power systems have demonstrated numerous advantages (and some unique capabilities) over other types of power supplies at power levels up to several kilowatts (electric) for long lived, unattended applications in outer space and in remote terrestrial locations. Many especially challenging power applications can be satisfied by radioisotope power systems by proper selection, design, and integration of the isotope heat source and the power conversion technologies which are now available or which can be developed. Such power systems are rugged, compact, highly reliable, relatively independent of the operating environment and can be safely produced and used with minimal risk to operating personnel, the general public, and the Earth's environment.

Although many radioisotope power systems have been successfully demonstrated and used in terrestrial applications (such as in arctic weather stations, light buoys, light houses, floating weather buoys, and mountain-top communication relays) over the past thirty years, the technology got its start and has received its strongest support in the United States for its unique capabilities as a power system for various space missions.

Since before the first Sputnik was launched into space, the Department of Energy (and its predecessor agencies) has been developing the technology to fabricate and deliver radioisotope power systems for use on U.S. military and civilian space missions. Radioisotope Thermoelectric Generators (RTGs) have been used on several major National space programs. The Apollo astronauts placed five RTGs on the surface of the Moon to power the Lunar Science Experiment Packages they left there. The Pioneer 10 and 11 spacecraft which flew by the outer planets before leaving the solar system are powered by RTGs. RTGs also provided the power for the two Viking Landers on Mars. The Voyager spacecraft that sent back the first close-up pictures of the planet Neptune during a tour of the outer planets is powered by RTGs. The Galileo and Ulysses spacecraft on their way to orbit Jupiter and explore the polar regions of the Sun, respectively, are also powered by RTGs. None of these space missions would have been possible without RTGs. Because of RTGs, the United States has established its position as the world leader in outer planetary and space science exploration.

RADIOISOTOPE FUELS

Selection of a suitable radioisotope fuel for use in space radioisotope power systems is the key to their acceptance and use. The radioisotope characteristics of an acceptable fuel include: a long half-life (that is, the time it takes for one-half of the original amount of fuel to decay) compared to the operational mission lifetime; low radiation emissions; a high power density and high specific power; and a stable fuel form with a high melting point suitable for the application. The fuel must be producible in useful quantities and at a reasonable cost (compared to its benefits). It must be capable of being produced and used safely, including in the event of potential launch accidents.

Radioisotope decay is a totally predictable and unalterable process involving the emission of particles and/or photons, including alpha, beta, neutron, and gamma radiation. Energy absorbed from these emissions by the fuel itself or its containment is transformed into useful heat. Any penetrating radiation which escapes the heat source becomes a direct radiation source of concern to nearby personnel and equipment. Alpha particles can produce energetic neutrons from alpha-neutron reactions with light elements in the fuel. Beta particles produce bremsstrahlung (or x-rays) when they are being slowed down. Heavy elements can also decay by spontaneous fission releasing energetic fission neutrons and gammas as well as fission fragments. The characteristics of the decay products from the radioisotope fuel must also be considered.

For radioisotopes which deposit similar amounts of heat energy per disintegration, the size and weight of the heat source are directly proportional to the half-life of the fuel. If the half-life is too long, the radioactive decay rate is slower and the amount of heat produced per unit time is low. This results in a fuel loading which is too large and too heavy for space missions. If the fuel has a shorter half-life, it decays at a faster rate (producing a higher heat output), but excess fuel must be added initially to produce the amount of heat required at the End-of-Mission (EOM). Therefore, the half-life of the radioisotope fuel should be at least as long, or longer than the mission lifetime to reduce the heat variation over the mission and to provide contingency for mission scheduling flexibility. However, for projected National Aeronautics and Space Administration (NASA) missions with operating lifetimes up to 25 years, a radioisotope fuel with a half-life of over 100 years is not acceptable, and would adversely impact the power system performance.

For shorter-lived fuels (such as for use in short duration, man-tended space missions), excess fuel inventory must be loaded into the system at Beginning-of-Mission (BOM) in order to compensate for the more rapid decay of the fuel. For most energy conversion systems, this excess heat must be dissipated directly from the heat source (by-passing the conversion system) to stay within design limits over the mission lifetime. This is called power flattening, and technology has been developed and tested which permits loading twice as much fuel at BOM as is required at EOM and provides a relatively constant heat input to the converter over the span of one half-life. The added complexity and weight of the power flattening equipment (as well as the added weight of the excess fuel and its containment) tends to offset the advantage offered by the higher power density of the shorter-lived fuel. If a mission is extended beyond its design life, or if the launch is unduly delayed, the smaller the ratio of fuel half-life to mission lifetime, the faster the electrical power output will decrease after the system has operated beyond its design life.

The levels of penetrating radiation (gamma, x-ray, neutron) emissions must be inherently low for any radioisotope fuel used in space applications. Some of the reasons for this include:

- o Weight is a critical consideration in space missions. The small spacecraft systems which use radioisotope power systems do not permit the use of heavy external shields such as those used with large reactor powered systems.

- o The small size of the fuel pellets and the thin encapsulation materials used in lightweight heat sources for space offer very little self-shielding.

- o Radiation exposures of operating personnel must be minimized during production, fabrication, and testing of the fuel, heat sources, and power systems prior to launch.

- o In the event of a launch accident, radioisotope heat sources are designed to return to the Earth intact; therefore, the probability of exposing the general populace to direct radiation from a heat source must be minimized.

- o The types of scientific experiments flown on missions to the unexplored regions of space include very sensitive particle and photon detectors, which are incompatible with radioisotope fueled power systems emitting higher levels of penetrating radiation.

All heat-to-electrical power conversion systems are heat engines in which their performance depends on operating the heat sources as hot as possible. Therefore, the radioisotope fuel must have a fuel form (for example, compound, alloy, or matrix) which has a high melting point suitable for use at high operating temperatures and which remains stable during postulated launch pad accidents (for example, fires) and accidental reentry heating in the Earth's atmosphere. The fuel form must be chemically compatible with its containment material (usually metallic cladding) over the operating life of the heat source. It is highly desirable that the fuel form have a low solubility rate in the human body and in the natural environment.

Daughter products must not adversely affect the integrity of the fuel form and the decay process should not degrade its properties. For example, helium gas build-up from the decay of an alpha emitting fuel must not create inhalable fines within the fuel or fuel cavity which might then be released during an accident. Also, the decay process must not destroy the chemical bonds within the fuel form.

The power density (watts/cubic centimeter) and specific power (watts/gram) of a radioisotope fuel form are directly proportional to the energy absorbed per disintegration and are inversely proportional to its half-life. Higher power density leads to smaller heat sources and higher specific power leads to lighter heat sources for comparable power levels. Both characteristics are highly significant for space power system fuels. In general, alpha decay energies are higher than beta decay energies, making the fuel inventory (in curries) smaller for an alpha emitting heat source. For radioisotope fuels with comparable half-lives, a beta emitting heat source will be larger and heavier than an alpha emitting heat source. However, management of the helium generated by an alpha emitting fuel must be considered. If the heat source is designed to contain the helium, then sufficient void volume and/or capsule wall thickness must be provided to contain the maximum pressure build-up over the life of the heat source. For space applications, thin-walled, vented capsules with minimal void volumes are used with alpha emitting fuels to minimize weight.

Very few radioisotopes can be produced without some fraction being composed of relatively stable isotopes of the same chemical element. These dilute the effective power density and specific power of the pure isotopic fuel and cannot be removed by chemical processing. Daughter products build up in the fuel as the fuel ages, but these can be removed by chemical separation. Since these daughter products also dilute the fuel's properties, it is desirable to chemically process the fuel as late as possible before it is to be used to maximize the power production of the fuel form.

Any radioisotope fuel selected for space power applications must be producible in sufficient quantities and on a schedule to meet mission requirements. Radioisotope heat sources require relatively large quantities of isotope material compared to other applications, such as radiation, photoluminescent, and medical sources. There are only two sources of radioisotopes in the quantities needed for thermal and electrical power applications:

1) Fission products and other by-products formed during the operation of a nuclear reactor and which must be recovered by reprocessing the spent reactor fuels; and,

2) Deliberate production of radioisotopes by irradiation of target materials in a nuclear reactor or a very high-powered accelerator facility.

Both require major investments in nuclear facilities capable of processing highly radioactive spent reactor fuel or irradiated targets. Appropriate target materials, designs, fabrication processes, and facilities are required to handle radioactive or non-radioactive targets. Sufficient reactor irradiation space at the appropriate neutron flux and energy levels is required to irradiate the targets. Chemical processing technology to produce the proper fuel compound with the necessary purity must be available. Fabrication processes and facilities to produce and encapsulate the final fuel form must also be provided. These radioisotope fuel facilities must be operated under the strictest safety and environmental standards and take into account the ultimate disposal of any radioactive wastes generated.

When it is determined that a proposed fuel form can be produced in adequate quantities and with the desired nuclear characteristics, the production quality fuel form must be extensively tested to qualify it for space use and to support the necessary launch safety reviews and approvals. Tests are conducted to determine the long-term compatibility between the fuel form and selected containment materials. Solubility rates of the fuel form in fresh water and sea water are also measured over extended periods of time. Physical properties of the fabricated fuel form (both fresh and aged) are determined. Helium retention characteristics are determined for alpha emitting fuel forms. Finally, impact response of the fuel form (in its proposed containment) is determined by actual tests at velocities somewhat higher than terminal velocities against unyielding targets such as steel, concrete, and granite as well as less demanding targets such as sand, soil, and water.

Development of a fuel production and fuel form fabrication capability for a new radioisotope fuel is very costly and time consuming. To qualify a new fuel form (and its containment) for flight use requires a large effort in terms of costs and schedule. In addition, there are only a limited number of radioisotope fuels which meet the requirements of half-life, radiation, power density, fuel form, and availability for use in space power system applications.

HISTORICAL FUELS PERSPECTIVE

Development of radioisotope power systems began in the early 1950s and since then a variety of radioisotopes have been evaluated for space and terrestrial applications. The isotope initially selected for development was Cerium-144 (Ce-144) because it was one of the most plentiful fission products available from reprocessing defense reactor fuel at the Hanford Site. Its short half-life (290 days) made Ce-144 compatible with the 6-month military reconnaissance satellite mission envisioned as the radioisotope power system application at that time. The cerium oxide fuel form and its heavy fuel capsule met all safety tests for intact containment of the fuel during potential launch abort fires, explosions, and terminal impacts. However, the high radiation field associated with the beta/gamma emission of Ce-144 caused handling and payload interaction problems as well as safety problems upon random reentry from orbit. Therefore, the Ce-144 fueled SNAP-1 (SNAP stands for Systems for Nuclear Auxiliary Power) power system was never used in space.

By the late 1950s, large amounts of Polonium-210 (Po-210) became available, also as a by-product of the nuclear weapons program. Po-210 is an alpha emitter with a very high power density (~1320 watts/cc) and low radiation emissions. It is made by neutron irradiation of Bismuth-209 targets in a nuclear reactor. It was used in Polonium-Beryllium neutron sources. Po-210 metal was used to fuel the small (5 watt(e)) SNAP-3 Radioisotope Thermoelectric Generator (RTG) to demonstrate the RTG technology and was first displayed at the White House in January 1959. Several SNAP-3 RTGs were fueled with Po-210 and used in various exhibits. However, the short half-life of Po-210 (138 days) makes it suitable for only limited duration space power applications.

In order to provide a longer-lived radioisotope fuel, Strontium-90 (Sr-90), an abundant fission product with a 28.6 year half-life, was recovered from defense wastes at Hanford. A very stable and insoluble fuel form, strontium titanate, was developed and widely used in terrestrial power systems. Sr-90 and its daughter Yttrium-90 emit high energy beta particles so they give off significant bremsstrahlung radiation which requires heavy shielding; but shield weights are not as critical in most terrestrial power systems as for space power systems.

By 1960, Plutonium-238 (Pu-238) had been identified as an attractive radioisotope fuel which could be made by irradiating Neptunium-237 (Np-237) targets in the defense production reactors. Its availability was extremely limited due to shortage of Np-237 target material which must be recovered from processing (and recycling) high burn-up, enriched uranium fuel. However, Pu-238 has all the necessary nuclear characteristics for a space power system fuel: long half-life (87.74 years), low radiation emissions, high power density and useful fuel forms (as the metal or the oxide). Therefore, after flight qualification of its heat source, a Pu-238 fueled SNAP-3 RTG was launched on the Transit 4A Navy navigation satellite in June 1961 - the first use of nuclear power in space.

The first heat sources used in space were relatively small and employed Pu-238 metal or plutonium-zirconium alloy fuel forms contained in tantalum lined superalloy (Haynes-25) fuel capsules. These heat sources withstood all postulated launch pad accident and downrange impact environments, but they were designed to burn-up and disperse throughout the upper atmosphere in the event they reentered from space. This type of accident happened during the fifth launch of an RTG (a SNAP-9A RTG) when the spacecraft failed to achieve orbit and the RTG burned up over the Indian Ocean in April 1964.

Subsequent Pu-238 fueled space power systems were designed to use progressively higher temperature fuel forms and containment materials with a progressively higher degree of containment of the fuel under all postulated accident conditions (including reentry). As the intact reentry heat source technology was developed, the fuel inventories (power levels) per launch also increased. A number of RTGs were launched on NASA and Navy missions with Pu-238 dioxide microsphere and plutonia-molybdenum-cermet (PMC) fuel forms in the late 1960s and early 1970s. Since the mid-1970s, pressed Pu-238 oxide fuel forms have been exclusively used in all radioisotope power systems launched into space.

The amount of Pu-238 which could be produced has always been a limiting factor in its use in space missions. Therefore, several other radioisotopes have been thoroughly evaluated for space use over the years. Sr-90 and Po-210 fuels were considered for use in higher powered military satellite constellations for which there were insufficient quantities of Pu-238 available. These programs were cancelled before they were completed, so these fuels were never used in space by the United States.

Curium-242 (Cm-242) was selected to fuel an isotope power system for the 90-day Surveyor mission to the Moon. Both the SNAP-11 RTG and the SNAP-13 thermionic generator were developed for the Surveyor mission. Cm-242 is produced by reactor irradiation of Americium-241 (Am-241) targets. Cm-242 has a short half-life of 162 days (which is acceptable for a 90 day mission) and a very high power density (which is necessary for a thermionic heat source). It also has a high melting point oxide fuel form capable of the high operating temperature necessary for thermionic energy conversion. A Cm-242 demonstration heat source was produced for the SNAP-13 engineering unit, but it was decided that the Surveyor program would not use isotope power units and Cm-242 fueled power systems have never been used in space. Due to its short half-life, Cm-242 is not suitable for use in longer duration missions required by most space programs.

However, short-lived radioisotope fuels can and should be considered for use in any program involving recurring human visits to the Moon. Besides Po-210 and Cm-242, Thulium-170 (Tm-170) with a half-life of 129 days can be considered for such missions where astronauts can replace the fuel periodically. Tm-170 is a low energy beta emitter which will require more shielding during handling, but it is unique in that it requires no chemical processing of radioactive materials. Natural non-radioactive Tm-169 metal or oxide targets can be irradiated in a reactor, loaded directly into a heat source and delivered to the Moon on a short schedule without ever handling unencapsulated radioactive fuel.

Curium-244 (Cm-244) was expected to become available in significant quantities from the U.S. breeder reactor fuel cycle and was investigated as a potential alternative long-lived fuel to Pu-238. However, this source has failed to materialize. Cm-244 was considered an attractive space fuel because it has a relatively long half-life (18.2 years), a power density five times greater than that of Pu-238 and has a very stable, high temperature oxide fuel form. However, higher neutron and gamma emissions due to the higher rate of spontaneous fission of Cm-244 would increase shielding requirements for handling and for protection of spacecraft instrumentation. The increased weight of shielding and power flattening equipment required with Cm-244 makes it less desirable than Pu-238, especially for long duration missions to outer space. Cm-244 is also more difficult to produce requiring successive neutron captures starting with Pu-239. Many years ago, several kilograms of Cm-244 were made as a target material for the Californium-252 program, but there is currently no practical production or processing capability for large quantities of Cm-244. Cm-244 has not been qualified for use in space.

In the final analysis, Pu-238 is clearly superior to other radioisotope fuels for use in long duration space missions, especially for deep space exploration. The technology for producing and processing Pu-238 fuel forms has been clearly demonstrated over the past thirty years. Pu-238 fueled heat sources have been through rigorous flight qualification testing and have performed safely and reliably in all of the radioisotope power systems employed in the U.S. space program to date.

The availability of Pu-238 fuel for future space missions is a continuing concern. For the past thirty years the production and processing of Pu-238 fuel has been accomplished as a by-product of the production of materials for nuclear weapons. Recent changes in the nation's nuclear weapons program will eliminate this traditional capability to produce Pu-238. Therefore, for Pu-238 fuel to continue to be available for use in the space program, a reliable source or sources must be established. Alternative sources of Pu-238 are currently being investigated by the Department of Energy including near-term purchases from Russia.

FUEL ENCAPSULATION

The various layers of material surrounding the radioisotope fuel form make up the rest of the isotope heat source and are called the fuel encapsulation. Each layer serves one or more important function(s) in the safe handling and use of radioisotope heat sources. The various encapsulation components include: a liner, strength member, oxidation cladding, helium vent, and reentry aeroshell.

The liner is a thin metallic capsule that is next to the fuel form and is chemically compatible with the fuel form at operating temperatures over the life of the heat source. The liner also serves as a decontamination container to prevent radioactive material from contaminating successive layers of encapsulation during assembly of the heat source. In the small burn-up design heat sources, a thin tantalum liner was used as the chemical containment for Pu-238 metal or plutonium-zirconium alloy fuel forms. In unvented Pu-238 fueled heat sources, an excess void volume equivalent to the volume of the fuel was provided to accommodate helium gas build-up over the life of the heat source.

The strength member provides the physical containment of the fuel under all operating and accident environments. It has the wall strength and toughness to withstand high velocity impacts, fragment impacts, explosion overpressure, and fire environments. It also contains the internal pressure generated by helium build-up in unvented fuel capsules. For smaller, lower-temperature heat sources designed to burn-up on reentry, superalloy strength members can be used. For larger, higher-temperature intact reentry heat sources, refractory strength members are required. In the latter case, a thin noble metal oxidation resistant cladding is required to protect the refractory metal strength member while it is in the Earth's atmosphere.

To minimize the size and weight of a heat source using an alpha emitting fuel, vents are provided that allow the helium gas to escape from the heat source without excessive pressure build-up. The vents are designed to permit the helium gas molecules to pass without releasing fuel particles or significant amounts of fuel vapor. Several different fuel capsule vents have been developed and successfully used in Pu-238 fueled heat sources. Ceramic vents have been used which selectively pass only the smaller molecules of helium and retain fuel particles and large fuel vapor molecules. "Pigtail" vents use a long coil of fine capillary tubing through which helium flows to limit the internal pressure. Due to the length of the capillary, any fuel vapor which might be entrained in the helium will condense within the tubing and back diffusion of oxygen (in an air environment) through the vent will be minimized. Another type of vent is the "frit" vent in which a tortuous path is provided within a metallic frit material which will only pass gaseous materials. All vents must be tested rigorously to make sure that they do not become clogged with condensed fuel or fuel contaminants during operation.

The reentry aeroshell is only used with a heat source designed for intact reentry through the atmosphere. The aeroshell must be made of a good ablative material, such as graphite, with a low sublimation/ablation rate and which can withstand the aerodynamic forces and high thermal stresses experienced during severe reentry heating pulses without coming apart. It is desirable for the aeroshell to be able to conduct the heat from the fuel during normal operation, but to prevent the metallic fuel capsule components from melting during reentry. Additional insulators can be used to design an optimum aeroshell to meet these conflicting requirements. The size and shape of the aeroshell is selected to minimize aerodynamic heating rates and to minimize terminal impact velocities.

As spacecraft power requirements and mission lifetimes have increased, radioisotope fuel inventories and safety concerns have also increased making it more difficult to design radioisotope heat sources within the increasingly important weight constraints. Fortunately, the early missions provided the experience and the time to develop new encapsulation materials for use in radioisotope heat sources which have kept up with mission requirements.

Noble metal alloys, such as iridium-tungsten and platinum-rhodium-tungsten, have been developed for use as the primary containment member. It also serves as a liner which is compatible with the fuel at very high temperatures and a partial strength member which will absorb physical loads and provide long-term containment of the fuel in an oxidizing atmosphere. To minimize the mass of the dense noble metal fuel clad, a frit vent is incorporated and a graphite impact shell made of 3-D composite material is used to absorb some of the impact energy to protect the fuel clad. A graphite aeroshell made of similar 3-D construction serves as the third level of physical containment as well as an excellent reentry body. Pyrolytic graphite coatings on the impact shell and a low density graphite insulation are included to protect the fuel clad from overheating during reentry and overcooling during the subsequent impact with the Earth's surface. (The iridium alloy exhibits brittle failure characteristics at high strain rates at relatively high temperatures.)

Improved composite graphite aeroshell and impact shell materials are under development to lower the fabrication cost and to enhance the safety performance of the heat source components (such as by designing an impact shell material which will absorb more energy on impact).

ENERGY CONVERSION SYSTEMS

The radioisotope heat source delivers its heat to some type of energy conversion system which converts part of the heat into useful electrical power. The heat source can be thermally coupled to the energy conversion system in any of several different ways as long as the required interface temperatures and heat fluxes can be met while at the same time proper considerations are given to minimizing thermal losses and power system weight.

Thermal coupling can be accomplished by conduction (such as by having thermoelectric hot shoes spring-loaded against the heat source), direct radiation, convection (for example, by use of a pumped fluid loop or heat pipes), or a combination of these.

Safety is a primary consideration during integration of the radioisotope heat source into the power system. The surrounding structure should help protect the heat source against explosion fragments and overpressure during launch pad aborts, but should readily disassemble upon atmospheric reentry to free the heat source aeroshells to reenter on their own. The power system must also be designed to reject the heat from the radioisotope heat source under all operating and failure conditions without exceeding temperature limits within the heat source.

After passing through the energy conversion system, the unconverted (or waste) heat must be rejected to the environment at much lower temperatures. For space power systems some of the waste heat can be utilized to control the temperature of the spacecraft equipment, but ultimately the waste heat must be dumped to the space vacuum environment by radiation.

Thus, the operating temperature boundary conditions for a radioisotope space power system are set on the hot side by heat source and conversion system material limitations (T_{HOT}) and on the cold side by the size, weight, and heat sink conditions of the radiator (T_{COLD}). The overall efficiency of the energy conversion system is limited to something less than the Carnot (or ideal) efficiency of $T_{HOT} - T_{COLD}/T_{HOT}$. Efficiency is an important consideration in selecting an energy conversion system because of its effect on the radioisotope inventory and its implications on cost, availability, size, weight, and safety.

Conversion system reliability is also important. Since mission success depends on having sufficient electrical power over the life of the mission, the selection of an energy conversion system must be consistent with mission power levels and lifetimes. Besides, it makes little sense to combine an unreliable or short-lived energy conversion unit with a 100% reliable, long-lived isotope heat source. Graceful power degradation over the life of a mission is acceptable as long as it is within predictable limits.

Other characteristics important in selecting an energy conversion unit for a radioisotope space power system are weight, size, ruggedness to withstand shock and vibration loads, survivability in hostile particle and radiation environments, scalability in power levels, flexibility in integration with various types of spacecraft (and launch vehicles), and versatility to operate in the vacuum of deep space or on planetary surfaces with or without solar energy inputs.

There are two general classes of energy conversion systems: static and dynamic. Static systems include thermoelectric, thermionic and thermophotovoltaic conversion devices which can convert heat to electricity directly with no moving parts. Dynamic systems involve heat engines with working fluids that transform heat to mechanical energy which in turn is used to generate electricity. Both rotating and reciprocating engines are used. Dynamic systems include Rankine, Brayton, and Stirling engines that operate on various types of working fluids.

RADIOISOTOPE THERMOELECTRIC GENERATORS (RTGs)

Of the various static energy conversion technologies, thermoelectric energy conversion has received the most interest, both in development and use, for radioisotope power systems. Thermoelectric converters are useful over a very wide range of power levels (from milliwatts to kilowatts) and their operating temperatures are ideally suited for radioisotope heat sources. Thermoelectric converters are highly reliable over extended operating lifetimes (tens of years), compact, rugged, radiation resistant, easily adapted to the application, and produce no noise, vibration or torque during operation. Thermoelectric converters require no start-up devices to operate. They start producing electrical power (direct current and voltage) as soon as the heat source is installed. Power output is easily regulated at design level by maintaining a matched resistive load on the converter. A limiting feature of thermoelectric conversion (especially for radioisotope power applications) is its relatively low conversion efficiency, typically less than 10%.

Thermoelectric materials, when operating over a temperature gradient, produce a voltage called the Seebeck voltage. When connected in series with a load, the internally generated voltage causes an electron current to flow through the load producing useful power. The Seebeck principle was discovered in 1825, but had little practical use, except in measuring temperatures with dissimilar metal thermocouples, until semiconductor materials became available in the 1950s. Good thermoelectric semiconductor materials have large Seebeck voltages in combination with a relatively low electrical resistance and a low thermal conductivity (in contrast to most metals). Power is produced in a thermoelectric element placed between a heat source and a heat sink. By proper doping, n and p type elements can be formed so that current will flow in the same or opposite directions as the heat. By electrically joining the n and p elements through a hot shoe, a thermocouple is formed which can be connected to other thermocouples at the cold shoe to form a converter with the desired output voltage and current. Thermocouples can be connected in a series-parallel arrangement to enhance reliability by minimizing the effect on total power due to an open circuit or short circuit failure in a single thermocouple. Typically, thermoelectric couples are low voltage, high current devices so a number of them must be connected in series to produce normal load voltages.

Thermoelectric converters can be operated in a shorted condition without any permanent damage. The short circuit current increases the Peltier cooling effect at the hot junction, so the thermoelectric elements run cooler than normal. They can also be operated under open circuit conditions (zero current flow) as long as the reduced Peltier cooling does not overheat the thermoelectric materials. RTGs are routinely stored and shipped while shorted, but are usually only momentarily subjected to open circuit conditions during operation.

A significant advantage of thermoelectric conversion systems is that they can be tested as individual couples or in small groups of couples called thermoelectric modules which are prototypical of flight systems. Long term performance testing of thermoelectric modules can be done much easier and at lower costs compared to testing complete systems.

The most widely used thermoelectric materials, in order of increasing temperature capability, are: Bismuth Telluride (BiTe); Lead Telluride (PbTe); Tellurides of Antimony, Germanium and Silver (TAGS); Lead Tin Telluride (PbSnTe); and, Silicon Germanium (SiGe). All of these (except BiTe) have been used in RTGs which have been flown on space missions. Many more materials have been, and are still being, investigated in hopes of finding that ideal thermoelectric material from which to produce higher efficiency, lower weight power systems with more stable performance over longer operating lifetimes.

The telluride materials are limited to a maximum hot junction temperature of 550°C. Due to the deleterious effects of oxygen on these materials and their high vapor pressures, the tellurides must be operated in a sealed generator with an inert cover gas to retard sublimation and vapor phase transport within the converter. Bulk-type, fibrous thermal insulation must be used due to the presence of the cover gas. Helium gas build-up within the converter (from vented alpha emitting heat sources) must be controlled by using a separate container around the heat source or permeable seals in the generator design. Gas management considerations in the generator housing design and the use of bulk insulation materials increase the size and weight of the generator. However, this type of RTG is equally useful for space vacuum or for planetary atmospheric applications.

SiGe materials can be operated at hot junction temperatures up to 1000°C. Their sublimation rates and oxidation effects, even at these higher temperatures, can be controlled by use of sublimation barriers around the elements and an inert cover gas within the generator during ground operation. A pressure release device, designed to open upon reaching orbital altitude, opens the generator to space vacuum for operation on deep space missions. This allows the use of multifoil thermal insulation and also vents the helium to space as it is generated. A SiGe RTG is usually smaller and lighter than is a telluride RTG of similar power level.

The overall efficiency of the two types of thermoelectric generators are comparable. Although the tellurides have a higher material efficiency than SiGe, the SiGe operates over a larger temperature gradient. Cold junction temperatures are determined more by radiator weight than by efficiency considerations for space RTGs and are normally in the range of 200-300°C. Although various convectively cooled (for example, heat pipe) radiator systems have been developed, conductively coupled finned radiators attached to the generator housing are normally more weight efficient for low-powered RTGs (up to 300 We).

DYNAMIC ISOTOPE POWER SYSTEMS (DIPS)

For higher power levels (such as 1000 to 10,000 We), the more efficient dynamic power conversion technologies provide for better use of the limited radioisotope fuel, offer systems with a higher power-to-weight ratio and make it easier to integrate the radioisotope power system with the spacecraft compared to the number of RTGs required to produce kilowatts of power. Dynamic heat-to-electricity conversion efficiencies of 20% (or more) are achievable which reduce the radioisotope inventory to one-third (or less) of that for RTGs. This reduces weight, cost, and potential safety risks for higher-powered radioisotope systems.

A Dynamic Isotope Power System (DIPS) is composed of the following components:

- o Isotope Heat Source
- o Heat Source Heat Exchanger (or Boiler)
- o Emergency Cooling System
- o Power Conversion Unit (Heat Engine)
- o Recuperator
- o Waste Heat Exchanger (or Condenser)
- o Radiator

- o Working Fluid(s)
- o Plumbing System
- o Thermal Insulation
- o Power Control and Conditioning System

The Power Conversion Unit (PCU) is the heart of the system where the thermal energy in the working fluid is partially transformed into mechanical work to drive an alternator to produce electricity and a pump (or compressor) to move the primary working fluid through the system. All of the moving (or dynamic) parts are contained within the PCU, except for the working fluid. (If auxiliary cooling loops are used, they could also include pumps.) The design of the PCU depends on the thermodynamic cycle employed.

Three different thermodynamic cycles have received the most attention for use in radioisotope power systems: Rankine, Brayton and Stirling cycles. The Rankine cycle is based on a two-phase working fluid which requires special vapor-liquid boiler and condenser designs for use in the microgravity of space. Development programs for radioisotope power units have been conducted on Rankine systems using liquid metal (for example, mercury) and organic working fluids (such as Dowtherm A and toluene). Brayton and Stirling cycles use a single-phase gas working fluid such as a helium-xenon mixture for Brayton and pure helium for Stirling. Both Brayton and Stirling engines are currently receiving development support for use with radioisotope heat sources because of their promise of very high conversion efficiencies.

In a Brayton or Rankine system, the gaseous working fluid from the Heat Source Heat Exchanger turns a turbine rotor which is mounted on a common shaft with the alternator rotor and the compressor (or liquid pump) rotor and is called the Combined Rotating Unit (CRU). The remainder of the PCU includes the turbine housing, turbine nozzles, bearings, alternator coils, pump housing, cooling ducts, etc. to form a compact energy conversion unit. To minimize size and weight, the CRU turns at very high speeds and the alternator produces high-frequency alternating current.

To start the unit, electrical power from a battery is used to spin-up the rotor until the system's temperature, pressure, and mass flow rates permit it to be self-sustaining. Start-up, operating speed, and electrical output is controlled by the Power Control and Conditioning System (PCCS). Since temperatures, pressures, mass flow rates, rotor speeds, heat exchanger areas, etc. are all inter-related design parameters for an optimized turbine-driven PCU, power output is sensitive to maintaining the system within its design limits once the hardware design is fixed. Its useful design life with a decaying radioisotope heat source is dictated by these built-in design limits. The Emergency Cooling System can be used to control the heat flow from the heat source to the working fluid over the design life of the system as well as to dump the heat prior to start-up or in case the working fluid is lost.

To achieve their high conversion efficiencies, both Brayton and Rankine cycles can include recuperators (highly effective heat exchangers) to transfer residual sensible heat from the fluid exiting the turbine to the fluid from the compressor (or pump) before entering the Heat Source Heat Exchanger. This reduces the amount of heat added from the Isotope Heat Source as well as the amount of heat that must be rejected through the radiator during each pass of fluid through the system.

There are two general areas of concern with the use of a turbine-driven electrical power system: reliability and spacecraft interactions. The main reliability issues have to do with bearings, loss of working fluid and failure of electronic parts in the control system. Foil bearings are used in the Brayton unit that allow the rotating shaft to ride on a thin film of working fluid gas during operation so that the metallic surfaces only rub during start-up and shut-down. Organic Rankine units can use either foil bearings or hydrodynamic thrust and journal bearings lubricated by the organic working fluid. Both have been shown to be highly stable and reliable once operating conditions are achieved and bearing temperatures are controlled. Since a puncture by a micrometeoroid or a crack caused by thermal or mechanical stresses during operation could cause the loss of the working fluid and total loss of power, this single-point failure mechanism must receive particular attention during the design. Conservative wall thicknesses, expansion joints, and meteoroid armor are usually employed to reduce the probability of failures of this kind. Totally redundant power conversion loops can also be used if the additional weight can be tolerated. The power conditioning and control system can be designed using standard approaches of redundancy and de-rating of electronic components to increase its reliability. However, this increases the parasitic power losses of the system and makes the DIPS less attractive at lower power levels.

The Organic Rankine system has an additional reliability concern due to thermal degradation of the organic fluids at elevated temperatures, that limits its hot side temperatures and its peak operating efficiency. The Brayton system temperatures are limited by the heat source and the hot side loop materials. Refractory metals can only be used if they can be protected from oxidation while in the Earth's atmosphere. Otherwise, DIPS Brayton cycle performance is limited by operating temperatures compatible with superalloy materials.

The high speed rotor produces an unbalanced torque which must either be reacted by the spacecraft's attitude control system or be negated by adding a compensating reaction wheel to the power system. The torque loads caused by the small turbine-driven power units are usually small enough that they can easily be handled by the larger spacecraft momentum control system.

Another difference between the Organic Rankine and Brayton cycles is in the heat rejection (or radiator) system. The Rankine working fluid changes phase (that is, condenses from a vapor to a liquid) in the cold side heat exchanger producing a relatively constant temperature profile which permits a more efficient, smaller radiator area. The actual condenser/radiator temperature depends on the vapor pressure properties of the working fluid. The vapor pressure must provide the necessary input pressure to the pump. The gas working fluid of the Brayton cycle requires a larger cold side heat exchanger with a larger temperature drop between the inlet and outlet. This non-uniform temperature produces a larger radiator and more complex designs to couple the cold side heat exchanger to the radiator via low-temperature heat pipes. The use of fluid loops for either system allows flexibility in thermally and mechanically integrating the DIPS radiator with the spacecraft. It is possible to integrate the DIPS radiator so that it can be used as part of the spacecraft structure and can provide a substantial degree of survivability protection for the spacecraft components.

Stirling cycle engines use a light working gas which expands by absorption of heat on the hot side and contracts by rejection of heat on the cold side causing rapidly changing pressure cycles across a piston forcing it to move in a reciprocating fashion. The movement of the piston can drive a linear alternator to produce electricity. Some Stirling engines use a rhombic drive mechanism to convert the reciprocating motion into a rotary motion to drive an ordinary (rotating) alternator. This requires lubrication of the gear box and seals to separate the working gas from the lubricating oil. The engine housing cannot be hermetically sealed because of the penetration of the rotary power shaft. However, such Stirling engines have been widely used in foreign Countries and have been studied for use with radioisotope heat sources in the U.S.. A more recent development is the Free Piston Stirling engine which requires no lubricating fluids and produces electricity by means of a linear alternator within the hermetically sealed engine housing. The piston moves back and forth at a resonant frequency on a cushion of working gas between it and the surrounding cylinder wall. Piston displacement is controlled by gas pressures across the piston. A permanent magnet is attached to the power piston and produces electrical currents in surrounding alternator coils as it vibrates back and forth. Since the reciprocating motion of the piston would cause unbalanced vibration loads, these Stirling engines usually are designed in pairs with dynamically opposed pistons so that no net load is transmitted to the engine mounts.

Heat is also exchanged between the hot and cold gas flowing from one side of the piston to the other to enhance the conversion efficiency. Due to the limited volume of working gas within the Stirling engine, heat transfer between the heat source and the heater head of the engine, between the hot and cold gas, and between the cold gas and a radiator system are the most challenging requirements for an optimum engine design. The Stirling cycle provides higher conversion efficiencies than the Rankine and Brayton cycles at the same cycle temperatures. Therefore, efficiencies of 30% (or more) are possible at operating temperatures achievable with isotope heat sources and oxidation-resistant superalloy structural materials. The Stirling engine also promises to retain its high performance characteristics at lower power levels compared to the Brayton and Rankine systems, which is also attractive for radioisotope power systems.

CURRENT STATE OF TECHNOLOGY

Since 1961, the United States has launched 41 RTGs on 24 spacecraft for various NASA and Department of Defense missions in high and low Earth orbit, on the surfaces of the moon and Mars, and fly-bys to and beyond the outer planets (See Table I). Power levels have varied from 2.7 to 288 We/RTG and one to four RTGs were used per spacecraft. All of these RTGs met or exceeded their predicted power performance. Many are still operating, including the RTGs on the Pioneer 10 spacecraft which are still sending data back to Earth after being in space over 20 years and travelling a distance of over 5 billion miles.

All of these RTGs have been fueled with Pu-238. The fuel form and heat source technologies have been steadily improved over the years to operate at higher temperatures and to meet the stringent aerospace nuclear safety requirements with increasingly larger fuel inventories. As the power levels of the RTGs have increased, improved thermoelectric materials and thermal insulations have been developed to increase long-term power stability and specific power characteristics of the RTGs commensurate with the needs of more ambitious space exploration missions. Concurrently, the Nation's capability to produce and process greater quantities of Pu-238 fuel and larger heat sources as a by-product of the operation of nuclear weapons facilities continued.

Technology advancement programs have also included the demonstration of the more efficient DIPS to expand the use of radioisotope power systems into the kilowatt power range for military satellites and special lunar and planetary surface applications.

GENERAL PURPOSE HEAT SOURCE (GPHS)

Following the launches of the RTG-powered Lincoln Experimental (Communications) Satellites and the Voyager Spacecraft in the late 1970s, employing the 2400 Wt Multi-Hundred Watt (MHW) heat sources, the Department of Energy has developed a new modular heat source called the General Purpose Heat Source (GPHS) to accommodate a range of radioisotope powered space missions.

The GPHS module is shown in Figure 1. Each module is designed to deliver up to 250 Wt at BOM and weighs 1.43 kg (3.16 lbs) for a specific power of 174.4 Wt/kg (79.1 Wt/lbs). This is a significant improvement over previously flown high temperature radioisotope heat sources designed for intact reentry from space. The module size and shape were selected to survive reentry through the atmosphere and impact the Earth at a modest terminal velocity of 50 m/s (164 ft/s). Each module is a rectangular parallelpiped with the overall dimensions of 9.72 cm x 9.32 cm x 5.31 cm (3.83 in x 3.67 in x 2.09 in).

Each GPHS module contains four pressed and sintered Pu-238 oxide fuel pellets of nominal thermal output of 62.5 W each. The cylindrical fuel pellet is 2.75 cm (1.08 in) in diameter and length. Each fuel pellet is individually encapsulated in a welded iridium alloy containment shell or cladding with a minimum wall thickness of 0.05 cm (0.02 in). The iridium alloy is capable of resisting oxidation in the post-impact environment while it also provides chemical compatibility with the fuel and graphite components during high temperature operation and postulated accident conditions. The iridium fuel cladding is equipped with a frit vent that allows release of helium produced by the decay of the Pu-238 without releasing plutonia particles. The combination of fuel pellet and cladding is called a fueled clad.

Two fueled clads are encased in a Graphite Impact Shell (GIS) made of Fine Weave Pierced Fabric (FWPF) carbon-carbon composite material. The GIS is designed to limit the damage to the iridium clads during free-fall or explosion fragment impacts. Two of these GISs are inserted into an aeroshell which is also made of FWPF graphite. A thermal insulation layer of Carbon Bonded Carbon Fiber (CBCF) graphite surrounds each GIS to limit the peak temperature of the iridium cladding during atmospheric reentry heating and to maintain its ductility during the subsequent impact. The aeroshell is designed to contain the two GISs under severe reentry conditions and to provide additional impact protection against hard surfaces at its terminal velocity. It also provides protection for the fueled clads against overpressures and fragment impacts during postulated missile explosion events. The aeroshell serves as the primary structural member to maintain the integrity and position of a stack of GPHS modules within a power system during normal operations, including testing, transportation, and launch.

The GPHS module has undergone an extensive safety analysis and test program over the last decade and is the first isotope heat source to be qualified for and used in a launch on the Space Shuttle (for example, the Galileo and Ulysses missions). It is the current standard heat source module for use in various radioisotope space power systems.

GPHS-RADIOISOTOPE THERMOELECTRIC GENERATOR

The current state-of-the-art in space RTGs is represented by the GPHS-RTG, so named because it employs the GPHS modules. The GPHS-RTG, shown in Figure 2, is the largest Pu-238 fueled, long-lived RTG built for use in space missions. It produces at least 285 We at launch from a Pu-238 heat source assembly containing a stack of 18 GPHS modules producing at least 4,410 Wt. The GPHS-RTG operates at a normal electrical voltage output of 28-30 V-DC. The overall dimensions of the GPHS-RTG are 42.2 cm (16.6 in) diameter (fin tip to fin tip) by 114 cm (44.9 in) long. The GPHS-RTG weighs 55.9 kg (123.3 lb) for a specific power (at launch) of 5.1 We/kg (2.3 We/lb).

The heat source is surrounded by 572 Silicon Germanium (SiGe) thermocouples or unicouples of the type used in the RTGs flown on the LES 8 and 9 and Voyager 1 and 2 spacecraft. The unicouples are individually bolted to and cantilevered from the aluminum alloy generator housing and are surrounded by a thermal insulation package consisting of 60 alternating layers of molybdenum foil and astroquartz fibrous insulation. The molybdenum silicon (MoSi) alloy hot shoes are radiatively coupled to the heat source. A gas management system is included to maintain an inert cover gas within the RTG prior to launch and to vent the generator prior to operation in space.

The unicouples are connected in two series-parallel electric wiring circuits in parallel to enhance reliability and provide the full output voltage. The RTG power output would be affected only by the loss of power from an individual unicouple in the event that unicouple failed in either the open or short circuit mode. The electrical wiring is also arranged to minimize the magnetic field of the RTG.

The SiGe unicouples operate between a hot junction temperature of 1273 K and a cold junction of about 573 K to produce a thermoelectric efficiency of about 9 percent. Waste heat is radiated from the finned RTG housing which is covered with a high emissivity coating.

The GPHS-RTG is equipped with coolant channels through its fin roots so that an auxiliary pumped-liquid cooling system can remove its heat while installed in the shuttle payload bay. The auxiliary cooling system is used prior to, and during, launch until the shuttle payload bay doors are opened on orbit and the spacecraft is deployed or during return of the payload to Earth, if necessary. These coolant channels do not need to be used within the payload fairing of an expendable launch vehicle where conditioned gas streams can be used to cool the RTG(s) prior to launch.

Two GPHS-RTGs were launched on the Galileo spacecraft in October 1989 and another was launched on the Ulysses spacecraft in October 1990. These GPHS-RTGs continue to perform as predicted. Three more GPHS-RTGs are scheduled to be launched on the Cassini spacecraft in October 1997 on a mission to orbit the planet Saturn.

FUTURE DEVELOPMENTS

Over the next decade, the U.S. Radioisotope Power Systems Program faces several programmatic and technological challenges if it is to continue to provide radioisotope power systems for use in NASA space missions that require them. NASA has identified a number of potential missions which can best be done or can only be done using radioisotope power and/or heat sources. There are three principal challenges:

First, the Department of Energy must provide a reliable and continuing supply of Pu-238 to fulfill the requirements of approved space (and terrestrial) applications. Efforts are underway to investigate alternative facilities in the U.S. to provide a continuing Pu-238 production and processing capability as well as the purchase of Pu-238 from foreign sources, such as Russia.

Second, technical developments will be required to design, fabricate and qualify lower powered RTGs (smaller than the present GPHS-RTG) for use in potential lunar and martian surface exploration missions, and perhaps smaller outer planetary spacecraft missions, such as a fly-by of Pluto.

And third, the development and qualification of a DIPS for use in support of the manned exploration of the Moon and Mars to provide a long-lived, reliable power source to bridge the gap between RTGs and space reactors.

Acknowledgments

This document describes work performed by and for the U.S. Government. The authors gratefully acknowledge the major contributions made in the preparation of this document by Robert T. Carpenter and Kathryn L. Yates of Fairchild Space Company. The preparation of this document was funded, in part, under contract to the U.S. Department of Energy.

TABLE 1. Summary of Space Radioisotope Power Systems Launched by the United States.

Power Source	Spacecraft	Mission Type	Launch Date	Status
SNAP-3B7	Transit 4A	Navigational	29 Jun 1961	RTG operated for 15 years. Satellite now shutdown but operational.
SNAP-3B8	Transit 4B	Navigational	15 Nov 1961	RTG operated for 9 years. Satellite operation was intermittent after 1962 high altitude test. Last reported signal in 1971.
SNAP-9A	Transit 5-BN-1	Navigational	28 Sep 1963	RTG operated as planned. Non-RTG electrical problems on satellite caused satellite to fail after 9 months.
SNAP-9A	Transit 5-BN-2	Navigational	5 Dec 1963	RTG operated for over 6 yrs. Satellite lost navigational capability after 1.5 years.
SNAP-9A	Transit 5-BN-3	Navigational	21 Apr 1964	Mission was aborted because of launch vehicle failure. RTG burned up on reentry as designed.
SNAP-19B2	Nimbus-B-1	Meteorological	18 May 1968	Mission was aborted because of range safety destruct. RTG heat sources recovered and recycled.
SNAP-19B3	Nimbus III	Meteorological	14 Apr 1969	RTGs operated for over 2.5 years (no data taken after that).
ALRH	Apollo 11	Lunar Surface	14 Jul 1969	Radioisotope heater units for seismic experimental package. Station was shutdown Aug 3, 1969.
SNAP-27	Apollo 12	Lunar Surface	14 Nov 1969	RTG operated for about 8 years (until station was shutdown).
SNAP-27	Apollo 13	Lunar Surface	11 Apr 1970	Mission aborted on way to moon. Heat source returned to South Pacific Ocean.
SNAP-27	Apollo 14	Lunar Surface	31 Jan 1971	RTG operated for over 6.5 years (until station was shutdown).
SNAP-27	Apollo 15	Lunar Surface	26 Jul 1971	RTG operated for over 6 years (until station was shutdown).
SNAP-19	Pioneer 10	Planetary	2 Mar 1972	RTGs still operating. Spacecraft successfully operated to Jupiter and is now beyond orbit of Pluto.
SNAP-27	Apollo 16	Lunar Surface	16 Apr 1972	RTG operated for about 5.5 years (until station was shutdown).
Transit-RTG	"Transit"	Navigational	2 Sep 1972	RTG still operating. (Triad-01-1X)
SNAP-27	Apollo 17	Lunar Surface	7 Dec 1972	RTG operated for almost 5 years (until station was shutdown).
SNAP-19	Pioneer 11	Planetary	5 Apr 1973	RTGs still operating. Spacecraft successfully operated to Jupiter, Saturn, and beyond.
SNAP-19	Viking 1	Mars Surface	20 Aug 1975	RTGs operated for over 6 years (until lander was shutdown).
SNAP-19	Viking 2	Mars Surface	9 Sep 1975	RTGs operated for over 4 years until relay link was lost.
MHW-RTG	LES 8*	Communications	14 Mar 1976	RTGs still operating.
MHW-RTG	LES 9*	Communications	14 Mar 1976	RTGs still operating.
MHW-RTG	Voyager 2	Planetary	20 Aug 1977	RTGs still operating. Spacecraft successfully operated to Jupiter, Saturn, Uranus, Neptune, and beyond.
MHW-RTG	Voyager 1	Planetary	5 Sep 1977	RTGs still operating. Spacecraft successfully operated to Jupiter, Saturn, and beyond.
GPHS-RTG	Galileo	Planetary	18 Oct 1989	RTGs still operating. Spacecraft in route to Jupiter.
GPHS-RTG	Ulysses	Planetary/Solar	6 Oct 1990	RTG still operating. Spacecraft in route to Solar Polar fly-by.

* Single launch vehicle with double payload.

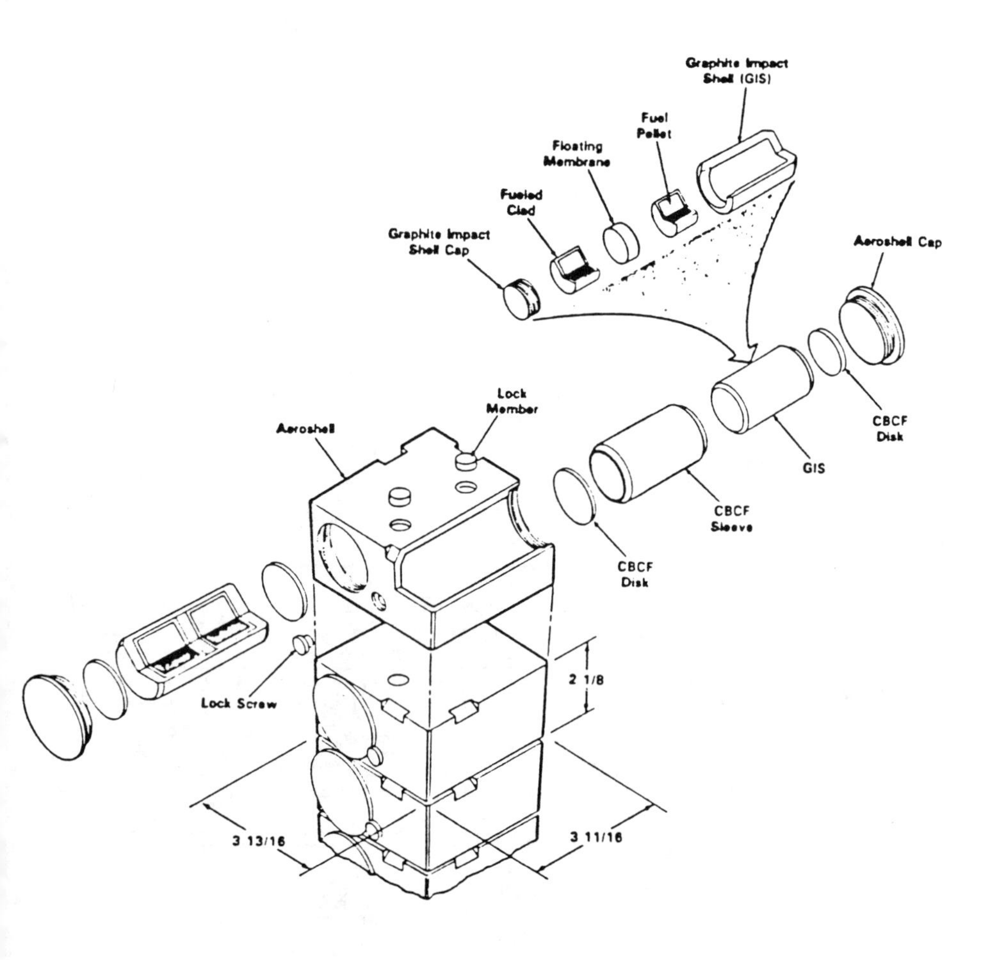

FIGURE 1. General Purpose Heat Source (GPHS) Module Assembly.

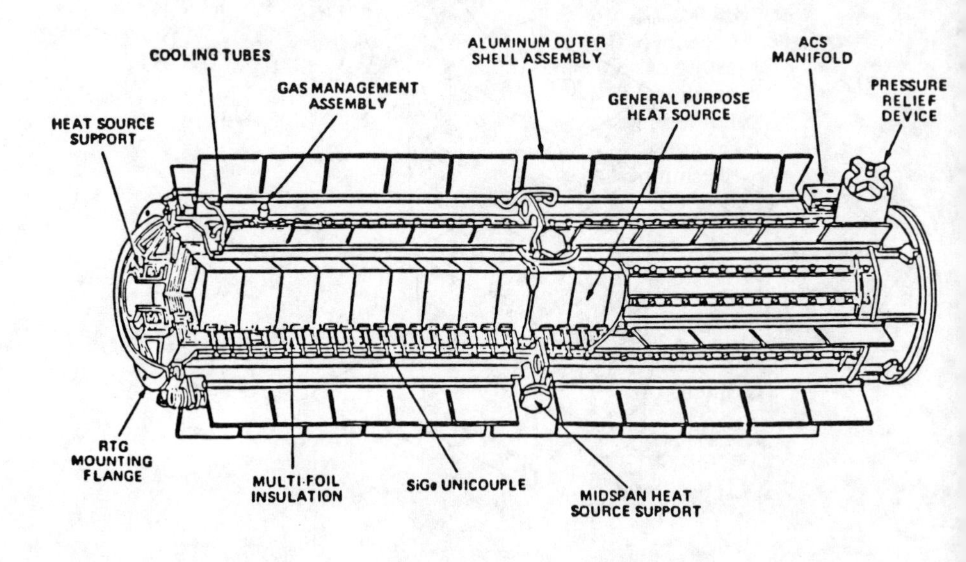

FIGURE 2. General Purpose Heat Source Radioisotope Thermoelectric Generator.

SUMMARY OF SPACE NUCLEAR REACTOR POWER SYSTEMS (1983-1992)

David Buden
Idaho National Engineering Laboratory
12108 Holly Ave., NE
Albuquerque, NM 87122
(505)846-7228

Abstract

Major developments in the last ten years have greatly expanded the space nuclear reactor power systems technology base. In the SP-100 program, after a competition between between liquid-metal, gas-cooled, thermionic, and heat pipe reactors integrated with various combinations of thermoelectric, thermionic, Brayton, Rankine, and Stirling energy conversion systems, three concepts were selected for further evaluation. In 1985, the high-temperature (1,350 K), lithium-cooled reactor with thermoelectric conversion was selected for full scale development. Since then, significant progress has been achieved, including the demonstration of a 7-y-life uranium nitride fuel pin. Progress on the lithium-cooled reactor with thermoelectrics has progressed from a concept, through a generic flight system design, to the design, development, and testing of specific components. Meanwhile, the U.S.S.R. in 1987-88 orbited a new generation of nuclear power systems beyond the thermoelectric plants on the RORSAT satellites. Two satellites using multicell thermionic Topaz I power plants operated six months and eleven months respectively in space at 5 kWe. An alternate single cell thermionic power plant design, called Topaz II, has been operated in a ground test for 14,000 h in one unit. The U.S. has continued to advance its own thermionic fuel element development, concentrating on a multicell fuel element configuration. Experimental work has demonstrated a single cell operating time of about 1 1/2-y. Technology advances have also been made in the Stirling engine; an advanced engine that operates at 1,050 K is ready for testing. Additional concepts have been studied and experiments have been performed on a variety of systems to meet changing needs, such as powers of tens-to-hundreds of megawatts and highly survivable systems of tens-of-kilowatts power.

INTRODUCTION—TEN YEAR TREND (1983-1992)

In 1983, the National Aeronautics and Space Administration (NASA), the Department of Energy (DOE), and the Department of Defense (DoD) entered into an agreement to fund a space nuclear reactor power program. This program, called SP-100, superseded the heat pipe reactor development program (also called SP-100) that had started in 1979. The goals of this program are to develop the technology to provide tens-to-hundereds of kWe of electric power with the specific initial design concentrating on 100 kWe at an operational time of 7 y and lifetime of 10 y. Starting with a broad range of candidates (Hylin and Moriarty 1985, Chiu 1985, Harty et al. 1985, Yoder and Graves 1985 and Terrill and Putnam 1985) including liquid-metal, gas-cooled, thermionic, and heat pipe reactors with various combinations of thermoelectric, thermionic, Brayton, Rankine, and Stirling energy conversion systems, three concepts were selected for further evaluation. These were: (1) a high-temperature, pin-fuel element reactor with thermoelectric conversion, (2) an in-core thermionic power system, and (3) a low-temperature pin-fuel element reactor with

Stirling cycle conversion. In 1985, the high-temperature pin-fuel element reactor with thermoelectric conversion was selected for development to flight readiness. The thermionic powerplant offered a more compact heat rejection subsystem and lower temperature structural materials, but issues of lifetime excluded its selection. The Stirling system offered higher energy conversion efficiency and lower temperature materials, but the technology risk was considered greater at that time because of the preliminary status of its development. Because sufficient merit was recognized in the other options, a technology program was started on in-core thermionic fuel elements, called the Thermionic Fuel Element Verification Program, and another program on developing a high-temperature Stirling engine that can be mated with the high-temperature pin-fuel reactor. Since 1985, the SP-100 power system has progressed from a concept, through a generic flight system design, to the design, development, and testing of specific components.

In the meantime, the U.S.S.R. launched a new generation of space reactors with flights of Cosmos 1818 and 1867 in 1987 and 1988. Details of these systems, known to the U.S. as Topaz I, were first introduced to the West at the Sixth Space Nuclear Power Systems Symposium in 1989.

During the past ten years, a number of activities occurred in developing multimegawatt-level power systems to support power needs for directed energy weapons and electric propulsion. The Multimegawatt Program evaluated systems for: (1) open loop, power levels of tens-of-megawatts for hundreds of seconds, (2) closed loop, power levels of tens-of-megawatts for one year, and (3) open loop, power levels of hundreds-of-megawatts for hundreds of seconds. Six concepts were considered during the Phase I preconceptual activities. The purpose of Phase I was to identify key technology feasibility issues. Phase II was to resolve the issues prior to technology selections. This program was terminated in 1990, because of funding constraints, before Phase II design contracts were awarded.

There is some interest in multimegawatt systems for use in electric propulsion (Doherty and Gilland 1992 and Gilland and Oleson 1992). The major candidate for this mission is a version of the SP-100 nuclear subsystems integrated with a Rankine cycle power conversion system.

In this report, emphasis is being given to the major system level developments. In the U.S., development concentrated on SP-100; in the U.S.S.R., development was on in-core thermionic Topaz power systems. In addition, all areas of activities where technology developments have occurred are summarized. The major U.S. technology efforts besides SP-100 have been in reactors with in-core thermionic power converters. The ten-year period ends with uncertainty as a result of changing directions and budget constraints.

HIGH-TEMPERATURE, LIQUID-METAL-COOLED POWER PLANT DEVELOPMENT (SP-100)

Requirements

SP-100 is being designed to provide tens-to-hundreds of kWe power for 7 y at full power and 10 y overall operation. Power plant components are to provide this wide range of power without significant requalification of the basic building blocks. These requirements were originally derived from projected DoD needs for robust surveillance systems, survivable communications including antijam capabilities, electric propulsion systems for reusable orbital transfer, and weapons applications (Redd and Fornoles 1985), and from projected NASA needs for nuclear electric propulsion for planetary missions, earth orbit tug, civilian anticollision aircraft radar, advanced direct broadcast satellites, growth space station power, and planetary bases (Ambrus and Beatty 1985). A specific reference system design point of 100 kWe was selected for demonstration of the power system technology. Table 1 summarizes the key requirements. The total system mass of <4,000 kg and safety requirements are key design drivers. In addition, the design is to decrease radiation at 25 m from the reactor to 1×10^{13} n/cm^2 and 5×10^5 rad. These criteria are to be met without servicing or maintenance in space. The size of the power plant must be compatible with the Space Shuttle. Recent modifications have incorporated the requirement for launch on a Titan IV with the operating orbit above 1,100 km. Survivability requirements for defense missions have been put on hold indefinitely.

TABLE 1. SP-100 Generic Flight System Design Major System Requirements.

Major Parameter	Values
System mass goal (kg)	4,000
Stowed length goal (m)	6.0
System power level (kWe)	100
Housekeeping power (We)	300
In-Orbit life/full power life	10/7
Orbit altitude (km)	≥1,100
Launch vehicle	Titan IV
Safety	No accidental criticality
Dose plane diameter (m)	4.5
Dose plane distance (m)	25
Radiation dose to user plane (n/cm^2)	10^{13}
Radiation dose to user plane (gamma) (rad)	5×10^5

Description

The Generic Flight System (GFS) design was completed in May 1988 and found in an independent System Design Review to be acceptable as the basis for technology and component development. The selected design is a fast spectrum reactor coupled to thermoelectric energy conversion units by liquid lithium, which is magnetically pumped through niobium alloy piping. Since then there have been major refinements to enhance reliability and to reduce mass. Currently, the major elements in the system shown in Figures 1 and 2 (Truscello and Rutger 1992) are: (1) a fast reactor using the niobium base alloy PWC-11 (niobium-1% zirconium with 1,000 ppm of carbon) for structures with a uranium nitride (UN) ceramic fuel, (2) control system using sliding neutron reflectors and safety rods with appropriate actuators, sensors, and controllers, (3) radiation attenuation shadow shield of lithium hydride and depleted uranium, (4) heat transport loops that transfer heat from the reactor to the power conversion units by means of pumped liquid lithium using self-actuated thermoelectric electromagnetic

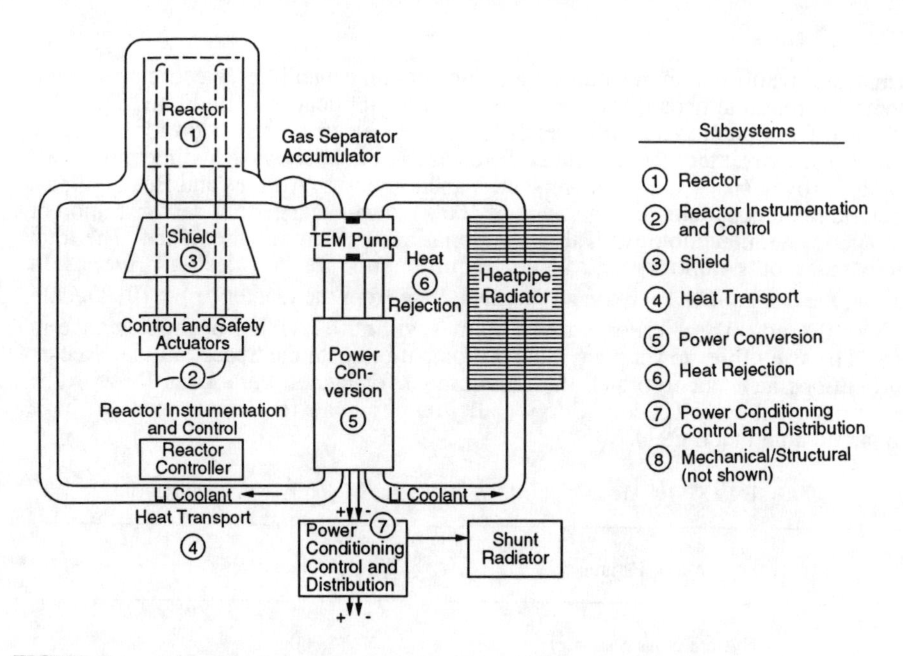

FIGURE 1. SP-100 Generic Flight System Schematic (Truscello and Rutger 1992).

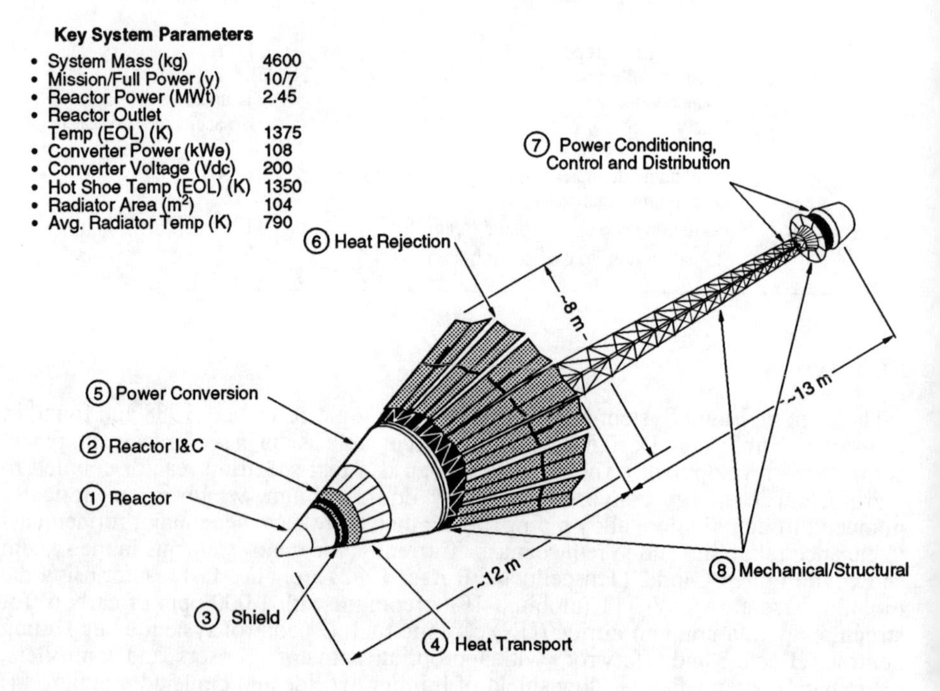

FIGURE 2. SP-100 Generic Flight System Configuration (Truscello and Rutger 1992).

pumps, (5) solid-state thermoelectric power conversion units using silicon germanium/gallium phosphide (SiGe/GaP) to convert thermal power to electricity, and (6) the heat rejection system consisting of a carbon/carbon matrix structure and armor with titanium/potassium heat pipes (Mondt 1989 and Josloff 1988).

The building blocks that form SP-100 can be configured in a number of arrangements depending on mission needs and desired power levels. The baseline configuration arrangement for a 100 kWe power plant has the reactor at the forward end of the power plant away from the payload. The largest segment of the power plant is the heat rejection area. The total length of the deployed power plant without the boom is 12 m. The boom, used to minimize the amount of shielding needed, makes the overall length 25 m. This can be stowed for launch.

The heart of the reactor design is the fuel pin, shown in Figure 3. The UN fuel is fabricated in the form of pellets, having a density of 94.5 percent theoretical and uranium enrichment of 97 percent. Approximately 50,000 pellets are needed for a 100 kWe thermoelectric system core. Cladding provides structural strength for the fuel pin, while a layer of rhenium acts as a barrier between the fuel and lithium coolant, prevents loss of nitrogen from the UN fuel, allows thinner cladding, and acts as thermal poison if the reactor is immersed in water. The fuel pin cladding is the niobium alloy PWC-11 refractory material tubing with rhenium (Re) tubing bonded to the internal surface. The UN fuel operates at a peak surface temperature at the beginning of life of 1,400 K and end of life of 1,450 K and a peak burnup of 6 atom percent. Surrounding the fuel pin is a wire wrap spacer that provides space for the lithium coolant to flow. The fuel pins have an upper plenum region to contain fission products.

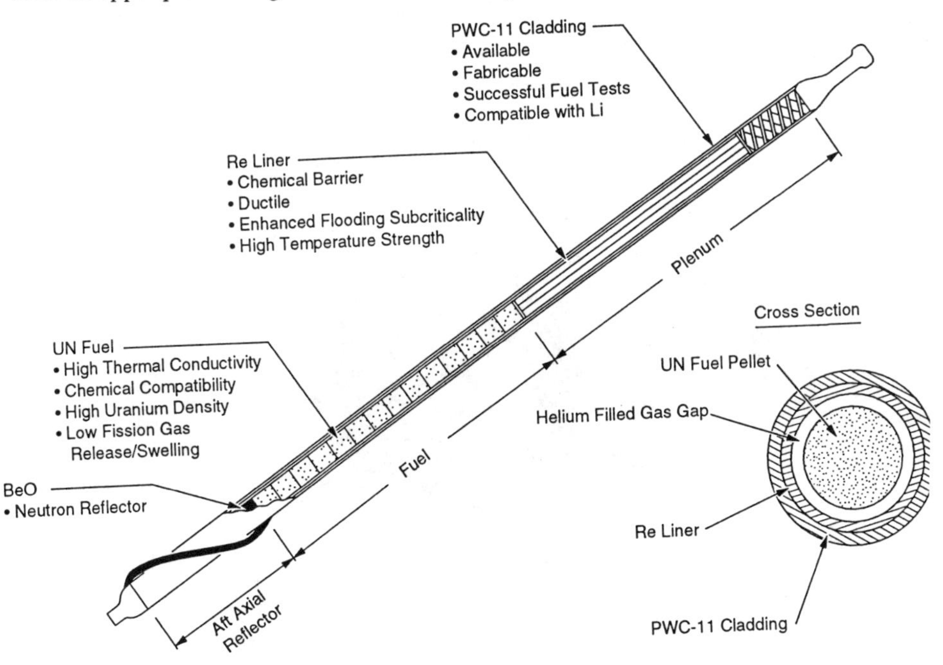

FIGURE 3. SP-100 Fuel Pin (Truscello and Rutger 1992).

The reactor core is separated into 10 hexagonal-shaped assemblies with approximately 61 fuel pins per assembly and six partial assemblies, each with approximately 50 pins, around the outer edge (similar to the configuration shown in Figure 4). The partial assemblies result in the hexagonal shape being closer to a circular form. The core also contains three safety rods for accidental criticality mitigation and, in case of loss-of-coolant flow, an auxiliary cooling system to prevent core melt down. The in-core safety rods provide a redundant shutdown system and help maintain the core subcritical in case of fire, explosion, water submersion, or compacting accidents. The spacing between the fuel pins is used as coolant paths for the lithium which transfers heat to the converters.

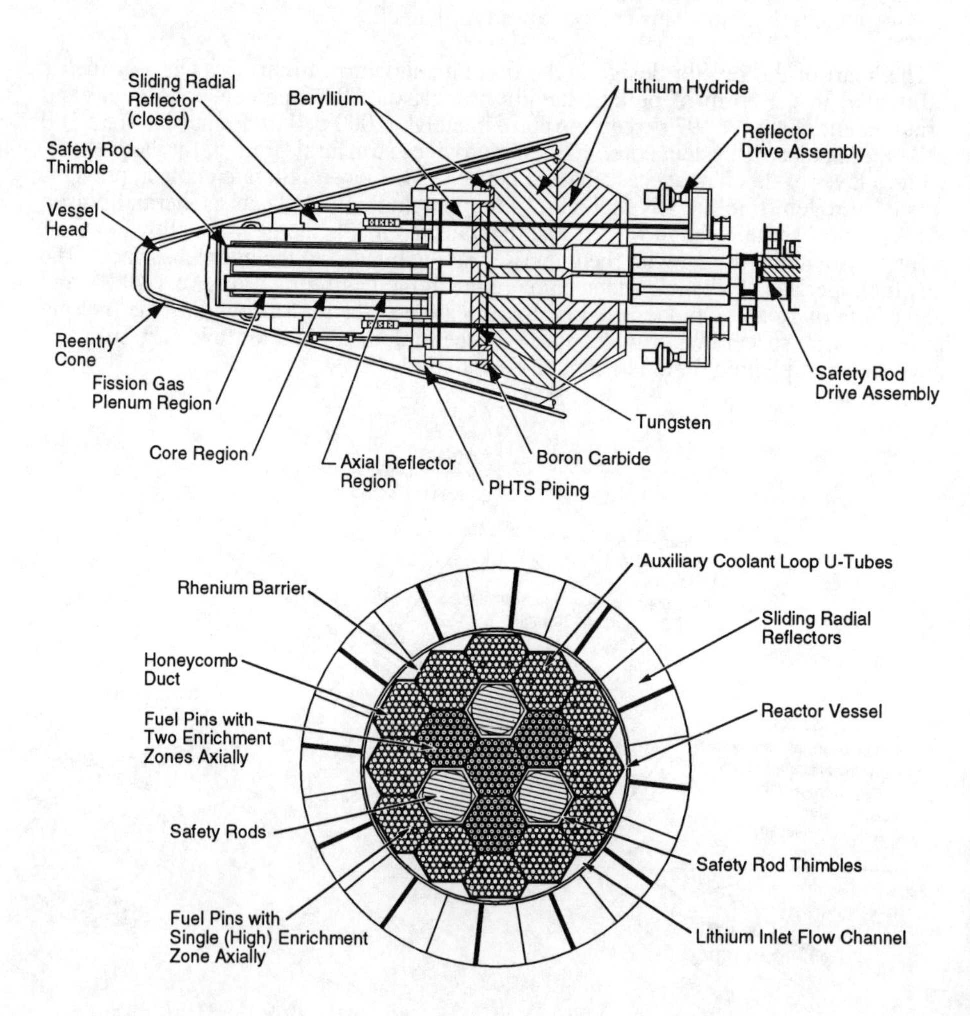

FIGURE 4. SP-100 Reactor Subsystem Components (Truscello and Rutger 1992).

The auxiliary coolant function option is designed to limit reactor fuel temperatures to less than 2,000 K in case of a loss-of-coolant accident. It provides coolant loops totally independent of the normal heat removal system. Heat removal is by 42 u-tubes located throughout the core pin bundles (Mondt 1992). The auxiliary lithium coolant passes through the core into a collection manifold, from which it is pumped to an independent radiator for rejection of heat to space.

The in-core safety rods are of boron carbide (B_4C) neutron absorbing material inside structural thimbles that can be located in or out of the core. The rods have a follower segment of beryllium oxide (BeO) that acts as moderator during operation.

Normal control is by means of twelve sliding tapered reflector segments. The segments control neutron leakage from the core. The positioning of the reflector segments is used to bring the reactor critical once the in-core safety rods are removed and to compensate for fuel burnup and swelling. The reflector segments consist of BeO contained within a Nb-1% Zr shell.

The radiation shielding approach for minimizing system mass resulted in a conical shadow-shield with a cone half-angle of seventeen degrees (Disney et al. 1990). Within the shield, the design must thermally isolate high performance/low mass shielding material with limited temperature capabilities from the adjacent high temperature components. The shield is fabricated from lithium hydride (LiH) pressed into a stainless-steel honeycomb to attenuate neutrons. Depleted uranium plates are added primarily to attenuate gamma radiation. Beryllium is used to transfer heat to the outer surface of the shield, where it is radiated to space. Stainless steel is used as a structural element and to contain the shield.

Surrounding the reactor and butting against the radiation shield is a carbon-carbon reentry heat shield. The purpose of the reentry shield is to ensure confinement of the core fission products if the reactor reenters the atmosphere. It is conical in shape, and the geometry protects the reactor vessel from overheating during reentry, keeping the temperature to 300 K even through the reentry shield might reach 3,200 K (Deane et al. 1989).

Heat is transferred to the converter by six interrelated pumped lithium loops (Atwell et al. 1989). Thermoelectric-electromagnetic (TEM) pumps are used to circulate liquid lithium in each loop (Figure 5). Each pump also circulates lithium coolant on the cold side of two heat rejection loops. These are self-energized, DC conduction, electromagnetic TEM pumps. Hot and cold ducts run parallel to each other along the active length of the pump. Sandwiched between them is a series of thermoelectric elements connected on each side by compliant pads. The temperature differential imposed across the thermoelectric converters generates an electrical current that travels at right angles to the lithium flow in the ducts and in closed paths around a magnetic center iron; as the current circulates, an induced magnetic flux is generated in the center iron that is directed through the lithium ducts perpendicular to the current. The interaction of the magnetic flux and the current produces a force on the molten lithium that drives it along the ducts. Hence, flow responds to the temperature in a self-regulated way.

During reactor operation, helium gas is generated in the coolant loop by neutron reactions with lithium. Enriched lithium in ^{7}Li (99.9% ^{7}Li, 0.1% ^{6}Li in contrast to natural lithium, which has an isotope ratio of 92.6/7.4) is used to minimize production

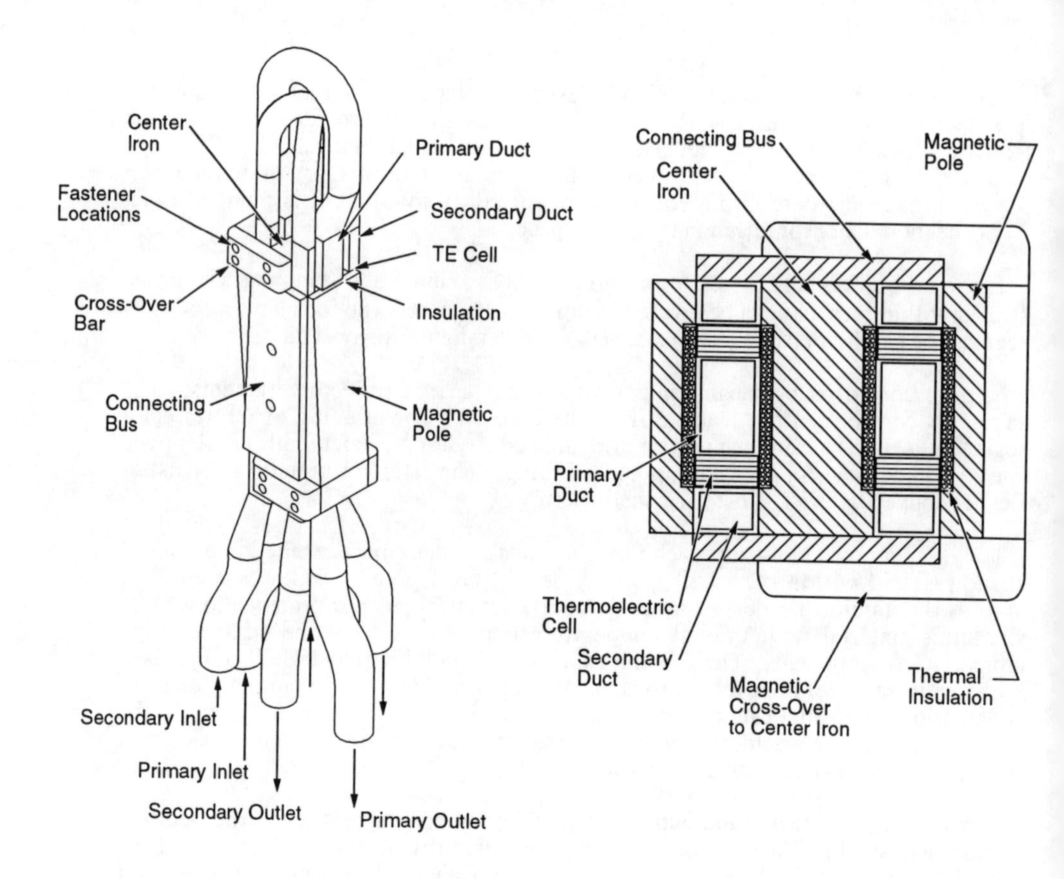

FIGURE 5. SP-100 TEM Pump Development (Truscello and Rutger 1992).

of helium. A gas separator/accumulator (Figure 6), used to separate the helium from circulating lithium, utilizes the principles of surface tension and centrifugal force.

The reactor is controlled based on temperature and neutronic measurements. Flow is inherently maintained by the TEM pumps (for example, higher reactor temperatures increase TEM pumping). Sensors are multiplexed in an analog multiplexer/amplifier located behind the reactor shield. These are designed to operate in a radiation environment of 4 x 10^{15} n/cm^2 and 2.4 x 10^8 rad (gamma) in 10 y of operation. N-type junction field effect transistors (JFET) semiconductor devices are being used.

Conversion of thermal energy from the reactor to electrical power is accomplished through the Power Conversion Assembly (PCA) (Figure 7) (Bond et al. 1993). Heat is conducted from hot lithium in a central heat exchanger, through TE cells on either side (which convert some of the heat to electricity), and then to a pair of heat exchangers where cooler lithium carries the waste heat to the heat rejection subsystem. The PCA building blocks can be packaged in any combination to generate the desired output. These building blocks are the thermoelectric cell and thermoelectric converter assembly (TCA). Each assembly has two arrays of 60 TE cells. The TCA consists of two cell arrays, one hot side heat exchanger, and two cold side heat exchangers. A

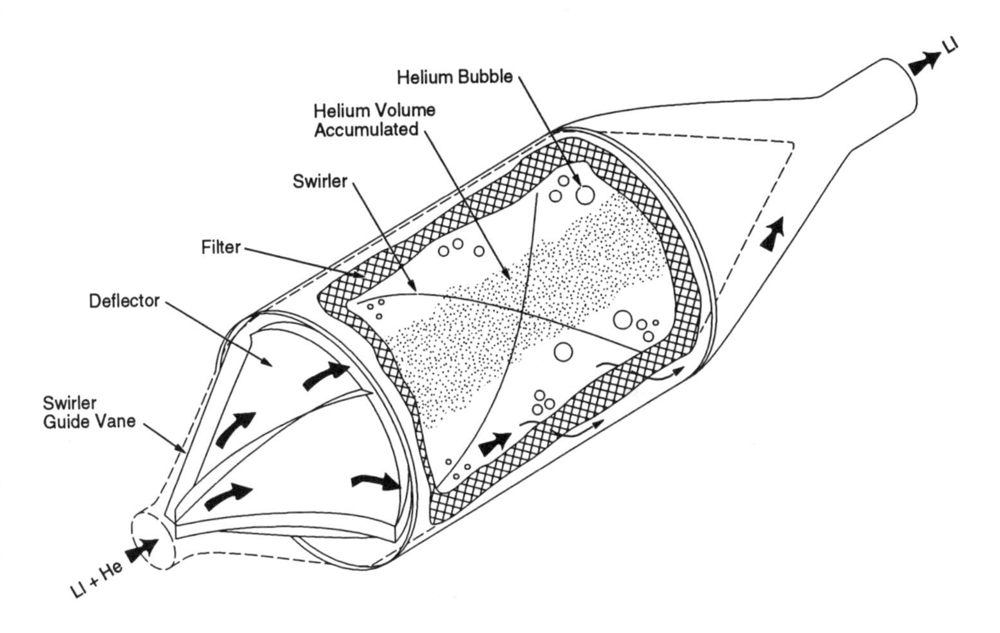

FIGURE 6. SP-100 Lithium/Helium Gas Separator-Accumulator (Truscello and Rutger 1992).

typical TCA configuration have 6 x 10 cell arrays. The thermoelectric (TE) converter assemblies are in a stack of six plate-and-frame configuration, each of which produces 1.5 kWe at 34.8 VDC. A total of 8,640 cells is required for a 100 kWe system.

The thermoelectric cells (see Figure 8) use silicon germanium/gallium phosphide (SiGe/GaP) materials. Each cell consists of a thermoelectric module, compliant pads to accommodate thermal stresses, electrical insulators to isolate the electrical power from the spacecraft, and conductive coupling to the heat exchangers. The cells are arranged in a parallel/series electrical network to provide the desired 200 V output. The thermoelectric material figure-of-merit is being increased to 0.85×10^{-3} K^{-1} by the addition of GaP to 80:20 SiGe from 0.67×10^{-3} K^{-1} for SiGe. A graphite-electrode-SiGe bond is used to make the electrical contact resistivity of that joint less than 25 $\mu\Omega$/cm. The compliant pad is used to prevent cracking of the TE elements from thermal expansion. It consists of niobium fibers bonded to niobium face sheets on both sides. The niobium face sheet matches the thermal expansion of both the heat source material (PWC-11) and the heat sink material (Nb-1% Zr). The insulators are single-crystal alumina, with 4,000 V/cm voltage gradient at 1,375 K.

The radiator will be tailored to the particular application. In the GFS, the heat rejection subsystem includes twelve radiator assemblies, each constructed of an array of varying lengths of potassium heat pipes brazed to a lithium duct-strongback structure at the center (see Figure 2). The lithium duct-strongback structure consists of the lithium supply and return ducts. Each of the twelve radiator panels have flexible joints for the supply and return ducts so that the radiator can be folded for launch and deployed for reactor operation. Accumulators are included within the heat rejection subsystem to accommodate the variable volume necessary to compensate for expansion

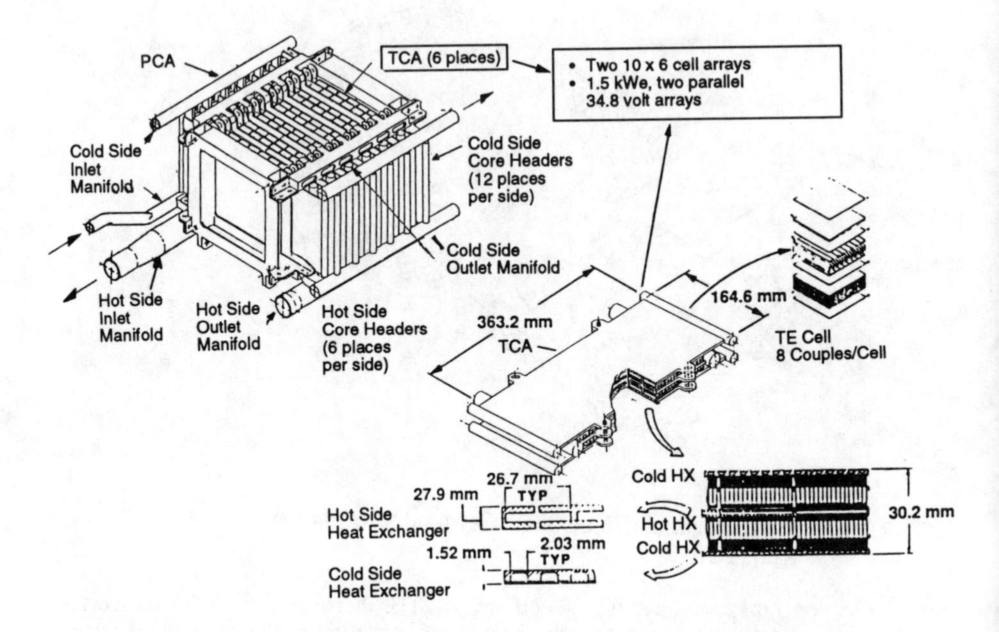

FIGURE 7. SP-100 Power Converter Subsystem Components (Truscello and Rutger 1992).

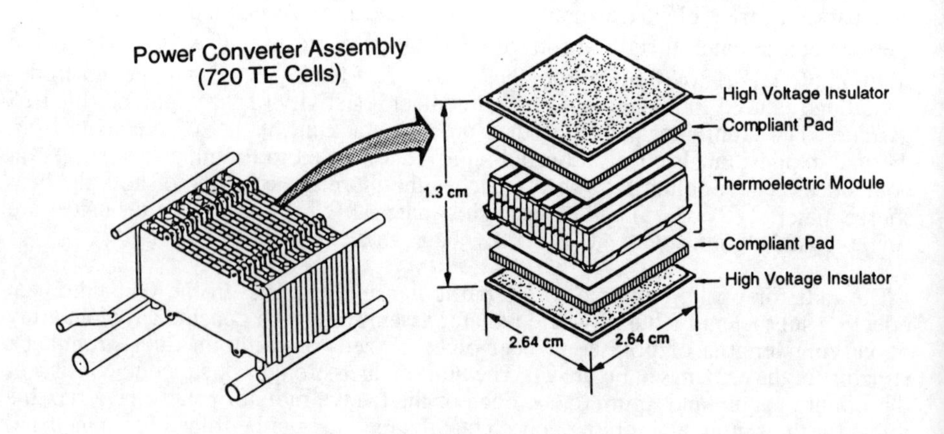

FIGURE 8. SP-100 Thermoelectric Cell (Truscello and Rutger 1992).

and contraction of the lithium coolant. The accumulators are fabricated out of titanium. Heat rejection area is 104 m^2.

For restarting the power plant in space, the reactor is used to heat an auxilliary liquid metal loop to melt the lithium throughout the system (Hwang et al 1993).

Table 2 summarizes some key power plant performance parameters.

TABLE 2. SP-100 GFS Design Performance Parameters.

Parameters	Values
Reactor power (MWt)	2.5
Peak reactor outlet temperature EOL (K)	1,375
Heat loop ΔT (K)	92
Heat loop mass flow (kg/s)	5.9
Reactor rejection loop ΔT (K)	48
Reactor rejection loop mass flow (kg/s)	10.4
Radiator inlet temperature (K)	840
Average radiator surface temperature (K)	790
Radiation black body area (m^2)	94
Radiation physical area (m^2)	104
Thermopile area (m^2)	5.5
Thermoelectric leg length (mm)	5.5
Gross power generated (kWe)	105.3
Subsystem Mass (kg)	
Reactor	650
Shield	890
Primary heat transport	540
Reactor instrumentation and controls	380
Power conversion	530
Heat rejection	960
Power conditioning, control and distribution	400
Mechanical/structural	250
Total	4,600

Performance and Scaleability

SP-100 is designed as a set of building blocks. Thus, the reactor power level can be varied by changing the number of fuel pins, changing the number of thermoelectric converters, or using different power conversion units. The radiator area can be adjusted according to the power level. This flexibility was one of the major reasons for selecting the concept in 1985. The mass of the system is a function of the various combinations selected. Concepts of 8, 10, 20, 30 40, 50 100, 200, 300, 1,000,

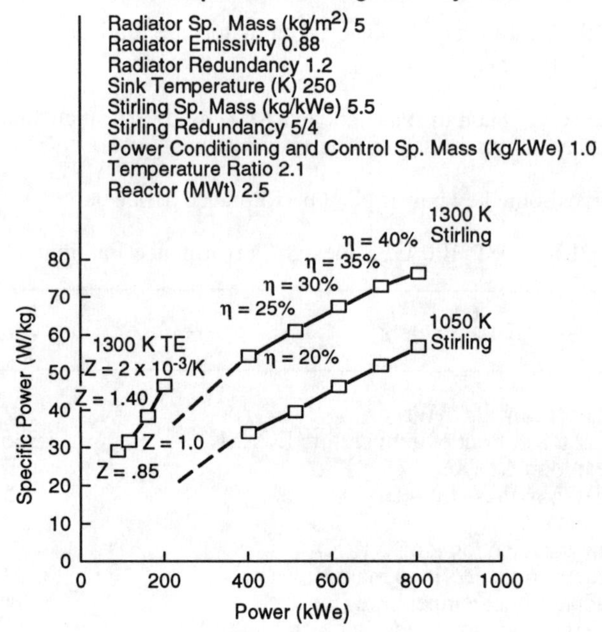

FIGURE 9. Extending SP-100 Reactor Power Systems Capability

5,000, 10,000 and 15,000 kWe have been configured. Figure 9 shows the specific power difference as a function of thermoelectric figure-of-merit and using Stirling engines at various efficiencies and peak operating temperatures. The same 2.5-MWt reactor, used in conjunction with thermoelectric conversion to generate 100 kWe, can be coupled to the Stirling conversion system to generate nearly 600 kWe.

Nearer term options are given in Figures 10 and 11 (Schmitz et al. 1992). The mass is shown to be a trade-off at the design thermoelectric conditions with Brayton or Stirling conversion systems. The Brayton conversion work was performed in the 1960s where 46,000 h of test data were accumulated on the rotating machinery. More details on the status of Brayton technology is given in Dudenhoefer et al. in this volume. The Brayton and Stirling conversion systems require much larger radiators than the thermoelectric conversion systems.

The SP-100 reactor can be reconfigured for higher thermal power levels and mated with a high-temperature Rankine cycle to provide even higher powers, such as several megawatts for nuclear electric propulsion. Also, SP-100 can be configured in a number of arrangements for lunar and Mars surface applications.

Development Status

Since 1985, substantial progress has been made in all key technology areas, and early feasibility issues have been resolved (Josloff et al. 1992a and Truscello and Rutger 1992). Fuel development includes the fabrication of the fuel pins and testing in an environment sufficient to develop high confidence of meeting the 7-y full-power operational lifetime requirement. The fabrication processes necessary to produce high

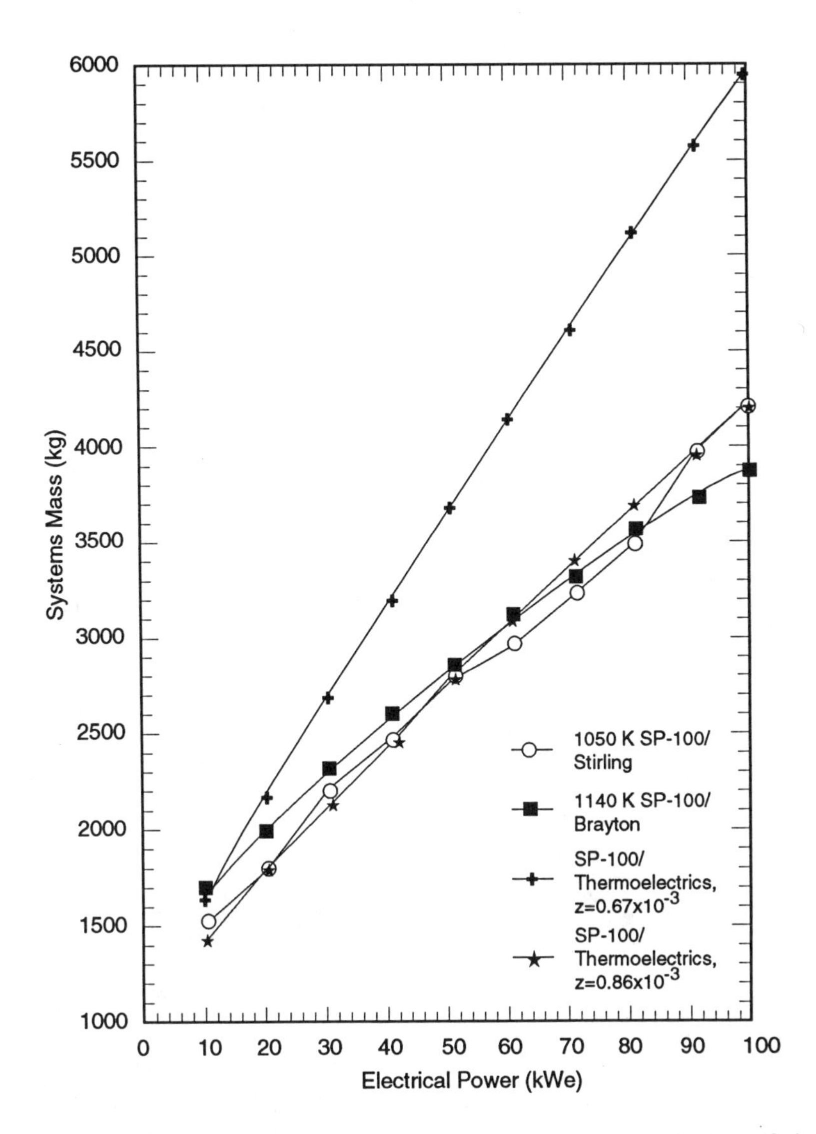

FIGURE 10. SP-100 System Mass Comparisons For Near Term Options (Schmitz et al. 1992).

quality UN fuel pellets have been successfully developed and the fuel irradiated in the Experimental Breeder Reactor II (EBR-II) and Fast Flux Test Facility (FFTF) test reactors. Approximately 75 fuel pins have been tested with some tests performed at three times the nominal power and at fuel pin surface temperatures as high as 1,500 K. All tests have met or exceeded expectations, and the goal of 6 atom-percent fuel burnup has been achieved. The pins tested used the Nb-1% Zr alloy cladding because PWC-11 was not yet qualified. This alloy is predicted to provide similar design margins, but at less mass. The tests show the fuel pins can meet the requirements of a 7-y system. Fuel pins testing with PWC-11 is planned to be completed in 1998 (Truscello and Rutger 1992).

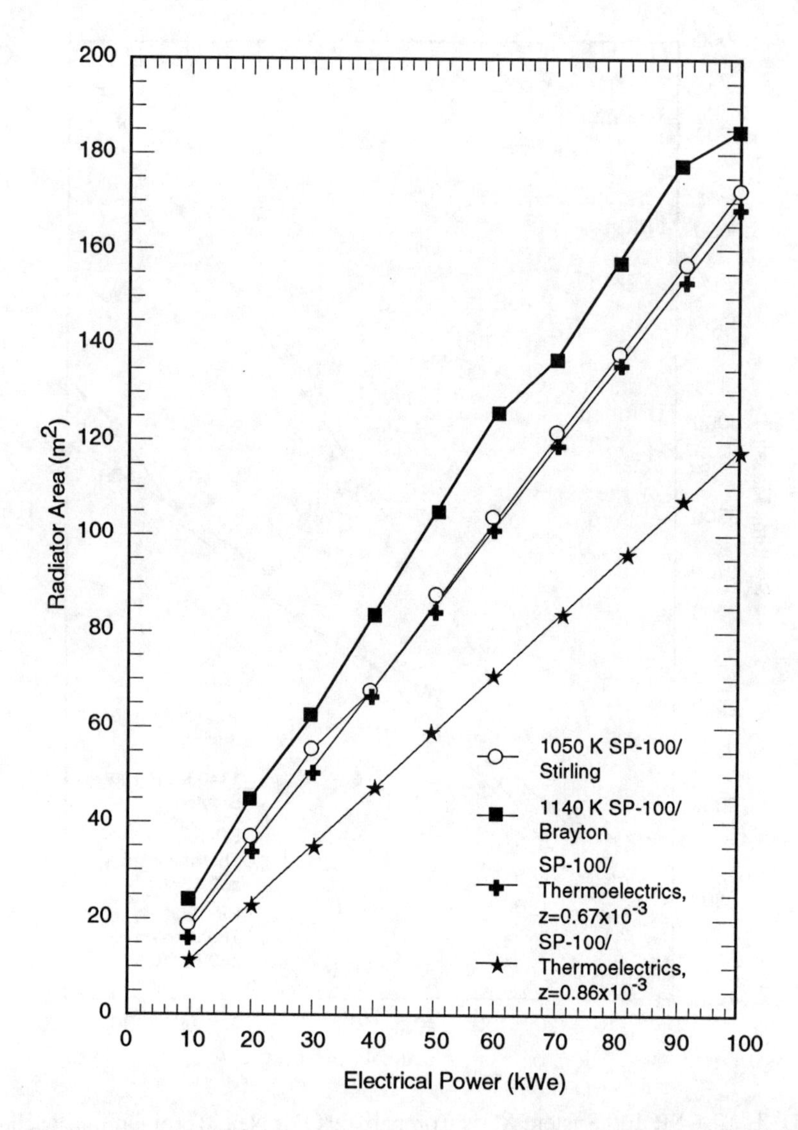

FIGURE 11. SP-100 Radiator Area Comparisons For Near Term Options (Schmitz et al. 1992).

A dual material tube has been qualified with an outer tube of niobium alloy and an inner tube of rhenium for strength as well as a barrier between the fuel and niobium alloy. Tubes 0.6 m long have been tested that show the bond is sufficiently rugged for 10 y of operation. Approximately 50 tubes have been fabricated with Nb-1% Zr cladding and another 10 were produced with PWC-11 cladding material.

The nearly 50,000 fuel pellets needed for a reactor core have been fabricated and are in storage awaiting the fabrication of the cladding. Production of the fuel pins for a reactor is being delayed until they are needed.

Materials are fundamental to the success of the development of a reactor that must operate for 7 y with a maximum coolant outlet temperature of 1,375 K. Sufficient materials data did not exist on the alloys of choice at the start of the program. Now, suppliers have been qualified for refractory alloys fabrication including Nb-1% Zr, PWC-11, and rhenium. Fabrication procedures include electron beam and gas tungsten arc welding, cold forging, drawing, hot isostatic pressing, diffusion bonding, vacuum sputtering, chemical vapor deposition, and high-temperature heat treatment. Compatibility testing with lithium has been performed in Nb-1% Zr and PWC-11 test loops at 1,350 K for thousands of hours without failures (Josloff et al. 1992b).

Reactor physics behavior has been experimentally verified in critical assembly tests. Measurements performed on the assembly showed close agreement with predictions under both normal design and postulated accident conditions. These tests provide assurance that the reactor will meet safety criteria imposed to protect the environment. From Mondt (1992), the experimental results verify that: (1) the internal shutdown rod reactivity met the shutdown requirement with margin, (2) reflector control worth versus position confirmed necessary control margin, (3) flooding reactivity worth was confirmed and subcriticality assured, and (4) buried reactor reactivity worth was confirmed and subcriticality was assured.

Control and safety drives are the only mechanical moving parts in the SP-100 power plant design. Low-temperature development tests have been completed to confirm the key mechanical design features. Environmental tests of control drive motor, clutch, and brake assemblies have demonstrated predicted performance of these components at 700 K. The self-aligning bearings of the reflector control drives have been successfully tested for 30,000 cycles at temperatures up to 1,170 K, and tests of safety rod bearings are underway at 1,570 K. Tribological coatings necessary to protect against self-welding, friction, and wear have been tested using refractory borides, carbides, and nitrides. These must operate at temperatures >1,600 K. Test data at 1,700 K indicate refractory carbides are the best material. Accelerated testing has been completed demonstrating that the equivalent life at SP-100 conditions is 50 y.

Long-life temperature and pressure sensors are needed to measure lithium coolant conditions. Temperature sensor concepts using a Johnson Noise Thermometer and W/Re thermocouples have been developed and fabricated units found to be mechanically robust. Sensor lifetime, accuracy, and stability are presently being established in a series of tests. In addition, the key features of a pressure transducer hydraulically coupled to the lithium coolant have been tested.

Multiplexers are located near the reactor where the nuclear radiation and temperature environments are severe (1.2×10^8 rad gamma, 1.6×10^{15} n/cm^2, and 800 K over 10 y). The temperature will be reduced to 375 K by the use of insulating blankets and radiators. Gamma testing indicates that radiation damage annealing will prevent unacceptable levels of drift in the circuit. Neutron testing is still underway.

Shadow shielding is used to attenuate the neutron and gamma radiation to acceptable levels for the spacecraft. Characterization of the LiH shield material for thermal conductivity and expansion, material compatibility experiments to confirm the long term behavior of the materials, and irradiation of LiH to establish swelling rates indicate no long-term issues with the current design. This includes irradiation testing at temperature. Material compatibility testing showed that LiH is not compatible with beryllium (Be). Therefore, the shield now has a stainless steel barrier between the Li and Be (Mondt 1992).

Work has been done on the major issues associated with the heat transport subsystem. For the gas separator, a feasibility experiment using air and water indicates that the design is sound. Experiments of a prototypical gas separator with helium and lithium are planned. A magnetic bench test has been performed on the TEM pump to demonstrate the ability to accurately predict the three-dimensional magnetic field. An electromagnetic integrated pump test has been performed to verify the calculated pumping forces.

Power conversion major issues included: (1) thermoelectric material figure-of-merit, (2) bonding of the cell between the heat exchangers to accommodate critical problems of thermal stress and electrical isolation, and (3) electrical insulation at 1,350 K while sustaining a voltage gradient of 8,000 V/cm for a period of 7 y in a deoxidizing environment created by the proximity of molten lithium. The SiGe material used in radioisotope thermoelectric generators (RTGs) has been improved so that the figure-of-merit has increased from 0.65×10^{-3} K^{-1} to 0.72×10^{-3} K^{-1}. The design goal of 0.85×10^{-3} K^{-1} must still be achieved.

A major technical achievement has been the development of a compliant pad to connect the thermoelectric cell to the heat source and sink. This was accomplished by use of a brush-like design that carries heat conductively while absorbing mechanical stresses due to the large temperature difference across the cell. The pad fibers are coated with a thin film of yttria to prevent self welding. Pads have performed under prototypic conditions for thousands of hours satisfactorily. Tests are continuing to demonstrate lifetime performance.

High voltage insulators are positioned at the top and bottom of the thermoelectric cell. Single crystal Al_2O_3 insulators, equipped with oxygen permeation barriers made of molybdenum, have been developed that will maintain the necessary electrical isolation for more than twice the lifetime required.

Very high electrical conductivity is needed to interconnect the TE couples. A multilayer electrode consisting of tungsten or niobium sandwiched between graphite layers has been developed. The tungsten provides good intercouple conductivity and strength, while the graphite isolates the tungsten from the TE material with which it reacts. Initial experiments have been performed, indicating that low resistivity bonds can be achieved, but long-term stability has yet to be demonstrated.

The process for fabricating the thermoelectric converter heat exchanger with the integral headers has been successfully demonstrated.

Cells can now be routinely fabricated using validated processes. The development has progressed through three phases. These all use SiGe thermoelectric materials. Progress in each phase includes (Don Matteo, May 25, 1993):

- PD-1 first demonstrated the basic concept of conduction coupling of thermoelectric cells to their heat source and heat sink heat exchangers. The PD-1 cell contained certain features (such as low temperature braze) which prevented driving the cell to full prototypic temperature levels, and the test fixture limitations resulted in thermal inefficiencies which prevented the cell from reaching maximum potential power output. The PD-1 cell delivered 4.0 We at 500 K temperature change for prototypic conditions and 4.8 We for 545 K ΔT at peak power (Matteo et al. 1992).

- The PD-2 cell improvements allowed operation of the cell at prototypic temperatures (1335 K hot side), but still was constrained by certain thermal inefficiences. The PD-2 cell produced 4.0 We at 500 K ΔT prototypic conditions and 8.7 We peak power at 730 K ΔT.

- New analytical techniques (Bond et al. 1993) were developed and applied to design a "fracture safe" configuration. The TA cells are near prototypic. These were subjected to prototypic and higher thermal conditions. The measured cells were 8.8 We (verus 8.9 We predicted) at the 500 K ΔT prototypic temperature conditions and 13.7 We under 660 K ΔT peak power temperature conditions. The two TA cells tested had efficiencies of conduction coupling of 75% and 79%, respectively, almost identical to pretest predictions. Lifetime testing has now reached 3,670 h on one cell, and is continuing.

Only limited work has been performed on the heat rejection system. A half-dozen titanium heat pipes, with potassium as the working fluid, were fabricated and have been life tested. A 0.9-m section of radiator duct was fabricated and tested using a low melting point liquid-metal (Cerrobend) that substituted for the lithium to demonstrate the ability of the lithium in the radiator to thaw during startup. Test results show that actual thaw rate was twice as fast as predicted.

The early SP-100 development issues are summarized in Table 3 (Mondt 1991). These challenges have been mainly resolved. To meet the safety challenges, an anuxiliary coolant loop has been designed to maintain the fuel pin clad below 2,000 K. This maintains the fuel structural integrity for disposal. The disposal location will be in space either at high Earth orbit or at its planetary destination. Thermoelectric cell design issues were resolved with development of the compliance pad and validation of the electrical insulator and electrical contact resistance. The fuel pin design and performance has been confirmed in reactor radiation testing. Heat transport hermiticity has been demonstrated in a lithium loop test. The major elements of the pump, including the magnetics, were demonstrated in element testing. The nitrogen loss that could limit lifetime has been resolved in testing that verified the success of the rhemium liner in the fuel pin to contain the nitrogen. System mass continues to be a challenge. Currently, the mass is at 4,600 kg for the GFS design and 3,900 kg for outer planet design; the specification is 4,000 kg. The gas separator to remove helium from the lithium loop has been demonstrated in air/water tests in Earth's gravity with further tests planned using lithium/helium. Enriched ^{7}Li is being used to minimize helium generation. To meet the low cost, low mass radiator heat pipe challenge, a design has been develped and successful operated for limited periods of time. For the shield temperature control, the foreward LiH material was replaced with Be and B_4C to reduce the radiation dose by a factor of 200 and the temperature of the LiH to 700 K in the LiH. This resolved the shield temperature control challenge. Remaining challenges for flight readiness will be discussed under the Flight Readiness section.

Flight Safety

The key in the design of SP-100 has been safety; the nuclear safety requirements are given in Table 4 (General Electric Co. 1989a). Figures 12 and 13 summarize key safety features built into the SP-100 design (General Electric 1989b). The reactor is

TABLE 3. SP-100 Top Ten Challenges in FY 1987.

1. Safety
   - Core coolability with loss of coolant
   - Reactor control and safety drives
   - End of mission disposal
2. Thermoelectric cell technology
   - Electrical insulator development and performance
   - Electrical contact resistance
3. Fuel pin design and performance validation
   - Fuel pellet development and scheduled
   - Fuel clad liner development
   - Fuel pin clad creep strength
4. Thawing coolants
   - Startup from frozen lithium
5. Highly reliable heat transport loop
   - Hermetic
   - TEM pump development and performance
6. System lifetime
   - N_2 loss from fuel elements
7. System mass
   - Compaliance with specification
8. Gas accumulator/separator
   - Li^7 versus natural lithium
9. Heat pipe design and manufacture
   - Transient performance/rethaw
10. Radiation shield temperature control

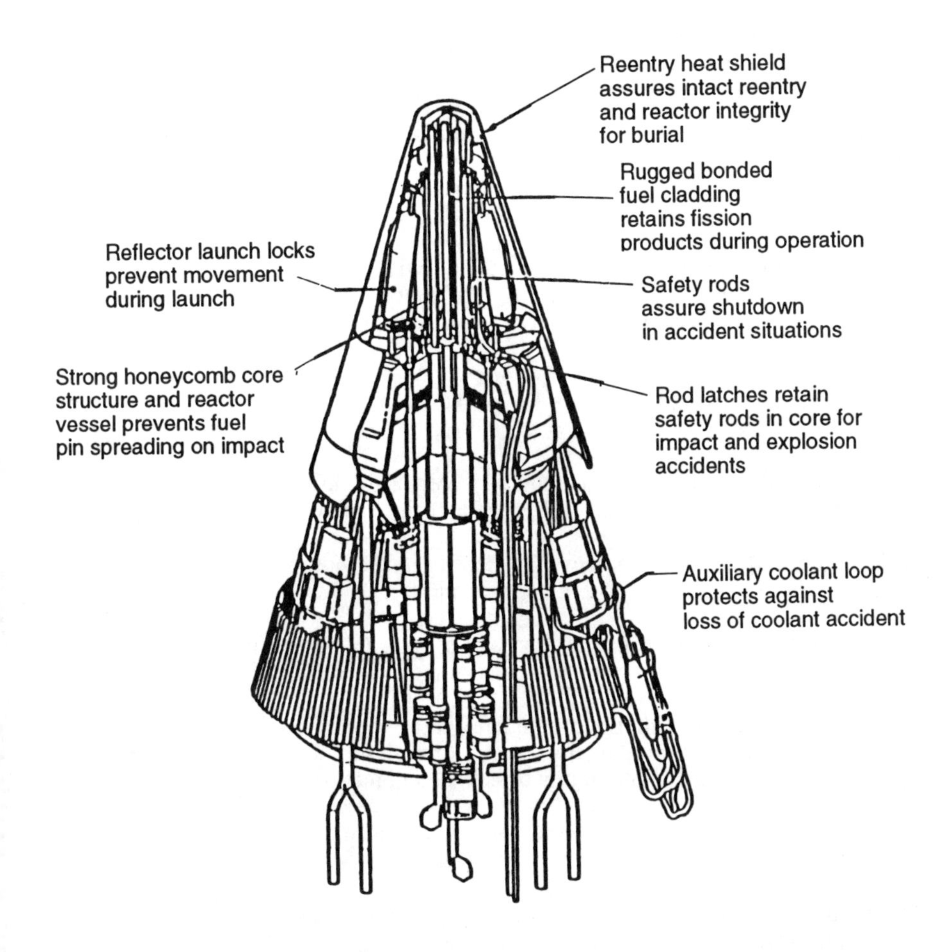

FIGURE 12. SP-100 Key Safety Features (General Electric 1989b).

designed to prevent inadvertent criticality during handling or in accident situations. This is accomplished by including two independent control elements that are physically locked in their shut down positions during ground transport, handling, launch, ascent, and final orbit acquisition (assuming Shuttle launch). These cannot be released until two independent signals are given. A large reactivity shutdown margin is provided by the safety rods to ensure that the reactor remains shutdown should any accident occur. This includes accidents involving severe fires, core compacting, projectile impacts, overpressure, and immersion/flooding environments. At the beginning of the mission, opening of any 7 of 12 radial reflectors or the insertion of any 1 of 3 safety rods will shut the reactor down. As the mission progresses, the number of reflectors required for shutdown reduces, becoming 3 after 7 y of full power operation. The reactor is designed with a prompt negative reactivity coefficient to ensure stable reactor control and enhance shutdown if a loss-of-coolant should occur.

TABLE 4. SP-100 Key Safety Requirements (General Electric 1989a).

---

1. Reactor Operation

Not started and operated (except for zero power testing) until operational orbit achieved.

2. Subcriticality

Remain subcritical to environments associated with credible failures or accidents during assembly, transportation, handling, prelaunch, launch, ascent, deployment, orbit acquisition, shutdown, and transfer to high permanent storage orbit. Minimum situations:

A. Core internal structure and vessel generally intact, all exterior components removed for:

a. All possible combinations of soil and water surrounding core.
b. Reactor vessel exposed to solid propellant fire for 1,000 s.

B. Core internal structure and vessel generally intact, compaction along the pitch line of the pins to produce pin-to-pin contact, and for:

a. All exterior components removed and all possible combinations of soil and water filling and surrounding the core.

b. Normal exterior components and reflectors compressed around the core; exterior absorber material, if any, in its normal shutdown position; and core containing its original coolant or any possible combination of soil and water.

c. All exterior components removed and aluminum surrounding the core containing its original coolant.

C. Reactor vessel as designed, core fuel pins spread radially apart to the maximum distance allowed within the fuel channel lattice design. All exterior components removed:

a. Water filing and surrounding the core.
b. Saturated soil filling and surrounding the core.
c. Dry soil filling and surrounding the core.

Calculated effective reactivities for these conditions, with margin for modeling and calculational uncertainties, shall be <0.98.

3. Response to Fires

Reactor, without reflector elements, neutron shield and reentry heat shield, remain subcritical in the liquid and solid propellant fire environments. Limited melting and creep deformation allowed, however, the as-built geometry essentially maintained.

TABLE 4. SP-100 Key Safety Requirements (General Electric 1989a) (continued).

---

4. Structural Response to Explosions

Reactor remain subcritical in launch vehicle explosion environments.

5. Reentry

For inadvertent reentry following reactor operation, reactor designed to reenter through the earth's atmosphere sufficiently intact to prevent the dispersion of fuel and fission products. Essentially intact defined as:

A. Reentry structural and thermal loads not cause loss of effective fuel/safety rod alignment.

B. If after reactor operation, the reactor structure remains sufficiently intact to allow effective burial on impact.

C. Reentry not breach the reactor vessel nor impair the predictability of its structural response on impact.

6. Burial

Intact reentry through the earth's atmosphere, the reactor capable of producing effective burial as it impacts on water, soil, or pavement-grade concrete. Effective burial defined that the fuel, reactor vessel, and internal components are within the formed impact crater and below normal grade level.

7. Transfer to Permanent Storage Orbit

Designed that reactor can be transferred to high permanent storage orbit at end-of-mission. Reactor designed to prevent such core disruption and structural degradation during and following operation that compromise structural integrity and predictability of desired reactor behavior during final shutdown and transfer to high permanent storage orbit.

8. Final Shutdown

Reactor designed to ensure high-confidence permanent subcriticality at the final shutdown to preclude further production of fission products and activated material and ensure subsequent reduction of radioactive inventory. Final shutdown activated automatically, irreversible, and not initiated or rendered inoperable by any credible single failure or initiating event.

9. Final Shutdown Clock

Final shutdown clock irreversibly interrupt the supply of power to all in-core safety rods and control reflector clutches when preset final shutdown time is reached.

TABLE 4. SP-100 Key Safety Requirements (General Electric 1989a) (continued).

---

10. Loss of Primary Coolant

Reactor designed to accommodate an instantaneous complete loss of main loop primary coolant followed by scram during operational phases without,

A. Exceeding fuel design cladding temperature limits.
B. Impairment of capability to achieve final shutdown.
C. Loss of structural integrity sufficient to impair the capability to boost to permanent storage orbit.

11. Core Heat Removal Capability

Coolability assured with high confidence for all credible accident conditions to maintain the structural integrity and thereby the predictability of desired reactor behavior during final shutdown and transfer to high permanent storage orbit.

12. Nuclear Feedback Calculations

Reactor core and associated coolant systems designed over entire power operating range, net effect of the prompt inherent nuclear feedback characteristics tends to compensate for the rapid increases in reactivity.

13. Power Oscillations

Nuclear subsystems designed to assure that thermal power oscillations which can result in conditions exceeding acceptable fuel design limits are not possible or can be reliably and readily detected and suppressed.

14. Reactivity Control Redundancy

Two means of reactivity control provided. Suitable independence and diversity provided to assure adequate protection against common cause failures. Each of these means capable of performing its nuclear safety function with a single active failure.

15. Inhibits

Until operational orbit achieved, startup shall be precluded by three independent inhibits, one of which precludes startup by radio frequency energy.

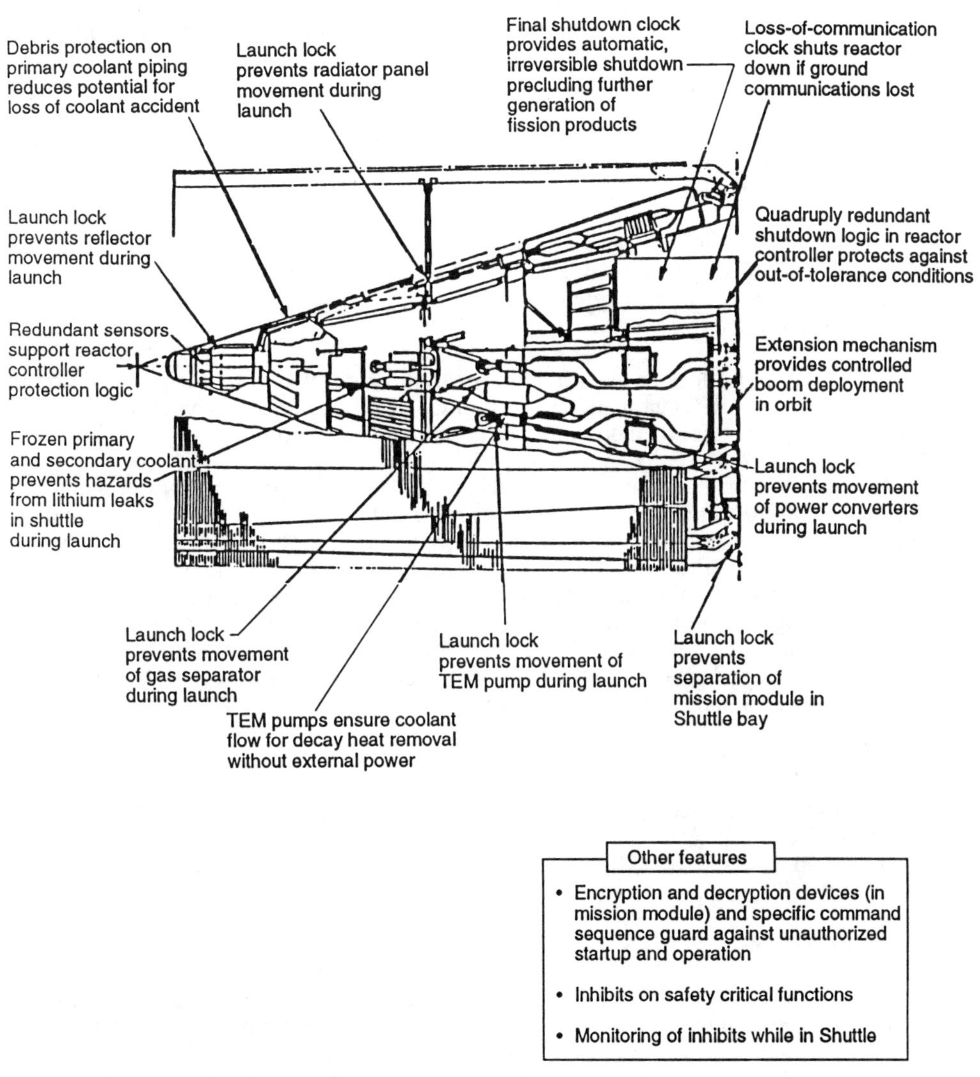

FIGURE 13. Additional SP-100 Safety Features (General Electric 1989b).

The reactor fuel pins include a rhenium poison that acts as thermal neutron absorption in case of water flooding.

While in space, the reactor is protected against impacts of micrometeorites and orbital debris by bumpers. If the debris would cause a loss-of-coolant, the reactor would automatically shut itself down. A loss-of-coolant auxiliary cooling loop is included in some configurations to ensure adequate core heat removal under accident conditions.

The reactor is protected by a reentry heat shield to ensure it remains intact if the reactor should reenter the atmosphere. It is also designed to bury itself on impact, and to remain subcritical if it falls into the ocean.

The SP-100 uses lithium as the working fluid. During launch, it is in the frozen state for added safety so that any launch-induced accident that might cause piping rupture will not endanger crew or equipment. Reactor energy is used for thaw once the operational altitude is reached.

In the event of a loss of electrical power to the control system during operation in space, the reactor will automatically shut down. Reflector elements and safety rods are spring actuated to their shutdown positions upon a loss of power.

As a result of the stringent design features, the reactor is predicted to offer much less radiation hazard to the public than transcontinental airline flights, diagnostic medical examinations, or therapeutic medical services. A mission risk analysis performed indicates no outstanding public safety issues. The analysis quantifies risk from accidental radiological consequences for a reference mission. The total mission risk based on expected population dose is estimated to be 0.05 person-rem based on a 1 mrem/y as the threshold for radiological consequences; this is a negligible amount in absolute terms and relative to the 1.5 billion person-rem/y that the world population experiences from natural radiation sources.

Flight Readiness

A summary of the technical status and challenges as seen in 1991 is given in Table 5 (Mondt 1991). The progress to date on each of the FY 91 technical challenges provides an assessment of flight readiness. Much of this information was prepared by J. Mondt (1993).

System Level

The power-verus-lifetime prediction codes are available and the verification of these codes is dependent on the completion of all component lifetime performance tests being conducted under the subsystems. With regards to verification of a 10-y system, all of the critical components, their failure modes, and the failure mechanisms have been identified. The failure mechanisms have been well defined and analyzed. The failures have been placed in three categories: i.e. 1) design margins determined to be adequate based on existing data, 2) addditional test data needed, which is included in the SP-100 planned effort, and 3) additional test data need which will be obtained during the flight development phase. System start-up and restart from frozen lithium in zero G has been conceptually designed based on component thaw tests and will be verified with system level ground tests. Flight system acceptance tests are still being defined.

TABLE 5. FY-91 Technical Challenges (Mondt 1991).

System Level

- Verified power versus lifetime prediction codes
- Verified reliable 10-y system-design margin codes
- Startup from frozen lithium in zero-gravity
- Flight system acceptance tests

Subsystems

1. Reactor

    - Verified prediction of fuel pin behavior
    - Verified 10-y creep strength of PWC-11
    - Verified transient behavior

2. Reactor Instrumentation and Controls

    - Reflector control drive actuator insulators and electromagnetic coil lifetime
    - Temperature sensors lifetime
    - Radiation hardened multiplexer amplifiers lifetime

3. Shield

    - Verified LiH swelling properties

4. Heat Transport Subsystem

    - Gas separator performance and plugging lifetime
    - TEM pump (TE/Busbar) bond performance and lifetime
    - TEM pump (Cu/Graphite Bus Duct) bond performance and lifetime

5. Converter Subsystem

    - Electrodes and bonds to TE legs lifetime
    - TE cell assembly low cost fabrication
    - High figure-of-merit TE material performance and lifetime ($Z = 0.85 \times 10^{-3} K^{-1}$)
    - Cell to heat exchanger bond performance and lifetime

6. Heat Rejection Subsystem

    - Carbon-carbon to titanium bond performance and lifetime
    - Low cost and low mass heat pipe lifetime

Reactor Subsystem

The fuel pin swelling and fission gas release predictions have been verified up to the 10-y mission lifetime (6% burnup) based on inpile accelerated fuel tests. The 10-y creep strength of Nb-1Zr has been verified by uniaxial and biaxial long term creep test. The very high creep strength of PWC-11 is dependent on the material processing and requires the results of existing long term creep tests, which are now scheduled to be complete in FY 98. The reactor transient behavior will be verified by the first operating SP-100 reactor. The SP-100 reactor component development is complete and technology ready for a flight system.

Reactor Instrumentation And Controls Subsystem

The actuator insulator and EM coil lifetime tests and analyses verify that these components will operate for the 10-y mission lifetime. The temperature sensors are verified for 5-y missions. The lifetime of the multiplexers in a very high radiation field still needs to be verified.

Shield Subsystem

The LiH swelling properties as a function of temperature, radiation dose, and radiation dose rate are verified. The SP-100 shield development is complete and technology ready for a flight system.

Heat Transport Subsystem

The gas separator performance and plugging is scheduled to be verified in FY 94 in a flowing lithium test. The TEM pump bond performance has been verified and the pump lifetime is primarily dependent on the TE cell lifetime. The pump TE cell performance is scheduled to be verified in FY93 and the lifetime by the end of FY98.

Converter Subsystem

The TE cell electrodes and bonds have been developed and incorporated into a prototype TE cell. Three of these prototype TE cells are now on test at design conditions, with one operated for 3,670 h, another for 3,100 h and one just started (from a telephone converstion with D. Matteo of Martin Marietta on 25 May 1993). The low cost fabrication of TE Cells has been factored into the design and will be verified in FY 98 with the manufacture of a large number of cells. The present converter manufacturing process uses hot isostatic pressure bonding of niobium-to-niobium for the cell-to-heat-exchangers bond. Small scale fabrication has shown this bonding to be very successful The high figure-of-merit TE material is progressing and should be available by the end of FY 98.

Heat Rejection Subsystem

An intregral carbon-carbon tube and fin has been developed and successfully bonded to a thin wall (0.085 mm) Nb-1Zr potassium heat pipe, 25 mm diameter by 376 mm long and operated as a radiator heat pipe. Since the Nb-1Zr tubes can be manufactured and are leak tight with such a thin wall it may not be necessary to go to a titanium liner. A high conductivity integral carbon-carbon tube and fin 25 mm diameter by 1 m long has also been fabricated. Nb-1Zr potassium screen-wick heat pipes have been tested for long times (>10,000 hr) with no failures or degration. The heat rejection FY 91 technical challenges are nearly resolved.

Design Growth—Stirling Engine Development

As shown in Figure 9, Stirling engines offer an attractive power conversion system in the hundreds of kWe when mated with an SP-100 reactor. For instance, a 600 kWe power plant using Stirling engines that operate at 1,050 K has a specific power over 45 W/kg. Therefore, in 1985, it was decided to develop Stirling engine technology along with thermoelectric power conversion for SP-100. The Stirling engine development program (Slaby 1987) is based on a free-piston design that features only two moving parts (displacer and power piston), close clearance, noncontacting seals (no wear of mating parts), hydrostatic gas bearings for dynamic members (no surface contact of dynamic components and no oil lubrication necessary), dynamic balancing, and the potential for a hermetically sealed power module. The free-piston Stirling concept utilizes gas springs, which have hysteresis losses.

The program started with a 650-K Space Power Demonstrator Engine (SPDE) technology development and is proceeding with the development of common designs for 1,050-K and 1,300-K (Dudenhoefer 1990). The 1,050-K engines provide the means to demonstrate that the engine technology issues have been solved using easier-to-work-with superalloy materials before demonstration of the refractory or ceramic materials version. The plan is shown in Figure 14 with the goals and specifications for the 1,050-K design in Table 6. A schematic of the 1,050-K engine is shown in Figure 15.

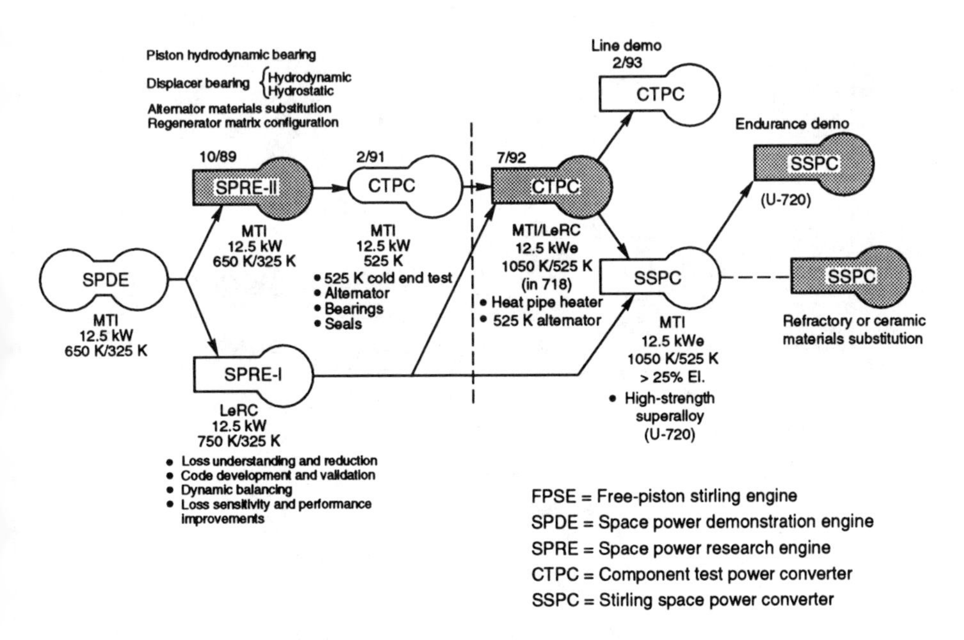

FIGURE 14. Evolution of a High Temperature Stirling Engine (Dudenhoefer 1990 and Dudenhoefer and Winter 1991).

TABLE 6. 1,050 K Stirling Space Engine Goals and Specifications.

Balanced opposed configuration total power output (kWe)	50
End of Lift power (kWe/piston)	25
Efficiency (percent)	>25
Life (h)	60,000
Hot side interface	Heat Pipe
Heater temperature (K)	1,050
Cooler temperature (K)	525
Vibration - casing peak-peak (mm)	<0.04
Bearings	Gas
Specific mass (kg/kWe)	<6.0
Frequency (Hz)	70
Pressure (MPa)	15

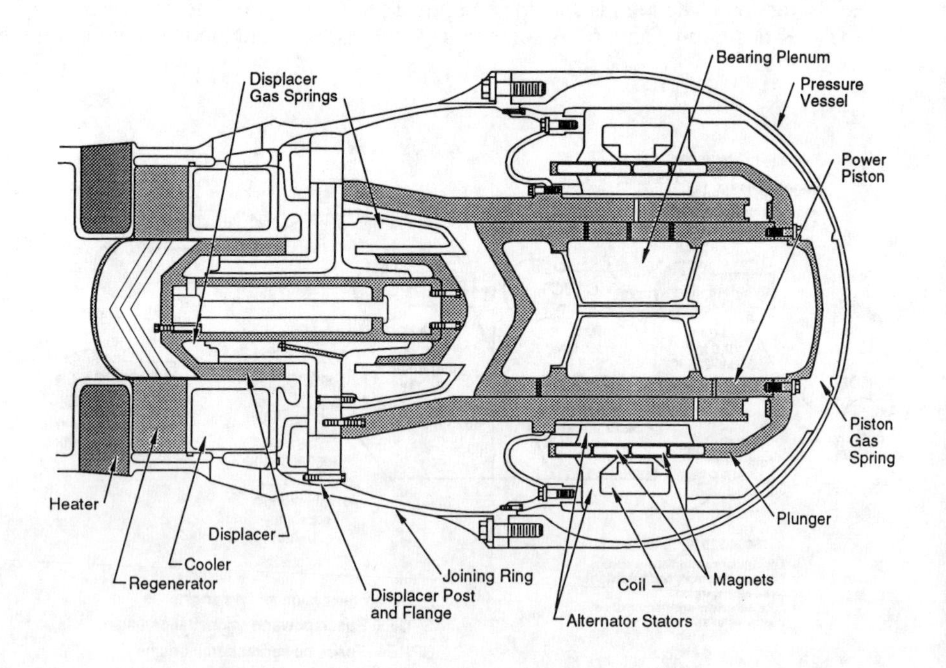

FIGURE 15. Preliminary Design of 1,050 K Stirling Space Power Converter (Dudenhoefer 1990).

In October 1986, the 650-K SPDE demonstrated 25 kWe (Dudenhoefer and Winter 1991). The SPDE was a dual-opposed configuration consisting of two 12.5-kWe converters. After this successful demonstration, the engine was cut in half to serve as test beds for evaluating key technology areas and components, now called Space Power Research Engines (SPRE). The electrical output has been measured as 11.2 kWe at overall efficiency of about 19%. The goal is 12.5 kWe and efficiency greater than 20%. Sensitivity of the engine performance to the displacer seal clearance and the effects of varying the piston centering port area are under study.

The Component Test Power Converter (CTPC) is a 12.5 kWe cylinder technology engine for the Stirling Space Power Converter (SSPC). Inconel 718 is being used as the heater head material to permit early testing for short terms (100-1,000 h at 1,050 K). This testing has demonstrated 12.5 kWe at a 1,050 K operation temperature and greater than the 20% goal. A heater for the SSPC of Udimet 40 L1 superalloy that will have a design life of 60,000 h still needs to be fabricated. The CTPC is being used to evaluate critical technologies identified as: bearings, materials, coatings, linear alternators, mechanical and structural issues, and heat pipes. The impact of temperature on close-clearance seal and bearing surfaces in the cold end of the power converter has been tested, with no problems observed. The CTPC linear alternator uses an alternator that can reach a peak temperature of 575 K, close to the upper operating limits of samarium cobalt magnets. Test results indicate sufficient design margins (Dudenhoefer et al. 1992).

Endurance testing is underway on a 2-kWe free-piston Stirling engine called EM-2 that operates at a heater temperature of 1,033 K. At the end of 5,385 h, only minor scratches were discovered due to the 262 dry starts/stops, and no debris was generated. The heater head of Stirling power conversion systems is the major design challenge because heater head creep is predicted to be the life-limiting mechanism for Stirling engines. The difficulty in creep analysis stems from inadequate knowledge of elevated temperature material behavior and inadequate knowledge of inelastic analysis techniques. Typically, the high operating temperatures and long operating periods of Stirling engines are taxing the ultimate capabilities of even the strongest superalloys.

THERMIONIC DEVELOPMENT PROGRAMS

U.S.S.R. Space Reactors

Since 1967, the U.S.S.R. has orbited approximately 33 thermoelectric reactor power systems as a power source for ocean surveillance radars in satellites called RORSAT. This includes nine in the decade starting in 1983, with the last one on March 14, 1988. Power levels ranged from several hundred watts to a few kilowatts. Limited information is available on the details of the RORSAT power system. We know that the RORSAT power systems are fast reactors using SiGe thermoelectric conversion system. The general characteristics (Bennett 1989) are summarized in Table 7.

In 1987–1988, the U.S.S.R. tested a different type of reactor power system using thermionic power conversion. Two space tests were performed, with one operating six months (Cosmos 1818) and the other operating 346 days (Cosmos 1867). These power plants are designated in the U.S. as Topaz I. Topaz I design output is 10 kWe. The flight-tested units used a multicell thermionic fuel element with an output power of approximately 5 kWe, one with a molybenum emitter and the other with a tungsten emitter. The power system with the tungsten emitter operated for the longer period of time; degradation of performance occurred, with the thermal power increased to compensate for this degradation.

TABLE 7. RORSAT Power System (Bennett 1989).

Thermal Power (kWt)	≤100
Conversion System	Thermoelectric
Electrical Power Output (kWe)	≤5 (~1.3 to 2)
Fuel Material	U-Mo (≥3wt% Mo)
Uranium-235 Enrichment (%)	90
Uranium-235 Mass (kg)	≤ 31 (~20 to 25)
Burnup (fissions/g of U)	$\leq 2 \times 10^{18}$
Specific Fuel Thermal Power (W/g of U)	~5
Core Arrangement	37 cylindrical elements (probably 20 mm dia)
Cladding	Possibly Nb or stainless steel
Coolant	NaK
Coolant Temperature Outlet (K)	≥970
Core Structural Material	Steel
Reflector Material	Be (6 cylindrical rods)
Reflector Thickness (m)	0.1
Neutron Spectrum	Fast (~ 1 MeV)
Shield	LiH (W and depleted U)
Core Diameter (m)	≤0.24
Core Length (m)	≤0.64
Control Elements	6 in/out control rods composed of B_4C with LiH inserts to prevent neutron streaming and Be followers to serve as the radial reflector
Overall Reactor Mass (kg)	<390

A schematic of the fuel element (Bennett 1989), shown in Figure 16, portrays the fuel element as divided into five fuel cells. Urania fuel is used, the cathodes are made from a tungsten or molybdenum alloy, anodes are made from the niobium alloy, insulators are beryllia, outer casings are stainless steel, and cesium vapor is used in the interelectrode gap. A cutaway of the Topaz I reactor showing the principal subsystems and design features is shown in Figure 17. During operation, some hydrogen leaks from the moderator; this is continuously removed by the cesium. During a year's operation about a kilogram of cesium is used.

A second form of thermionic reactor, called Topaz II, has been purchased by the U.S. for testing and evaluation. Topaz II also has a 6 kWe design output, but a single thermionic cell fuel element has replaced the multicell fuel element. A significant advantage of Topaz II is the ability to test the entire system at full temperature in an electrically heated configuration (Nicitin et al. 1992). The single-cell fuel element makes this possible.

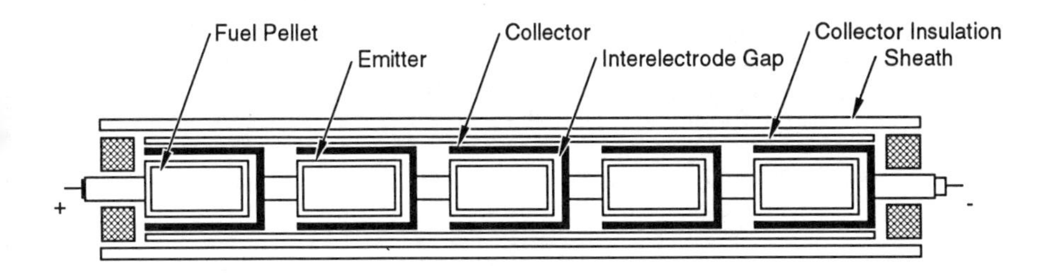

FIGURE 16. Basic Arrangement of the Multicell TOPAZ Thermionic Fuel Element (TFE) (Bennett 1989).

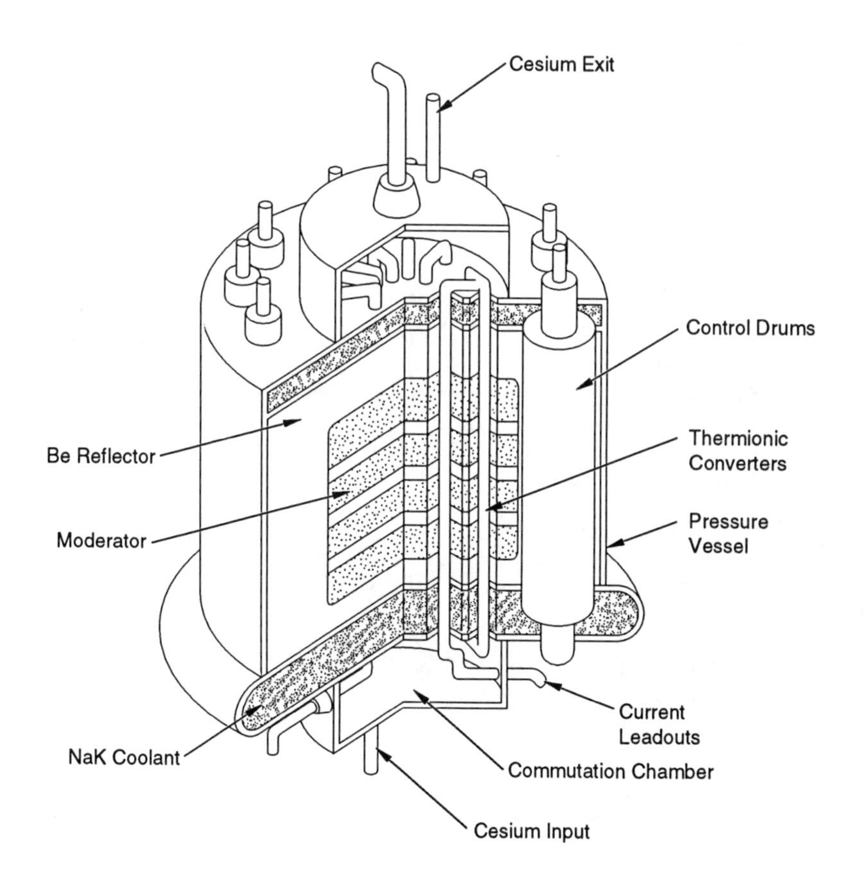

FIGURE 17. Configuration of the TOPAZ I Reactor (Bennett 1989).

The characteristics of the Topaz series of space power systems are summarized in Table 8 (Bennett 1989, updated using information from the TSET/NEPSTP Workshop 1992). The following discussion focuses on Topaz II because it is the system now being tested in the U.S., with plans to use a Topaz II in a U.S. space test in 1996 or 1997, and because more information is available.

TABLE 8. TOPAZ Reactor Characteristics.

	TOPAZ I	TOPAZ II
Electrical Power (kWe)	5 to 6	6
Voltage at Lead (V)	5 to 30	28 to 30
Thermal Power (BOL/EOL)(kWt)		115/135
Number of TFEs	79	37
Cells/TFE	5	1
System Mass (kg)		1,061
Emitter Diameter (cm)	1.0	1.73
Core Length (cm)	30	37.5
Core Diameter (cm)	26	26
Reactor Mass (kg)		290
Moderator	$ZrH_{1.8}$	$ZrH_{1.85}$
Emitter	Mo/W	Mo/W
Emitter Temperature (K)	1,773	1,800 to 2,100
Coolant	Pumped NaK	Pumped NaK
Number of Pumps	1	1
Type of Pump	Conduction	Conduction
Pump Power	19 TFEs	3 TFEs
Reactor Outlet Temperature (BOL/EOL)(K)	773/873	560/600
Coolant Flow Rate (kg/s)		1.5
Cesium Supply	Flow Through	Flow Through
Axial Reflector	Be Metal	Be Metal
Radial Reflector	Be Metal	Be Metal
Number Control Drums	12	12
Shield Mass (kg)		390
Radiator		Tube and Fin
Radiator Area (m^2)		7.2
Radiator mass (kg)		50

Topaz II Description

A schematic of Topaz II is shown in Figure 18. At the beginning of life, the reactor produces approximately 115 kWt with a conversion efficiency of 5.2%; and at the end of life, the reactor produces 135 kWt with a conversion efficiency of 4.4%. The Topaz

II is cooled by a liquid metal of eutectic sodium-79% potassium-21% (NaK) that remains liquid during all phases of the Topaz II lifetime, excluding the end-of-mission shutdown. The NaK coolant removes the waste heat from the reactor and transports it to the radiator, where it is rejected to space.

The Topaz II core is a right circular cylinder 260 mm in diameter and 375 mm high (Space Power Inc. 1990). Thirty-seven cylindrical thermionic fuel elements (TFEs) are arranged in an approximately triangular pitch within a block of moderator. The

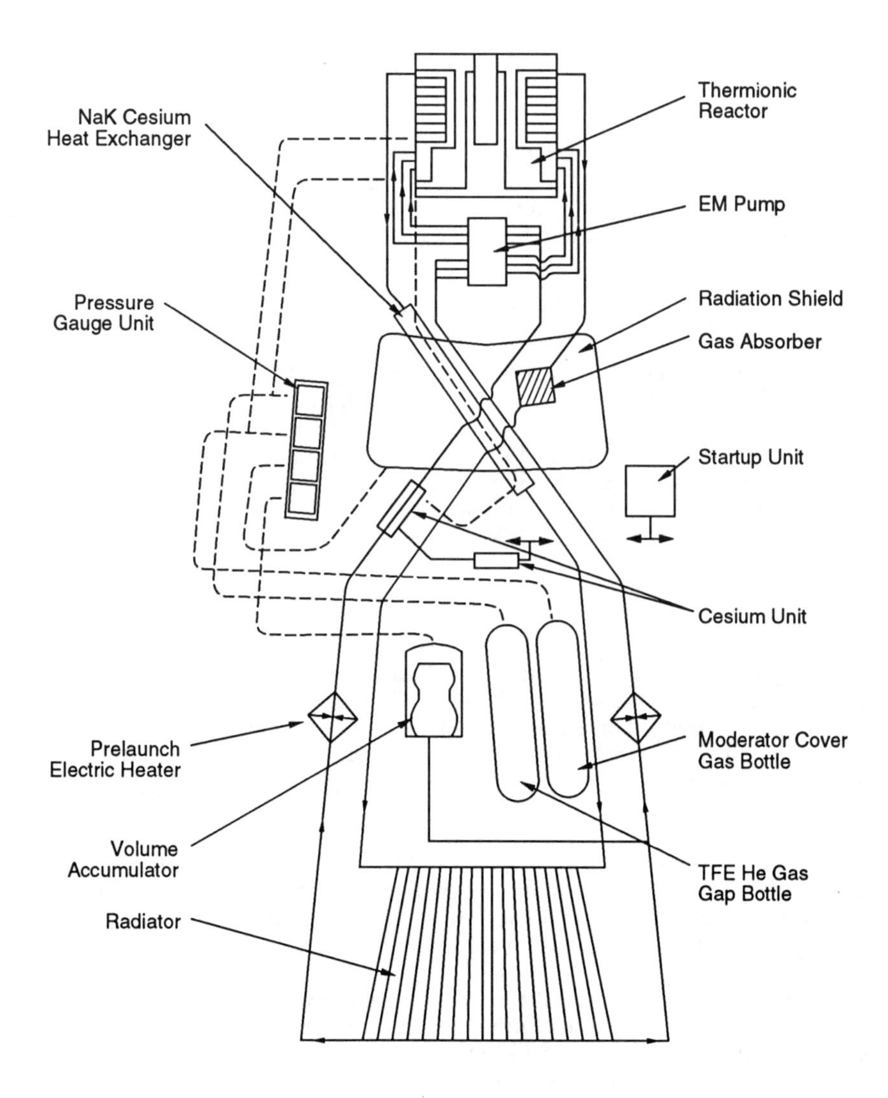

FIGURE 18. TOPAZ II Schematic (Topaz CoDR 1992).

moderator is epsilon phase zirconium hydride (ZrH_{2-x}) with hydrogen stoichiometry in excess of 1.8. The moderator is canned in a stainless steel calandria that has 37 circular channels in it to accommodate the TFEs and NaK coolant. The coolant channel gap between the TFE outer sheath and the calandria wall is a grooved surface.

A gas mixture of approximately 50% CO_2, 50% helium, and other trace gases is maintained within the moderator/axial reflector region to help inhibit the release of hydrogen from the $ZrH_{1.85}$, and to increase the heat transfer from the $ZrH_{1.85}$ to the outer surface of the vessel and to the coolant channels.

The thin-walled, stainless steel cylinder reactor vessel encloses the TFEs, moderator calandria, and axial beryllium metal neutron reflectors. It supports the core and TFEs and provides plena for the cesium vapor, helium gas, and NaK coolant.

Outside the reactor vessel within the radial reflector are 12 rotating control drums. Three of these drums are used in the safety system. The remaining nine are for control, and are driven by a common mechanism. The radial reflector assembly is held together with two fused tension bands. The control drum bearings and drive trains are lubricated with MoS_2.

Figure 19 is a cross section of the Topaz II TFE. The fuel contained within each thermionic fuel element is highly enriched (96%) UO_2 in the form of pellets stacked within the cavity of the emitter. Each fuel pellet is ~8 mm high with an outer diameter of 17 mm. The fuel pellets possess a central hole with a diameter of 4.5 mm or 8 mm, depending upon the position within the core, to help flatten the power profile in the radial direction. The fuel height is 355 to 375 mm, where the height of the fuel can be varied at the time of loading to compensate for variations in fabrication that can affect core excess reactivity. The maximum fuel temperature is ~1,775 to 1,925 K, and the end temperatures are ~1,575 K.

Topaz II has regenerating cesium supplies and vents fission gases outside the reactor. The emitter contains the fuel and fission products and serves as the source of thermal electrons. Emitter strain limits system life, so the mechanical properties of the emitter are extremely important. The emitter is made from monocrystal molybdenum alloy substrate with a chemical vapor deposition (CVD) coating of tungsten. The tungsten coating is for improved thermionic performance. This coating is deposited from chloride vapor, and is also monocrystalline. The outer tungsten layer is enriched in the isotope ^{184}W in order to limit the adverse neutronic effect. The emitter temperature is 1,873 K.

Beryllium oxide (BeO) pellets on both ends of the fuel stack provide axial reflection. The BeO pellets have central holes that match up with the holes in the fuel. The pellets are stacked to a height of 55 mm. They are used to compensate for variations in the fuel loading. The total height of the core, including the BeO end reflector, is 485 mm.

The monocrystal molybdenum collector tube is coaxial with the fueled emitter. High collector temperatures are desirable to reduce radiator size, while lower temperatures reduce thermionic back-emission and keep the dissociation pressure of hydrogen in the moderator within bounds. A temperature of 925 K is used at the outlet of the reactor; the temperature is about 100 K lower at the inlet.

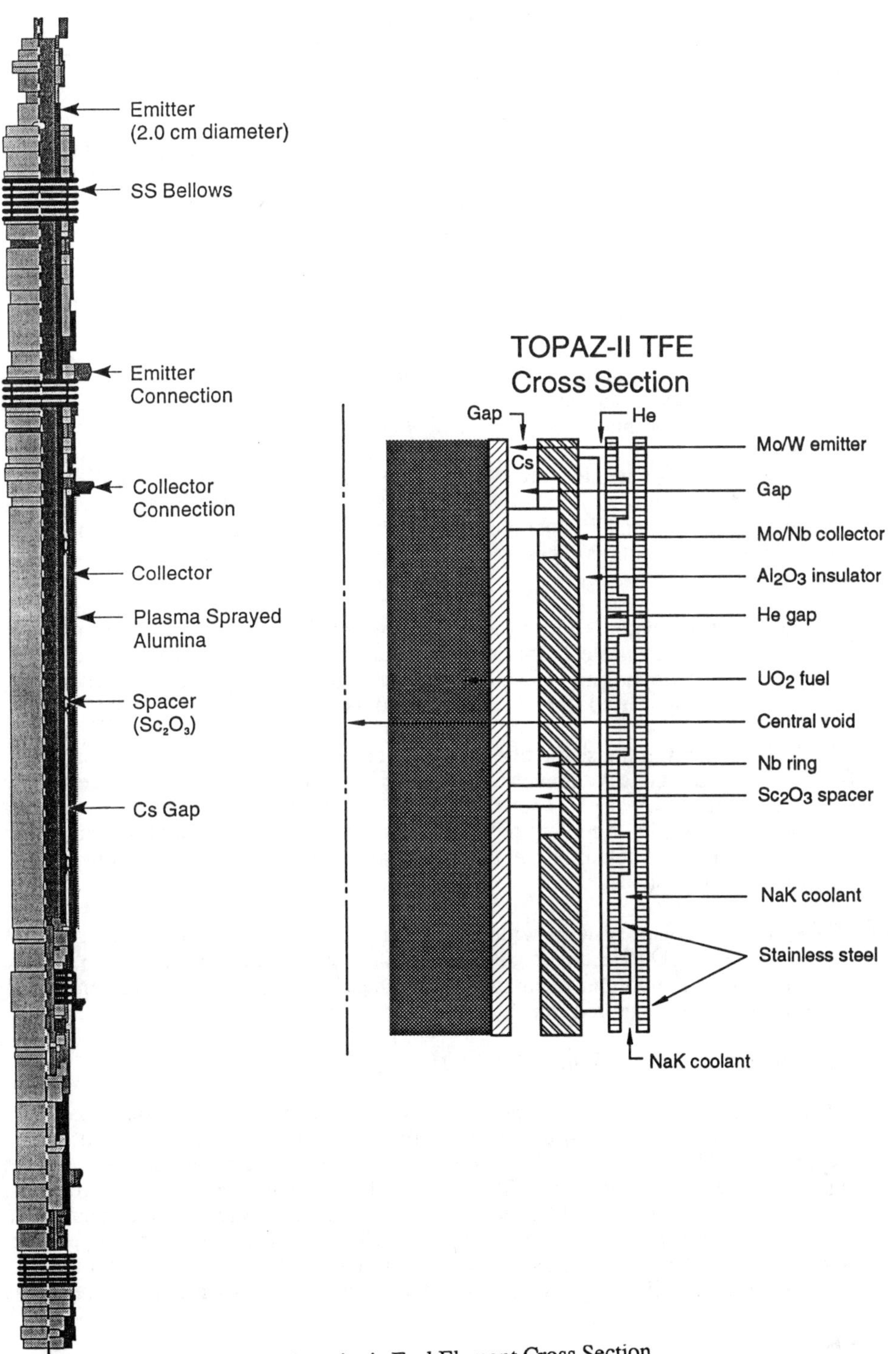

FIGURE 19. TOPAZ II Thermionic Fuel Element Cross Section.

Between the collector and emitter is the interelectrode thermionic gap. This gap is 0.5 mm (U.S. Topaz II Flight Safety Team 1992). There are scandium-oxide (Sc_2O_3) spacers between the emitter and collector surfaces in the interelectrode gap to prevent the emitter from shorting to the collector as a result of emitter distortion caused by fuel swelling.

External to the collector is a helium gap between the collector insulator and inside diameter of the inner coolant tube. The helium provides a good thermal bond for the transfer of heat to the coolant, while maintaining electric insulation of the thermionic fuel element. A helium bottle is located in the radiator region that maintains helium in the gap over the lifetime of the system.

The stainless steel sheath completes the TFE structure. The sheath supports the TFE and provides a heat transfer surface to the NaK coolant.

The fuel is vented from its interior directly to space. The cesium supply is a still that feeds cesium vapor to the interelectrode gap and vents any gases to space. Cesium venting is 0.5 g/d for this system (Marshall et al. 1993).

The Topaz II uses eutectic NaK to remove heat from the reactor. The NaK must be kept liquid during launch. A single DC conduction pump powered by a group of three dedicated TFEs connected in parallel is used. The coolant piping is divided into two groups of three channels as it passes through the pump. An on-board current source is used for startup.

A tube-and-fin radiator is employed in the shape of a truncated cone. The surface of this cone is formed from steel tubes welded to circular manifolds at the top and bottom of the radiator. Copper fins are welded to these tubes. To improve emissivity of the fins, a glass coating is used that adheres well and has good thermal resistance.

A shadow shield in the shape of a truncated cone is used for radiation attenuation. Both end caps are concave downward, spherical, and thick walled. The sides of the shield are thin-walled steel. The space between the end caps is filled with LiH for neutron shielding. Four coolant pipes pass through the shield at angles designed to minimize radiation streaming. A stepped channel through the shield contains the control drum drive shaft.

The gas and coolant systems within the Topaz II include the NaK coolant, cesium supply system, CO_2/He cover gas, the He thermionic fuel element gas gap, the argon/He gas in the volume accumulator, and the helium gas in the radiation shield. The last four systems are fed from pressurized bottles.

Technology Status of Topaz I and II

Topaz I demonstrated 1 y operation in space, while Topaz II has demonstrated 1.5 y of nuclear ground testing. A claim of 3-y lifetime is based on component life data. Experience with the multicell TFE indicates that swelling and intercell leakage are significant life-limiting problems. The single cell TFE tends to correct these problems, partially by using a high void fraction. A primary life limiting element appears to be the loss of hydrogen from the ZrH moderator. The rate is about one percent per year. Also, with only 65 cents of excess reactivity at design (Topaz CoDR 1992) the fuel burnup can be life limiting, especially if the reactor cools down before a restart is achieved Another issue is the oxygen getter. Modifications may be needed in the cesium supply, but these do not appear to be life limiting.

Topaz I

For the thermionic fuel elements in Topaz I, the chloride-deposited tungsten layers on Mo alloy monocrystalline substrates have been subjected to in-pile tests of up to 17,340 h duration and high temperature (1,900 K) tests of over 13,000 h. For the collector, after 1.5 y of in-pile testing opposite a tungsten emitter, a condensed mass transfer layer has reached a thickness of 300 nm. The layer consists primarily of tungsten, but also contains some oxygen, carbon, and cesium. The UO_2 fuel behavior under irradiation is very complex. The high temperature at the outer edge (~1,900 K), combined with the relatively large diameter of the fuel pin (~10 to 15 mm), implies high temperatures at the center of the fuel pin. Within a matter of hours after startup, the fuel column restructures so that the original stack of pellets has become a single fused structure. The interior of this structure is an isothermal void, which extends the length of the TFE. The 25 to 40 mm gap between the fuel pellets and the interior of the emitter closes due to sublimation and redistribution of the fuel. This void and the micropores originally present in the UO_2 fuel pellet are swept to the interior void. As burnup proceeds, xenon, krypton, and a few volatile fission products are similarly swept from the fuel into the interior. This central void is vented by a passage in a screw hold-down plug fitting in the end of the TFE.

Extensive testing on components has been performed by the U.S.S.R. Coated zirconium hydride material to reduce hydrogen loss has been tested in a reactor with losses of hydrogen being less than 1% of the initial inventory. Zirconium hydride swelling data from neutron irradiation show a volume change of 2 % at 823 K for a fluence of 1.5 x 10^{21} n/cm^2. The temperature in the system is actually 50 to 75 K less during design operation. Cladding the hydride improves performance by six orders of magnitude, and the use of a cover gas was found to provide another factor of five improvement.

Topaz II

For Topaz II, extensive component and systems testing has been performed. Table 9 provides summary information for some of the major components including the number of components tested, the type of testing, and the time at test. During the development phase, component testing was done in two stages. The first preliminary assurance of functional sufficiency and identified potential problems in an informal manner. Later, more formal tests were performed to ensure that the components met the defined acceptance criteria.

Table 10 (prepared by Susan Voss, Los Alamos National Laboratory) summarizes the power plant testing data on Topaz II. The longest test was 14,000 h. Transient behavior, shown in Figure 20, shows stable operation.

A new automatic control system must be designed that meets U.S. qualification standards for space applications and that is compatible with U.S. launch systems. Also, it is uncertain whether the nuclear fuel will be procured from Russia or fabricated in the U.S. If fabricated in the U.S., the fuel will need to meet both U.S. and Russian quality standards.

TABLE 9. TOPAZ II Component Testing.

Single Component Tests				System Test		
Component	Test Description	Number of Tests	Time on Test (h)	Test Description	Number of Test	Time on Test (h)
1. Reactor	Thermo-Physical	3	12,000	Electrical Tests	7	12,500
	Mechanical (Dynamic)	2		Mechanical		
				• Ground Transport	4	
				• Static	2	
				• Dynamic	4	
				Cold Testing	4 & 1	
				Nuclear Tests	6	
2. Control Drum Drive	Operational	5	13,000	Electrical Tests	6	14,000
	Lifetime	6		Mechanical		
				• Ground Transport	1	
				• Static	1	
				• Dynamic	3	
				Cold Testing	3	
				Nuclear Tests	4	
3. Safety Drive	Mechanical	3		Mechanical		
	Thermo-Physical	3		• Ground Transport	12	
	Climatic	3		• Static	6	
	Operational	3		• Dynamic	12	
				Cold Testing	12 & 3	
4. Radiation Shield	Mechanical (Static)	2		Electrical Tests	7	14,000
	Lifetime	2		Mechanical		
	Characteristics and Material Changes			• Ground Transport	4	
				• Static	2	
				• Dynamic	4	
				Cold Testing	4 & 1	
				Nuclear Tests	6	
5. Cesium Unit	Thermal Lifetime	7	26,400	Electrical Tests	7	14,000
	Operational Lifetime	3		Mechanical		
				• Ground Transport	4	
				• Static	2	
				• Dynamic	4	
				Cold Testing	4 & 1	
				Nuclear Tests	6	
6. Radiator	Thermal Lifetime	1	10,000	Electrical Tests	7	14,000
				Mechanical		
				• Ground Transport	4	
				• Static	2	

TABLE 9. TOPAZ II Component Testing (continued).

Component	Single Component Tests: Test Description	Number of Test	Time on Test (h)	System Test: Test Description	Number of Tests	Time on Test (h)
				• Dynamic	4	
				Cold Testing	4 & 1	
				Nuclear Tests	6	
7. Volume Compensator	Thermal Lifetime	7	40,000	Electrical Tests	7	14,000
	Operational	5		Mechanical		
				• Ground Transport	4	
				• Static	2	
				• Dynamic	4	
				Cold Testing	4 & 1	
				Nuclear Tests	6	
8. Ionization Chamber	Thermal Lifetime	3	26,300	Electrical Tests	7	14,000
				Mechanical		
				• Ground Transport	4	
				• Static	2	
				• Dynamic	4	
				Cold Testing	4 & 1	
				Nuclear Tests	6	
9. Pressure Sensor Unit (4)	Lifetime	1	13,500	Electrical Tests	6	2,000
	Operational	1		Mechanical		
				• Ground Transport	3	
				• Static	2	
				• Dynamic	3	
				Cold Testing	3 & 1	
				Nuclear Tests	6	
10. Start-up Unit	Operational	3		Electrical Tests	3	
	Mechanical			Mechanical		
	• Ground Transport	6		• Ground Transport	3	
	• Dynamic Launch	2		• Static	1	
	Thermal Lifetime	6		• Dynamic	3	
				Cold Testing	1 & 1	
				Nuclear Tests	1	
11. Thermal Cover	Mechanical (Static)	2		Mechanical		
	Mechanical (Dynamic)	1		• Ground Transport	4	
	Operational	2		• Static	2	
				• Dynamic	4	
				Cold Testing	4 & 1	

TABLE 10. TOPAZ II Systems Test Data.

Test	Date	Time at Test	Findings
Plant 23	1975-76	2,500 h at 6 KWe 5,000 h total test	Ground Demonstration Unit. Significant degradation of power due to TFE.
Plant 31	1977-78	4,600 h	Flight Demonstration Unit. Same as TFEs as Plant 23 Startup with Automatic Control System (ACS). 4,600 h was planned test time.
Plant 24	1980	14,000 h	Ground Demonstration Unit. Same TFEs as Plant 23 Startup with ACS Provided life testing of many components.
Plant 81	1980	12,500 h at 4.5 to 5.5 kWe	Ground Demonstration Unit. New TFE design. Did not complete 1000-h electric testing prior to nuclear testing. NaK leaked 150 h into nuclear test. Fixed NaK leak and continued testing.
Plant 82	1983	8,300 h at 4.5 to 5.5 kWe	Ground Demonstration Unit. Loss of flow progressing to a loss-of-coolant accident. Final shutdown of the reactor due to loss of H from the $ZrH_{1.85}$. No major structural damage to the reactor.
Plant 38	1986	4,700 h at 4.5 to 5.5 kWe	Flight Demonstration Unit. Test ended due to a NaK leak in upper radiator collector First ground test with the temperature regulator as part of the ACS.

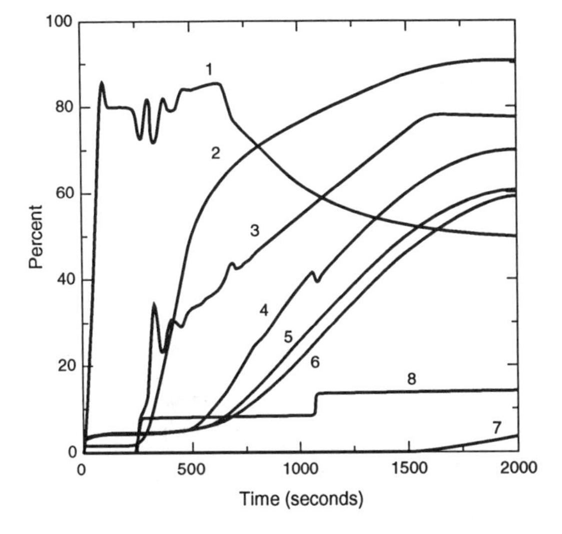

Parameter	Range
1. Drive position	0 to 180 deg
2. Temperature at the fuel element center	0 to 2275 K
3. Reactor power	0 to 150 kW
4. Coolant temperature at the reactor outlet	0 to 1075 K
5. Average moderator temperature	0 to 1075 K
6. Vessel temperature	0 to 1075 K
7. Load current	0 to 400 A
8. Coolant flow rate	0 to 6 kg/s

FIGURE 20. TOPAZ II Regular Power Plant Startup (Topaz CoDR 1992).

Safety

Topaz II was originally designed for operation in geosynchronous orbit. Therefore, the Russians were not concerned with dispersal after operation. If reentry occurs, it is doubtful that complete break up at sufficiently high altitudes will achieve full reactor dispersal (Topaz CoDR 1992).

A number of design modifications are being considered to meet U.S. safety philosophy. These include the inclusion of a reentry thermal shield to avoid breakup if reentry occurs and the addition of removable poison in the annulus of a number of TFEs or removal of fuel during launch to ensure against reactor supercriticality when immersed and flooded with water. For the planned U.S. flight in 1996 -1997, the initial operational altitude is planned to be well above a sufficiently high orbit for fission product decay.

Another safety concern is the delayed positive temperature coefficient in the reactor core. This coefficient has been confirmed experimentally, and the Topaz reactors have been proven to be experimentally controllable. The delayed positive temperature coefficient effect is due primarily to moderator spectrum hardening when temperature is increased. This leads to fewer moderator captures and more fuel captures. The positive feedback time constant is very long (~330 s) relative to the control system. The long time constant is due to a large heat capacity and high thermal resistance. An effect of having a delayed positive feedback coefficient is to reduce the amount of excess reactivity required to very low levels (65 cents). This results from no negative temperature defect and a minimum of burnup reactivity loss to compensate for. The reactor has the unique feature that startup prompt disassembly accidents are not probable.

If the reactor has a loss of coolant accident and the control system does not shut down the reactor, Topaz II has a built-in safety feature to shut down the reactor. The ZrH moderator will heat up, releasing the hydrogen, shutting down the reactor.

Flight Readiness

A launch of Topaz II is being planned by the U.S. in 1996-1997. It will use information from two unfueled reactors that the U.S. purchased for electric heating testing in the Thermionic System Evaluation Test (TSET) program (TSET/NEPSET Workshop 1993). Goals of this program include: learning from the Russians, evaluating performance within the design limits of Topaz II, and training a cadre of U.S. experts on space nuclear power systems. By using purchased Topaz II power plants, the U.S. obtains insights into Russian technology, insights into complete non-nuclear satellite qualification and acceptance methodology, knowledge of Russian safety methods, and reduced cost to develop U.S. thermionic systems. First heat-up testing of TSET got underway in November 1992 (Thome 1992).

Additional reactors are being purchased for the space program. The space program will demonstrate the feasibility of launching a space nuclear power system in the U.S. and evaluate in-orbit performance of the Topaz II power system. It will also demonstrate electric propulsion options and measure the nuclear electric propulsion self-induced environments. Modifications to the Topaz II power plant will include a new automatic control system, acquiring the nuclear fuel, possibly adding the thermal reentry shield, and adding removable poison in the TFEs or removeable fuel for launch.

Thermionic Fuel Element (TFE) Verification Program

In 1986, the Thermionic Fuel Element Verification Program was initiated to resolve the technical issues identified with thermionic reactors during the SP-100 Phase I concept selection. The program's objectives are to resolve technical issues for a multicell TFE suitable for use in thermionic space reactors with electric power output in the 0.5 to 5 MWe range and a full-power lifetime of 7 to 10 y. The main concerns are fuel/clad swelling, insulator integrity for long irradiation times, and demonstration of performance and lifetime of TFEs and TFE components (Brown and Mulder 1992). The program has been restructured to better address the performance and lifetime requirements currently of interest to the DoD (i.e., 5 to 40 kWe, 1.5 to 5 y lifetime).

System Concept

Thermionic power systems are an attractive option for providing electric power in space because their radiators tend to be smaller than for other concepts (higher heat rejection temperatures and/or efficiencies); outside of the fuel and emitter, the temperatures are sufficiently low that refractory metals are not needed; and scaleability is from a few kilowatts to megawatts (Homeyer et al. 1984, Snyder and Mason 1985 and Holland et al. 1985). The heart of these systems is the thermionic fuel element

(TFE). A single thermionic fuel cell is shown schematically in Figure 21 (Strohmayer and Van Hagan 1985). The TFE is a building block, as seen in Figure 22.

One of the objectives of the TFE Verification Program is to size the cell for a 2 MWe space nuclear power system with a 7-y operating life. Another objective is to demonstrate scaleability from 500 kWe to 5 MWe. Table 11 summarizes systems parameters for the 2 MWe system; Table 12 defines the TFE design parameters (General Atomics 1988).

TABLE 11. 2 MWe System Parameters (General Atomics 1988).

Parameter	Value
Net electrical power (MWe)	2
System efficiency (percent)	8.9
Lifetime at full power (y)	7
Shield cone half-angle (deg)	12
Reactor thermal power (MWt)	22.5
Coolant	Lithium
Heat rejection (MWt)	20.3
Radiator temperature (K)	1,020
Radiator area (m^2)	426
System Mass (kg)	
Reactor	8,720
Shield	2,803
Coolant loop	1,788
Main radiator	6,390
Power conditioning and shielding	1,230
Power conditioning radiator	262
Cables	506
Structure and miscellaneous	791
Total	22,490

Technology Developments

TFE cells were extensively developed during the 1960s and early seventies with the operation of elements out-of-pile of over 47,000 h and over 12,000 h in-pile (Samulelson and Dahlberg 1990). The major issues with developing a 7-y TFE are fuel/clad dimensional stability, seal insulator integrity, and sheath insulator integrity under temperature, irradiation, and applied voltage. Since 1986, component and TFE cell verification testing has been underway through accelerated and real-time irradiation testing and analytical modeling. The results of these activities are summarized in Tables 13 and 14 (Begg et al. 1992).

Significant progress in the TFE Verification Program include:

- Higher burn up for fuel emitters to 3% at 1800 K and 4% at 1700 K equivalent to over 5 y operation
- Sheath insulators tested under 10-12 V for 8 mo (the test were stopped by a water leak with the test samples still in good condition)

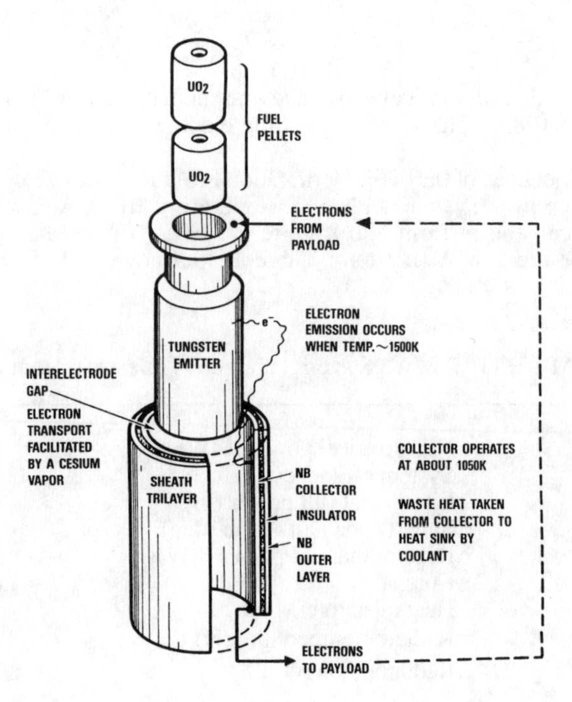

FIGURE 21. Schematic of a Thermionic Fuel Cell (Strohmayer and Van Hagan 1985).

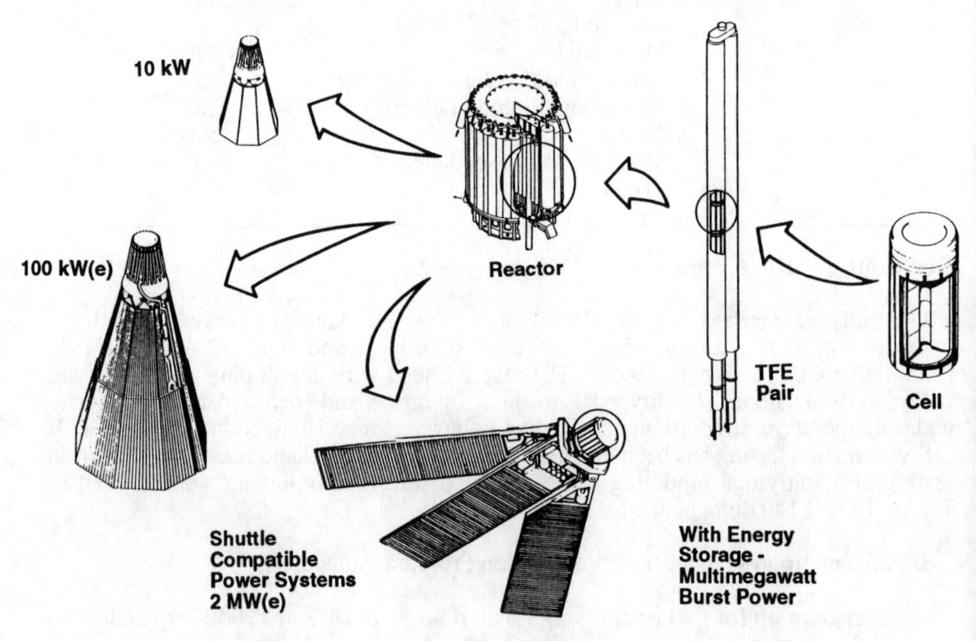

FIGURE 22. Thermionic Cell is the Building Block for Space Nuclear Power Systems to All Levels (Samuelson and Dahlberg 1990).

TABLE 12. TFE Design Definition (General Atomics 1988).

Performance	Values
Overall TFE:	
Output electrical power (We)	705
Efficiency (%)	8.9
Maximum voltage (v)	5.9
U-235 burnup (a/o)	4.1 average, 5.3 peak
Fluence (n/cm^2)	2.7 x 10^{22} average, 3.5 x 10^{22} peak
Converter:	
Converter power (Wt/We)	658/58.8
Thermal power/length (Wt/cm)	137.8
Emitter power flux (We/cm^2)	2.9
Diode current density (A/cm^2)	7.0
Thermionic work function (eV)	4.9
Emitter temperature (K)	1,800
Collector temperature (K)	1,070
Cesium pressure (Pa)	2.7
Converter output voltage (V)	0.49
Converter current (A)	140
Configuration	
Overall TFE:	
TFE length (active core) (cm)	100.6
TFE length (overall)	TBD
Sheath tube O.D. (cm)	1.8
Lead O.D. (cm)	2.2
Lead length (cm)	10.2
Converters per TFE	12
Converter	
Emitter O.D. x L x t (cm)	1.3 x 5.1 x 0.1
Emitter stem L x t (cm)	1.1 x 0.05
Diode gap (cm)	0.025
Trilayer thickness:	
collector(cm)	0.07
insulator (cm)	0.04
outer cylinder(cm)	0.07
Fuel specification	93% enriched UO_2; variable volume fraction
Intercell axial space (cm)	1.88

TABLE 13. TFE Verification Program Key Components Demonstrated Lifetimes (Begg 1992).

Key Components	Tested To	Life in 40 kWe Design (y)
Sheath insulators	4×10^{20} n/cm^2 (in-core with voltage)	0.75
Insulator seals	2.3×10^{21} n/cm^2	4.3
Graphite reservoir	3×10^{22} n/cm^2	>>10
Fueled emitters	>3% at 1800 K	>5
TFEs	13,500 h	1 to 1.5

TABLE 14. TFE Test Results (Begg 1992).

TFE Designation	Time at Power (h)	Cause of Performance Degradation	Comments
1H1	12,000	In PIE	First TFE test in 16 y
1H3	13,500	Fission gas related interelectrode space (planned)	Mixed fission gasses into
1H2	8,800	Internal short caused by over voltage (external)	First graphite reservoir
3H1	13,000	Under evaluation	Still at power, output performance started declining at 8,000 h)
3H5	3000	N/A	Good output, end of February 1993
6H1	N/A	N/A	Planned start in March 1993

- Insulator seals in converters still functional after testing for 39,000 h
- Graphite reservoirs radiation tested to equivalent of over 10 y operation
- TFEs tested to 13,500 h.

Forty Kilowatt Thermionic Power Systems Program

In 1992, a 40-kWe thermionic power plant program was initiated. The design life goal is 10 y; however, the initial lifetime requirement is 1.5 y. Currently, there are two concepts selected. One, called S-Prime Thermionic Nuclear Power System, builds on the multicell TFE Verification Program as well as Topaz I technology. The second concept, called the SPACE-R Thermionic System, builds on the single cell Topaz II technology. Initial mass calculations at 40 kWe indicate both systems have a specific power of 18 W/kg and growth capabilities above 100 kWe. This program is just getting underway with a plan to complete preliminary designs and demonstrate key technologies and components supporting these designs by the end of 1995 (Phillips Laboratory 1992).

RADIATIVELY COUPLED THERMIONIC SYSTEMS

Radiative coupling allows physical separation of the nuclear fuel from the thermionic energy converters. Heat is conducted from the fuel to the surface of the fuel and radiatively coupled to the thermionic emitters. This eliminates traditional thermionic fuel element constraints (such as fuel swelling effects on emitter dimensions) and permits the separate development of the thermionic converters and the fuel. Also, radiatively coupled systems, because of the reactor arrangement, permit the elimination of the primary coolant loops and the development of systems without single failure points. However, the fuel temperatures are higher in these configurations, operating over 2,000 K. Radiatively coupled systems can meet the DOD's interest for very compact, highly survivable, ~ 40 kWe systems. SEHPTR and STAR-C are two possible configurations that have received some funding to meet these needs.

Small Externally-Fueled Heat Pipe Thermionic Reactor (SEHPTR)

The SEHPTR power reactor concept uses hexagonal fuel elements of UO_2 clad in tungsten. The outer surfaces of the fuel elements radiate heat to thermionic heat pipe module energy conversion elements. The elements are single cell core-length thermionic energy converters configured with a cylindrical emitter sleeve on the outside, surrounding a central collector on the inside. The external cylindrical emitter is a tungsten annular heat pipe that receives the radiated heat from the tungsten fuel surface and distributes it isothermally both axially and circumferentially. The inner surface of the heat pipe emitter forms the converter cathode surface. The converter anode is an internal cylindrical sleeve bonded to a central collector heat pipe through an electrically insulating material. The insulator electrically isolates the thermionic converter cell from the central collector heat pipe. The collector heat pipe transports the waste heat from the converter anode sleeve to the radiator. The emitter and collector heat pipes are separated by the interelectrode gap, which is maintained by ceramic centering spacers. Sixty-two thermionic heat pipe units are employed to provide 40 kWe.

In addition, the SEHPTR power system can be configured so that hydrogen can pass between the fuel elements and thermionic modules (Figure 23, Zubrin et al. 1992). Using this configuration, a 40 kWe/400 kWt power plant would have a mass of 2,600 kg and generate 111 N of thrust at a specific impulse of 7,140 m/s. The combination of nuclear thermal propulsion and power permits the use of less expensive launch vehicles for many missions. This ability to combine thrust and power may be especially useful for defense satellites where it allows launch on demand of satellites when needed, rapid development and redeployment of satellites, and higher powers for surveillance, communications, electronic jammers and illumination missions.

SEHPTR was designed to eliminate single failure modes and to provide for electrical testing with electric heaters substituted for the fuel. This greatly simplifies the development and qualification process. The major design concerns with SEHPTR development are: higher fuel temperatures, long term stability of the fuel element including fission product management, long term emissivity of the fuel to converter, and lifetime of the converter heat pipes. Dimensional stability is not nearly as critical in this configuration as in the fueled thermionic fuel element configurations. However,

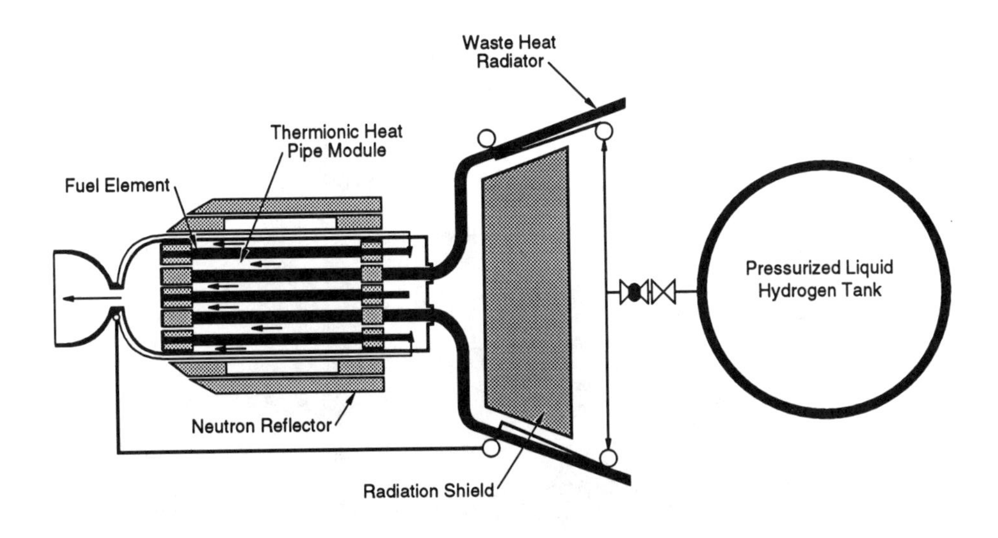

FIGURE 23. Schematic of Propellant Flow for Bimodal Power and Propulsion SEHPTR Operation (Zubrin et al. 1992).

the tungsten cladding must provide enough ductility at 2,100 K to allow for about 8% creep. While creep rupture values for tungsten are much higher than this, data are needed to indicate what effect irradiation will have on ductility. The tungsten clad for the SEHPTR fuel is designed to operate 300 degrees higher than the TFE emitters; data from the 710 Program indicates that chemical compatilility should not be a problem if the UO_2 is stablized. A heat pipe of W-Li has operated for 1.5y; if the heat pipes all fail, there is a 20% degradation in power output. In the tested heat pipe, the mass flow was ten times that needed in the SEHPTR design. This indicates that the 7 to 10-y life is possible. Data on emissivity coatings is now being developed.

A technology demonstration program is underway on the SEHPTR thermionic converter, called Thermionic Heat Pipe Module (THPM) (Horner-Richardson et al. 1992). A schematic is shown in Figure 24. Emitter and collector heat pipes of a preprototype (THPM 1127A) were successfully designed and built in 12 weeks. The converter produced a maximum output of 33.4 W with an emitter temperature of 1,810 K, collector temperature of 1,000 K, and cesium reservoir temperature of 550 K. The emitter heat pipe developed a leak after 50 h due to corrosion of the molybdenum by the lithium working fluid. It is hypothesized that this corrosion was driven by residual nitrogen and/or oxygen in the lithium. Another THPM (1127B) has also been fabricated with a thick emitter sleeve of tungsten. This device has been tested up to temperatures of 2,050 K and has produced 180 W of power. Full length, full power THPMs are planned for later in 1993.

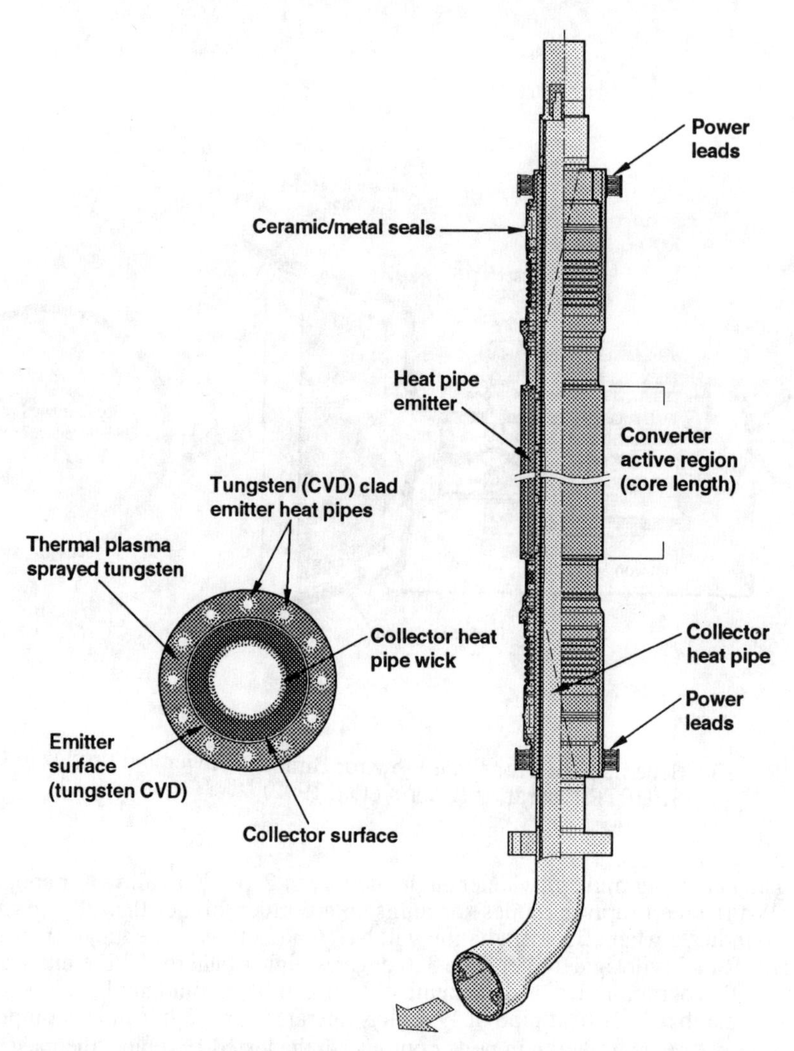

FIGURE 24. Schematic of SEHPTR Thermionic Heat Pipe Module (Horner-Richardson et al. 1992).

Space Thermionic Advanced Reactor-Compact (STAR-C)

In STAR-C (Begg et al. 1992 and Allen et al. 1991), illustrated in Figure 25, the solid core is composed of annular plates of UC_2 fuel supported in graphite trays. Heat is radiated from the radial core surface to a surrounding array of thermionic converters where electric output is generated. Planar thermionic converters operate in the ignited, high-pressure cesium mode with an interelectrode gap of 0.1 mm. The emitter is tungsten, collector niobium, and insulator-seal alumina tri-layer. Reject heat is taken from the thermionic converter collectors through the radial core reflector by heat pipes to an extended surface immediately on the outside of the reactor, from which heat is radiated directly to space. Key parameters are given in Table 15. Design studies are the extent of the funded activities on this concecpt.

STAR-C has the advantages of relatively simple thermionic converters located outside the core, a simple fuel design, no pumped loops, and being a compact power plant overall. The major concerns relate to the quantity and containment of fuel swelling, sublimation of graphite, venting of fission gases without loss of fuel, and limited power system growth.

TABLE 15. STAR-C Key Performance Features.

Thermal Power (kWt)	340
Reactor Output (kWe)	42.8
Net Electrical Power (kWe)	40.9
Net System Efficiency (%)	12.0
Peak Fuel Temperature (K)	2,150
Core Surface Temperature (K)	2,000
Emitter Temperature (K)	1,854
Collector Temperature (K)	1,031
Main Radiator Area (m^2)	5.9
Mass (kg)	2,502

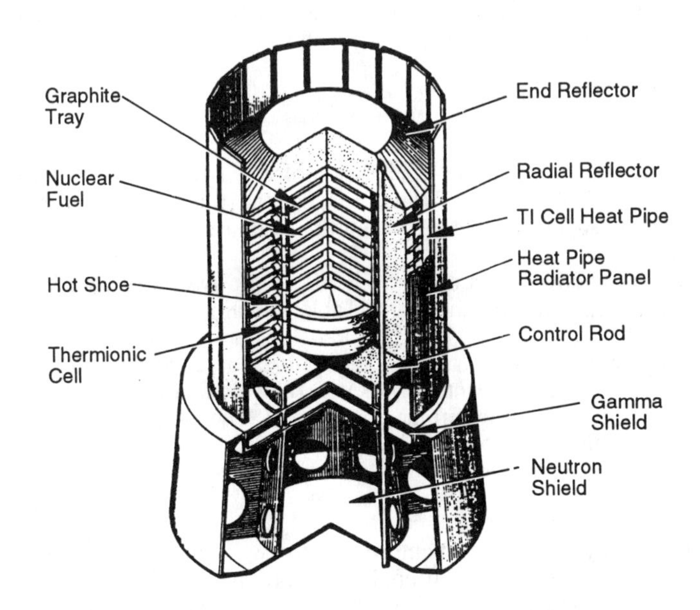

FIGURE 25. STAR-C Configuration (Allen et al. 1991).

MULTIMEGAWATT PROGRAM

Beginning in 1985 and until 1990, a program was underway to develop electrical power in the multimegawatt range for neutral particle beams, free electron lasers, electromagnetic launchers, and orbital transfer vehicles. The program was discontinued because of a shift in emphasis within the Strategic Defense Initiative. The power requirements were grouped into three categories, as seen in Table 16.

TABLE 16. Multimegawatt Space Power System Requirements

	Category I	Category II	Category III
Power requirements (MWe)	10s	10s	100s
Operating time (s)	100s	100s + 1 y of total life	100s
Effluents allowed	Yes	No	Yes
1 Orbit Recharge	No	Yes	No

System Concepts

Six concepts were selected for Phase I studies: three for Category I, two for Category II, and one for Category III. The program was terminated prior to Phase II awards. A brief summary of these concepts follows (much of the information is curtesy of Richard Shutters, Multimegawatt Project Office, Idaho National Engineering Laboratory).

For Category 1:

(1) A fast-spectrum, cermet-fuel, gas-cooled reactor derived from the 710 program (Angelo and Buden 1985) drives twin counter-rotating open Brayton cycle turbines coupled to super-conducting generators. Table 17 summarizes key parameters.

(2) A fast-spectrum, gas-cooled reactor with a two-pass core heats hydrogen and drives twin gas turbo-generators. Table 18 summarizes key parameters.

(3) A gas-cooled, nuclear rocket derivative based on the Nuclear Engine Rocket Vehicle Application reactor (NERVA) (Angelo and Buden 1985) drives two open-cycle, counter-rotating turbine generators (Schmidt et al. 1988 and Chi and Pierce 1990). Table 19 summarizes key parameters.

For Category 2:

(1) A space thermionic advanced reactor with energy storage system (STARS) consisting of a liquid-metal-cooled, in-core thermionic reactor coupled to alkaline fuel cells for burst power. Table 20 summarizes key parameters.

(2) A lithium-cooled, cermet-fuel, fast reactor drives a potassium-vapor Rankine cycle with a Na/S battery storage system. Table 21 summarizes key parameters.

TABLE 17. Category I, MMW, Fast-Spectrum, Cermet-Fuel Reactor (710 Program Derivative) with Twin Counter-Rotating Open Brayton Cycle Turbines Coupled to Super-conducting Generators.

Concept Details	
Reactor Inlet Temperature (K)	164
Reactor Outlet Temperature (K)	800
Turbine Inlet/Exit Temperature (K)	1,400/800
Reactor	
Type (derivative of 710 reactor)	Gas-cooled fast reactor
Fuel (86 to 97% enriched U-235)	Cermet ($U0_2$-Mo),
Materials	Reactor structure, W-Re; pressure vessel, Inconel-750X
Maximum Fuel Temperature (K)	1,083
Coolant	Hydrogen
Burnup	Negligible
Power Conversion	
Turbine/Compressor Type	Conventional axial turbine
Generator Type	Superconducting
Recuperation	No
Materials	Single-crystal Rene N5 for first blades; second stages A286; shaft and rotor astrology; casing HS188; bearings M50 99.999% aluminum stator
Heat Rejection Method	
Main	Effluent
Decay Heat Removal	Hydrogen flow
System Mass (kg)	8,136

For Category 3:

(1) A gas-cooled, particle-bed reactor drives a turbine generator (Powell and Horn 1985). Table 22 summarizes key parameters.

Technology Developments

The major Multimegawatt Program development activities were concerned with fuels. Scoping tests were performed to evaluate the compatibility of UN fuels with W-Re and Mo-Re alloys. The test results showed some problems at high temperatures, but these could be mitigated through control of the UN stoichiometry. Thermodynamic analyses were performed to estimate the chemical compatibility of UC fuels with these alloys. Also, a testing program was performed on two particle bed fuel elements. The elements did not perform as expected. Post-irradiation examinations indicated power/flow matching problems exhibited by nonuniform flow distributions, particle-frit chemical and mechanical interactions, and cycling problems.

TABLE 18. Category I, MMW, Fast-Spectrum Reactor with a Two-Pass Core and with Gas Turbo-generators.

Concept Details	
Reactor Inlet Temperature (K)	150
Reactor Outlet Temperature (K)	1,200
Turbine Inlet/Exit Temperature (K)	1,200/800
Reactor	
Type	Gas-cooled fast reactor with two-pass core
Fuel	UC pins, 316 S.S. clad first pass, Mo-41 Re for second pass
Materials	Vessel is 316 stainless steel
Maximum Fuel Temperature (K)	1,650
Coolant	Hydrogen
Burnup	Insignificant
Power Conversion	
Turbine/Compressor Type	Twin axial turbine counter-rotating; 15,000 rpm
Generator Type	Wound field, non-salient pole, hydrogen cooled
Recuperation	Yes
Heat Rejection Method	
Main	Effluent
Decay Heat Removal	Hydrogen flow
System Mass (kg)	12,887

Some materials development activities were undertaken on Ta-, Mo-, and W-based alloys. Several material lots were produced, but the efforts were terminated before conclusive data were obtained on the alloys.

FOREIGN CONCEPTS

A number of international participants have presented concepts during the Space Nuclear Power Systems Symposiums in Albuquerque, NM in addition to the U.S.S.R. concepts described above. Rolls-Royce of the United Kingdom participated in the Multimegawatt Program. The most active foreign participation at the Symposia has been by France. In addition, Japan presented a concept of a reactor that could be launched with a fuel cartridge separate from the reactor (Yasuda et al. 1990). The fuel is coated particle and the reactor is cooled using heat pipes.

The French space nuclear power program has entered a period of lower activity. They have concentrated on power levels in the 10 to 30 kWe range. The power conversion system selected is the Brayton cycle. Several reactor concepts have been reported, including gas-cooled, liquid-metal-cooled, and water moderated reactors (Tilliette et al. 1990, Proust et al. 1990, Carre et al. 1990 and Tilliette and Carre 1990). The selected reference system is a 930 K, NaK-cooled, fast spectrum reactor (Figure 26). This selection was based on available or near term technologies.

TABLE 19. Cateorgy I, MMW, Gas-Cooled, NERVA-Derivative Reactor (NDR) with Two Open-Cycle, Counter-Rotating Turbine-Generators.

Concept Details	
Reactor Inlet Temperature (K)	35
Reactor Outlet Temperature (K)	1,150
Turbine Inlet Temperature (K)	1,150
Reactor	
Type	NERVA derivative, graphite-moderated near thermal
Fuel	Pyrolitic carbon and SiC-coated $UC_{1.7}$ particles in graphite matrix
Materials	Titanium vessel, Zr-C coated graphite internals
Maximum Fuel Temperature (K)	1,200
Coolant	Hydrogen
Burnup	Negligible
Power Conversion	
Turbine/Compressor Type	Two counter-rotating, high pressure axial turbines exhaust to a pair of counter-rotating, low pressure axial turbines. Pressure provided by LH_2 turbopumps.
Generator Type	Hyper-conducting
Recuperation	None except by jacketing the reactor outlet line
Materials	99.999% Aluminum conductors
Heat Rejection Method	
Main	Effluent
Decay Heat Removal	Hydrogen flow through tie tubes to a radiator
System Mass (kg)	10,513

The initial version of the liquid-metal-cooled reactor consisted of a tight lattice of 780 UO_2 fuel pins arranged with a pitch to diameter ratio of 1.07 (9.1/8.5 mm). This leads to a fuel inventory of 75 kg of 93% enriched uranium with active core height of 270 mm and diameter of 290 mm. Alternative core designs, including an SP-100 derivative architecture using a honeycomb fuel supporting structure, are being considered, and the mass penalty associated with the lesser fuel volume fraction within the core is being evaluated. The reduction of core size and reactor mass afforded by the use of UN fuel is also being assessed. The reactor control system has 12 rotating drums, and the core contains 7 safety rods. A shadow shield consists of B_4C and LiH elements fitted in a stainless steel honeycomb structure for neutron attenuation and tungsten for gamma ray attenuation are included in the design. The Brayton systems are designed with a single recuperated turbogenerator directly coupled to an armored gas cooled radiator as a heat

TABLE 20. Category II, MMW, Space Thermionic Advanced Reactor with Energy Storage System (STARS) and with a Liquid-Metal-Cooled, In-Core Thermionic Reactor Coupled to Alkaline Fuel Cells for Burst Power.

Concept Details	
Reactor Inlet Temperature (K)	230
Reactor Outlet Temperature (K)	1,130
Main Radiator Inlet/Exit (K)	1,073/1,023
Reactor	
Type	Liquid-metal-cooled, in-core thermionic
Fuel	UO_2 pellets
Materials	W-HfC
Maximum Fuel Temperature (K)	2,520
Coolant	Lithium
Burnup	Negligible
Power Conversion	
Type	Thermionic
Emitter	W-HfC, 2,200 K
Collector	Nb, 1,200 K
Energy Storage	
Type	Fuel cell
Heat Rejection Method	
Main	Heat-pipe panel radiators for steady-state system, expandable radiator for fuel cell system with water.
Decay Heat Removal	Heat pipe radiators
System Mass (kg)	58,152

sink. Alternative design options under study include using redundant dual Brayton engines and a heat pipe radiator. The mass of a 20 kWe system is calculated to be 2319 kg.

ADDITIONAL CONCEPTS

A number of other concepts deserve discussion. Space limits mentioning all of the concepts that have appeared in the Space Nuclear Power Systems Symposium over the last ten years.

TABLE 21. Catgegory II, MMW, Li-cooled, Cermet-Fuel, Fast Reactor with a Potassium-Vapor Rankine Cycle and Na/S Battery Storage System.

Concept Details	
Reactor Inlet Temperature (K)	1,500
Reactor Outlet Temperature (K)	1,660
Turbine Inlet/Exit Temperature (K)	1,500/1,255
Radiator Inlet/Exit (K)	1,040 average
Reactor	
Type	Liquid-metal-cooled fast reactor Cermet (UN in W-Re)
Materials	Pressure vessel and piping, ASTAR 811-C
Maximum Fuel Temperature (K)	1,870
Coolant	Lithium
Burnup	~7%
Power Conversion	
Working Fluid	Potassium
Turbine/Compressor Type	Axial turbine
Generator Type	Wound rotor
Energy Storage	
Type	Sodium-sulfur batteries
Heat Rejection Method	
Main	Carbon-carbon heat pipes
Decay Heat Removal	Heat pipe radiators
System Mass (kg)	54,544

Heat Pipe Reactors

Heat-pipe-cooled reactors were being pursued before the SP-100 program started in 1983. These used heat pipes to cool the reactor and transfer the heat to the electric power conversion equipment. Significant features of such a system are that all heat transport is by passive means and there is multiple redundancy throughout the system. The disadvantages of such a system are that no heat pipe reactors have actually been built, there are technical questions as to whether the redundancy is actually achieved, and the technology base for 1,400 K, long-life heat pipes is limited (El-Genk et al. 1985 and Koenig 1985). More recent studies have suggested a heat pipe reactor with operating temperature of 1,125 K to reduce development risks (Ranken 1990).

TABLE 22. Category III, MMW, Gas-Cooled, Particle-Bed Reactor with a Turbine Generator.

Concept Details	
Reactor Inlet Temperature (K)	220
Reactor Outlet Temperature (K)	1,050
Turbine Inlet Temperature (K)	1,050
Reactor	
Type	Gas-cooled, particle-bed reactor
Fuel	UC_2 coated with ZrC and pyrolytic C particles
Materials	Al vessel, Mo-Re hot frit
Maximum Fuel Temperature (K)	1,240
Coolant	Hydrogen
Burnup	Negligible
Power Conversion	
Turbine/Compressor Type	Axial turbines on counter-rotating shafts, direct coupled generator
Generator Type	Cryo-cooled, wound rotor alternator, 3-phase output at 20 kV
Recuperation	None
Heat Rejection Method	
Main	Effluent
Decay Heat Removal	H_2 flow via multiple independent channels
System Mass (kg)	42,000

Gas Vapor Cores

Gas vapor cores have been studied for hundreds of MWe of power. The concepts are futuristic and could find applications in large power plants for beaming power back to Earth. The Ultra-High Temperature Vapor Reactor (UTVR) is an externally-moderated (with BeO), circulating fuel reactor with highly enriched (>85%) UF_4 fuel (Kahook and Dugan 1991). The working fluid is in the form of a metal fluoride. A side view schematic is shown in Figure 27. The reactor has two fissioning core regions: (1) the central ultrahigh temperature region contains a vapor mixture of highly-enriched UF_4 and a metal fluoride working fluid at an average temperature of ~3,000 K and a pressure of ~5 MPa., and (2) the boiler columns which contain highly enriched UF_4 fuel. This reactor has symmetry about the midplane, with identical tandem vapor cores and boiler columns separated by the midplane BeO slab region and the magnetohydrodynamic ducts where power is extracted. The walls are maintained at about 2,000 K by tangential injection of the metal fluoride. A schematic is shown in Figure 28 with representative values of some parameters (Diaz et al. 1991). This is a research program with recent results: UF_4 is the fuel of choice above 1,800 K; UF_4 is

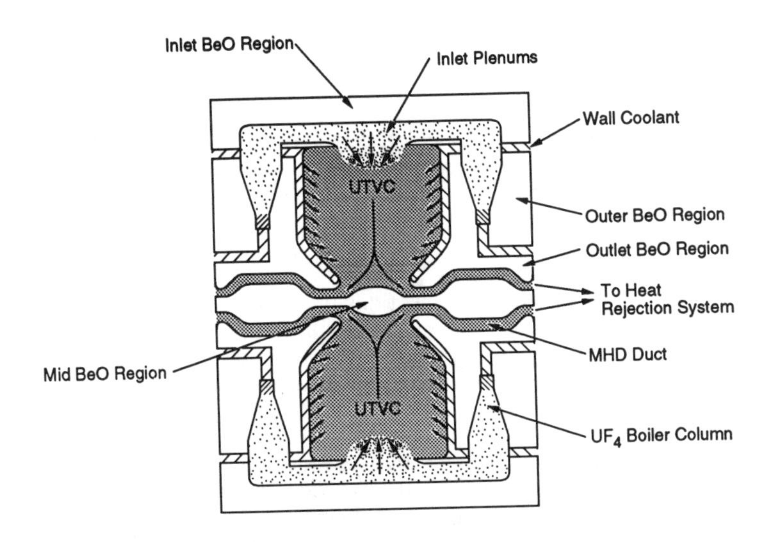

FIGURE 27. Side View Schematic of the UTVR (Diaz et al. 1991).

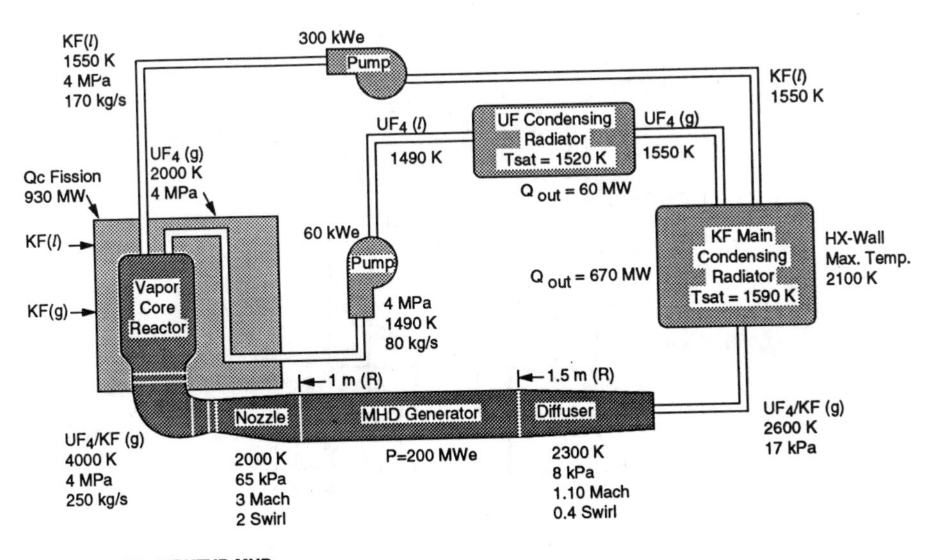

FIGURE 28. Schematic Diagram of a 200 MWe UTVR-MHD Conceptual Design with Specific Mass of ~ 1 kg/kWe (Diaz et al. 1991).

compatible with tungsten, molybdenum, and glassy carbon to 2,200 K for up to 2 h; specialized gaseous core neutron data were developed; and nuclear-induced ionization enhances electrical conductivity by factors greater than 10.

SUMMARY

Table 23 summarizes major potential applications of space nuclear power. The table divides applications into near-Earth, solar system exploration (Yen and Sauer 1991), Lunar-Mars exploration, and Near-Earth resources. The requirements presented are representative values, there are many possible optimizations for each specific application.

In the near term, most applications would involve missions to support planet Earth, and most of those would be defense missions that provide unique capabilities. There is a need for rapidly deployable satellites from launch site storage in times of conflict and for redeployment of satellites already in space for better coverage of key geographic areas. Improved surveillance, communications, battlefield illumination, and electronic jammers are some of the unique systems that would be enabled by nuclear power.

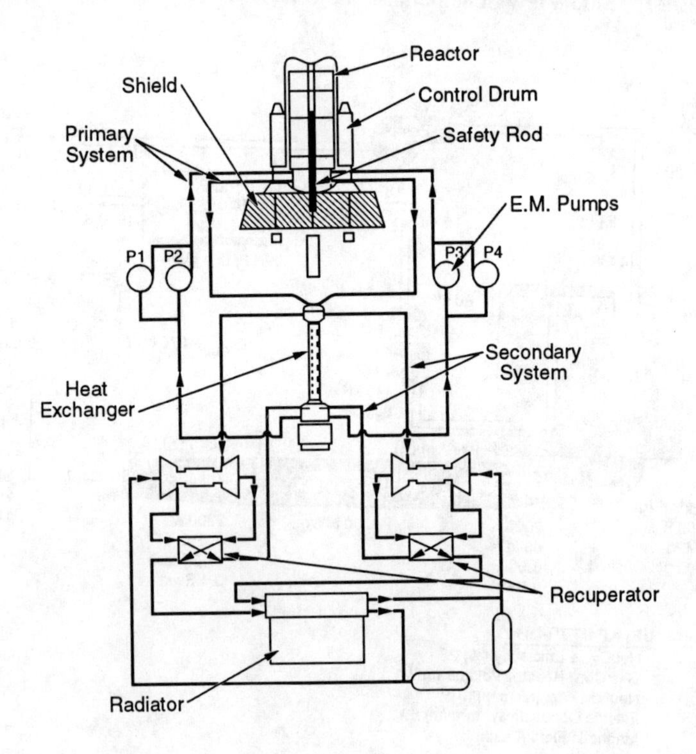

FIGURE 26. Double Loop 20 kWe Nuclear Brayton Power System Concept proposed by French (Tilliette et al. 1990).

These require power levels of 10 to 40 kWe with lifetimes of 7 to 10 y. The orbits are sufficiently high to satisfy concerns about safety and possible interference with gamma ray observatories. Other possible missions include performing underground measurements using ground penetrating radar or microwaves and chemical, nuclear and biological effluent monitoring using laser spectroscopy for non-proliferation and treaty verification. The Federal Aeronautics Administration has a need for oceanic anti-collision aircraft radar. Environmental monitoring needs world-wide measurements of ozone and pollution and also upper air turbulence. Commercial use could include a satellite component of the data information superhighway for remote and mobile sites and high definition television satellites.

Robotic exploration of the solar system and piloted exploration of the Moon and Mars are given less weight in the next decade. However, if nuclear systems enable low cost orbiters to the outer planets, these might be planned at the beginning of the next century. Near-Earth resource recovery from comets and asteroids is another possible application that might become a driver, but only at the beginning of the next century. Precursor missions are needed to establish the viability of this approach.

During the last ten years, many challenges have been overcome in the development of space nuclear power systems. For instance, SP-100 has demonstrated a 7-y fuel pin that can operate at 1,400 K. Based on the work already performed, there is high confidence that the liquid metal cooled, fuel pin SP-100 reactor, with thermoelectric or some other converter, can be successfully developed. The SP-100 program has progressed to the hardware demonstration of most of the key components. NASA has tended to favor the development of the SP-100 to meet their mission needs.

Thermionic power systems have received a major boost with the change in relations with the former Soviet Union and the accessibility to its technology. Defense missions have tended to favor thermionic systems because of being more compact, with smaller radiators. Themionic power plants have been demonstrated by the Russians in flight tests. There are plans for the U.S. to space test a Topaz II power system, modified to meet U.S. safety standards, in 1996 or 1997. Follow-on thermionic developments in both the U.S. and Russia are planned to develop a flight-ready 40 kWe system. Thermionic fuel element lifetime is still the key issue.

Changing mission emphasis is leading to emerging interest in other forms of nuclear space systems. This interest includes a possible combination of electric power and thermal propulsion in a single power plant.

Acknowledgments

The author wishes to thank the excellent inputs provided by a large number of people for this review. In particular, Dr. Jack Mondt (Jet Propulsion Laboratory), Dr. Michael Schuller (Phillips Laboratory), Susan Voss (Los Alamos National Laboratory), Mike Jacox and Richard Strutters (Idado National Engineering Laboratory), and James Dudenhoefer (Lewis Research Center) provided invaluable information. The paper reviewers and Dr. Mohamed El-Genk (University of New Mexico) also added much to the paper.

Work supported by the U.S. Department of Energy, DOE Idaho Operations Office Contract DE-AC07-76IDO01570. Neither the United States Government nor any agency thereof, nor any of their employees, makes any warranty, express or implied,

TABLE 23. Representative Potential Space Nuclear Power Missions.

Mission	Key Requirements
Near-Earth	
Defense	
Wide-area surveillance	20 to 40 kW, lifetime 7 to 10 y, rapid deployment, high elliptical Earth orbit (HEEO)/medium Earth orbit (MEO)
Battlefield communications	10 to 20 kW, lifetime 7 to 10 y, rapid deployment, geosynchronous Earth orbit (GEO) orbit
Battlefield illumination	10 to 40 kW, lifetime 7 to 10 y, rapid deployment, HEEO/MEO orbits
Electronic jammers	>10 kW, lifetime 7 to 10 y, rapid deployment, HEEO/MEO orbits
Submarine communications	10 to 40 kW, lifetime 7 to 10 y, rapid deployment, HEEO orbit
Non-Proliferation and Treaty Verification	
Under ground measurements	to 40 kW, lifetime 7 to 10 y
Moveable surface sensors	to 40 kW, lifetime 7 to 10 y
CBN effluent monitoring	to 40 kW, lifetime 7 to 10 y
Federal Aeronautics Administration	
Anti-collision aircraft radar	20 to 40 kW, lifetime 7 to 10 y, HEEO/MEO orbits
Commercial	
Electronic information highway	25 to 100 kWe
Direct broadcast television	25 to 100 kWe
Environmental Monitoring	
Earth observations	>10 kW, lifetime 7 to 10 y
Upper air turbulence	>10 kW, lifetime 7 to 10 y
Solar System Exploration	
Neptune orbiter/probe	Payload 1.8 Mg, 100 kW, power system mass 3.7 Mg
Pluto orbiter/probe	Payload 1.4 Mg, 56 kW, power system mass 2.8 Mg
Uranus orbiter	Payload 1.4 Mg, 100 kW, power system mass 3.7 Mg
Jupiter grand tour	Payload 1.4 Mg, 58 kW, power system mass 2.9 Mg
Rendezvous	Payload 1.4 Mg, 40 kW, power system mass 2.35 Mg
Comet Sample/Return	Payload 1.8 Mg, 100 kW, power system mass 3.7 Mg
Lunar-Mars Exploration	
First lunar outpost	>12 kW
Enhanced outpost (ISRU)	>200 KW
Mars transportation	Flight time <180 d, payload 52 MT
Mars stationary (600 d)	75-150 kW
Mars in situ resources	> 200 kW
Mars comsats	20 KW
Near-Earth Resources	
In situ probes	30 kW, payload 1.5 Mg, lifetime 3 y
Transportation	20 kN

or assumes any legal liability or responsibility for accuracy, completeness, or usefulness of any information herein. The views and opinions of the author expressed herein do not necessarily state or reflect those of the United States Government or any agency thereof.

References

Allen, D. T., G. O. Fitzpatrick, and D. M. Ernst (1991) "The STAR-C Nuclear Power With High-Efficiency Transparent Thermionic Converters," in *Proceedings Eighth Symposium on Space Nuclear Power Systems*, Conf-910116, M. S. El-Genk and M. D. Hoover, eds., American Institue of Physics, New York, AIP Conf. Proc. No. 217, 3:1093–1099.

Ambrus, J. H. and R. G. Beatty (1985) "Potential Civil Mission Applications for Space Nuclear Power Systems," in *Space Nuclear Power Systems 1984*, Proceedings of the First Symposium on Space Nuclear Power Systems, held in Albuquerque, NM, Jan. 1984, Chapter 10, 43–51.

Angelo, J. A., Jr. and D. Buden (1985) *Space Nuclear Power*, Orbit Book Company, Malabar, FL.

Atwell, J. C., et al. (1989) "Thermoelectric Electromagnetic Pump Design For The SP-100 Reference Flight System," in *Transactions Sixth Symposium On Space Nuclear Power Systems,* held in Albuquerque, NM, Jan. 1989, 280-283.

Begg, L. L., T. J. Wuchte, and W. D. Otting (1992) "STAR-C Thermionic Space Nuclear Power System," in *Proceedings Ninth Symposium on Space Nuclear Power Systems* , held in Albuquerque, NM, Jan. 1992, 1:114–119.

Bennett, G. (1989) "A Look At The Soviet Space Nuclear Power Program," in *Proceedings 24th Intersociety Energy Conversion Engineering Conference*, held in Washington, D.C., August 1989, 2:1187-1194.

Bond, J. A., D. N. Matteo, and R. J. Rosko (1993) "Evolution of the SP-100 Conductivity Coupled Thermoelectric Cell," in *Proceedings Tenth Symposium Space Nuclear Power and Propulsio*n, held in Albuquerque, NM, January 1993, 2:753-758.

Brown, C. E. and D. Mulder (1992) "An Overview of the Thermionic Space Nuclear Power Program," presented at the *1992 Space Nuclear Power Systems Symposium*, held in Albuquerque, NM, Jan. 1992.

Carre, F., et al. (1990) "Update of the ERATO Program and Conceptual Studies on LMFBR Derivative Space Power Systems," in *Proceedings Seventh Symposium on Space Nuclear Power Systems*, held in Albuquerque, NM, Jan. 1990, 1:381–386.

Chi, J. W. H. and W. L. Pierce (1990) "NERVA Derivative Reactors and Space Electric Propulsion Systems," in *Proceedings Seventh Symposium on Space Nuclear Power Systems*, held in Albuquerque, NM, Jan. 1990, 1:208-213.

Chiu, W. S. (1985) "Impact of Stirling Engines to a Space Nuclear Power System," in *Proceedings Second Symposium on Space Nuclear Power Systems*, held in Albuquerque, NM, Jan. 1985, EC-2.

Deane, N. A. et al. (1989) "SP-100 Reactor Design and Performance," in *Proceedings 24th Intersociety Energy Conversion Engineering Conference* , held in Washington, DC, Aug. 1989, 2:1225–1226.

Diaz, N. J. et al. (1991) *Ultrahigh Temperature Reactor and Energy Conversion Research Program, Annual Report For Period July 1990-Sept. 1991*, Innovative Nuclear Space Power & Propulsion Institute Report INSPI-91-110, Oct. 1991.

Disney, R. K., J. E. Sharbaugh, and J. C. Reese (1990) "Shielding Approach And Design For SP-100 Space Power Applications," in *Proceedings Seventh Symposium On Space Nuclear Power Systems*, held in Albuquerque, NM, 1:121–124.

Doherty, M. P. and J. H. Gilland (1992) "NEP Systems Engineering Efforts In Fy92 Plans and Status," in *Proceedings Nuclear Technologies For Space Exploration NTSE-92*, American Nuclear Society Meeting held in Jackson, WY, August 1992, 1:42-49.

Dudenhoefer, J. E. (1990) "Programmatic Status of NASA's CSTI High Capacity Power Stirling Space Power Converter Program," in *Proceedings 25th Intersociety Energy Conversion Engineering Conference*, held in Reno, Nevada, Aug. 1990, 1:40.

Dudenhoefer, J. E. and J. M. Winter (1991) "Status of NASA's Stirling Space Power Converter Program," in *Proceedings 26th Intersociety Energy Conversion Engineering Conference*, held in Boston, MA, Aug. 1991, 2:38-43.

Dudenhoefer, J. E., D. Alger, and J. M. Winter (1992) "Progress Update of NASA's Free-Piston Stirling Space Power Converter Technology Project," in *Proceedings Nuclear Technologies For Space Exploration, NTSE-92*, American Nuclear Society Meeting held in Jackson, WY, August 1992, 1:294–303.

El-Genk, M. S. et al. (1985) "Review of the Design Status of the SP-100 Space Nuclear Power System," in *Space Nuclear Power Systems 1984*, Chapter 23, Orbit Book Co., Malabar, FL, 1985, 177–188.

General Atomics (1988) *TFE Verification Program Semiannual Report for the Period Ending September 30, 1987*, General Atomics Documents GA-A19115, LaJolla, CA, March 1988.

General Electric Co. (1989a) "SP-100 Mission Risk Analysis," in *Report No. GESR-00849*, San Jose, CA, Aug. 1989.

General Electric Co. (1989b) "SP-100 Safety Feature," presented at the *Sixth Symposium On Space Nuclear Power Systems*, held in Albuquerque, NM, Jan. 1989.

Gilland, J. H. and S. R. Oleson (1992) "Combined High and Low Thrust Propulsion For Fast Piloted Mars Missions," *in Proceedings Nuclear Technologies For Space*

Exploration NTSE-92, American Nuclear Society Meeting held in Jackson, WY, August 1992, 1:22-31.

Harty , R. B., M. P. Moriarty, and T. A. Moss (1985) "Capabilities of Brayton Cycle Space Power Systems," in *Proceedings of the Second Symposium on Space Nuclear Power Systems*, held in Albuquerque, NM, Jan. 1985, EC-3.

Holland, J.W. et al. (1985) "SP-100 Thermionic Technology Program Status," in *Space Nuclear Power Systems 1985*, Chapter 24, Orbit Book Co., Melabar, FL, 179–188.

Homeyer, W. G., et al. (1984) "Thermionic Reactors for Space Nuclear Power," in *Space Nuclear Power Systems 1984*, Chapter 25, Orbit Book Co., Melabar, FL, 197-205.

Horner-Richardson, D., J. R. Hartenstine, D. Ernst, and M. G. Jacox (1992) "Fabrication and Testing of Thermionic Heat Pipe Modules for Space Nuclear Power Systems," in *Proceedings 27th Intersociety Energy Conversion Engineering Conference*, Paper No. 929075, San Diego, CA, August 1992, 6:6.35-6.41.

Hwang, Choe et al. (1993) "SP-100 Lithium Thaw Design, Analysis, and Testing," in *Proceedings Tenth Symposium on Space Nuclear Power and Propulsion Systems*, held in Albuquerque, NM, Jan. 1993, 2:697-702.

Hylin, E. C. and M. P. Moriarty (1985) "Impact of Stirling Engine Characteristics upon Space Nuclear Power Systems Design," in *Proceedings Second Symposium on Space Nuclear Power Systems*, held in Albuquerque, NM, Jan. 1985, EC-1.

Josloff, A. T. (1988) "SP-100 Space Reactor Power System Scaleability," in *Proceedings 23th Intersociety Energy Conversion Engineering Conference*, held in Denver, CO, Aug. 1988, 1:263–265.

Josloff, A. T., H. S. Bailey, and D. N. Matteo (1992a) "SP-100 System Design and Technology Progress," in *Proceedings Ninth Symposium on Space Nuclear Power Systems*, held in Albuquerque, NM, Jan. 1992, 1:363–371.

Josloff, A. T., D. N. Matteo, and H. S. Bailey (1992b) "SP-100 Technology Development Status," in *Proceedings Nuclear Technologies For Space Exploration, NTSE-92*, American Nuclear Society Meeting held in Jackson, WY, August 1992, 1:238–246.

Kahook, S. D. and E. T. Dugan (1991) "Nuclear Design Of The Burst Power Ultrahigh Temperature UF4 Vapor Core Reactor System," in *Proceedings Eighth Symposium on Space Nuclear Power Systems*, CONF-910116, American Institute of Physics, 1:205–210.

Koenig, D. R. (1985) "Heat Pipe Reactor Designs for Space Power," in *Space Nuclear Power Systems 1984*, Chapter 27, Orbit Book Co., Malabar, FL, 1985, 217–228.

Marshall, A., et al (1993) "Topaz II Preliminary Safety Assessment," in *Proceedings Tenth Space Nuclear Power and Propulsion Systems Symposium*, held in Albuquerque, NM, Jan. 1993, 1:439-445.

Matteo, D. N., J. A. Bond and R. J. Rosko (1992) "SP-100 Thermoelectric Converter Technology Development," in *Proceedings of the 27th IECEC*, held in San Diego, CA, August 1992, 6:73-77.

Matteo, D. N. (1993) Personal Communication, Martin Marietta, Valley Forge, PA, 25 May, 1993.

Mondt, J. F. (1991) "Overview of the SP-100 Program," in *AIAA/NASA/OAI Conference on Advanced SEI Technologies*, AIAA 91-3585, held in Cleveland, OH, September 1991.

Mondt, J. (1992) *SP-100 Accomplishments*, in SP-100 Project Review held June 15, 1992, JPL D-9650.

Mondt, J. F. (1989) "Development Status Of The SP-100 Power System," in *Proceedings AIAA/ASME/SAE/ASEE 25th Joint Propulsion Conference,* Paper No. AIAA-89-2591, held in Monterey, CA, July 1989.

Mondt, J. F. (1993) *SP-100 Project Memorandum No. 8501-93-004*, 25 March 1993.

Nicitin, V. P., et al. (1992) "Special Features And Results Of The "Topaz II" Nuclear Power System Tests With Electric Heating," in *Proceedings Ninth Symposium on Space Nuclear Power Systems*, Conference 920104, American Institute of Physics, 1992, 1:41–46.

Phillips Laboratory (1992) *Thermionic Space Nuclear Power Technical Program Review,* held in Albuquerque, NM, 9-10 November 1992.

Powell, J. R. and F. L. Horn (1985) "High Power Density Reactors Based on Direct Cooled Particle Beds," in *Space Nuclear Power Systems 1985*, Chapter 39, Orbit Book Co., Melabar, FL, 319–329.

Proust, E., S. Chaudourne, and J. Tournier (1990) "Performance Modeling and Optimization of Design Parameters for the Space Nuclear Brayton Power Systems Considered in the ERATO Program," in *Proceedings Seventh Symposium on Space Nuclear Power Systems* , held in Albuquerque, NM, Jan. 1990, 1:426–431.

Ranken, W. A. (1990) "Low Risk Low Power Heat Pipe/Thermoelectric Space Power Supply," in *Proceedings Seventh Symposium on Space Nuclear Power Systems*, held in Albuquerque, NM, Jan. 1990, 1:488–496.

Redd, F. J. and E. V. Fornoles (1985) "Emerging Space Nuclear Power Needs," in *Space Nuclear Power Systems 1984*, Proceedings of the First Symposium on Space Nuclear Power Systems, held in Albuquerque, NM, Jan. 1984, Chapter 9, 41–42.

Samulelson, S., L. and R. C. Dahlberg (1990) "Thermionic Fluid Element Verification Program - Summary of Results," in *Proceedings Seventh Symposium on Space Nuclear Power Systems*, held in Albuquerque, NM, Jan. 1990, 1:77–84.

Schmidt, J. E., J. F. Wett, and J. W. H. Chi (1988) "The NERVA Derivative Reactor and a Systematic Approach to Multiple Space Power Requirements," in *Space Nuclear Power Systems 1988*, Chapter 19, Orbit Book Co., Melabar, FL, 95–100.

Schmitz, P., H. Bloomfield, and J. Winter (1992) "Near Term Options For Space Reactor Power," in *Proceedings of Nuclear Technologies For Space Exploration*, American Nuclear Society Meeting held in Jackson, WY, August 1992, 2:313-321.

Slaby, J. G. (1987) "Overview of NASA Lewis Research Center Free-Piston Stirling Engine Technology Activities Applicable to Space Power Systems," in *Space Nuclear Power Systems 1987,* Chapter 5, 35–39.

Snyder, H., J. and J. H. Mason (1985) "Thermionic Space Power System Concept," in *Space Nuclear Power Systems 1985*, Chapter 29, Orbit Book Co., Melabar, FL, 223–229.

Space Power Inc. (1990) "*Characteristics Of The Soviet Topaz II Space Power System,*" Report SPI-52-1, San Jose, CA, October 1990.

Strohmayer, W.H. and T. H. Van Hagan (1985) "Parametric Analysis of a Thermionic Space Nuclear Power System," in *Space Nuclear Power Systems 1985*, Orbit Book Co., Melabar, FL, Chapter 20, 141–149.

Terrill, W. and L. R. Putnam (1985) "SP-100 Thermocouple Assembly, Design and Technology Development," in *Proceedings of the Second Symposium on Space Nuclear Power Systems*, held in Albuquerque, NM, Jan. 1985, EC-5.

Thome, F. (1992) "TSET," presented at the *Thermionic Space Nuclear Power Technical Program Review*, Phillips Laboratory, Albuquerque, NM, Nov. 9-10, 1992.

Tilliette, Z. P. and F. O. Carre (1990) "French Investigations on an Alternative, Longer Term, 20-kWe, Direct Cycle, Gas-Cooled Reactor, Space Power System," in *Proceedings Seventh Symposium on Space Nuclear Power Systems*, held in Albuquerque, NM, Jan. 1990, 1:508–513.

Topaz CoDR (1992) *Flight Topaz Conceptual Design Review, U.S. Air Force Phillips Laboratory*, review hold in Albuquerque, NM, Sept. 16-17, 1992.

Truscello, V. C. and L. L. Rutger (1992) "The SP-100 Power System," in *Proceedings Ninth Symposium on Space Nuclear Power Systems*, held in Albuquerque, NM, Jan. 1992, 1:1–23.

TSET/NEPSTP Workshop (1993) held in the Albuquerque Convention Center, sponsored by Phillips Laboratory (AFSC), 15 January 1993.

U.S. Topaz II Flight Safety Team (1992), *Topaz II Flight Program Safety Assessment,* Draft report Aug. 24, 1992.

Yasuda, H. et al. (1990) "Conceptual Study of a Very Small Reactor With Coated Particle Fuel," in *Proceedings Seventh Symposium on Space Nuclear Power Systems*, held in Albuquerque, NM, Jan. 1990, 1:103–107.

Yen, C. L. and C. G. Sauer (1991) "Nuclear Electric Propulsion for Future NASA Space Science Missions," in *AIDAA, AIAA/DGLR/ JSASS 22nd International Electric Propulsion Conference,* Paper No. IEPC-91-035, Viareggio, Italy, October 1991.

Yoder, G. L. and R. R. Graves (1985) "Analysis of Alkali Liquid Metal Rankine Space Power Systems," in *Proceedings of the Second Symposium on Space Nuclear Power Systems*, held in Albuquerque, NM, Jan. 1985, EC-4.

Zubrin, R. M. et al. (1992) "The Integrated Power and Propulsion Stage: A Mission Driven Solution Utilizing Thermionic Technology," in *Proceedings Ninth Space Nuclear Power Systems Symposium*, held in Albuquerque, NM, Jan. 1992, 3:1259–1267.

SILICON-GERMANIUM: AN OVERVIEW OF RECENT DEVELOPMENTS

Cronin B. Vining and Jean-Pierre Fleurial
Jet Propulsion Laboratory/
California Institute of Technology
M/S 277-212
4800 Oak Grove Drive
Pasadena, CA 91109-8099
(818) 354-9374
(818) 354-4144

Abstract

Over the last decade an impressive variety of innovative techniques has been applied to prepare silicon-germanium alloys with improved thermoelectric figure of merit values. Today, technologically important improvements appear nearly ready for device-level testing. Some approaches have favored the reduction of the lattice thermal conductivity through smaller grain size or addition of inert particulates to enhance phonon scattering. Other approaches sought to improve the electrical properties of the materials by carrier concentration and mobility enhancement. Although most of the experimental work has been conducted on sintered/hot-pressed materials, more sophisticated and complex techniques have been developed to strive for more control over microstructure and composition. The difficulty of understanding the physical mechanisms involved and of predicting the magnitude of possible improvements has led to the development of theoretical tools based on solid-state physics and physical chemistry. This paper attempts to provide an overview of the experimental and theoretical efforts to improve the thermoelectric properties of silicon-germanium with a particular emphasis on some of the less well publicized results. Future prospects and open questions are also briefly discussed.

INTRODUCTION

Particularly well suited for space power applications, heavily doped silicon-germanium (SiGe) alloys have been the exclusive choice for NASA's radioisotope thermoelectric generator (RTG) needs since the launch of Voyager I and II and has performed with distinction in this role. The development of SiGe thermoelectrics, from the early research efforts at RCA's Princeton Laboratory to use in recent space applications, has been amply and capably reviewed by Rosi (1968), Bhandari and Rowe (1980), Rowe (1987), Wood (1988), Rowe (1989), Cody (1990), and Rosi (1991). Slack (1991) provides a very detailed discussion of the theoretical maximum efficiency of silicon-germanium thermoelectric converters. Equally significant, but beyond the scope of this overview, is the extensive effort on SiGe heterostructures, recently reviewed in (Pearsall 1989). The Voyagers, employing a unicouple thermoelectric device configuration and SiGe prepared by hot-pressing, have established a staggering reliability record criteria of 16 years without a single unicouple failure.

After the launch of Voyagers I and II, advantage was taken of the expected lapse before the next RTG would be needed and an ambitious RTG program was undertaken at 3M which included development of a new class of thermoelectric materials based on selenium. This program, unfortunately, encountered serious technical difficulties and the SiGe manufacturing facilities had already been shut down at RCA's Harrison Laboratories. With the re-establishment of a SiGe manufacturing capability, now at General Electric's Valley Forge Space Center (GE-VF), and the approximately concurrent beginnings of the SP-100 space nuclear reactor program, there began a period of renewed interest in SiGe the United States in the early 1980's.

The unfortunate experience with the selenides and the now proven performance of SiGe encouraged a re-examination of the possibility of achieving further improvements in this trusted material. Over the past 10-15 years a variety of sponsors and programs, mostly concerned with the US space program, have contributed significantly to this effort. There will be no attempt to discuss in detail the various sponsors, goals and programs involved. Nevertheless, there are certain general points to be made.

Through a long-standing arrangement the United States Department of Energy (DOE) has provided the United States National Aeronautics and Space Administration (NASA) with RTG's. As such, DOE has sponsored a variety of RTG and SiGe related development activities. NASA, largely through the Jet Propulsion Laboratory/California Institute of Technology (JPL), also maintains a high level of interest in these areas because JPL utilizes these technologies. DOE, NASA and the Strategic Defense Initiative Office (SDIO) share interests in the development of a space nuclear reactor capability and each contribute to the SP-100 program for that reason. At a somewhat lower level of effort, the U.S. Army Electronics Technology and Devices Laboratory (LABCOM) has also expressed a serious interest in SiGe for terrestrial applications.

The primary purpose of this overview is to attempt to follow the evolution of the technical ideas to improve SiGe examined in these programs, with special emphasis on less well known results. In some cases, results discussed here may not have appeared elsewhere in print, but every effort has been made to give appropriate credit.

The majority of this overview is organized along the lines of the two main approaches for improving any thermoelectric material: 1) reduce in the thermal conductivity and/or 2) improve the electrical properties. Two contributions warrant particular note due to their influence on subsequent developments. Raag, then at Syncal, initiated work involving gallium phosphide (GaP) additions to SiGe. Generally the notation used to describe this is "SiGe/GaP." This work (Pisharody 1978), which appeared to result in a lower thermal conductivity and a higher figure of merit ($Z = \sigma\alpha^2/\lambda$), has had a major influence on all the following efforts related to SiGe.

A second body of work, which remains a major influence today, is the work of Rowe (1974) and others on the grain boundary scattering of phonons sometimes simply referred to as "fine grain SiGe." More than anything else, the two ideas of "SiGe/GaP" and "fine grain SiGe," have in one way or another dominated all the following work in this area and the story of SiGe technology after the Voyagers is largely the story of these two ideas.

One final comment is in order before proceeding. Several terms, such as "standard SiGe," "SiGe/GaP," and "fine grain SiGe," will be used here in much the same way as they are commonly used throughout the community. "Standard SiGe" means something very like the materials used, say, in the Voyagers. "SiGe/GaP" means it has GaP in it and "fine grain SiGe" means the grains are more or less fine (say, 5 μm or less). Strictly speaking, however, there is no "standard SiGe," for example, only a family of closely related specifications which, religiously executed, produce a product with a narrow range of properties.

The situation is even more obscure for the terms "SiGe/GaP" and "fine grain SiGe," as these terms have been applied to materials prepared by tremendously different methods, and with concomitantly variant properties. Further, they are not exclusive terms. A "fine grain SiGe" may or may not contain GaP, and a "SiGe/GaP" sample may or may not have small grains. And GaP content, grain size, Si/Ge ratio and doping level can vary substantially. Still, with some care, these terms do have some meaning and are convenient for categorization purposes.

THERMAL CONDUCTIVITY REDUCTION METHODS

Grain Boundary Scattering of Phonons

Early work on the effect of grain boundary scattering on the thermal conductivity of fine grain SiGe has been reviewed through 1980 by Bhandari and Rowe (1980), so only a brief discussion will be given here. Goldsmid and Penn (1968) first pointed out that boundary scattering of phonons could be an important effect in the thermal conductivity of alloys at high temperatures, even though the effect is negligible in pure materials. Parrott (1969) extended these calculations and predicted significant reductions in the thermal conductivity of SiGe alloys. Experimental results by Savvides and Goldsmid (1973) on silicon and by Meddins and Parrott (1976) on SiGe alloys tend to support the prediction that fine grain SiGe should have a lower thermal conductivity, particularly in undoped materials.

Rowe, in a series of collaborative papers (Rowe 1974, 1981a,b, Bhandari 1978), has examined the effect of fine grain size and predicted significant improvements in Z. This body of work, both experimental and theoretical, was well known when work on SiGe began to increase in the United States in the early 1980's. Particularly influential was the repeated assertion and widespread expectation that the thermal conductivity could be significantly reduced by fine grain effects, long before any significant reduction in the electrical properties would be observed (Rowe 1974, 1981a,b, Meddins 1976).

While several studies had achieved Z values in sintered or fine grain materials similar to previous results on zone leveled SiGe (Rosi 1968), none had reported the expected improvements. To address this discrepancy as well as other thermoelectric materials issues, DOE supported the Improved Thermoelectric Materials (ITM) program at GE-VF beginning in 1984 (ITM Final Report 1989). These studies examined a wide variety of hot pressing and particle comminution methods and achieved thermal conductivity reductions in fully dense (>97%) materials of up to 50% (Vining 1991a), as shown in Figure 1. While qualitatively confirming the expected reduction in λ with grain size, the temperature dependence of the reduced thermal conductivity, shown in Figure 2, was not quite as expected from theory and a highly correlated reduction in the electrical mobility resulted in no net improvement in Z compared to similar zone leveled materials, as indicated in Figure 3.

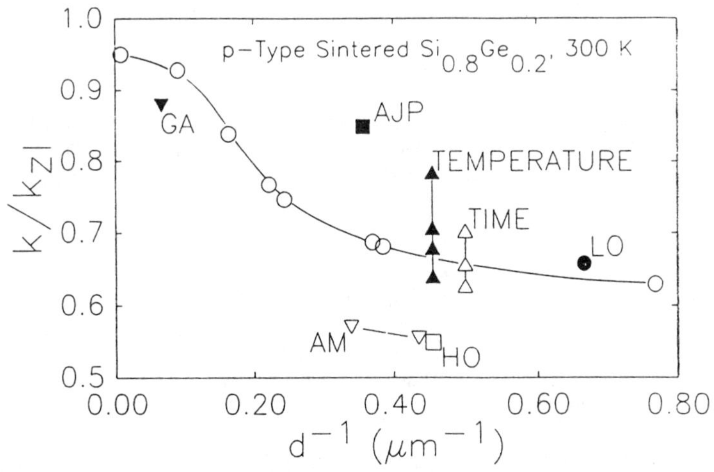

FIGURE 1. Thermal Conductivity of Sintered, p-type SiGe as a Function of Particle Size, Normalized to Zone Leveled Material. The Labels Indicate Processing Variations, Discussed in (Vining 1991a).

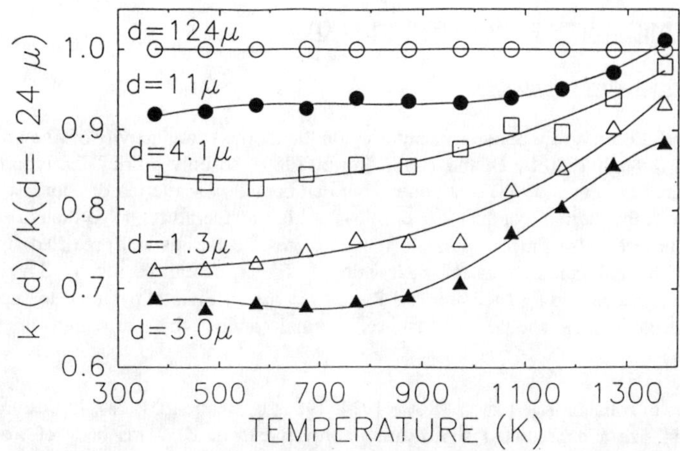

FIGURE 2. Thermal Conductivity for Five Samples of Sintered, p-type SiGe, Normalized to the Thermal Conductivity of the Largest Particle Size Sample (Vining 1991a).

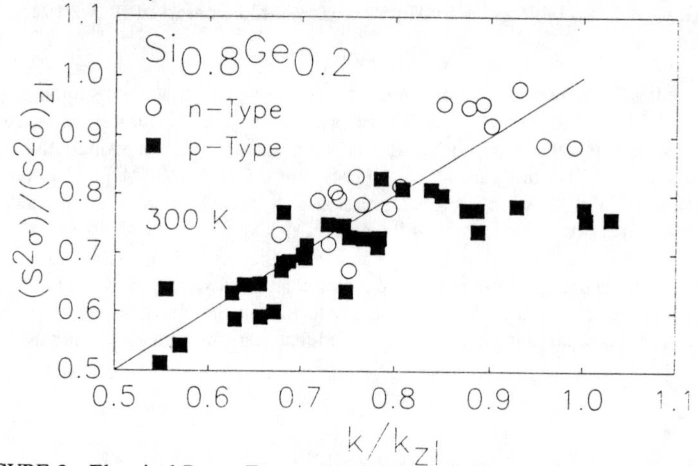

FIGURE 3. Electrical Power Factor as a Function of Thermal Conductivity, Where Each Property has been Normalized to Zone-Leveled Materials. The Solid Line Represents a Figure of Merit Equivalent to Zone-Leveled Materials (Vining 1991a).

A wide variety of powder metallurgical processes were examined in this study, including pulverization in a shatter box, gas atomization from the melt, planetary ball milling in various media, attrition milling, and air jet pulverization. Experiments were designed to both increase and decrease the oxygen content of samples in order to examine the role of oxygen. A fine grain sample (labeled LO in Figure 1) prepared with special attention to keep the oxygen content (0.23 wt. %) comparable to large grain size material (typically 0.2 wt. %), still showed a substantial thermal conductivity reduction, but no improvement in Z.

Each method produces distinct particle size and shape distributions as well as variations in the amounts and types of impurities introduced. Generally, smaller particle sizes produce smaller thermal and electrical conductivities. But virtually every other processing variable can also result in changes on the same order of magnitude. Particles produced by different methods or hot pressed under different conditions result in different thermoelectric properties, even when of nominally the same size. Increasing the hot pressing temperature, for example, by only 45 K resulted in a 2% increase in density (from 97% to 99% dense), but a 20% increase in thermal conductivity.

An important result from this study is the need for tight control over virtually all of the experimental variables and the need to measure all the thermoelectric properties on a single sample. These effects are expected to be superposed on any other study involving sintered SiGe and can easily obscure the primary intention of the experiment. It seems clear that straightforward variations of hot pressing conditions alone offers little overall advantage in terms of Z. But equally important is the conclusion that over a wide range of preparation conditions, excellent Z values essentially identical to zone leveled values can be achieved, in spite of large changes in the individual thermoelectric properties.

While several of the particle comminution methods examined under the ITM program have distinct advantages, one process developed at General Electric's Corporate Research and Development (GE-CRD) center deserves particular note. The gas atomization process, described by Miller (1983), has several advantages as applied to SiGe. This rapidly quenched process can readily produce industrial quantities of highly uniform particles of doped SiGe on the order of 15 μm in diameter directly from the starting elements. Actually, significant compositional segregation occurs within these particles, but on a length scale that allows complete homogenization during ordinary hot pressing. These powders can then be hot pressed without any further processing.

Mechanical Alloying

Mechanical alloying, which produces homogeneous sub-micron SiGe alloys at room temperature, has been studied at Ames for both n-type (Cook 1989) and p-type (Cook 1991, 1993 and Harringa 1992) materials. In addition to very fine microstructures which may produce low thermal conductivity values, this unique process may be suitable for incorporating fine dispersals of inert additions or for incorporating volatile materials difficult to handle by other methods. While reductions in thermal conductivity were reported for both n-type materials (which also contained GaP) and p-type materials, Z values for the n-type materials were similar to previous results while the p-type materials appear to be up to about 10% improved. In this case, however, the improvement may be due to the higher doping level of the mechanically alloyed samples compared to standard materials, rather than the lower thermal conductivity values.

A very similar method of preparation has been reported by Gogishvilli (1990). This study emphasizes the advantages for preparing homogeneous materials by a low temperature process, avoiding the unpleasant consequences of the SiGe phase diagram. No transport properties were reported.

GaP Additions - Indications of λ Reductions

The first efforts to incorporate GaP into SiGe were performed under a JPL sponsored program at Syncal Corporation under Raag (Pisharody 1978). In this report, the thermal conductivity of both n-type and p-type SiGe/GaP is reported to be 40-50% lower than materials without the GaP. A significantly enhanced Z was reported in spite of some increase in electrical resistivity.

Pisharody's paper on SiGe/GaP was extremely influential on future work on SiGe. Eventually, several independent measurements seemed to support these results. Samples of SiGe/GaP, prepared by Syncal, were sent to the Thermophysical Properties Research Laboratory (TPRC) at Purdue University for thermal conductivity measurements (Taylor 1980) and Ames Laboratory, USDOE (Ames) for Seebeck coefficient and electrical resistivity measurements (Raag 1984). This data set, widely referred to as the "Ames/Purdue" data, appears to confirm the lower thermal conductivity and higher resistivity and suggests an overall improvement in SiGe/GaP compared to standard SiGe of about 23% in Z. See Schock (1983) for the only open publication of these data.

Based partly on these results, a unique device geometry called a 'bicouple' was developed at GE-VF and the new SiGe/GaP material was tested at the device level (Cockfield 1984), considered a more definitive test than direct thermoelectric property measurements. Four bicouples were built at GE-VF from SiGe/GaP prepared by Thermo Electron Corporation (now called, and hereafter referred to as, TTC) (SiGe RTG Report 1984) and four more bicouples were built from a type of fine grain SiGe then being developed at GE-VF. It should be emphasized that the fine grain SiGe was "expected to exhibit both Seebeck and electrical resistivity no different than standard SiGe, but to have a thermal conductivity reduced by 18%" (Cockfield 1984), but had not been fully characterized. This assumption, though reasonable at the time, is hardly consistent with the studies on fine grain SiGe discussed above.

These eight bicouples were tested simultaneously at Fairchild Space and Electronics Company (Fairchild), as reported by Cockfield (1984). Analysis of the results required estimates of the temperatures at the device junctions, which are different for the two types of devices, and this in-turn required certain assumptions about material and junction properties. The electrical properties of the fine grain SiGe devices varied considerably amongst themselves and a few of the test fixture thermocouple readings were judged unreasonable and had to be disregarded. The various assumptions and analyses of these tests have been extensively discussed in the community, some of which is documented in a contract monthly report (SiGe RTG Report 1984), but most of which appears to be entirely undocumented.

The most widely known analysis is due to Eck (1984) and is widely regarded as an excellent analysis of the available data. Among Eck's conclusions regarding this test are: 1) both the fine grain SiGe and SiGe/GaP bicouples produced less electrical power than expected, probably due to higher electrical and thermal contact resistances within the devices; 2) the SiGe/GaP has a distinctly lower thermal conductivity and a distinctly lower electrical power factor ($\sigma\alpha^2$) than the fine grain SiGe; 3) based on preliminary inferential calculations, the Z of the SiGe/GaP material appeared to be 23% higher than that of the fine grain SiGe. Some of these conclusions will be discussed in the summary in context with more recent results.

The combination of the Syncal (Pisharody 1978), the Ames/Purdue (Taylor 1980, Raag 1984) material test results and the bicouple device test results (Cockfield 1984, SiGe RTG Report 1984, Eck 1984) resulted in a concerted effort to determine the precise mechanism by which GaP additions resulted in a lower thermal conductivity. Another consequence of these results was the adoption of n-type SiGe/GaP for the MOD-RTG program (MOD-RTG Program 1991), an advanced, multicouple oriented RTG development program based on a Fairchild design (Schock 1983). A type of fine grain SiGe was adopted for the p-type materials due to concerns over the stability of p-type SiGe/GaP.

It should be pointed out that the processing specifications for n-type SiGe/GaP adopted by the MOD-RTG program result in thermoelectric properties no better than standard n-type SiGe (deduced from data in ITM Program Final Report 1989). Indeed, atomic concentrations of Ga and P had to be changed drastically to insure reliable manufacturing of SiGe/GaP material with improved Z values using the MOD-RTG process (Vandersande and Fleurial 1992a). Even today further effort is required before SiGe/GaP with improved Z can be manufactured with sufficiently reproducible performance for a flight program. Nevertheless, by building and testing devices manufactured from a current generation SiGe/GaP material, the MOD-RTG program will presumably resolve any bonding or materials compatibility issues which may arise.

GaP Additions - λ Reduction Mechanism Studies

A significant impetus had developed, therefore, to determine the precise mechanism by which GaP additions lower the thermal conductivity of SiGe. The ITM program at GE directed considerable effort in this direction, much of which was aimed (at least initially) at reproducing the microstructure observed in SiGe/GaP samples produced by TTC and used in the bicouple tests described above. These efforts were successful in that the microstructures of the GE samples did resemble the TTC samples, and the GE samples did exhibit reduced thermal conductivity values, although the Z values were not improved.

A variety of measurements were performed on these low thermal conductivity GE samples. Extended X-ray absorption fine structure (EXAFS) measurements on the GE samples indicated that the bonding environment of the Ga atoms in the n-type material is identical to the bonding environment of Ga atoms in GaP, while in the p-type material, Ga was bonded in an entirely different (and as yet undetermined) way (ITM Program Final Report 1989). Transmission electron microscopy (TEM) performed on these samples at the University of Virginia (Owusu-Sekyere 1989) revealed a rich microstructure with a distinct bimodal distribution of grain sizes. Fine grains (0.1 to 1.5 μm), not readily observable with optical techniques, were found to occupy as much 55% of the sample. The remainder of the materials were composed of coarser particles (1.5 to 10 μm).

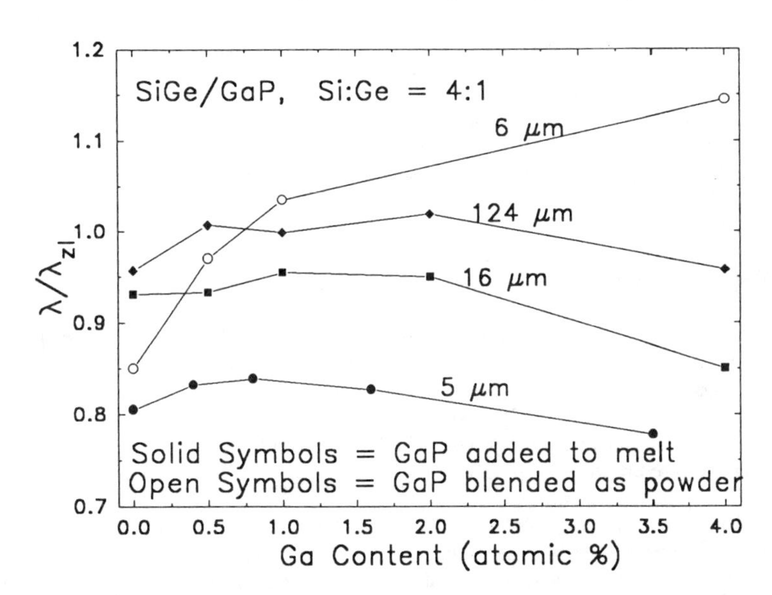

FIGURE 4. Thermal Conductivity of SiGe/GaP, Normalized for Carrier Concentration Effects, as a Function of Ga Content and Preparation Method.

Several calculations performed independently in the mid-1980's suggested GaP alone should not lower the thermal conductivity. Rosi's calculations (the only result published) indicate addition of 5% GaP to $Si_{0.8}Ge_{0.2}$ should increase the thermal conductivity by about 8% (Rosi 1991). In retrospect, the reason for this is simple: SiGe has a low thermal conductivity due to the large difference in mass between Si and Ge. But the masses of Ga and P are both between the masses of Si and Ge, so any addition of GaP actually decreases the average mass difference. This is not in discord with either the EXAFS results, which indicate (at least for n-type) that the GaP was not in solution in any case or the TEM results, which provided ample microstructural reasons for the low thermal conductivity.

Further experiments at GE involved preparation of samples with various amounts of GaP. Some samples were prepared by blending powders of SiGe with powders of GaP and hot pressing the mixtures. Others were prepared by adding GaP directly to a melt of SiGe (in an effort to get more GaP in solution), followed by pulverization and hot pressing. Although widely discussed in programmatic reviews in the mid-1980's, these results are not documented in the open literature, and the raw data must be extracted from Appendix A of the ITM program final report (1989).

Figure 4 shows the effect of additions of GaP on the thermal conductivity, normalized by the thermal conductivity of a comparable zone leveled sample (λ_{zl}) to account for variations in doping level. Four variations of preparation conditions are considered. Most of the thermal conductivity reduction observed in this study is due to grain size effects. Indeed, except for the very highest concentrations studied (where the normalizations used may be less reliable), the normalized thermal conductivity actually increases with increasing GaP content, in qualitative agreement to the theoretical calculation of Rosi discussed above.

An important point regarding Figure 4 is that a quite different conclusion might be reached if the normalization is neglected. Typically, adding more GaP results in higher carrier concentrations due to the higher solubility of P compared to Ga. Higher carrier concentrations result in lower thermal conductivities, as is well known (Rosi 1968). Also, the processes generally used to prepare SiGe/GaP tended to produce much finer grain sizes than had previously been studied, which, as discussed above, certainly lowers the thermal conductivity.

Thus, it seems clear today that the mechanisms responsible for the lower thermal conductivity of SiGe/GaP materials are the fine grain size and high carrier concentration, both of which are a natural consequence of the processes adopted to introduce GaP, as first suggested in (Amano 1987). In a sense, then, GaP works, but not by the mechanisms originally supposed. In analogy to the fine grain studies, it might be expected that GaP additions would not improve Z, at least compared to similar zone leveled materials.

P-Type Materials: Boron and III-V Additions

Several early SiGe/GaP results had raised concerns about the stability of the p-type materials, which required additions of large quantities of boron to overcompensate for the inherent n-type tendencies of GaP additions. Studies were directed at the cross-doping effects (Gunther 1982) and after the SiGe/GaP bicouples (SiGe RTG Report 1984) experienced relatively rapid degradation with time (presumably associated with the p-type SiGe/GaP materials), much of the experimental efforts shifted focus to n-type SiGe/GaP.

One small program involving LABCOM, TTC and JPL examined the effects of additions of BP, GaP and GaSb on p-type SiGe (McLane 1986). At the time, the mechanism by which GaP lowers the thermal conductivity was still in serious question. It seemed plausible that regardless of how GaP additions worked, other III-V compounds might function similarly and one needed only to avoid poisoning the p-type electrical properties. Preliminary results seemed encouraging, but reproducibility of the initial measurements proved difficult and no real improvement seems to have been achieved by this approach.

A second program involving the same laboratories examined the effects of really large additions of boron to SiGe, up to 20 atomic percent boron, again based on encouraging preliminary results (Vining 1988). Figure of merit improvements were not immediately forthcoming, mostly due to the formation of relatively high thermal conductivity second phases. But these experiments did serve to emphasize that very low resistivity values could be achieved by boron additions alone due to the high solubility of boron. Although even the earliest SiGe studies indicated boron is sufficiently soluble to achieve greater than optimum hole concentrations (Rosi 1968), only recent studies appear to confirm that standard SiGe is in fact under-doped. Improvements in Z of 10-15% have been reported merely by increasing the atomic boron concentration from 0.23 at% (used in the Voyager RTGs) to 0.6-1.3 at% (Bajgar 1991, Loughin 1993, and Fleurial 1993a).

Column IV Additions: Sn and Pb

An attractive idea that recurs from time to time is the question of alloys of Si and/or Ge with one of the other column IV elements: C, Sn or Pb. From the point of view of lower lattice thermal conductivity, one desires the greatest difference between the masses of a two-component alloy as possible and generally heavier masses. Assuming the only difference between a Si-Pb and a Si-Ge alloy were the mass of the atom, for example, a recent calculation suggests the Si-Pb material would have twice the figure of merit (Vining 1990) of the SiGe material. It seems natural, then, to investigate alloys involving the heavy elements Sn and Pb.

A patent (Duncan 1985) has been granted for Sn and Pb additions to SiGe which indicates a major reduction in thermal conductivity and improvement in Z, as described in (Duncan 1989). Unfortunately, more recent results (Elsner 1990) strongly suggest some measurement error in the original data. Recent thermodynamic calculations of several Si-Ge-Metal systems (Fleurial 1990c) have shown that the maximum solid solubilities of Sn and Pb in $Si_{80}Ge_{20}$ are about 0.12 at% and less than 0.001 at% respectively. In this case, neither Sn nor Pb are soluble enough in either Si or Ge to have any observable decrease in the thermal conductivity of a SiGe alloy.

Neutral Donor Scattering

White (White 1990, 1992) has examined theoretically the effect of neutral impurities on the lattice thermal conductivity. Shallow, neutral impurity states perturb the electronic structure in the vicinity of the impurity, which strains the lattice and results in phonon scattering. This contribution to phonon scattering is estimated to lower the thermal conductivity by about 8% at high temperatures, which may be observable in some cases. Unfortunately, further increases in this phonon scattering mechanism appear to be ineffective at lowering the thermal conductivity at high temperatures in these heavily doped alloys.

Inert Particulates

For some years it has been recognized that alloying is effective at scattering short wavelength phonons and free carriers (and/or grain boundaries) are effective at scattering long wavelength phonons. A substantial amount, perhaps the majority, of the heat is therefore attributable to a range of intermediate wavelength phonons with (relatively) low scattering rates. A scattering mechanism directed at these intermediate wavelength phonons might, therefore, be quite effective at lowering the thermal conductivity.

One possible scattering mechanism is to introduce a fine, randomly distributed dispersion of very small (say, 50-100 Å) particulates into an otherwise undisturbed single crystal. This general idea has been recognized as early as 1966 and is described in a Monsanto patent (Henderson 1966), summarized in Sittig's (1970) review of the patent literature. This report claimed to improve Z of SiGe by additions of aluminum oxide, calcium oxide, boron nitride, and a long list of other additions. Up to 14% of particulates were added to SiGe, with particulate sizes ranging from 50 to 500,000 Å and inter-particulate distances from 50 to 500 Å. Z values as high as $1.1 \times 10^{-3}\ K^{-1}$ were reported.

Serious experimental and theoretical investigations are more recent and still active. Slack, under the ITM program, has attempted to grow suitable particulates in-situ by precipitation of boron (ITM Program Final Report 1988). An SDIO sponsored program with JPL and Los Alamos pursued a plasma deposition process with similar goals, but encountered difficulties before the concept could be verified. This same program examined the possibility of producing extended defects by neutron radiation (Vandersande 1990a,b). While not strictly particulates, the neutron induced defects were supposed to cluster in a manner which would scatter intermediate wavelength phonons. Reductions in thermal conductivity were in fact observed, but the effect annealed out rather quickly with time and the reduction of the electrical mobility was greater than the reduction of the thermal conductivity. Still, this is an important result because the properties rapidly returned to their pre-radiation values, demonstrating the radiation resistance of SiGe.

The most substantive effort on particulate additions is currently underway at TTC, with support from JPL. Calculations performed by Klemens (1991), White (1990, 1992), Vining (1991b), and Slack (1991) suggest significant improvements in Z (10-40%) may be possible by incorporating fine particulates in SiGe. Slack (1991) discusses the importance of achieving a large grain size in the final material in order to avoid degradation of the electrical conductivity.

TTC has developed a remarkable spark erosion preparation method (Beaty 1990) to produce ultra-fine particles of inert, phonon scattering centers, introducing them into B-doped SiGe and evaluating their thermal conductivity after processing. For this effort, conducted under a JPL/TTC program, particles of SiGe 20 to 120 Å in size were produced and methods were developed for handling, mixing, hot-pressing and heat-treating these powders (Beaty 1991a). After demonstrating Z values typical of standard p-type SiGe/B using appropriate high temperature heat-treatments, attempts were made to introduce inert scattering centers. Suitable scattering center materials were to be electrically, thermally and chemically inert, but electrically conductive enough to be spark eroded.

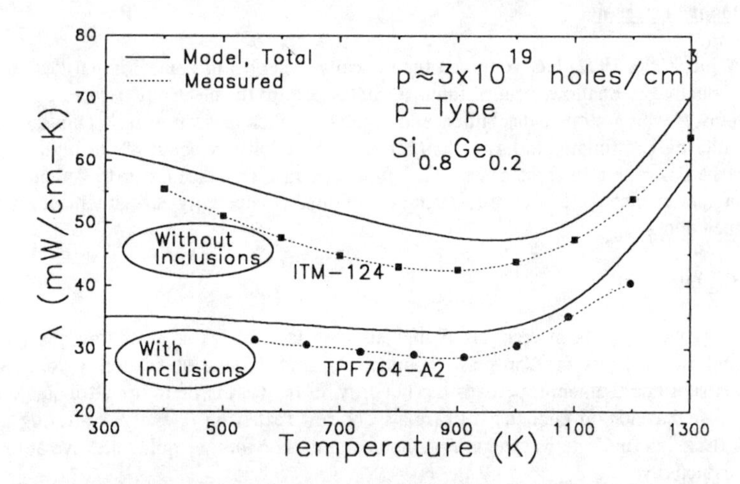

FIGURE 5. Thermal Conductivity for SiGe With and Without Particulates. Solid Lines Were Calculated as Described in (Vining 1991b). Dashed Lines Represent Data From (ITM Final Report 1989) and (Beaty 1991b).

Several materials have been investigated to date (JPL/TTC Workshop 1991): 1) SiO_2, which dissociated within the SiGe matrix; 2) aluminum oxides, which proved to be only metastable when formed by spark erosion; 3) Si_3N_4, which reacted with the B dopant to form BN, and 4) BN which appeared stable within the SiGe matrix. Samples prepared with Si_3N_4 and BN particles exhibited much lower thermal conductivities than standard SiGe, even after several high temperature anneals. Figure 5 shows the good agreement between experiment and theory for the change in thermal conductivity (at equivalent doping level) due to particulate scattering centers (Beaty 1991b). TEM analysis of several such samples showed particles containing B and ranging in size from 50 to 200 Å were embedded in the SiGe grains (Vandersande 1992b). Several samples with BN inclusions succeeded in improving Z up to 0.7×10^{-3} K^{-1} but reproducible results have been hampered by difficulties in the manufacturing process.

Minimum Lattice Thermal Conductivity

Common to all of these thermal conductivity reduction concepts is a limiting principle proposed by Slack (1979) and recently reviewed by Cahill and Pohl (1991). This principle is the idea of a minimum thermal conductivity. In simplistic terms, a phonon is a wave-like disturbance of the atoms in a crystal. If phonon scattering rates became too large, conventional perturbation theory will eventually break down. One way around this breakdown is to impose a lower limit on the phonon mean free path. A natural cutoff is to require the phonon mean free path to be greater than one interatomic distance. Slack proposes a somewhat more restrictive limit to cutoff the scattering distance at one wavelength, arguing that one cannot really speak of a phonon until it travels at least one wavelength.

Recent estimates now place the minimum lattice thermal conductivity for Si at about 10 mW/cm-K (Cahill 1989). Slack has estimated that if the thermal conductivity of SiGe could be reduced to the minimum value, the Z would more than double over current values, as shown in Figure 6 (ITM Final Report 1989).

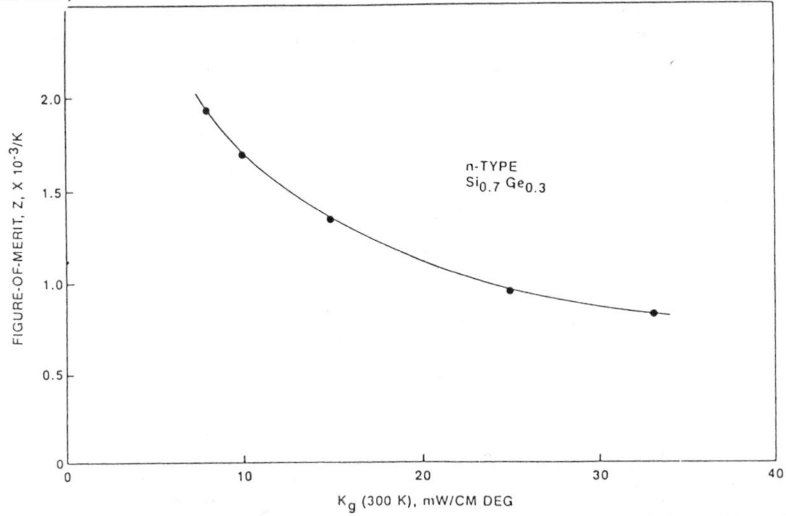

FIGURE 6. Average Thermoelectric Figure of Merit From 773 to 1273 K as a Function of the Lattice Thermal Conductivity at 300 K (ITM Final Report 1989).

IMPROVEMENT OF THE ELECTRICAL PROPERTIES

GaP Additions - Carrier Concentration Enhancement

Through about 1987, most studies focused on thermal conductivity reductions and little attention was given to the possibility of optimizing the electrical properties of n-type $Si_{80}Ge_{20}$ by increasing the carrier concentration. But since the work done in the 1960's at RCA (Rosi 1968) and in the Soviet Union (Erofeev 1966), it has been known that, unlike boron used for p-type materials, phosphorus (P) could only produce room temperature carrier concentrations up to 2.2×10^{20} cm^{-3} in n-type $Si_{80}Ge_{20}$, due to its limited solubility. Only recently, under the ITM program at GE, have new results on SiGe/GaP sparked renewed interest in high carrier concentrations in n-type materials (ITM Final Report 1989). Although the ITM studies were originally intended to characterize the influence of GaP additions on the thermal conductivity, as discussed in the preceding sections, contradictory results seemed to show that the GaP additions did alter the transport properties but not for the reasons originally expected.

Indeed, regardless of the theoretical suggestion of thermal conductivity reduction by small particulates, it retrospectively seems that the possibility of having stable GaP particles in a Si+Ge environment was quite limited. Since 1968, the work of Glazov (1968) on doping Si and Ge using III and V elements had shown there was no pseudo-binary equilibrium between Si or Ge and the III-V compounds investigated, as the ratio of the solid solubilities of the III and V elements was not retained upon incorporation into Si or Ge. Glazov also found that the solubility values depended on the stability of the corresponding III-V compounds.

The ITM work on heavily doped SiGe/GaP showed: 1) the sample composition was lower in Ga and Ge after hot-pressing and subsequent heat treatments; and 2) higher carrier concentrations could be obtained, up to 2.8×10^{20} cm^{-3}. Although a different interpretation was given at the time, these ITM samples were the first ones to exhibit carrier concentrations substantially higher than for standard SiGe material.

The first encouraging results on improvement of Z came in 1987 when JPL reported Z values 30% higher than standard SiGe could be achieved with proper GaP doping and high temperature heat treatments in air (Vandersande 1987). Shortly thereafter, TTC reported confirmation of these results using $Si_{80}Ge_{20}$ material doped with 2 mole% GaP (Vandersande 1988a). The improvement in Z due to heat treatments was widely reported in dramatic plots such as shown in Figure 7, which is due to Draper (1987).

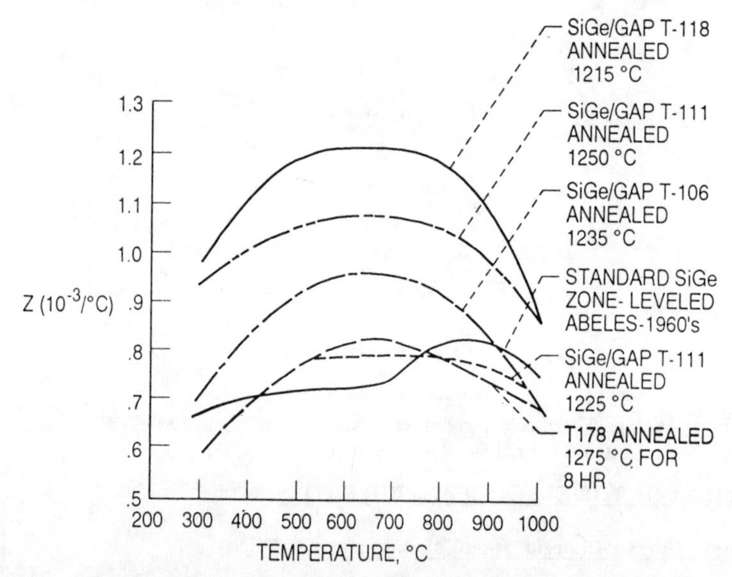

FIGURE 7. Figure of Merit of Annealed n-type SiGe/GaP as a Function of Annealing Temperature, after (Draper 1987).

The improvement in Z was attributed to a lower electrical resistivity brought about by higher carrier concentrations, while the Seebeck coefficient and the thermal conductivity barely changed. A preliminary model for the transport properties of n-type SiGe due to Vining (1988) supported the initial experimental findings and suggested increases in Z of 40% to 50% beyond these improved samples might be possible by increasing the doping level alone. This model predicted an improvement would result from an increase in power factor and reductions in the ambipolar and lattice contributions to the thermal conductivity. Moreover, a somewhat refined version of this model (Vining 1991c) indicates the experimental results on improved SiGe/GaP, shown in Figure 8, are entirely consistent with previous results on zone leveled materials, except that the SiGe/GaP has a higher carrier concentration. Calculations performed by Slack, using a somewhat distinct approach, reach generally similar conclusions (Slack 1991).

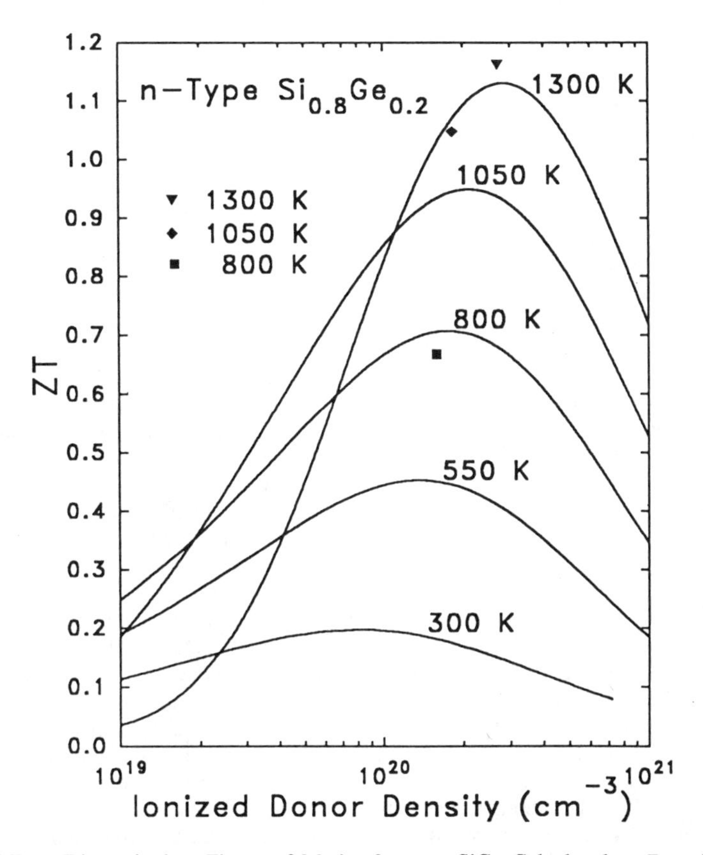

FIGURE 8. Dimensionless Figure of Merit of n-type SiGe Calculated as Described in (Vining 1991c). Solid Points Represent Experimental Results on T373, a Sample of SiGe/GaP.

However, because of the complexities of the manufacturing process for hot-pressed SiGe/GaP, it was difficult to control the amount of dopants introduced, the size of the grains and more generally the homogeneity of the samples. As previously pointed out, every processing variable affects the thermoelectric properties, obscuring the true relationships between transport properties and microstructure composition of these heavily-doped samples. Indeed, the first improved samples with integrated average Z values (between 873 and 1273 K) close to 1.0×10^{-3} K^{-1} were very difficult to reproduce. But comforted by the few experimental results and the theoretical predictions, an extensive effort was initiated at TTC, GE and JPL to obtain and study high Z SiGe/GaP samples (Vandersande 1988b). Development plans in 1988 called for determining the Ga+P amounts necessary for the improvement, understanding the nature of Ga+P doping in terms of dopant solid solubility changes and determining the role of the microstructure of the samples.

Experimental programs at GE-VF, TTC and JPL on hot-pressed SiGe/GaP concentrated on varying systematically the Ga/P ratios and the total Ga+P concentrations introduced in the samples using additions of SiP and GaP (Nakahara 1990). Microstructure analysis of the improved SiGe/GaP samples had shown the formation of Si-rich and Ge-rich SiGe areas within the samples, with substantial changes in Ga and P contents. One major finding was that the resulting Ga/P ratio in the SiGe matrix was always lower than 1 (Fleurial 1989a). Due to the lack of success in reproducing the Z improvements and the poor understanding of the mechanism(s) involved, a research effort directed towards the synthesis of high quality, homogeneous SiGe alloys was started at JPL, together with an intensive theoretical approach for directing the experiments, using models based on physical chemistry (Borshchevsky 1989) and solid state physics (Vining 1989). Studies of the effect of processing methods on homogeneity in Si-Ge-GaP materials have also been performed at the University of Virginia (Kilmer 1991).

Zone-leveling (ZL) and liquid phase epitaxy (LPE) were developed at JPL in order to grow bulk large grain, homogeneous ingots and single crystalline layers of doped SiGe. Also, experiments to diffuse P and/or As dopants into ZL SiGe materials were set up. The ZL method developed in the mid-60s at RCA (Dismukes 1965), produced ingots doped either by P, As or B only which, because of furnace limitations, were generally of a higher Ge content (typically $Si_{70}Ge_{30}$). The technique was redeveloped at JPL to produce $Si_{80}Ge_{20}$ doped with several elements simultaneously (Borshchevsky 1990a). Microprobe analysis of the grown ingots with various III and V dopant combinations demonstrated that good quality materials with a uniform Si/Ge ratio and dopant(s) content throughout the samples could be obtained. The addition of III and V elements together in the form of III-V compounds (such as BP or GaP) led to a very different result from the original equiatomic ratio of 1:1. This result clearly manifested the strong dissociation of III-V compounds such as GaP in the Si-Ge melt. Unfortunately, the processing temperature of the zone-leveling process is too close to the $Si_{80}Ge_{20}$ melting point to achieve high dopant concentrations, except for B, and the introduced amounts remained well below the optimum values (Borshchevsky 1990b).

Zone-leveled materials produced at JPL, together with original ingots manufactured by RCA, were used for diffusion experiments (Fleurial 1989b,c). Diffusion of P, As and P+As were conducted on ZL $Si_{70}Ge_{30}$, $Si_{80}Ge_{20}$ and $Si_{85}Ge_{15}$ samples doped with either P or As. Higher carrier concentrations were obtained by combining P and As as dopants. These higher values were reached exclusively in samples displaying localized inhomogeneities in SiGe composition together with second-phase inclusions similar in composition to compounds of P and As with Si and Ge. P-only diffusion did not succeed in increasing the carrier concentration of P-doped samples, demonstrating that the maximum P solid solubility had already been obtained in these samples. As-only doping was found to achieve carrier concentrations substantially higher than P-only doping, a result in contradiction with previous RCA work (Rosi 1968). This difference was accounted for by the high operating temperatures of the ZL process, incorporating the dopants into a Si-Ge melt, compared to the solid state diffusion process.

Actually, the JPL results correlated with work on heavily-doped Si (Fleurial 1990a) where As solid solubility and carrier concentrations are higher than for P. The large gap in carrier concentration between the P and the As diffusion experiments had been explained (Nobili 1982) by the difference in formation of dopant defects and complexes: 1) a precipitation mechanism for P, with the carrier concentration saturating at values independent of the total doping concentration; and 2) a clustering model for As, with the carrier concentration continuing to increase with the total doping concentration up to the solid solubility limit. This illustrated the fact that the actual carrier concentration obtained from a dopant is always lower than the dopant concentration, due to the polytropy of the impurities (Fistul 1967).

However, not only did P doping yield better carrier mobility values than As doping for similar carrier concentrations, but the difference in maximum carrier concentrations tended to decrease as the

Si/Ge ratio increased from 70/30 to 85/15. Thus, As doping eventually proved unattractive, but this study was instrumental illuminating several points: 1) P-only doping did not yield carrier concentrations higher than already achieved on standard SiGe; 2) Ga+P was not the only combination to enhance carrier concentration; 3) carrier concentration enhancement is linked to impurity-rich second-phase inclusions and localized Si/Ge inhomogeneities; 4) doping heavily n-type SiGe materials was a complex problem and modeling of the doping mechanisms was required.

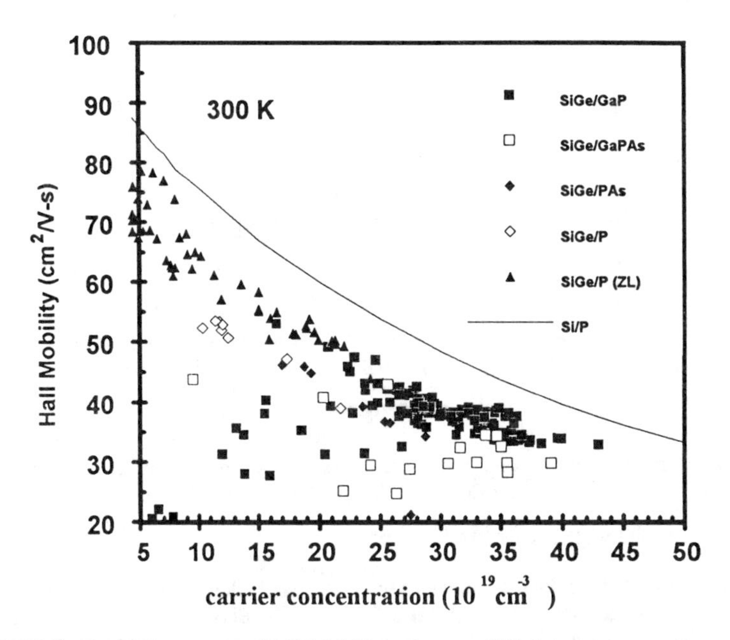

FIGURE 9. Room Temperature Hall Mobility of n-type SiGe Materials as a Function of Carrier Concentration, Redrawn From (Fleurial 1991b).

To further evaluate the potential of multi-doping for enhancing carrier concentrations, similar experiments were conducted at JPL on hot-pressed SiGe/P and SiGe/GaP materials obtained from TTC. Using 3 at% P and 1 at% Ga additions in $Si_{80}Ge_{20}$, high carrier concentration values above $3.5x10^{20}$ cm^{-3} have now been obtained repeatedly. Figures 9 and 10 display compilations of room temperature Hall mobility and electrical resistivity versus carrier concentration for n-type hot-pressed $Si_{80}Ge_{20}$ doped with P, P+As, Ga+P and Ga+P+As. For comparison, values obtained for ZL P-doped $Si_{80}Ge_{20}$ (Rosi 1968) and polycrystalline P-doped Si (Fleurial 1991b) samples are also reported.

These Figures demonstrate the large carrier concentration increases achieved by multidoping, from $2.2x10^{20}$ cm^{-3} for SiGe/P to $3.9x10^{20}$ cm^{-3} for SiGe/GaPAs and up to $4.3x10^{20}$ cm^{-3} for SiGe/GaP. Although, combining P and As did enhance the carrier concentration value by 25%, lower mobilities resulted in higher electrical resistivities. By comparing the mobility of zone-leveled SiGe/P material with the best values of the hot-pressed SiGe/GaP material, it seems clear that the mobility decrease is due only to the increase in carrier concentration, not to a poorer quality of the samples. Comparison with P-doped Si material tends to confirm this finding, showing that the top mobility curve of $Si_{80}Ge_{20}$ is shifted only due to the alloying effect, the shift becoming smaller for the highest carrier concentration. Thus, this top mobility curve representing the best electrical properties for a given carrier concentration, significantly extended the range of carrier concentrations accessible for Z optimization studies.

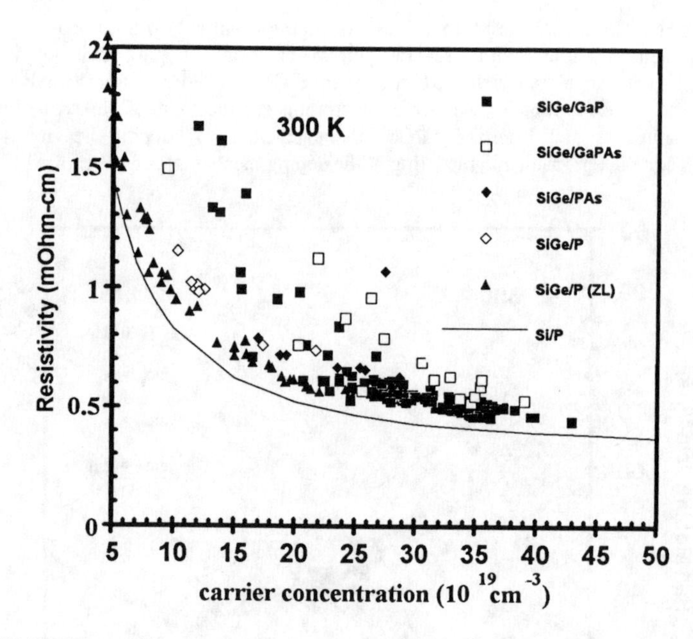

FIGURE 10. Room Temperature Electrical Resistivity of n-type SiGe Materials as a Function of Carrier Concentration, Redrawn From (Fleurial 1991b).

GaP Additions - Understanding the Mechanisms

During the course of these doping studies to produce heavily doped SiGe materials, it was found absolutely necessary to develop a thermodynamic model of the Si+Ge+Metal interaction at different temperatures. In other words, to calculate the relevant ternary and quaternary phase diagrams and predict the dopant(s) concentration(s) in Si-Ge alloys when multi-doping. Computation of these phase diagrams using a thermodynamic equilibrium model permitted calculation of liquidus temperatures and solvent solid solubility curves required for the growth and doping conditions of Si-Ge alloys from solution in metallic solvents such as Al, Ga, In, Sn, Pb, P, Sb and Bi (Fleurial 1990c). These results were then extended to quaternary solutions of Si-Ge with combinations of solvents such as Ga-In and Ga-P, supposing no thermodynamic interaction between the metallic solvents. Figure 11, redrawn from (Fleurial 1990c), shows the Ga solid solubility curves as a function of temperature for several Si-Ge alloys compositions. Similar results were obtained for several other systems of interest.

Building on the results achieved in Si-Ge growth using ternary systems, experiments were conducted at JPL involving more than one metallic solvent. SiGe thin films were grown by liquid phase epitaxy (LPE) out of Ga+In, In+P and Ga+In+P melts (Fleurial 1991c, Borshchevsky 1991). Attempts to grow Si-Ge material out of Ga+P melts proved unsuccessful, although high Ga/P(l) were used (values like 9 or 16), because of the difficulty in dissolving the GaP additions into the excess Ga, even at temperatures higher than 1373 K. Instead, as the solid solubility of In is much lower than Ga in $Si_{80}Ge_{20}$ and the liquidus temperatures for the two Si-Ge-Ga and Si-Ge-In ternary systems are very close in the range of interest, and as the melting point of InP (1343 K) could be reached by the LPE furnace, additions of InP were made into the excess Ga solvent.

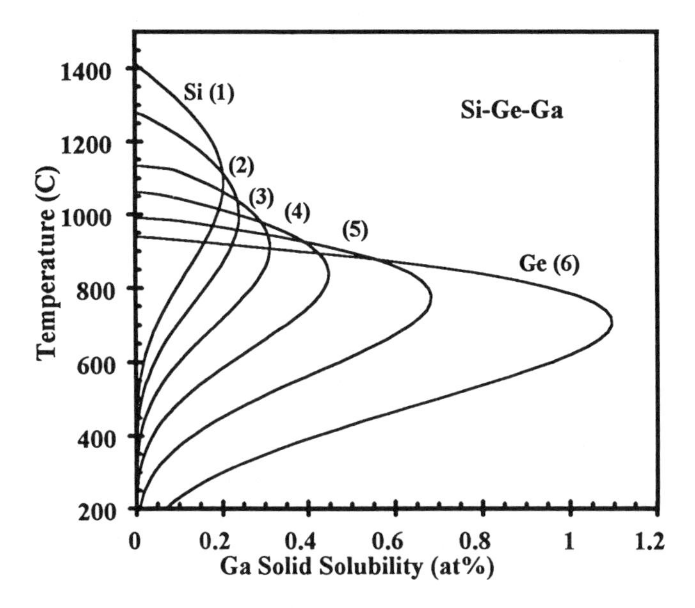

FIGURE 11. Calculated Ga Solid Solubility Curves as a Function of Temperature for Various Si-Ge Alloy Compositions in at%: (1) Si, (2) 80% Si, (3) 60% Si, (4) 40% Si, (5) 20% Si, (6) Ge.

Dopant solid solubilities in $Si_{80}Ge_{20}$ layers grown out of melts consisting of only column III solvents were very close to the values determined from the thermodynamic calculations based on the limiting binary systems, neglecting any additional multicomponent interaction. These results were predictable as no association in the melt exists between these elements. However, in the case of III-V combinations, the existence of compounds such as InP and GaP should bring some kind of association in the melt between these elements. This additional multicomponent interaction in the liquid phase was expected to modify dopant solid solubility behavior depending on the strength of this association (Glazov 1968). Neglecting a multicomponent interaction translated to systematically higher calculated liquidus temperatures than the observed by differential thermal analysis (DTA). A correction could be made to the model, but because of insufficient data a satisfactory recalculation of the quaternary phase diagram could not be carried out.

Microprobe analysis of the (III+V)-doped Si-Ge layers showed substantial differences between the experimental and calculated findings. The experimental Ga and P solid solubilities were more than fivefold higher than the values calculated neglecting multicomponent interactions. Such substantial enhancements indicate the Ga-P interaction could not be neglected. Additional experiments to grow bulk single crystals of SiGe by the traveling solvent method at JPL confirmed such changes of Ga and P solid solubilities (Fleurial 1991c, Borshchevsky 1993). These results also demonstrated that Ga additions to P-doped SiGe could significantly enhance the P solid solubility and thus increase their carrier concentrations.

The absence of similar effects when using only In+P melts shows that the In-P interaction is much weaker. This was expected because the strength of the III-V interaction, or ion pairing, is somewhat related to the melting point and stability of the corresponding III-V compound. Similar reasoning could be applied to the other III-V couples such as B-P, Al-P or Ga-Sb.

The schematic in Figure 12 has been presented numerous times during programmatic presentations although it was never published (Fleurial 1990b). It displays an enlarged portion of a section of the

Si-Ge-Ga-P phase diagram, illustrating how the increase in carrier concentration is obtained when double-doping by enhancing the P solid solubility and reducing the polytropy of P for certain Ga/P ratios.

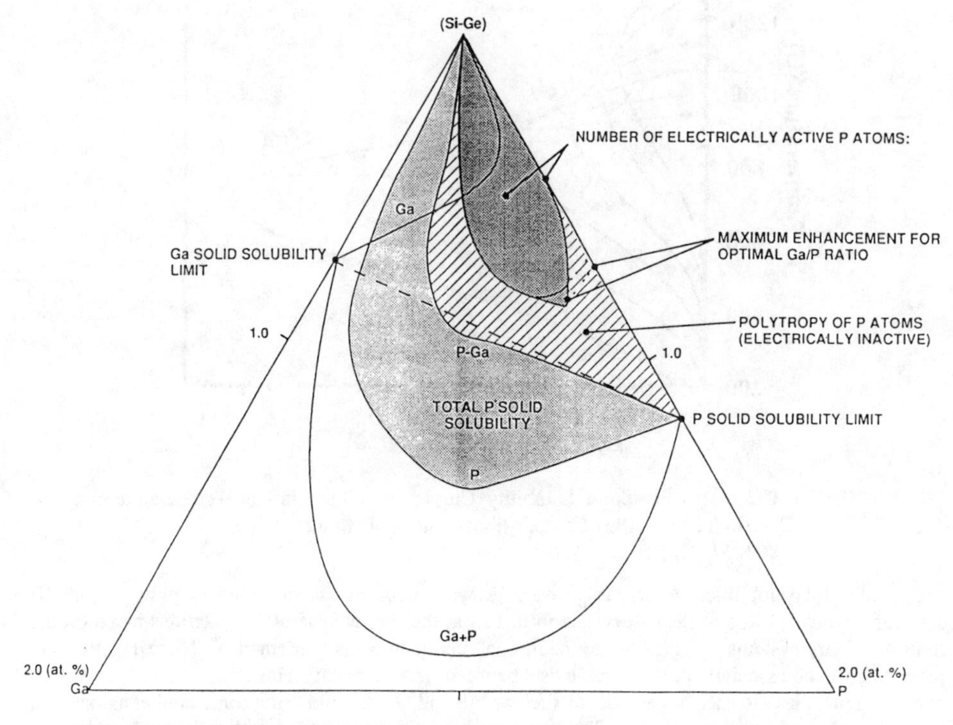

FIGURE 12. The (SiGe)-Rich Region of the (Si-Ge)-Ga-P Pseudoternary Phase Diagram Indicating the Region of Enhanced Carrier Concentrations (Fleurial 1990b).

The substantial changes in P solid solubility in Si-Ge alloys observed when Ga is added, demonstrated without any ambiguity that significant carrier concentration increases could be achieved for n-type hot-pressed $Si_{80}Ge_{20}$ thermoelectric materials. However, this also meant that understanding and controlling the doping mechanisms in these materials is a complex process, illustrated by the difficulty in reproducing high thermoelectric figures of merit.

Comparing the microprobe analysis of heavily-doped zone-leveled and hot-pressed materials, it was concluded that the high carrier concentrations reached in these samples were due to the formation of a complex multi-phase structure. The micrograph shown in Figure 13 is characteristic of this structure in hot-pressed SiGe/GaP materials (Fleurial 1991a). Si-rich and Ge-rich SiGe areas are mostly present, with inclusions rich in SiO_x, SiP_y (darkest spots) and other rich in Ge, GaP (brightest spots). Also, the Ge-rich SiGe areas have a much higher dopant concentration than the Si-rich SiGe areas, acting as reservoirs of dopants during subsequent temperature cycles. Although the mechanisms of carrier concentration enhancement were similar for diffused zone-leveled samples and diffused or annealed hot-pressed samples, the much smaller grain size of the latter enabled the development of a much larger number of dopant-rich phases. Thus, the capacity for dopant absorption of hot-pressed materials was far larger, accounting for the substantially higher carrier concentrations.

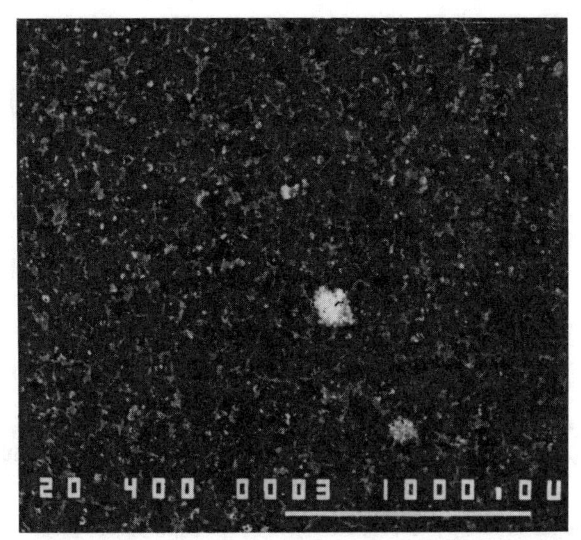

FIGURE 13. Micrograph of T-428, Hot-Pressed SiGe/GaP, After Various Heat Treatments (40x).

While at room temperature SiGe/GaP can exhibit carrier concentration values as much as twice the values for SiGe/P, high temperature Hall effect measurements conducted at JPL indicate this gap narrows rapidly above 1073 K, with no more than a 15 to 20% difference remaining at 1273 K. Similar effects had been observed before on pure Si doped with both III and V elements. An interpretation of these results expressed at JPL was that: 1) room temperature carrier concentration values are higher because of ion pairing between the Ga and P atoms in the form of electrically active Ga-P_n complexes, preventing the P from forming electrically inactive SiP_4 complexes and SiP precipitates; and 2) for temperatures higher than 1073 K, an increasing fraction of the Ga-P_n complexes redissolve into electrically inactive Ga-P pairs and SiP precipitates, reverting to standard SiGe/P electrical behavior. This reversible process would account for the very low resistivity values obtained on improved SiGe/GaP materials, due to high carrier concentration and good mobility values, the Ga dopant not acting as a carrier compensator.

GaP Additions - Reproducing Improvements in Z

Because high carrier concentrations had only been obtained in hot-pressed materials, crystal growth techniques were largely dropped at JPL. Efforts then focused on the conditions necessary to reproducibly achieve a high Z. In 1990, a JPL/TTC program performed systematic high temperature annealing of supersaturated (Ga+P)-doped SiGe materials at various temperatures and for different amounts of time (Fleurial 1991a,e and Scoville 1991a,b). These experiments indicated the formation of a multi-phase structure with Si-rich and Ge-rich Si-Ge regions was responsible for substantial increases in carrier concentrations, sometimes close to a factor of two at room temperature compared to standard P-doped SiGe materials. Optimization of the electrical properties of n-type materials throughout the extended temperature range was started, guided by the JPL theoretical work described in the preceding section.

Results reported on both large cylindrical samples and thin disks showed that substantial variations in microstructure and composition brought by the different annealing temperatures and times led to significant differences in electrical properties. It was particularly demonstrated how the changes in Ga and P concentrations correlated with annealing condition and electrical properties.

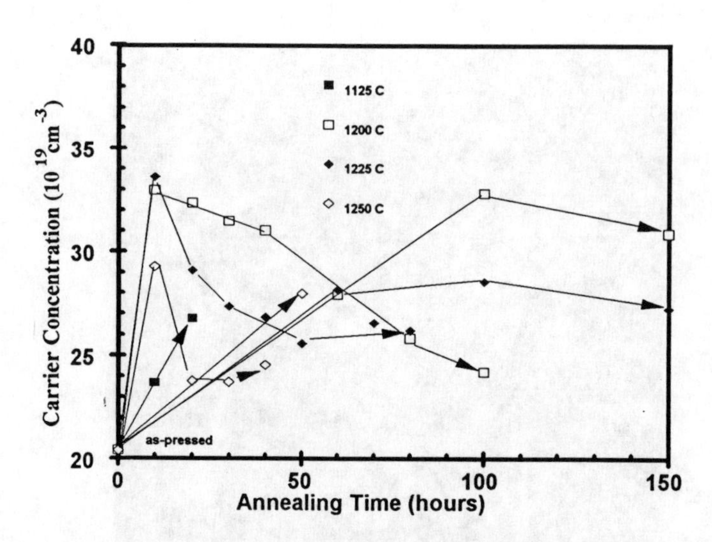

FIGURE 14. Changes in Room Temperature Carrier Concentrations of SiGe/GaP Samples With Annealing Time and Temperature, Redrawn From (Fleurial 1991b).

Figures 14 and 15, show changes in carrier concentration with annealing time and changes in mobility with carrier concentration for various annealing temperatures. The initial increases in carrier concentration (Figure 14, after the first anneal) appear typical of the dopant redistribution occurring in SiGe/GaP. Actually, the mobility values also increase dramatically after the first anneal (Figure 15), sometimes translating into electrical resistivity reductions of 60% over the entire temperature range (298-1273 K). Subsequent anneals generally lower the carrier concentration, but at a slow rate, with corresponding changes in mobility.

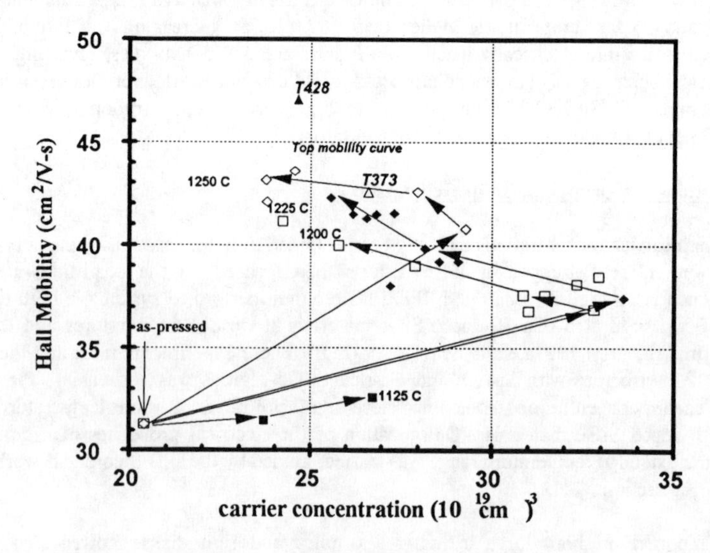

FIGURE 15. Room Temperature Hall Mobility Versus Carrier Concentration for SiGe/GaP. Curves Show Changes in Electrical Properties for Various Annealing Temperatures With Successive Heat-Treatments, Redrawn From (Fleurial 1991b).

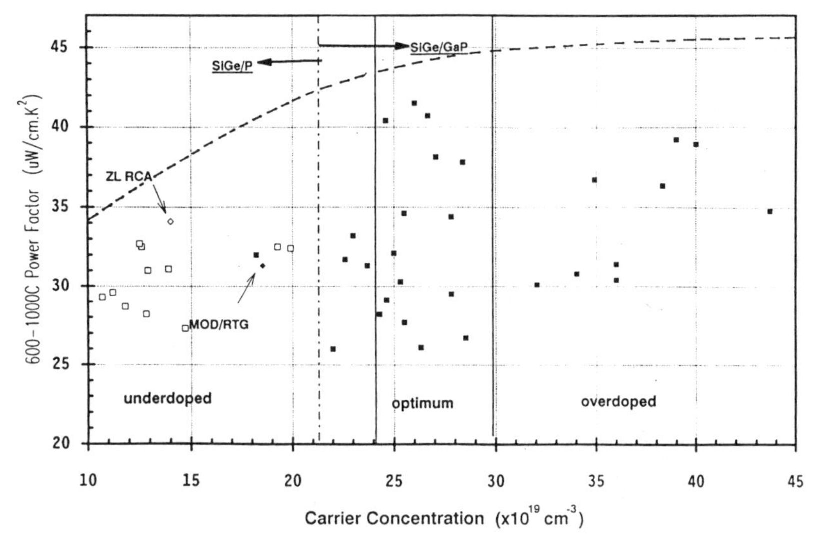

FIGURE 16. Integrated Average Power Factor (873-1273 K) of n-type Hot-Pressed $Si_{80}Ge_{20}$ Versus Room Temperature Carrier Concentration, Redrawn From (Fleurial 1991b).

Figure 14 also displays several curves for a same annealing temperature, but for different annealing times. These curves have shown that the loss of dopants due exposure to high temperatures depends on sample geometry (surface to volume ratio), an effect attributed to formation of a coherent Ga_2O_3 layer on the outer surface of the overdoped samples inhibiting carrier concentration loss and grain growth.

The power factor (S^2/ρ) of hot-pressed SiGe/P and SiGe/GaP is presented on Figure 16, where the integrated average of the power factor over the 873 - 1273 K temperature range is plotted as a function of room temperature carrier concentration. The only available RCA data on P-doped ZL $Si_{80}Ge_{20}$ (Rosi 1968) and the current standard SiGe materials are also shown for comparison.

The dotted line in Figure 16 represents the maximum power factor curve as calculated using the model described in (Vining 1991b). Note that the calculation reproduces the relatively steep increase in power factor observed experimentally between 1 and 3×10^{20} cm^{-3}. Both the calculation and the experimental results qualitatively indicate the power factor changes little for carrier concentrations above $n=3 \times 10^{20}$ cm^{-3}, which is reasonable agreement considering the uncertainties and extrapolations involved.

It was also clear that the P-doped only SiGe was underdoped, and that all samples showing low mobility values on Figure 9 were situated well below the maximum power factor curve of Figure 16, stressing the value of measuring the room temperature mobility for monitoring the improvement. The systematic preparation at TTC of hot-pressed samples with P:Ga atomic ratios ranging from 6:1 to 1:1 with a total P concentration varied between 2 and 3 at.% resulted in the determination of optimum conditions for achieving best electrical properties of n-type SiGe/GaP material. The experimentally found optimal room temperature electrical properties consist in a) Hall mobility ranging from 45 to 40 cm^2.V^{-1}.s^{-1}; b) carrier concentration between 2.5 and 3.0×10^{20} cm^{-3}; c) electrical resistivity between 0.55 and 0.52×10^{-3} Ω-cm. To obtain these properties, the P:Ga ratio must be close to 3:1 with a Ga concentration on the order of 0.75 at.%. The minimal grain size necessary to achieve these high mobility values has been found to be about 20 to 30 μm.

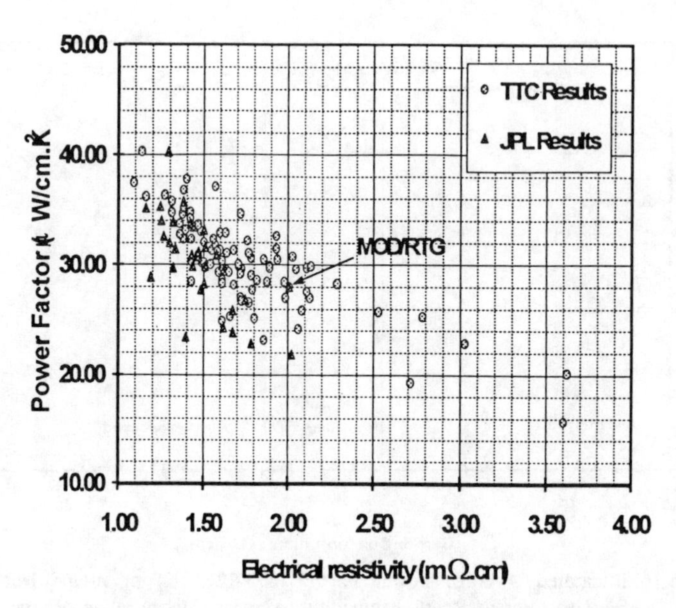

FIGURE 17: 600-1000 C Integrated Average Power Factor of recent n-Type SiGe/GaP Samples Versus Integrated Average Electrical Resistivity, Redrawn From (Fleurial 1993a).

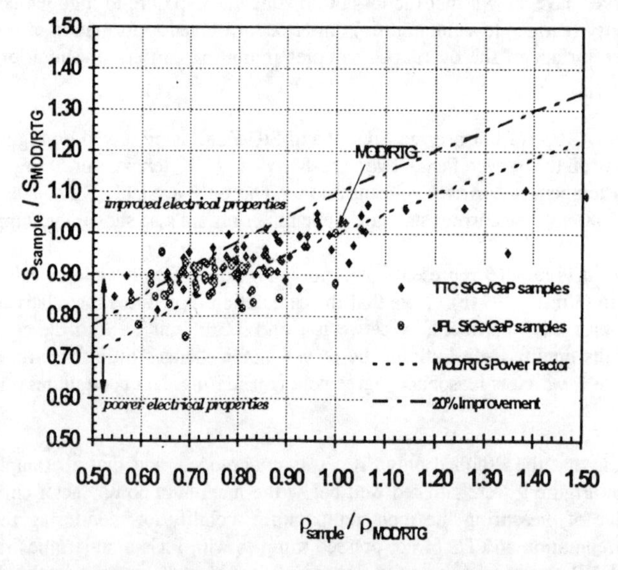

FIGURE 18: 600-1000 C Integrated Average Properties Normalized to MOD/RTG Values: Power Factor of Recent n-Type SiGe/GaP Samples Versus Electrical Resistivity, Redrawn From (Fleurial 1993a).

The large carrier concentration increases obtained by heavily doping hot-pressed SiGe samples with several dopants such as Ga, P and As have been instrumental in the improvement of the power factor. By doubling the range of room temperature carrier concentrations attained with P-only doping (Figure 16), optimization of the electrical properties was made possible. Simultaneous doping with adequate concentrations of Ga and P resulted in up to 30% increase in power factor over n-MOD/RTG materials (Figure 17). This is mostly due to a sharp decrease in the electrical resistivity over the 300-1000 C temperature range (up to 45%) combined to a much smaller decrease in the Seebeck coefficient (up to 15%), as seen in Figure 18.

The decrease of mobility with carrier concentration for these heavily doped SiGe/GaP samples was not affected by the additions of Ga, a potential acceptor impurity (Fleurial 1991b). This is attributed this to the nature of the Ga-P interaction resulting in the creation of Ga-P_n complexes, thus preventing Ga to act as a compensator and to degrade the mobility. Recent results on heavily P-only doped samples prepared by mechanical alloying and subsequent hot-pressing have confirmed this theoretical explanation (Han 1992). The authors reported carrier concentration as high as 3.0×10^{20} cm^{-3}; but no difference in Hall mobility values with our SiGe/GaP samples with similar doping levels. This demonstrated that point defect scattering is solely responsible for the lower mobility values obtained in optimized heavily doped SiGe/GaP compared to P-doped Si.

In Figures 17 and 18, it is clear that low electrical resistivity values systematically resulted in high power factor values. Also, the magnitude of the reproducible improvement is now about 20%, as larger values can be obtained but are hard to duplicate with the current processing conditions. Such improvement have been recently confirmed on mechanically alloyed SiGe samples, using similar P and Ga doping concentrations and with 20-50 μm grain size (Cook 1992). Indeed, the successive high temperature heat treatments tend to eventually degrade the electrical properties, but they are still necessary to ensure sufficient grain growth and dopant redistribution in the samples. To remedy these difficulties, change to the current hot-pressing parameters are investigated to obtain substantial grain growth in situ. This can be achieved through longer pressing time and/or higher pressing temperature. Preliminary results indicate that the loss of dopant in vacuum remained at reasonable levels, and that these new experimental conditions are successful at reproducing good quality samples with improved electrical properties (Fleurial 1993b).

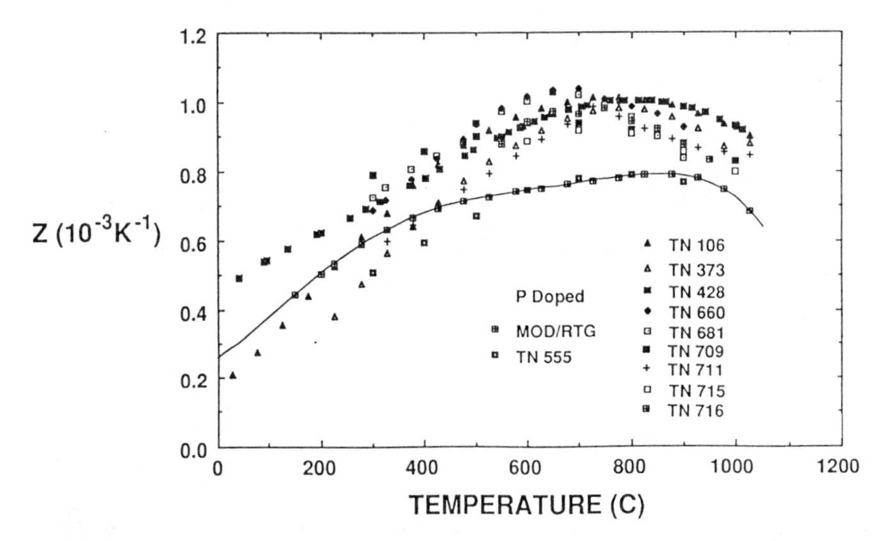

FIGURE 19. Figure of Merit Results for n-type SiGe/GaP Compared to Samples Doped With Phosphorus Only, Labeled MOD/RTG and TN 555.

Figure 19 displays Z versus temperature for the improved SiGe/GaP samples obtained over the past year, including previous results first obtained in 1987 (samples TN 106 and TN 373). Comparison with current standard materials (represented by the labels MOD-RTG and TN 555) show that improvements of about 30% in 873-1273 K integrated average Z have been obtained: from 0.78×10^{-3} K^{-1} up to 1.0×10^{-3} K^{-1}. This substantial increase in Z represents the maximum improvement to be reached by optimization of the electrical properties alone (Fleurial 1991b). To reach the goal of 1.1×10^{-3} K^{-1} set by the SP-100 program for n-type SiGe, thermal conductivity reductions by introduction of inert scattering centers, similar to the program described above for p-type SiGe, are to be applied.

The carrier concentration enhancement mechanisms appear to be understood, the optimum electrical properties and the maximum Z values have been identified. The desired microstructure and composition have been determined. Current efforts at TTC, GE and JPL have developped a reliable SiGe/GaP manufacturing procedure for reproducibly obtaining average Z values of 0.9×10^{-3} K^{-1} in the 300-1000°C tempoerature range, a 15% improvement over baseline MOD/RTG materials. Work is in its final stages, but it seems probable that Z values of $0.95\text{-}1.0 \times 10^{-3}$ K^{-1} will also be reproducibly demonstrated in the near future.

SiGe/GaP Solubility and Precipitation Effects

Preliminary studies of the long term effects of GaP in SiGe have been performed at Ames (Tschetter 1990), JPL (Vandersande 1989), and Cardiff (Rowe 1990, Rowe 1991, Min 1991). The Ames study reproduced Raag's results on dopant precipitation (Raag 1978) in SiGe and extended these studies to n-type SiGe/GaP as prepared by GE-VF. As shown in Figure 20, the time and temperature dependencies are significantly altered in SiGe/GaP, presumably due to some combination of solubility and kinetic coefficient variations. Indeed while acknowledging a non-standard behavior, it has been found by several authors that SiGe/GaP MOD/RTG material was inferior to standard P-doped SiGe in the high temperature range (Fu 1992).

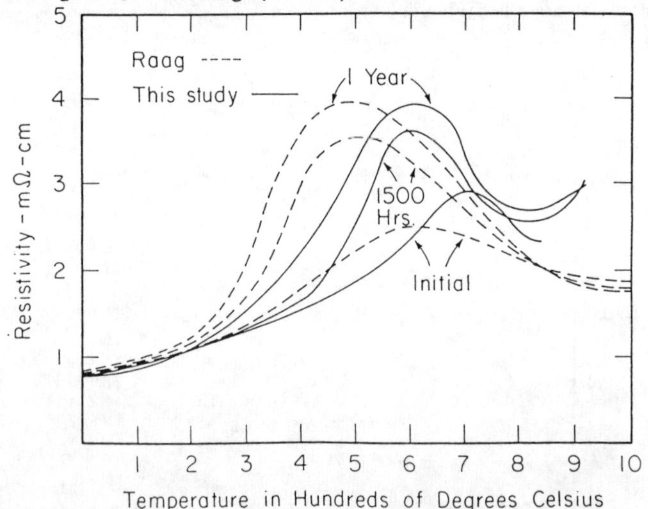

FIGURE 20. Electrical Resistivity of n-type SiGe (Dashed Line) and SiGe/GaP (Solid Line) as a Function of Temperature and Time (Tschetter 1990).

However, baseline MOD/RTG materials do not present any substantial improvement in Z over standard samples due their unoptimized atomic concentration of Ga and P dopants (2% of each). A more realistic test needed to be done using high power factor improved SiGe/GaP samples, which have a much different microstructure and composition (high P/Ga ratio). It was expecteded that the same mechanisms that resulted in increased P solid solubility and P electrical activity would also improve the dopant precipitation data, provided that optimal P:Ga ratio be met. Figure 21 (a), 21 (b),

21 (c) and 21 (d) display the results of long term resistivity measurements done in isothermal furnaces at respective temperatures of 450°C, 600°C, 800°C and 1000°C. The changes of resistivity over time for improved SiGe/GaP samples and standard SiGe/P samples demonstrate after more than 1850 hours of continuous operation that the SiGe/GaP samples retained their much lower electrical resistivity (except at 600 C where values are very close together). Moreover, at 450 C and 1000 C the rates of resistivity increase are substantially larger for standard SiGe/P samples. At 600 C and 800 C these rates are almost identical. These results confirmed that the substantial improvements in power factor over the entire temperature range were retained, especially for the most heavily-doped samples (Fleurial 1993a).

The long term dopant precipitation test has now been running for more than 5600 hours without any significant departure from the results obtained at 1850 hours (Fleurial 1993b).

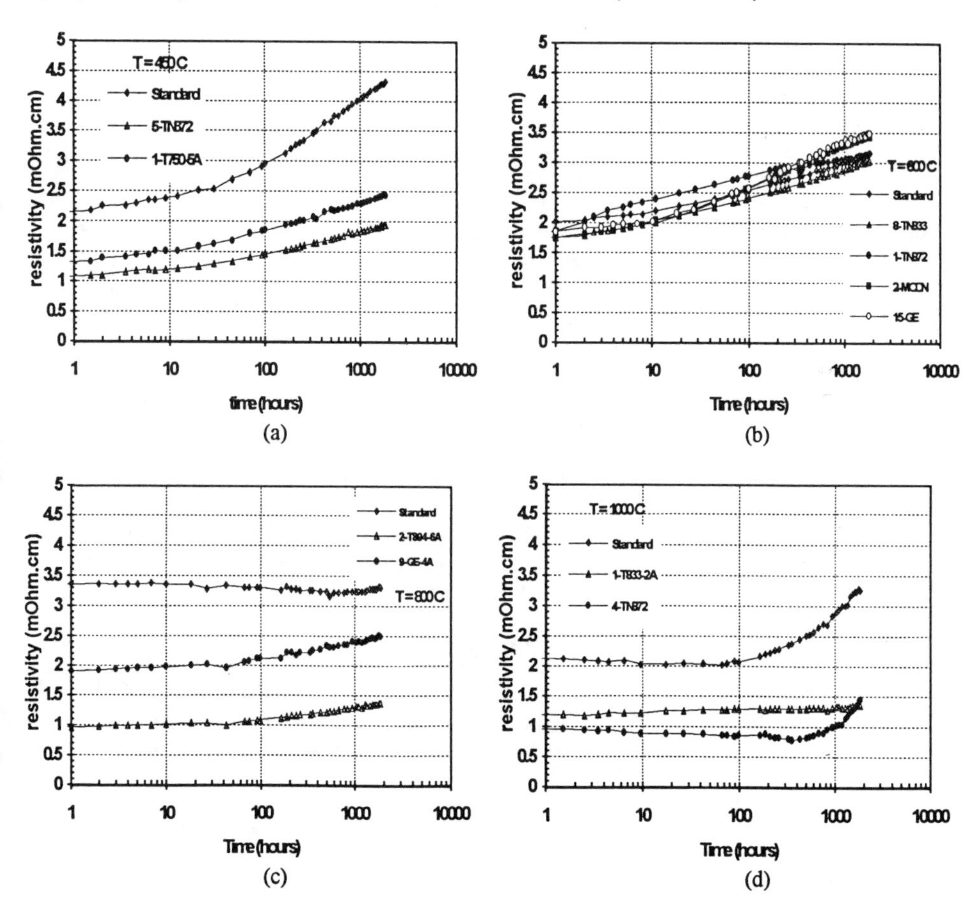

FIGURE 21: Dopant Precipitation Study of Improved n-Type SiGe/GaP Samples Compared to Standard P-Doped Only SiGe Samples: Data After 1850 hours at Temperatures of 450 C (a), 600 C (b), 800 C (c) and 1000 C (d), Redrawn from (Fleurial 1993a).

SUMMARY AND OPEN QUESTIONS

The variety of preparation methods, characterization techniques and theoretical tools applied to SiGe over the past 15 years testifies to the vitality of current thermoelectric materials development activities. The "fine grain" ideas for lowering thermal conductivity have evolved into a quite different approach of inert particulate additions. The idea of adding GaP has evolved tremendously, to the point that today improvements are attributed not to thermal conductivity reductions, as originally supposed, but to electrical property enhancements.

Yet in spite of the changes, both approaches appear nearly ready for practical applications and are entering the final development stages. The inert particulate approach for lowering thermal conductivity still requires some attention to the electrical properties. And both the particulate and carrier concentration enhancement approaches still require long term studies and device manufacturing processing development on improved materials.

But have those early ideas really been exhausted? The fine grain idea in fact remains something of an open question. The correlation between reduced mobility and reduced thermal conductivity reported for fine grain SiGe in (Vining 1991a) is at this point largely empirical. It is still possible that this correlation results from entirely different mechanisms acting independently on the phonons and on the carriers (such as described by Slack 1991), but are coincidentally similar in magnitude. After all, the grain boundary phonon scattering mechanism described by Rowe is probably not the mechanism responsible for the observed thermal conductivity reductions, as evidenced by the unexpected temperature dependence shown in Figure 2. Some clever trick may yet be found to avoid degradation of electrical properties in fine grain SiGe and realize the predicted improvements in Z.

And what was wrong with those early results on SiGe/GaP, which indicated such large gains in Z due to thermal conductivity reductions? Even extensive efforts have not reproduced the early material properties, which may simply be erroneous. The usual problems of performing all the important measurements on precisely the same material, changes in sample properties between (and during!) measurements and propagation of experimental errors may be to blame.

The bicouple results, however, seemed very persuasive to many in the field. But this test actually could at most conclude that the "SiGe/GaP" *devices* were better than the "fine grain SiGe"*devices*. Relatively less could be said about the absolute Z values of the materials, for a variety of reasons. Contributing factors include: 1) the unique bicouple device geometry employed; 2) the significant device-to-device variability observed in the "fine grain SiGe" devices; 3) uncertainties about internal resistances (both electrical and thermal); 4) non-optimal device geometries; and 5) test instrumentation difficulties. Given the experience of recent years it seems unlikely that the "SiGe/GaP" used in the bicouples was better than standard SiGe, although it may have been better than the particular "fine grain SiGe" materials actually used in that test for comparison.

Still, the efforts to understand fine grain SiGe and SiGe/GaP have been useful. Because of the need for higher carrier concentrations to achieve optimum electrical properties, the fine grain approach has been dropped because of accelerated dopant precipitation effects even during the course of high temperature thermoelectric properties measurements. Additions of GaP are now exclusively used for enhancing the carrier concentration of n-type materials by double doping mechanisms. Systematic experimental and theoretical studies on n-type and p-type materials have resulted in determining the range of microstructure, composition, doping level and electrical properties necessary to achieve improved Z values. Optimization of the hot-pressing process to approach these optimal parameters with a minimal amount of heat treatments is in its final stages. Currently reproducible average Z values in the 300-1000°C temperature range are 0.9×10^{-3} K^{-1} for n-type SiGe/GaP and 0.59×10^{-3} K^{-1} for n-type SiGe/B, a substantial combined 18% increase over the performance of baseline MOD/RTG materials. Addition of ultra fine BN particles to lower the lattice thermal conductivity could possibly bring another 15% improvement.

While many workers have made significant contributions, the efforts of Rowe and Raag deserve special note. Both made contributions in the 1970's, during a period of extremely limited resources for thermoelectric materials work, and most of the work today can be traced fairly directly to those early studies. Some future contribution may yet allow major improvements in SiGe, beyond the levels described here, but the standard of comparison has been raised significantly and any such new concept must certainly go well beyond the efforts described in this overview.

Acknowledgments

This paper was prepared at the Jet Propulsion Laboratory, California Institute of Technology, under contract with the National Aeronautics and Space Administration. The authors would like to express their gratitude to all those who have contributed to SiGe development over the years, who are far too numerous to mention. While every effort has been made to be complete, the authors would also like to apologize for any inadvertent omissions or misstatements of fact or opinion attributed to others.

References

Amano, T., B. J. Beaudry, K. Gschneidner, Jr., R. Hartman, C. B. Vining, and C. A. Alexander (1987) "High-Temperature Heat Contents, Thermal Diffusivities, Densities, and Thermal Conductivities of N-Type SiGe(GaP), P-Type SiGe(GaP), and P-Type SiGe Alloys," *J. Appl. Phys.*, 62(3): 819-823.

Bajgar, C., R. Masters, N. Scoville, and J. Vandersande (1991) "Thermoelectric Properties of Hot Pressed P-Type SiGe Alloys," in *Proc. of 8th Symposium on Space Nuclear Power Systems*, CONF-910116, M. S. El-Genk and M. D. Hoover, eds., American Institute of Physics, New York, NY: 440-445.

Beaty, J. S., J. L. Rolfe, and J. W. Vandersande (1990) "Properties of 100 Å P-Type Atomclusters and Thermoelectric Material," in *Proc. of 25th Intersoc. Energy Conv. Eng. Conf.*, P. A. Nelson, W. W. Schertz, and R. H. Till, eds., American Institute of Chemical Engineers, New York, NY, 2: 379-381.

Beaty, J. S., J. L. Rolfe, and J. W. Vandersande (1991a) "Thermoelectric Properties of Hot-Pressed Fine Particulate Powder SiGe Alloys," in *Proc. of 8th Symposium on Space Nuclear Power Systems*, CONF-910116, M. S. El-Genk and M. D. Hoover, eds., American Institute of Physics, New York, NY: 446-450.

Beaty, J. S., J. L. Rolfe, and J. W. Vandersande (1991b) "Thermoelectric Properties of Hot-Pressed Ultra-Fine Particulate SiGe Powder Alloys with Inert Additions," in *Modern Perspectives on Thermoelectrics and Related Materials, Mat. Res. Soc. Symp. Proc. 234*, D. D. Allred, C. B. Vining, and G. A. Slack, eds., Materials Research Society, Pittsburgh, PA: 105-110.

Bhandari, C. M. and D. M. Rowe (1978) "Fine Grained Silicon Germanium Alloys as Superior Thermoelectric Materials," in *2nd International Conf. on Thermoelectric Energy Conversion*, IEEE Cat. No. 78CH 1313-S Reg 5, K. R. Rao, ed., The University of Texas at Arlington, Arlington, TX: 32-35.

Bhandari, C. M. and D. M. Rowe (1980) "Silicon-Germanium Alloys as High-Temperature Thermoelectric Materials," *Contemp. Phys.*, 21(3): 219-242.

Borshchevsky, A., and J. P. Fleurial (1989) "Computation of Ternary Phase Diagrams for LPE Single Crystal Growth of SiGe Alloys," in *Eighth International Conf. on Thermoelectric Energy Conversion*, ISBN 2-905267-15-1, H. Scherrer and S. Scherrer, eds., Institut National Polytechnique de Lorraine, Nancy, France: 87-90.

Borshchevsky, A., J. P. Fleurial, J. W. Vandersande, and C. Wood (1990a) "Preliminary Results on Zone-Leveling of Multidoped SiGe Thermoelectric Alloys," in *Proc. of 7th Symposium on Space Nuclear Power Systems*, CONF-900109, M. S. El-Genk and M. D. Hoover, eds., American Institute of Physics, New York, NY: 229-233.

Borshchevsky, A., J. P. Fleurial, and J. W. Vandersande (1990b) "Experimental Approaches for Improving SiGe Thermoelectric Efficiency at JPL," in *Proc. of 25th Intersoc. Energy Conv. Eng. Conf.*, P. A. Nelson, W. W. Schertz, and R. H. Till, eds., American Institute of Chemical Engineers, New York, NY, 2: 397-401.

Borshchevsky, A. and J. P. Fleurial (1991) "Growth of SiGe from Metallic Solutions," in *10th International Conf. on Thermoelectrics*, D. M. Rowe, ed., Babrow Press, Cardiff Wales, UK: 19-26.

Borshchevsky, A. and J. P. Fleurial (1993) "Growth of Heavily-Doped SiGe from Metallic Solutions," *J. Crystal Growth*, 128: 331-337.

Cahill, D. G., H. E. Fischer, T. Klitsner, E. T. Swartz, and R. O. Pohl (1989) "Thermal Conductivity of Thin-Films - Measurements and Understanding," *J. Vac. Sci. Technol.*, A7: 1259.

Cahill, D. G. and R. O. Pohl (1991) "Is There a Lower Limit to the Thermal Conductivity of Solids?," in *Modern Perspectives on Thermoelectrics and Related Materials, Mat. Res. Soc. Symp. Proc. 234*, D. D. Allred, C. B. Vining, and G. A. Slack, eds., Materials Research Society, Pittsburgh, PA: 27-38.

Cockfield, R. D. and P. D. Gorsuch (1984) "The Bicouple - An Alternative Approach for Thermoelectrics," in *Proc. of 19th Intersoc. Energy Conv. Eng. Conf.*, W. Dodson, ed., American Nuclear Society, New York, NY: 2229-2234.

Cody, G. D., B. Abeles, J. P. Dismukes, and F. D. Rosi (1990) in "Research and Development of the Silicon-Germanium Thermoelectric Power Generator: The Long Term Payoff of Basic Research," *Proc. of IX International Conf. on Thermoelectrics (USA)*, C. B. Vining, ed., Jet Propulsion Laboratory internal document number D-7749, Pasadena, CA:. 182-205.

Cook, B. A., B. J. Beaudry, J. J. Harringa, and W. J. Barnett (1989) "The Preparation of SiGe Thermoelectric Materials by Mechanical Alloying," in *Proc. of 24th Intersoc. Energy Conv. Eng. Conf.*, W. D. Jackson, ed., Institute of Electrical and Electronics Engineers, New York, NY, 2: 693-700.

Cook, B. A., B. J. Beaudry, J. J. Harringa, and W. J. Barnett (1991) "Thermoelectric Properties of Mechanically Alloyed P-Type $Si_{80}Ge_{20}$ Alloys," in *Proc. of 8th Symposium on Space Nuclear Power Systems*, CONF-910116, M. S. El-Genk and M. D. Hoover, eds., American Institute of Physics, New York, NY: 431-439.

Cook, B.A., J. L. Harringa, B. J. Beaudry, and S. H. Han (1992) "Relationship between Thermoelectric Properties and the P/Ga Ratio in Mechanically Alloyed N-Type Si-20 at.% Ge Alloys," in *Proc. of XI International Conf. on Thermoelectrics*, K. R. Rao, ed., The University of Texas at Arlington, Arlington, TX: 28-32.

Cook, B. A., J. H. Harringa, and B. J. Beaudry (1993) "A Solid State Approach to the Production of Kilogram Quantities of $Si_{80}Ge_{20}$ Thermoelectric Alloys," in *Proc. of 10th Symposium on Space Nuclear Power Systems, AIP Conf. Proc. 271*, M. S. El-Genk and M. D. Hoover, eds., American Institute of Physics, New York, NY: 777-783.

Dismukes, J. P. and L. Ekstrom (1965) "Homogeneous Solidification of Ge-Si Alloys," *Transactions of the Metallurgical Society of AIME*, 233: 672-680.

Draper, S. L. (1987) "The Effect of High Temperature Annealing on the Microsctructure and Thermoelectric Properties of GaP doped SiGe," NASA Technical Memorandum 100164.

Duncan, W. and A. J. Barlow (1985) "Thermoelectric Alloy Composition," *European Patent 0185499*, 25 June 1986.

Duncan, W. (1989) "Properties of SiGe," in *Eighth International Conf. on Thermoelectric Energy Conversion*, ISBN 2-905267-15-1, H. Scherrer and S. Scherrer, eds., Institut National Polytechnique de Lorraine, Nancy, France: 96-97.

Elsner, N. B., and J. H. Norman (1990) "SiGe Alloys with Sn and Pb Additons," in *Proc. of IX International Conf. on Thermoelectrics (USA)*, C. B. Vining, ed., Jet Propulsion Laboratory internal document number D-7749, Pasadena, CA: 227.

Eck, M. (1984) Unpublished Results, Fairchild Space and Electronics Company, Germantown, MD.

Erofeev, R. S., E. K. Iordanishvilii, and A. V. Petrov (1966) "Thermal Conductivity of Doped Si-Ge Solid Solutions," *Sov. Phys.-Sol. St.*, 7: 2470-2476.

Fistul', V. I. (1967) *Heavily Doped Semiconductors*, Plenum Press, New York, NY: 252-266 and 273-280.

Fleurial, J. P. (1989a) "Thermodynamical Model for SiGe," Unpublished Results, *5th SiGe Integration Meeting*, Thermoelectron Technology Corporation, Waltham, MA.

Fleurial J.P. (1989b) "Double Doping of SiGe," Unpublished Results, *5th SiGe Integration Meeting*, Thermoelectron Technology Corporation, Waltham, MA.

Fleurial, J. P., C. B. Vining, and A. Borshchevsky (1989c) "Multiple Doping of Silicon-Germanium Alloys for Thermoelectric Applications," in *Proc. of 24th Intersoc. Energy Conv. Eng. Conf.*, W. D. Jackson, ed., Institute of Electrical and Electronics Engineers, New York, NY, 2: 701-705.

Fleurial, J. P. and A. Borshchevsky (1990a) "Mechanisms of Carrier Concentration Enhancement in Multidoped SiGe," in *Proc. of IX International Conf. on Thermoelectrics (USA)*, C. B. Vining, ed., Jet Propulsion Laboratory internal document number D-7749, Pasadena, CA: 206-221.

Fleurial, J. P. (1990b) "Dopants Behavior in n-Type SiGe," Unpublished Results, *6th SiGe Integration Meeting*, GE Valley Forge, PA.

Fleurial, J. P., and A. Borschevsky (1990c) "Si-Ge-Metal Ternary Phase-Diagram Calculations," *J. Electrochem. Soc.*, 137(9): 2928-2937.

Fleurial, J. P., A. Borshchevsky, and J. W. Vandersande (1991a) "Improved N-Type SiGe/GaP Thermoelectric Materials," in *Proc. of 8th Symposium on Space Nuclear Power Systems*, CONF-

910116, M. S. El-Genk and M. D. Hoover, eds., American Institute of Physics, New York, NY: 451-457.

Fleurial, J. P., A. Borshchevsky, and J. W. Vandersande (1991b) "Optimization of the Thermoelectric Properties of Hot-pressed n-type SiGe Materials by Multiple Doping and Microstructure Control," in *10th International Conf. on Thermoelectrics,* D. M. Rowe, ed., Babrow Press, Cardiff Wales, UK: 156-162.

Fleurial, J. P., A. Borshchevsky, and D. Irvine (1991c) "LPE Growth of Doped SiGe Layers Using Multicomponent Phase Diagrams Calculations," in *Modern Perspectives on Thermoelectrics and Related Materials, Mat. Res. Soc. Symp. Proc. 234,* D. D. Allred, C. B. Vining, and G. A. Slack, eds., Materials Research Society, Pittsburgh, PA: 123-134.

Fleurial, J. P. (1991d) Unpublished Results, Jet Propulsion Laboratory, Pasadena, CA.

Fleurial, J. P., A. Borshchevsky, and J. W. Vandersande (1991e) "Effect of Microstructure and Composition on the Improvement of N-Type SiGe/GaP Material," in *Proc. of 26th Intersoc. Energy Conv. Eng. Conf.,* D. L. Black, ed., American Nuclear Society, La Grange Park, IL, 3: 202-207.

Fleurial, J. P., J. W. Vandersande, N. Scoville, C. Bajgar, and J. Beaty (1993a) "Progress in the Optimization of N-Type and P-Type SiGe Thermoelectric Materials," in *Proc. of 10th Symposium on Space Nuclear Power Systems, AIP Conf. Proc. 271,* M. S. El-Genk and M. D. Hoover, eds., American Institute of Physics, New York, NY: 451-457.

Fleurial, J. P. (1993b) Unpublished Results, Jet Propulsion Laboratory, Pasadena, CA.

Fu, L. W., D. M. Rowe, and G. Min (1992) "Long Term Electrical Power Factor Stability of High Temperature Annealed SiGe/GaP Thermoelectric Alloys," in *XI International Conf. on Thermoelectrics,* K. R. Rao, ed., The University of Texas at Arlington, Arlington, TX: 83-86.

Glazov, V. M. and V. S. Zemskov (1968) *Physico-chemical Principles of Semiconductor Doping,* IPST, Jerusalem, Isreal.

Gogishvili, O., Sh., I. P. Lavrinenko, S. P. Lalykin, T. M. Melashvili, and I. D. Rogovoy (1990) "Highly Homogeneous Silicon-Germanium Thermoelectric Alloys Produced by Using Mechano-Chemical Synthesis," in *Proc. of IX International Conf. on Thermoelectrics (USA),* C. B. Vining, ed., Jet Propulsion Laboratory internal document number D-7749, Pasadena, CA: 271-277.

Goldsmid, H. J. and A. W. Penn (1968) "Boundary Scattering of Phonons in Solid Solutions," *Phys. Lett.,* 27A(8): 523-524.

Gunther (1982) "Solitons and Their Role in the Degradation of Doped Silicon-Germanium Alloys," in *IV International Conf. on Thermoelectrics,* K. R. Rao, ed., The University of Texas at Arlington, Arlington, TX: 104-114.

Han, S.H., Cook, B.A. and Gschneidner, K.A. (1992) "Effect of Doping Process on Phosphorous Solubility in $Si_{80}Ge_{20}$," in *XI International Conf. on Thermoelectrics,* K. R. Rao, ed., The University of Texas at Arlington, Arlington, TX: 57-61.

Harringa, J. L., B. A. Cook, and B. J. Beaudry (1992) "Effects of Vial Shape on the Rate of Mechanical Alloying in $Si_{80}Ge_{20}$," *J. Mater. Sci.,* 27(3): 801-804.

Henderson, C. M., E. R. Beaver, Jr. (1966) *U.S. Patent 3,285,017.*

Improved Thermoelectric Materials, Development of (1989) Final Technical Progress Report for the period from 28 September 1984 through 17 April 1989, US-DOE Contract Number DE-AC01-84NE-32123.

JPL/TTC SP-100 Technical Workshop (1991) August.

Kilmer, R. J., W. A. Jesser, and F. D. Rosi (1991) "Effect of Melt-Growth Processing Methods on Si-Ge-GaP Homogeneity," *Appl. Phys. Lett.*, 58(22): 2529-2531.

Klemens, P. G. (1991) "Thermal Conductivity of N-Type Si-Ge," in *Modern Perspectives on Thermoelectrics and Related Materials, Mat. Res. Soc. Symp. Proc. 234*, D. D. Allred, C. B. Vining, and G. A. Slack, eds., Materials Research Society, Pittsburgh, PA: 87-94.

Loughin, S., D. X. Centurioni, A. G. Robison, J. J. Maley, and J. P. Fleurial (1993) "High-Boron P-Type Silicon Germanium Thermoelectric Material Prepared by the Vacuum Casting and Hot Pressing Method," in *Proc. of 10th Symposium on Space Nuclear Power Systems, AIP Conf. Proc. 271*, M. S. El-Genk and M. D. Hoover, eds., American Institute of Physics, New York, NY: 747-752.

McLane, G., C. Wood, J. Vandersande, V. Raag, and L. Danielson (1986) in *Proc. of VI International Conf. on Thermoelectrics*, K. R. Rao, ed., The University of Texas at Arlington, Arlington, TX: 311-321.

Meddins, H. R. and J. E. Parrott (1976) "The Thermal and Thermoelectric Properties of Sintered Germanium-Silicon Alloys," *J. Phys. C: Solid State Phys.*, 9: 1263-1276.

Miller, S. A. (1983) in *Amorphous Metallic Alloys*, F. E. Luborsky, ed., Butterworth, London, UK.

Min, G., and D. M. Rowe (1991) "The Effect of High-Temperature Heat-Treatment on the Electrical-Power Factor and Morphology of Silicon Germanium-Gallium Phosphide Alloys," *J. Appl. Phys.*, 70: 3843-3847.

Modular Radioisotope Thermoelectric Generator (RTG) Program (1983-1991) US-DOE contract number DE-AC01-83NE-32112.

Nakahara, J. F. and M. J. Platek (1990) "Development of N-Type SiGe/GaP Alloys for the SP-100 Program," in *Proc. of IX International Conf. on Thermoelectrics (USA)*, C. B. Vining, ed., Jet Propulsion Laboratory internal document number D-7749, Pasadena, CA: 222-226.

Nobili, D., A. Armigliato, M. Finetti, and S. Solmi (1982) "Precipitation as the Phenomenon Responsible for the Electrically Inactive Phosphorus in Silicon," *J. Appl. Phys.*, 53: 1484-1491.

Owusu-Sekyere, K., W. A. Jesser, and F. D. Rosi (1989) "Characterization of SiGe and SiGe-GaP Thermoelements," *Materials Science and Engineering*, 3(3): 231-240.

Parrott, J. E. (1969) "The Thermal Conductivity of Sintered Semiconductor Alloys," *J. Phys. C: Solid State Phys.*, 2: 147-151.

Pearsall, T. P. (1989) "Silicon-Germanium Alloys and Heterostructures - Optical and Electronic Properties," *Critical Reviews in Solid State and Materials Sciences*, 15(6): 551-600.

Pisharody, R. K. and L. P. Garvey (1978) "Modified Silicon-Germanium Alloys with Improved Performance," in *Proc. of 13th Intersoc. Energy Conv. Eng. Conf.*, American Institute of Chemical Engineers, New York, NY: 1963-1968.

Raag, V. (1978) "Comprehensive Thermoelectric Properties of N- and P-Type 78% Si-22% Ge alloy," in *2nd International Conf. on Thermoelectric Energy Conversion*, IEEE Cat. No. 78CH 1313-S Reg 5, K. R. Rao, ed., The University of Texas at Arlington, Arlington, TX: 5-10.

Raag, V. (1984) "Thermoelectric Properties of SiGe Alloys," in *Third Working Group Meeting on Thermoelectrics*, Jet Propulsion Laboratory internal document number JPL D-1335, Pasadena, CA: 353-367.

Rosi, F. D. (1968) "Thermoelectricity and Thermoelectric Power Generation," *Solid-State Electronics*, 11: 833-868.

Rosi, F. D. (1991) "The Research and Development of Silicon-Germanium Thermoelectrics for Power Generation," in *Modern Perspectives on Thermoelectrics and Related Materials, Mat. Res. Soc. Symp. Proc. 234*, D. D. Allred, C. B. Vining, and G. A. Slack, eds., Materials Research Society, Pittsburgh, PA: 3-26.

Rowe, D. M. (1974) "Theoretical Optimization of the Thermoelectric Figure of Merit of Heavily Doped Hot-Pressed Germanium-Silicon Alloys," *J. Phys. D: Appl. Phys.*, 7: 1843-1846.

Rowe, D. M., V. S. Shukla, and N. Savvides (1981a) "Phonon Scattering at Grain Boundaries in Heavily Doped Fine-Grained Silicon-Germanium Alloys," *Nature*, 290: 765-766.

Rowe, D. M. and V. S. Shukla (1981b) "The Effect of Phonon-Grain Boundary Scattering on the Lattice Thermal Conductivity and Thermoelectric Conversion Efficiency of Heavily Doped Fine-Grained, Hot-Pressed Silicon Germanium Alloy," *J. Appl. Phys.*, 52(12): 7421-7426.

Rowe, D. M. (1987) "Recent Advanced in Silicon Germanium Alloy Technology and an Assessment of the Problems of Building the Modules for a Radioisotope Thermoelectric Generator," *Journal of Power Sources*, 19: 247-259.

Rowe, D. M. (1989) "United States Activities in Space," in *Eighth International Conf. on Thermoelectric Energy Conversion*, ISBN 2-905267-15-1, H. Scherrer and S. Scherrer, eds., Institut National Polytechnique de Lorraine, Nancy, France: 96-97.

Rowe, D. M., and G. Min (1990) "High-Temperature Heat-Treatment of Silicon-Germanium-Gallium Phosphide Alloys," *J. Phys. D: Appl. Phys.*, 23: 258-261.

Rowe, D. M., G. Min, and S. G. K. Williams (1991) "Silicon Germanium-Gallium Phosphide Research at ELSYM, Cardiff," in *10th International Conf. on Thermoelectrics*, D. M. Rowe, ed., Babrow Press, Cardiff Wales, UK: 181-185.

Savvides, N. and H. J. Goldsmid (1973) "The Effect of Boundary Scattering on the High-Temperature Thermal Conductivity of Silicon," *J. Phys. C: Solid State Phys.*, 6: 1701-1708.

Schock, A. (1983) "Revised MITG Design, Fabrication Proceedure, and Performance Predicitons," in *Proc. of 13th Intersoc. Energy Conv. Eng. Conf.*, American Institute of Chemical Engineers, New York, NY: 1093-1101.

Scoville, A. N., C. Bajgar, J. W. Vandersande, and J. P. Fleurial (1991a) "High Carrier Concentration Improved N-Type SiGe/GaP Alloys," in *Modern Perspectives on Thermoelectrics and Related Materials, Mat. Res. Soc. Symp. Proc. 234*, D. D. Allred, C. B. Vining, and G. A. Slack, eds., Materials Research Society, Pittsburgh, PA: 117-122.

Scoville, A. N., C. Bajgar, J. W. Vandersande, and J. P. Fleurial (1991b) "High Figure of Merit N-Type SiGe/GaP Alloys," in *Proc. of 26th Intersoc. Energy Conv. Eng. Conf.*, D. L. Black, ed., American Nuclear Society, La Grange Park, IL, 3: 224-229.

"Silicon Germanium (SiGe) Radioisotope Thermoelectric Generator (RTG) Program for Space Missions" (1984) Fifty-Sixth Technical Progress Report, US-DOE contract number DE-AC01-79ET-32043.

Sittig, M. (1970) *Thermoelectric Materials*, ISBN: 8155-0313-X, Noyes Data Corporation, Park Ridge, NJ: 189-191.

Slack, G. A. (1979) "The Thermal Conductivity of Nonmetallic Crystal," *Solid State Physics*, 34: 1-71.

Slack, G. A. and M. A. Hussain (1991) "The Maximum Possible Conversion Efficiency of Silicon-Germanium Thermoelectric Generators," *J. Appl. Phys.*, 70: 2694-2718.

Taylor, R. E. and H. Groot (1980) "Thermophysical Properties of Silicon Germanium/Gallium Phosphide," Unpublished Results, Thermophysical Properties Research Laboratory report #TPRL 221, Purdue University, West Lafayette, IN.

Tschetter, M. J., and B. J. Beaudry (1990) "Effect of GaP on the Resistivity Versus Time at Elevated Temperatures of N-Type $Si_{80}Ge_{20}$ Alloys," in *Proc. of 25th Intersoc. Energy Conv. Eng. Conf.*, P. A. Nelson, W. W. Schertz, and R. H. Till, eds., American Institute of Chemical Engineers, New York, NY, 2: 382-386.

Vandersande, J. W., C. Wood, and S. L. Draper (1987) "Effect of High Temperature Annealing on the Thermoelectric Properties of GaP Doped SiGe," in *Novel Refractory Semicondutors, Mat. Res. Soc. Symp. Proc. 97*, D. Emin, T. L. Aselage, and C. Wood, eds., Materials Research Society, Pittsburgh, PA: 347-352.

Vandersande, J. W., C. Wood, S. L. Draper, V. Raag, M. Alexander, and R. Masters (1988a) in *Trans. of 5th Symp. Space Nuclear Power,* M. S. El-Genk and M. D. Hoover, eds., American Institute of Physics, New York, NY: 1: 629-632.

Vandersande, J. W. and C. Wood (1988b) Unpublished Results, *CSTI High Capacity Power Project Review*, Lewis Research Center, Cleveland, OH..

Vandersande, J. W. (1989) Unpublished Results, Jet Propulsion Laboratory, Pasadena, CA.

Vandersande, J. W., J. McCormack, A. Zoltan, and J. Farmer (1990a) "Effect of Neutron Irradiation on the Thermoelectric Properties of SiGe Alloys," in *Proc. of 25th Intersoc. Energy Conv. Eng. Conf.*, P. A. Nelson, W. W. Schertz, and R. H. Till, eds., American Institute of Chemical Engineers, New York, NY, 2: 392-396.

Vandersande, J. W., J. McCormack, A. Zoltan, and J. Farmer (1990b) "Neutron Irradiation of SiGe Alloys," in *Proc. of IX International Conf. on Thermoelectrics (USA)*, C. B. Vining, ed., Jet Propulsion Laboratory internal document number D-7749, Pasadena, CA: 228-233.

Vandersande, J. W. and J. P. Fleurial (1991) Unpublished Results, *GE-JPL Technical Meeting on Thermoelectric Material Development*, GE AstroSpace, Valley Forge, PA.

Vandersande, J. W. and J. P. Fleurial (1992a) Unpublished Results, *MOD-RTG Materials Development Meeting*, GE Astrospace, Valley Forge, PA.

Vandersande, J. W., J. P. Fleurial, J. S. Beaty, and J. L. Rolfe (1992b) in *Proc. of XI International Conf. on Thermoelectrics*, K. R. Rao, ed., The University of Texas at Arlington, Arlington, TX: 21-23

Vining, C. B. (1988) Unpublished Results, Jet Propulsion Laboratory, Pasadena, CA.

Vining, C. B. (1989) "Electrical and Thermal Model," Unpublished Results, *5th SiGe Integration Meeting*, Thermoelectron Technology Corporation, Waltham, MA.

Vining, C. B. (1990) "High Figure of Merit Thermoelectrics: Theoretical Considerations," in *Proc. of 25th Intersoc. Energy Conv. Eng. Conf.*, P. A. Nelson, W. W. Schertz, and R. H. Till, eds., American Institute of Chemical Engineers, New York, NY, 2: 387-391.

Vining, C. B., W. Laskow, J. O. Hanson, R. R. Van der Beck, and P. D. Gorsuch (1991a) "Thermoelectric Properties of Pressure-Sintered $Si_{0.8}Ge_{0.2}$ Thermoelectric Alloys," *J. Appl. Phys.*, 69: 4333-4340.

Vining, C. B. (1991b) "A Model for the High Temperature Transport Properties of Heavily Doped P-Type Silicon-Germanium Alloys," in *Modern Perspectives on Thermoelectrics and Related Materials, Mat. Res. Soc. Symp. Proc. 234*, D. D. Allred, C. B. Vining, and G. A. Slack, eds., Materials Research Society, Pittsburgh, PA: 95-104.

Vining, C. B. (1991c) "A Model for the High Temperature Transport Properties of Heavily Doped P-Type Silicon-Germanium Alloys," *J. Appl. Phys.*, 69,: 331-341.

White, D. W. (1990) "Phonon Scattering and Thermal Conductivity in Semiconductors with Thermoelectric Applications," PhD Thesis, University of Connecticut, Storrs, CN.

White, D. P. and P. G. Klemens (1992) "Thermal Conductivity of Thermoelectric $Si_{0.8}Ge_{0.2}$ Alloys," *J. Appl. Phys.*, 71: 4258-4293.

Wood, C. (1988) "Materials for Thermoelectric Ènergy Conversion," *Reports on Progress in Physics*, 51: 459-539.

REVIEW OF THERMIONIC TECHNOLOGY: 1983 TO 1992

Richard C. Dahlberg
Lester L. Begg
Joe N. Smith, Jr.
General Atomics
P. O. Box 85608
San Diego, CA 92186-9784
(619) 455-2998

G. Laurie Hatch
John B. McVey
Ned Rasor
Rasor Associates, Inc.
253 Humboldt Court
Sunnyvale, CA 94089
(408) 734-1622

Gary O. Fitzpatrick
Daniel T. Allen
Advanced Energy Technologies
P. O. Box 327
La Jolla, CA 92038
(619) 455-4310

David P. Lieb
Gabor Miskolczy
ThermoTrex Corporation
85 First Avenue
Waltham, MA 02254
(617) 622-1000

Abstract

This review summaries important research and development that occurred in the U.S. from 1983 to 1992 in the area of thermionic power conversion technology for space applications.

INTRODUCTION

Thermionic technology programs and system design efforts for space applications were essentially terminated in the U.S. at the end of 1973, the issue being a lack of a foreseeable application. In 1983, however, the growing utilization of space for both civil and defense missions led to the initiation of the SP-100 program to study the design and technology of alternative space nuclear power systems. The major thermionic programs undertaken since 1983 in the U.S. are summarized in Table 1, and briefly described below.

SP-100 Phase I (1983 to 1985): This was a 3 year effort to design a 100 kW thermionic space nuclear power system. The fast spectrum reactor core used 180 6-cell TFEs. The extensive design effort completed in this program is a starting point for the design of fast spectrum systems at any power level. Funding was provided by DOD, DOE and NASA.

SP-100 Technology Program (1984 to 1986): The following technology issues were studied in this program:

- o Emitter distortion in 9 fueled emitters irradiated in the General Atomics (GA) TRIGA reactor;
- o Sheath insulator integrity in a reactor with a voltage applied across the sheath; and
- o Insulator seal integrity in a cesium vapor environment.

Thermionic Technology Program (1984 to 1986): The following thermionic issues were studied in this DOE funded program:

- o The effect of oxygenated electrodes on thermionic efficiency;
- o The effect of structured electrodes on thermionic efficiency; and
- o The performance of advanced insulator materials and insulator configurations.

TFE Verification Program (1986 to Present): The TFE-VP was established by DOE and SDIO to demonstrate the performance and long life of thermionic components and TFEs for megawatt-class systems.

Multiple SBIR Contracts (1986 to Present): Small business contracts are funded by DOE, NASA and DOD (SDIO) to investigate advanced thermionic components, technologies and systems.

Multimegawatt Design Effort (1987-1989): Several technologies were evaluated in a DOE/SDIO design effort for their ability to provide many megawatts of burst power for space based weapons. A technology down selection was made but the program did not go forward due to changing mission priorities.

Advanced Thermionic Initiative (1989 to Present): Several advanced technology efforts are focused on the development and test of an advanced TFE. The basis for the TFE is one single long cell. The effort is managed by Wright Laboratories and funded largely by SDIO-IST.

USAF Design and Technology Programs (1990 to Present): In 1991, the Air Force Phillips Laboratory funded the design and evaluation of different thermionic systems to meet potential Air Force requirements in the 5 to 40 kW range. Several technology programs were also undertaken, and these efforts continue.

TSET Program (1991 Ongoing): In 1991, SDIO working with the Air Force Phillips Laboratory, University of New Mexico and Sandia National Laboratory acquired a Russian TOPAZ thermionic power system for use in a non-nuclear systems test program at the University of New Mexico. The use of single cell thermionic fuel elements permits electrical heating of the TFEs. System performance will be measured as a function of various operating parameters.

DOE/SDIO/USAF Systems Program (1992 On-going): In 1992, the government initiated a major thermionics program whose goal was to develop by 1995 the design and technology information necessary to make a flight program decision. The demonstration flight would occur in the year 2000, or before.

From these thermionic programs, the following specific technology efforts are briefly reviewed in the following sections:

Section 2: Thermionic theory and converter models.

Section 3: Thermionic SP-100 technology program.
Section 4: Thermionic technology program.
Section 5: Characteristics of cesium-graphite compounds
Section 6: High performance with close spaced electrodes.

A large and extensive thermionic program has been carried out in Russia over the last 25 years, and a review of that effort is not included herein.

THERMIONIC THEORY AND CONVERTER MODELS

The TECMDL computer subroutine package is a widely used tool in the U.S. for simulation of ignited mode thermionic converter performance. It is simple and fast and has reasonably good accuracy over the practical range of converter operation. It has been used as part of numerous computer codes which model thermionic devices and power systems. Development of the code was begun in 1982 as the computer implementation of the "phenomenological model" described in Rasor (1982). It has been continuously upgraded since that time.

TECMDL is based on a simplified physical model in which all quantities are described by easily solved algebraic equations. These equations contain a set of "effective plasma parameters," which are related to fundamental properties and cross-sections. Their values have been adjusted slightly within their ranges of uncertainty to optimize the agreement between calculated results and experimental data. However, they are not simply fitting parameters, since all are calculable to within 20% of their values from known properties.

Derivation of Plasma Model

As shown in Figure 1, the ignited mode volt-ampere characteristic consists of two regions: the obstructed region and the saturation region. The obstructed region is characterized by a "double-valued sheath barrier," or virtual cathode, at the emitter. The formulation of Rasor (1982) is primarily concerned with the obstructed region and secondarily with the saturation region. The electron motive diagrams for the two regions are shown in Figure 2.

Basic Obstructed Mode Formulation

The basic obstructed mode formulations of Rasor (1982) combines a set of simplified boundary conditions with the fluid transport equation for electrons in the plasma to derive Equation (1):

$$\frac{J_E}{J} = 1 + \exp\left(-\frac{V_E}{kT_{eE}}\right)\left[\frac{3}{4}\frac{d}{\lambda_e} + R\right] , \tag{1}$$

where

$$R = \left[1 + \frac{J_C}{J}\right] \exp\left[\frac{V_C}{kT_{eC}}\right] - 1 . \tag{2}$$

The value of R is constant and equal to 4.5 in the absence of local thermodynamic equilibium (LTE) constraints,as will be discussed below.A global energy balance equation gives Equation(3)

$$\begin{aligned} V_d = \; & 2k(T_{eE} - T_E)(\frac{J_E}{J} - 1) + 2k(T_{eC} - T_E) \\ & + 2k(T_{eC} - T_C)\frac{J_C}{J} + \Delta V_{rad} . \end{aligned} \tag{3}$$

Ion conservation at the collector and a local energy conservation equation give Equations (4) and (5)

$$V_C = 3k(T_{eE} - T_{eC}) - 2k(T_{eC} - T_C)\frac{J_C}{J} \tag{4}$$

and

$$T_{eC} = \frac{3T_{eE} + 2T_C\frac{J_C}{J}}{\ln\left[\frac{H + 1/2}{1 + J_C/J}\right] + 2\frac{J_C}{J} + 3} . \tag{5}$$

The parameter H is related to the ratio of the electron and ion mean free paths and is nearly a constant. A value of 5 is used in the TECMDL code. The emitter sheath is related to the arc drop and collector sheath by

$$V_E = V_d + V_C . \tag{6}$$

A global ion balance gives Equation (7) for the emitter-side electron temperature:

$$T_{eE} = \frac{V_I}{2k \ln\left[B\frac{d}{\lambda_{ea}}\right]} , \tag{7}$$

where V_I is the effective ionization potential which characterizes the multistage ionization of cesium and B is a constant which characterizes the ability of the plasma to produce and retain ions. In TECMDL V_I is 3.2 eV and B is 30. The electron temperatures given by Equations (5) and (7) are subject to the LTE limitation

$$\bar{T}_e \geq \bar{T}_s = \frac{\bar{\phi}_n}{k \ln\left(\dfrac{120\bar{T}_s^2}{J}\dfrac{\lambda_e}{d}\right)} , \tag{8}$$

where the equilibrium plasma potential ϕ_n is approximated as a linear function of T_s/T_R,

$$\phi_n = 1.7 + .383\frac{T_s}{T_R} . \tag{9}$$

Finally, the output voltage is

$$V_{out} = \phi_E - \phi_C - V_d + \Delta V , \tag{10}$$

where the double-valued sheath barrier height ΔV is

$$\Delta V = kT_E \ln\left(\frac{J_s'}{J_E}\right) . \tag{11}$$

Improvements to Obstructed Mode Formulation

To improve accuracy, several improvements over the formulation in Rasor (1982) have been incorporated into TECMDL.

Improved LTE Constraints. In the original formulation the average electron temperature is constrained by LTE to be higher than T_s. In actual converter's LTE occurs first at the collector, where the electrons are coldest, and then spreads gradually to the rest of the interelectrode gap. The present version of TECMDL approximates this process by using two LTE constraints; a new collector LTE formulation and the bulk LTE formulation.

The LTE-limited electron temperature at the collector side of the plasma will be designated T_{sC}. If ϕ_{nc} is the local equilibrium plasma potential at the collector, then the LTE current density is defined by

$$J_{nC} = 120T_{eC}^2 \exp\left(\frac{-\phi_{nC}}{kT_{sC}}\right) . \tag{12}$$

The parameter H_s, defined by

$$H_s \equiv \frac{J_{nC}}{J} . \tag{13}$$

replaces the constant value H in Equation (5) above, to give the temperature T_{sC}. TECMDL requires that $T_{eC} \geq T_{sC}$.

Radiation Loss. At low current densities the power loss from the plasma by resonance and nonresonance radiation is significant. From Norcross (1962) the formula for power lost from the plasma by photons was derived as:

$$P_{rad} = 9.65\times10^{5}\,\frac{Pd}{T_a}\exp\left(\frac{-2}{kT_e}\right)\times \left[1 + 0.069\exp\left(\frac{0.58}{kT_e}\right)\left(\frac{\epsilon_\lambda}{\sqrt{d}} + \frac{1}{2}\right)\right] , \tag{14}$$

where ϵ_λ is the spectral emissivity of the electrodes at the resonance wavelength for cesium and d is the gap in centimeters. A value of 0.4 is typical for ϵ_λ. An additional contribution to the arc drop due to radiation loss is defined as

$$\Delta V_{rad} = \frac{P_{rad}}{J} , \tag{15}$$

which is added to the right-hand side of Equation (3):

Ion-Retaining Collector Sheath. Under conditions of very high collector back emission, the collector sheath changes polarity so that it limits the back emission entering the plasma and retards ion current from the plasma to the collector. The new set of equations for this case, along with Equations (1) and (6) through (10), is

$$R = \frac{J_C}{J}\exp\left(\frac{V_C}{kT_{eC}}\right) , \tag{16}$$

$$\begin{aligned} V_E = \; & 2k(T_{eE} - T_E)(\frac{J_E}{J} - 1) \\ & + 2k(T_{eC} - T_E) \\ & + 2k(T_{eC} - T_C)R + \Delta V_{rad} , \end{aligned} \tag{17}$$

$$0 = \frac{J_C}{J}\zeta^2 + (H + 1/2)\zeta - 2H , \tag{18}$$

where $\zeta \equiv \exp(V_C/kT_C)$, and

$$T_{eC} = \frac{3T_{eE} + 2T_C R}{2R + 3} , \tag{19}$$

Electron Mean Free Path. The ratio of the interelectrode gap to the electron mean free path is given by

$$\frac{d}{\lambda_e} = \frac{d}{\lambda_{ea}} + \frac{d}{\lambda_{ei}}$$
$$= 17 P_{Cs} d + 3.4 \times 10^7 \frac{Jd}{T_e^{5/2}} , \tag{20}$$

where P_{Cs} is in torr and d is in millimeters. TECMDL uses

$$T_{sC} = \frac{3T_{eE} + T_C(2H_s - 1)}{2(H_s + 1)} . \tag{21}$$

This expression takes into account the influence of temperature on the cesium neutral density.

$$\frac{d}{\lambda_e} = \frac{35000}{T_a} P_{Cs} d + 3.4 \times 10^7 \frac{Jd}{T_e^{5/2}} . \tag{22}$$

Saturation Mode. In the saturation region additional energy is dissipated in the interelectrode gap which is rejected to the emitter in the form of an excess ion current density J_{ix}. Adding this ion current into the global ion-conservation equations gives a new formula for the emitter-side electron temperature,

$$T_{eE} = \frac{V_I}{2k \ln\left[B \frac{d}{\lambda_{ea}} \right] - k \ln\left[1 - B' \frac{J_{ix}}{J} \right]} . \tag{23}$$

The parameter B′ is a constant (TECMDL uses B′ = 50). As current increases, the value of J_{ix}/J increases, causing T_{eE} to increase. Continuity of current gives

$$\frac{J_S}{J} = 1 + \exp\left[\frac{-V_E}{kT_{eE}} \right] \left[\frac{3}{4}\frac{d}{\lambda_e} + R - \frac{1}{2}\frac{J_{ix}}{J} \right] + \frac{J_{ix}}{J} , \tag{24}$$

where R is given by Equation (2). Energy conservation gives

$$\begin{aligned} V_d = SCALESYM200\{&2k(T_{eE} - T_E)(\frac{J_S}{J} - 1) + 2k(T_{eC} - T_E) \\ &+ 2k(T_{eC} - T_C)\frac{J_C}{J} + \Delta V_{rad} \\ &- (V_C + V_i + 2kT_{eE})\frac{J_{ix}}{J} SCALESYM200\} \left[\frac{1}{1 + J_{ix}/J} \right] . \end{aligned} \tag{25}$$

Finally, an equation for the Schottky effect derived from Hansen (1967) gives:

$$J_S = J_S' \exp\left[\frac{612}{T_E} \left(-J_{ix} \sqrt{V_E} \right)^{1/4} \right] . \tag{26}$$

Similar equations can be derived for the saturation mode in the case of an ion-retaining collector sheath.

Electron Cooling. TECMDL calculates the electron-cooling and ion-heating portions of the heat loss from the emitter. Heating of the emitter by photons emitted from the plasma is also accounted for. Equations for the "effective electron cooling" are:

$$q_{el} = J(\phi_E + \Delta V + 2kT_{eE}) - J_E 2k(T_{eE} - T_E) - \frac{P_{rad}}{2} \tag{27}$$

for the obstructed mode, and:

$$q_{el} = J(\phi_S + 2kT_{eE}) - J_S 2k(T_{eE} - T_E) + J_{ix}(V_E + 3.89 + 2kT_{eE}) - \frac{P_{rad}}{2} \tag{28}$$

for the saturation mode.

Work Function Models

The simplified expressions for cesiated emitter and collector work functions used in Rasor (1982) have been replaced by more rigorous models using a formulation given in Rasor (1964) for the emitter. Inputs are the emitter temperature, the cesium reservoir temperature, and the bare work function of the emitter surface. Excellent agreement between calculated T_C/T_R curves and experimental data for CVD tungsten and single-crystal rhenium Pigford (1969) is shown in Figure 3.

The formulation of Rasor (1964) is not valid at the high cesium coverages characteristic of collectors. Therefore, a curve fit based on experimental data has been used in TECMDL. The T_C/T_R curve is shown in Figure 4. Only the dependence on T/T_R is taken into account for the collector work function, as experimental data indicate that the work function is relatively insensitive to electrode material under typical collector conditions. Comparisons of TECMDL output with experimental data indicate that the curve fit is reasonably accurate for niobium and molybdenum collectors.

Comparison with Experimental Data

The values of the effective plasma parameters in the TECMDL formulation (V_I, B, H, B′, and the linear fit for ϕ_n), which have been given in the discussion above, were obtained by adjusting them within their ranges of uncertainty to give the best possible fit to a particular set of experimental data in the following parameter range:

Emitter temperature:	1600-2000 K
Collector temperature:	800-1100 K
Cesium pressure:	1-12 torr
Interelectrode gap:	0.064-1.0 mm

The primary source for comparison is Thermoelectron Technologies Corporation's converter #46, which was chosen as the standard for the Thermionic Fuel Element Verification Program. This was a variable spacing planar converter with an active guard ring. The emitter was chloride deposited tungsten and the collector niobium. Due to the lack of collector temperature variation in the converter 46 data, data from PD-6 (tested by Rasor Associates in the Thermionic Technology Program) and TTC #42 (similar to #46 but with a single crystal tungsten emitter) have been used to check TECMDL's accuracy with collector temperature.

Comparisons between planar converter data from TTC #46 and TECMDL calculations are shown in Figures 5, 6, and 7 for a cesium pressure family, a spacing family, and an emitter temperature family respectively. The spacing family is particularly interesting as it is the only parameter variation which allows checking the plasma model independently of the surface work function model. Reasonable agreement is found for spacings from 0.254 to 1.016 mm. Accuracy begins to diminish for spacings below 0.127 mm. Inaccuracy at small values of pressure spacing product $P_{Cs}d$ is expected from any model based on a fluid approximation for the plasma. The value of $P_{Cs}d$ corresponding to 0.127 mm gap and a 577 K cesium reservoir is about 0.25 torr-mm, which is approximately the lower limit of the fluid description.

The behavior of output voltage with collector temperature is compared for TECMDL, TTC #46, and the PD-6 converter in Figure 8. TECMDL has a tendency to predict a lower optimum collector temperature than shown by experimental data. However, as seen from Figure 8 the total inaccuracy is no worse than the discrepancy in performance between similar experimental converters.

One of the primary difficulties with TECMDL is illustrated in Figure 9. Here the comparison between the shape of a calculated I-V curve with data from converter TTC #42 indicates that accuracy degrades sharply for portions of the I-V curve which are far from the transition point. The inaccuracy deep in the saturation region is attributed to the neglect of the plasma electric field in deriving Equation (23), and the neglect of the voltage drop within the plasma in Equation (6). The inaccuracy in the lower part of the obstructed mode is not presently understood at this time, and is exhibited in some form by all existing thermionic converter models. One speculation is that the thickening of the emitter sheath in this region invalidates the model of distinct sheath and plasma regions. For the present, the TECMDL subroutine returns a warning if calculated points are too far away from the transition point for accuracy.

Computer Implementation and Use

TECMDL is implemented as a subroutine package which users can call from their applications. TECMDL input parameters consists of the following:

- The emitter temperature;

- The collector temperature;
- The liquid cesium reservoir temperature;
- An integer value which selects from a menu of emitter surface bare work functions;
- The value of the cesiated collector work function;
- The interelectrode gap spacing; and
- The operating current density.

Output parameters are:

- The output voltage; and
- The emitter electron cooling.

Recent and Planned Improvements

Efforts are underway to combine TECMDL with the UNIG unignited mode model. This combination is capable of producing a reasonable approximation to the overall volt-ampere curve, rather than just the ignited mode portion. This avoids the nonphysical effect where, due to Equation (13), the voltage calculated by TECMDL tends toward large negative values at very small current densities.

For the long term, kinetic models of the thermionic converter interelectrode plasma are under development. Checking the validity of the basic TECMDL equation set against detailed results from such codes could lead to significant improvements in TECMDL's accuracy and range of validity, while maintaining its computational simplicity.

THERMIONIC SP-100 TECHNOLOGY PROGRAM

The technology program supporting the SP-100 design effort began in 1984 and consisted of three tasks:

1. Irradiation of 9 fueled emitters to study emitter distortion as a function of time;
2. Test of sheath insulators in-reactor with a voltage applied across the sheath; and
3. Evaluation of the degradation at high temperature of seal insulators in contact with cesium vapor.

These three tasks are discussed more fully below. More detail is provided in General Atomics (1985-1990).

Fueled Emitter Irradiations

Test Article Description: Nine thermionic emitters fueled with uranium oxide were tested in the TRIGA Mark F reactor to measure emitter deformation as a function of emitter temperature, emitter thickness, and operating profile to determine dominant deformation mechanisms. The deformation data obtained with neutron radiography can be used to validate fuel performance and lifetime models.

Each emitter was fabricated into a thermionic cell with the design configuration shown in Figure 10. Three cells were assembled into each irradiation capsule as shown in Figure 11. Each cell has its own cesium reservoir with heaters for temperature control and a large plenum with a pressure tap to collect and monitor the fission gas released from the fuel. The test specimen was contained within a helium filled primary containment which was in turn contained within a secondary containment.

Emitter temperature was controlled in part by adjusting the electron cooling, such as, the current flow from the emitter across the gap filled with cesium vapor. It was also controlled by adjusting the argon/helium mixture in the gas gap between the primary and secondary containments.

Instrumentation for the capsule included thermocouples for temperature measurement of the emitter flange, the collector and cesium reservoir, voltage taps for the emitter and collector, and power leads for the cesium reservoir heaters.

The test matrix is shown in Table 2. Cell 1.2 is, for example, the middle of the 3 cells in capsule 1. The matrix contains three emitter temperatures (1650, 1750 and 1850 K) and three emitter thicknesses (.04, .07 and .10 inches). These values span the region of interest for a conservative emitter design.

Irradiation History: The three fueled emitter capsules were charged into the TRIGA as shown in Table 3. Total irradiation times are also shown.

Degradation of the test specimens occurred with irradiation for two major reasons:

1. Electrical shorts in the instrumentation outside of the cell; and

2. Failure of the insulator seal due to thermal shock (from reactor scrams), mechanical loads from the heavy cells, and the effect of fast neutrons on the strength of the ceramic insulator. The effect of fast fluence on the alumina insulator component of the seal also weakened the seal.

Seal failure allowed helium gas to enter the interelectrode space and reduce electron cooling. Electrical shorts compromised diagnostics (for example, current-voltage (I-V) sweeps) and prevented a good assessment of emitter temperature. An accurate knowledge of the emitter temperature is essential to an evaluation of the data.

The deformation of the emitter was measured periodically from neutron radiographs, and the results of these measurements are shown on Table 4. The dates at which the radiographs were taken are shown, along with the corresponding hours of irradiation and the radial deformation in mils.

The maximum irradiation times for the 3 capsules are shown. However deformation data are not presented toward the end of the irradiation because either a short had occurred or the seal had cracked.

The deformation data are valid provided the emitter temperature is well known. The emitter temperatures were calculated from TRIGA operational data as described in General Atomics (1987), and the amount of valid data is shown on Table 5. Qualifying comments are provided below.

Cell 1.1. An early external short did not prevent thermionic operation of the cell but prevented a good estimate of emitter temperature.

Cell 1.2. Valid data were obtained for about 10,000 hours.

Cell 1.3. Valid data were obtained for about 6000 hours.

Cell 2.1. Cell 2.1 had an external short from the inception of irradiation, and the emitter temperature could not be accurately estimated. However, partial ignition could be achieved for about 34,000 hours.

Cell 2.2. Valid data were obtained for 20,225 hours, at which point the I-V curves showed a degradation in performance. After about 26,000 hours (June, 1988), a short occurred, electron cooling was lost and the emitter temperature increased to 2000 K or more.

Cell 2.3. Valid data were obtained for 16,000 hours. After 19,000 hours (July, 1987), there was indication of helium in the interelectrode gap.

Cell 3.1. Valid data were obtained for 14,000 hours.

Cell 3.2. Valid data were obtained for 14,000 hours.

Cell 3.3. Valid data were obtained for about 16,000 hours.

Sheath Insulator Irradiation

Test Article Description: The design concept of the incore sheath insulator capsule is illustrated in Figure 12. The sheath insulator samples are stacked along a common axis inside a vacuum chamber formed by a stainless steel containment tube. The insulators are heated from two sources: gamma radiation within the bulk material of each insulator, and by thermal radiation from a separate gamma-heated tantalum cylinder inside the stack of sheath insulators. A cylindrical stainless-steel water cooled cold finger is located inside the tantalum gamma heater and a gas gap exists between the two. The temperature of the tantalum heater and thus the temperature of the insulators is controlled by changing the gas mixture in this gap.

Chromel-alumel thermocouples are provided for temperature measurement of the samples and tantalum heater. The sheaths on four of these are used as insulator voltage probes. Isolated wires serve as current leads for each electrically loaded sample or guard ring.

Six sheath insulator samples were installed for testing, five to be electrically loaded and one to be irradiated without an electric field. Four of the insulator samples were yttria, two were alumina. The electrically loaded samples were qualified in out-of-core thermal and electrical stability tests prior to installation. Figure 13 shows the general configuration of the samples and Table 6 describes the samples.

Irradiation History: Sheath insulator testing was initiated on September 27, 1985. With argon in the gas gap between the water cooling cold finger and tantalum heater, insulator temperatures were about 1030 K.

Resistivity as a function of time is shown on Figure 14. All samples initially had high electrical conductivities. The resistivity of samples 2 through 5 increased slowly during the first 2000 hours of operation and then reached a stable steady state for the next 3000 hours. The resistivity of alumina sample 1 increased by two orders of magnitude during this same period and also reached and sustained a stable steady state operating point for the next 1000 hours. These resistivity changes are probably associated with end losses and the cleanup of the ends with time.

A change occurred in the resistivity of samples 2, 4, and 5, all Y_2O_3, at about 5000 hours when a drop in sample temperature of about 50°C was observed. Several means of increasing the temperature were tried to no avail. Both the temperature and resistivity of these samples were stable, at the new values, throughout the rest of the irradiation.

Samples 1 and 3, both alumina, did not exhibit a change in resistivity during the same time period. Sample 3 remained stable at a relatively low resistivity of about 5 M /cm. Sample 1 showed several sharp changes in resistivity between about 230 M /cm and 10 M /cm. This could be indicative of some foreign material that could have gotten lodged in the sample during one of the neutron radiography outages when the capsule was moved.

After about 6450 hours of irradiation time, neutron radiographs revealed that the gas gap between the tantalum heater and the water coolant cold finger was filled with some foreign substance. It appeared that coolant water from the cold finger had reacted with the tantalum heater forming the foreign substance, probably TaOH, which filled the gap. At this point it was decided to remove capsule 4 from service of irradiation testing. All sheath insulator samples were functioning about as expected at that time.

Seal Cesium Compatibility

The function of the metal ceramic seal is to provide a hermetic seal between the cesium and fission gas spaces in individual converters within a TFE. The seal must electrically isolate the emitter from the collector. Hence, the seal ceramic must be compatible with cesium at high temperature.

The ceramic-cesium compatibility testing examined yttria (Y_2O_3), yttrium-aluminum garnet (YAG), polycrystalline alumina (Lucalox) and alumina single crystal (sapphire). Two cermets, Y_2O_3-Nb and YAG-Nb, were also tested. All samples were exposed to 60 torr of cesium vapor for 200 h at temperatures of 1100, 1250, and 1400 K. To differentiate between the effect of exposure to high temperatures and to cesium vapor, similar samples were also tested at 1400 K, without cesium vapor. After completion of test, the samples were examined for weight change (chemical reaction), resistance change, change in visual appearance, change in microscopic structure as revealed in a scanning electron microscope, and change in chemical composition determined by EDX or ESCA.

None of the samples, cesiated or uncesiated, after having been exposed at any of the three test temperatures for 200 h showed a significant weight change (to $\pm$0.5 mg) or a loss of insulation at 300 K (resistance >10 M). The only significant effect was an observed discoloration of Y_2O_3 and YAG. This discoloration occurred also in uncesiated samples, precluding cesium attack as a possible cause.

No difference in the microstructure of the cesiated and uncesiated YAG samples was observed.

THERMIONIC TECHNOLOGY PROGRAM

The goals of the Thermionic Technology Program (TTP) were to investigate alternative sheath insulator technologies, to determine the performance enhancement from using oxygen additives on the electrodes and to determine the performance enhancement of structured electrodes. Program summaries are provided in Cone and Dunlay (1987), Thermo Electron (1990) and Hatch (1988).

Sheath Insulator Development

The sheath insulator electrically insulates the series connected collectors of the thermionic fuel element from the outer metallic sheath tube. In this application it is subjected to the irradiation flux of the reactor at collector temperature of typically 1000 K. The insulator must withstand the voltage of the series string and at the same time furnish a high thermal conductivity path for the reject heat from the collector. Typical performance requirements are:

Fast fluence:	1 to $3x10^{22}$ nvt, E>.1 mev
Temperature:	900 to 1100 K

Electrical potential:	15 to 25 volts
Resistivity:	10^5 ohm-cm
Lifetime:	5 to 10 years.

Studies in the early 1970s (Proier 1972) had shown that cermets composed of niobium metal embedded in a matrix of alumina powder formed a strong insulating layer which bonded well to the inner and outer niobium sheath walls. A slip casting technique was developed using a slurry containing both metal and ceramic powder. In some cases about 0.2 percent of an organic deflocculent was added to maintain a uniform suspension with high solids content. It was also observed that yttria at about 2 volume percent was an effective deflocculent for alumina slurries.

Investigations of yttria alumina garnet (YAG) trilayers were carried out because this material exhibits minimal radiation induced swelling. Application of YAG to the sheath insulator trilayer caused difficulties probably due to either the thermal expansion mismatch with niobium or to a reaction between niobium and YAG.

Alumina/niobium and YAG/niobium cermet trilayers were fabricated for testing. The alumina samples were slip cast with 25 volume percent niobium spheres and the YAG samples were filter pressed. Finite element stress analyses indicated that stress relieving the alumina samples by cutting back the ends of either the inner or outer niobium sheath layer would aid in preventing cracking from the radiation induced swelling. The YAG samples were not treated this way because of the small swelling expected. Irradiation tests occurred in October 1986.

Ceramic matrix composites were also studied, the composites consisting of a ceramic matrix with a dispersion of a second fibrous or powder material. The composites offer the possibility of strengthening and toughening the ceramic matrix through increasing the amount of energy required for crack propagation (Agarwal 1980), Liu 1988). Fibers or particulates can be helpful but they must be chemically and mechanically compatible. Table 7 summarizes the composite and cermet materials evaluated.

Analysis of the alumina-chromia samples revealed the presence of chromium particles near the niobium sheaths and the existence of radial cracks. Additional samples with modified fabrication techniques eliminated the cracking but the free chromium was still present. The oxygen originally contained in the chromia has apparently been taken up by the niobium sheaths probably as dissolved oxygen. The possible effects of such oxygen must be evaluated. The chromia reduction appears to be processing temperature dependent.

Thermionic Performance of Converters with Oxygen Additive

Test Description: The output power density of a thermionic converter can be enhanced by increasing the bare work function of the emitter surface. Chemically vapor deposited (CVD) tungsten is a suitable emitter material. Tungsten derived from WCl_6 yields a (110) crystal orientation and a bare work function about 4.9-5.0 eV; tungsten derived from WF_6 yields a (100) crystal orientation and a bare work function of about 4.7 eV. The fluoride coating has sufficient mechanical strength for fuel cladding, but to take advantage of the high bare work function of the chloride coating, it is generally deposited onto the stronger fluoride-deposited substrate. This combination of coatings is referred to as duplex tungsten.

It has been shown that oxygen adsorbed onto the surface of tungsten can result in a large increase in bare work function. (Early work in this area is presented in Lieb and Rufeh (1970), Levine (1965), Lieb and Kitrilakis (1966), Rufeh etal. (1967), Lieb and Rufeh (1972), Alleau and Bacal (1970), and Dunlay and Rufeh (1972).) Techniques for introducing oxygen into the converter have included oxide coatings deposited on various components of a thermionic converter. In this work the oxide is formed on the niobium collector as molybdenum oxide.

Studies of the adsorption of oxygen on fluoride and chloride CVD surfaces show that the resulting bare work function appears to be relatively independent of the crystal orientation of the substrate. The validity of this result in a thermionic converter is important to test since the fluoride coating is much easier to manufacture.

The series of diodes given in Table 8 was tested to establish how the performance of a diode with an oxide collector was affected by crystal orientation of the emitter and the concentration of oxygen in the oxide layer.

Two nonadditive diodes were included to compare the fluoride and chloride CVD emitters. Finally, two cylindrical diodes, one additive and one nonadditive, were tested to establish the relationship between the performance of planar and cylindrical geometry diodes. Diodes PD-2 and PD-6 were the reference nonadditive diodes with tungsten (100) and (110) emitter crystal orientation, respectively. Diodes PD-3 through PD-5 contained various concentrations of oxygen with tungsten (100) emitter, while PD-7 and PD-8 contained various oxygen concentrations for a tungsten (110) emitter.

Test Results: The test diodes are shown schematically in Figures 15 and 16. Performance-mapping included a wide range of emitter and collector temperatures and interelectrode spacings. The mapping consisted of cesium vapor pressure families taken at the diode parameters listed in Table 9. Because of the large number of J-V curves planned for this study, a data acquisition system was created which would record, display for selection, and store the curves efficiently.

Measured current-voltage curves are shown on Figures 17, 18 and 19.

Figure 17 is a cesium vapor family for the reference planar tungsten-niobium diode. Figure 18 is a cesium vapor family for the reference cylindrical diode with the same electrode materials and parameters. Figure 19 is representative of the oxygen additive effects on diode performance. Even at twice the spacing, the output power density is about twice that of the reference diode.

Diodes with an oxide collector exhibit virtually the same performance at oxygen concentrations from 800 ppm to 15,000 ppm, for both duplex and fluoride CVD tungsten emitter surfaces. Since the oxide collector has a lower work function than bare niobium, back emission degrades diode performance at high collector temperature.

At the reference parameters T_E=1800 K, J=7 A/cm^2, and d=10 mils, the oxide diodes have peak power density at T_c=800 K, low cesium pressure (~.5 torr), and pd=5 mil-torr, while the reference diodes (bare niobium collector) have peak power at $T_c \geq$1000 K, a typical cesium pressure (2 torr), and pd=20 mil-torr.

In the TFE application for the reference parameters given above, the duplex or fluoride tungsten emitter and bare niobium is the electrode combination yielding the maximum output power density.

Structured Electrodes

Structured electrodes, that is electrodes with a large number of uniform grooves to enhance the surface area, were studied as a second method to improve converter performance. Earlier work (Howard and Dunlay 1968, Shimada 1977), Shimada 1979) had reported modest improvements, on the order of 50 to 100 millivolts for converters with structured emitters or collectors, but there was always a question of normal fabrication parameter spreads producing the same results. In addition, several reports from the USSR (Gverdtsiteli et al. 1978, Gverdtsiteli et al. 1979, Atamasov et al. 1984, and Babanin et al. 1990) had indicated dramatic improvements of 100 to 200 millivolts and 1.2 to 1.9 times the power.

Four converters were constructed under the technology program. A reference converter with smooth electrodes, a grooved emitter-smooth collector, a smooth emitter-grooved collector and a grooved emitter and grooved collector. Emitters were duplex tungsten with a chloride <110> electrode surface. Grooves of 0.25 mm (0.010 in) depth and 0.0075 mm (0.003 in) width with equal separation were ground into the emitter face. The surfaces were finally electropolished. The niobium collectors had grooves with a depth of 0.75 mm (0.010 in), 0.01 mm (0.05 in) cut by a gang saw. Moloybdenium foil was used to clean and debur the grooves before final firing. Cesium family J-V curves were obtained for all four converters. Table 10 summarizes the test results for the converters. In each case the barrier index is measured at 6 A/cm^2 and optimum collector temperature and spacing.

The results show an improvement of as much as 100 millivolts and are consistent with the earlier U.S. work discussed above. The USSR studies used very low cesium pressure operating modes with the addition of barium vapor to adjust the work functions or flowing cesium with magnetic fields so that the results are not comparable. The effect of the grooving on effective emissivity was also examined but the differences observed were believed to be within the experimental error.

GRAPHITE-CESIUM RESERVOIR TECHNOLOGY

Characteristics of Cesium-Graphite Compounds

Cesium reacts with graphite to form layered intercalation compounds with the stoichiometry $C_{12n}Cs$ for $n \geq 2$. The number n is called the stage value and, in the classical model, gives the number of graphite layer planes between adjacent cesium-containing interplanar galleries (there are two first stage compounds with the formulas C_8Cs and $C_{10}Cs$). Alternatively, the Cs intercalation compounds can be described by an island model in which each interplanar gallery contains Cs atoms, but they accumulate in discreet islands, rather than being evenly distributed. The two models are compared in Figure 20. The island model is more suitable for describing the kinetics of the intercalation process, and it is uniquely useful in discussing the neutron irradiation response of these compounds.

Highly oriented pyrolytic graphite (HOPG) consists of large crystallites, with a common c-axis orientation. Isotropic graphite, such as POCO, consists of much smaller crystallites, with random orientation. The structure and thermodynamics of cesium compounds in the two materials are the same, whereas they respond quite differently to neutron bombardment. The cesium inventory, or loading, in grams of cesium per gram of carbon, is given in Table 11.

Figure 21 shows the vapor pressure of Cs over intercalated graphite as a function of graphite temperature for four different Cs loadings. These data were obtained with POCO graphite, however, the same vapor pressure is obtained with HOPG graphite. Experiments have verified that lower Cs pressures result if the loading is less than 278 mg Cs/g C and higher pressures are obtained when the loading exceeds 586 mg Cs/gC. Thus,in the range of temperatures appropriate to typical collector operation, Cs pressures in the range of 1 to 10 torr are easily obtained. One must be aware that the sensitivity of the Cs pressure to changes in loading depends upon the loading point. The staging is important in the design of graphite-cesium reservoirs.

Figure 22 displays the loading isotherms for cesium intercalated graphite. Although the isotherms shown were obtained using a pyrolytic graphite developed by Gulf General Atomic in the 1960s, modern specimens of POCO and HOPG display quantitatively the same behavior. As seen in the figure, at higher loading and lower temperature, more cesium may be lost to the system before the vapor pressure of Cs begins to fall off. In particular, below about 1073 K, the pressure will remain constant at a given temperature as long as the cesium inventory maintains the loading on the plateau between about 0.4 and 0.8 g Cs/g C. Operation in this range corresponds to having the cesium pressure determined by an equilibrium between $C_{10}Cs$ and $C_{24}Cs$. At lower loading and higher temperature, the reservoir response is more sensitive to cesium loss.

The response of cesium intercalated graphite to neutron irradiation has also been examined in the Thermionic Fuel Element Verification Program (General Atomics 1987-1990). Radical dimensional changes were observed in the case of HOPG: c-axis swelling of nearly 400% and a-axis contraction of about 60% were observed. Furthermore, the Cs pressure over HOPG was reduced by a factor of 10 after irradiation. The results obtained with POCO were quite different. Little or no dimensional change was observed with POCO. Figure 23 shows the pre- and postirradiation pressure-temperature isosteres for Cs-intercalated POCO graphite. As seen in the figure, there is very little difference between the data before and after neutron irradiation. Unlike HOPG, isotropic POCO graphite is a suitable choice for an integral reservoir material for thermionic systems.

System Considerations Associates with Cs-Graphite Reservoirs

There are two major considerations that drive the location and the space allocation of a cesium-graphite reservoir in a thermionic reactor-converter. In a static system, that is one in which the electric load requirements remain fixed, the operating Cs pressure and the range of reservoir temperatures available will uniquely determine the range of cesium-graphite loading that may be employed. In an integral reservoir configuration, the temperature range will be determined by the operating emitter and collector temperatures; the actual reservoir temperature will be determined by its passive thermal coupling to these electrodes.

In a dynamic power system, an additional constraint is imposed by the need to assure that the emitter is not subject to high temperature excursions due to low cesium pressure. An extreme example of the problem is represented by the startup of the power system. As the reactor power is increased, the emitter temperature increases, however, electron cooling does not occur until sufficient Cs vapor is released into the converter gap. The coupling between the reservoir and the emitter must be sufficient to raise the cesium pressure before the emitter temperature exceeds a safe value.

Once the temperature and loading are determined, the amount of graphite, and thus the volume of the reservoir, may be determined. The design criteria that must be considered are related to the desired lifetime of the power system. There must be a sufficient inventory of cesium available to accommodate loss without detriment to the converter performance. The loss could be a small leak in a seal insulator, consumption of cesium by reaction with impurities in the insulators, or other mechanisms. An analysis of an acceptable loss scenario will result in a calculated Cs loss rate. The amount of graphite required can be calculated so that the staging does not change, or changes by no more than an amount that is deemed acceptable, due to the calculated Cs loss.

An additional consideration in the design of an integral Cs-graphite reservoir results from the fact that Cs-intercalated graphite has been demonstrated to be fragile. The thermodynamics of the compounds are unaffected by the mechanical state of the graphite, but there is a possibility that fragments could migrate into lower temperature regions and become pressure-controlling. This problem may be eliminated by designing the reservoir so that fragments are confined to the isothermal region by baffles or other means.

HIGH PERFORMANCE WITH CLOSE SPACED ELECTRODES

It has long been recognized that near ideal thermionic converter performance would be achieved if the interelectrode gap could be reduced to a few micro meters. With such close spacing, the effects of atoms or ions in the interelectrode space are negligible because the distance from emitter to collector is less than the electron mean free path, and, as well, the interelectrode negative-space-charge effect is reduced in significance. The presence of cesium is still desirable, however, for its effect in lowering electrode work functions. The low pressure cesium diode with close spacing operates in what is called the "quasi-vacuum mode", since the thermionic performance is as if the interelectrode space were vacuum but the electrode work functions are as cesiated.

The potential benefit from such close spacing is shown in Figure 24 which shows lead efficiency as a function of emitter temperature and spacing. Efficiencies above 10% are possible at relatively low emitter temperatures.

Background

Some of the early thermionic devices achieved high efficiency with small interelectrode gaps,(Hatsopolous and Kaye 1958, Jacobson and Campbell 1969). Spacings in the range of 10 m or less resulted in lead efficiencies well above 10%. One of the problems with these early converters, however, was shorting between the electrodes. Converter fabrication and operation with small gaps were difficult. Other problems that were encountered in the early work with close spacings were the impact of uneven heating on configuration of the electrodes and the possibility that very small ceramic grains could gouge the electrode surfaces and reduce performance.

The Self Adjusting Versatile Thermionic Converter, SAVTEC, technology (Fitzpatrick 1987, Dick et al. 1983, NyRen et al. 1984) was invented in the mid-1980s to address these problems. Emitter sizes are small so that typical fabrication tolerances did not result in shorting; also, small electrodes are easier to keep parallel. Radiation coupling of the emitter to the hot shoe was employed to reduce further the sensitivity of converter performance to tolerances, and it also facilitates series-parallel connections within a common enclosed container.

The SAVTEC concept is shown schematically on Figure 25. The emitter is supported by the emitter lead. The key to obtaining a small spacing is to have the emitter and collector in contact in the cold condition. Then during heatup, differential expansion of the emitter lead will lift the emitter off the collector by a predetermined amount.

Experimental Data

The SAVTEC test program (Dick et al. 1983, NyRen et al. 1984, McVey 1984) results are shown in Table 12 and in Figure 26. Lead efficiencies are calculated values based upon the test data and the phenomenological model of converter performance (Rasor 1982) with an assumed value of 0.2 effective emissivity for thermal heat transport between electrodes.

Several converters were built and tested and good efficiency was achieved at low emitter temperature. It was also demonstrated that at higher emitter temperature (SAVTEC 15), an efficiency of 18% could be achieved.

Recent Results

Further development of the SAVTEC concept continues. Interest is in assuring a stable close spacing with gaps on the order of 5 m.

An alternative SAVTEC configuration was recently described in Fitzpatrick (1992) in which the emitter, collector and emitter lead and support are all made of ceramic, in particular sapphire. The emitter and collector electrode surfaces are provided with a thin film of metal over the structural sapphire. Electrical current path is provided both by deposited film on the sapphire and by a lead wire. These innovations have the potential to allow operation at small interelectrode gaps with reduced probability of electrode shorting and short circuits. Figure 26 is a schematic picture.

Because of the inherent electrical loss in the films, these converters optimize at lower current and higher voltage than conventional converters. In addition, optimum electrode diameters are 5 to 10 mm, and so large numbers of such small converters would be combined in modules for applications.

Experimental work with similar planar close-spaced converters at Scientific Industrial Association (NPO) LUTCH, in Podolsk, Russian Federation (Nikolaev 1990), has recently been published and is continuing. These converters use monocrystal alloys of tungsten and molybdenum for electrodes, which provide the exceptionally stable electrode surfaces required for close spacing. An interelectrode spacing of 3 to 6 m in the LUTCH converter is maintained by the placement of closely-machined ceramic spacers in recesses of the collector. The emitter is held against the spacers in the collector by a spring mechanism. Figure 27 reproduces a figure comparing their experimental results with calculated predictions.

Figure 28 illustrates the calculated lead efficiencies for the SAVTEC data points and one recent LUTCH result. Also drawn in Figure 28 are lines of predicted performance from the barrier index method and from the LUTCH paper.

References

Agarwal, B. D. and L. J. Broutman (1980) *Analysis and Performance of Fiber Composite,* Book, Wiley Interscience Publication, New York, NY, USA.

Alleau, T. and M. Bacal (1970) "Influence of Oxygen Adsorption and Oxygen-Cesium Coadsorption on (100) Tungsten Work Function," *Proceedings of the 1970 Thermionic Conversion Specialists Conference* held in Miami Beach, FL, IEEE, New York, 434-440.

Atamasov, V. P., S. A. Skvebkov, L. N. Tashchillin and I. A. Urtminstev (1984) "Experimental Study of Cesium Arc Thermionic Converters with Multicavity Emitter," *Sov. Phys. Tech. Phys.*, 29:1:40-42.

Babanin, V. I., A. Y. Ender, I. N. Kolshkin and V. I. Kuznetsov (1990) "Aspects of the Knudsen Thermionic Converter," in *Proceedings Seventh Symposium on Space Nuclear Power Systems*, CONF-91009 Albuquerque, NM, 7-10 January 1990.

Cone, V. P. and J. Dunlay (1987) *Thermionic Technology Program Fiscal Year 1986 and Final Technical Report*, Thermo Electron Report No. TE4400-227-87.

Dick, R. S., J. B. McVey, G. O. Fitzpatrick and G. J. Britt (1983) "High Performance, Close Spaced Thermionic Converters," *Proc. Eighteenth Intersociety Energy Conversion Engineering Conference,* Orlando, FL, AIChE, New York, 1:198-201.

Dunlay, J. B. Matsuda and V. Poirer (1972) "Evaluation of Sublimed Molybdenum Collector Coatings for Additive Diode Operation," in *Third International Symposium on Thermionic Electrical Power Generation* held in Julich, Germany: Kernforschungsanlage, F45:8.

Fitzpatrick, G. O. (1987) *Thermionic Converter,* U.S. Patent No. 4667126, May 16, 1987.

Fitzpatrick, G.O. et al. (1990) "New Thermionic Converter for Out-of-Core Space Reactor Power Systems," *Proc. Twenty-Seventh Intersociety Energy Conversion Conference,* held in San Diego, CA, IECEC, New York, NY.

General Atomics Report GA-A18780 (1987) *TFE Verification Program Semiannual Report for the Period Ending March 31, 1987,* by GA Technologies, Rasor Associates, Inc., Space Power, Inc. and Thermo Electron Corporation, GA Technologies, San Diego, CA, April 1987.

General Atomics Report GA-A19115 (1988) *TFE Verification Program Semiannual Report for the Period Ending September 30, 1987,* by General Atomics,

Rasor Associates, Inc., Space Power, Inc. and Thermo Electron Corporation, General Atomics, San Diego, CA, March 1988.

General Atomics Report GA-A19269 (1988) *TFE Verification Program Semiannual Report for the Period Ending April 30, 1988,* by General Atomics, Rasor Associates, Inc., Space Power, Inc. and Thermo Electron Corporation, General Atomics, San Diego, CA, June 1988.

General Atomics Report GA-A19412 (1989) *TFE Verification Program Semiannual Report for the Period Ending October 31, 1988,* by General Atomics, Rasor Associates, Inc., Space Power, Inc. and Thermo Electron Corporation, General Atomics, San Diego, CA, January 1989.

General Atomics Report GA-A19666 (1989) *TFE Verification Program Semiannual Report for the Period Ending April 30, 1989,* by General Atomics, Rasor Associates, Inc., Space Power, Inc. and Thermo Electron Corporation, General Atomics, San Diego, CA, September 1989.

General Atomics Report GA-A19876 (1990) *TFE Verification Program Semiannual Report for the Period Ending September 30, 1989,* by General Atomics, Rasor Associates, Inc., Space Power, Inc. and Thermo Electron Corporation, General Atomics, San Diego, CA, March 1990.

General Atomics Report GA-A20119 (1990) *TFE Verification Program Semiannual Report for the Period Ending March 31, 1990,* by General Atomics, Rasor Associates, Inc., Space Power, Inc. and Thermo Electron Corporation, General Atomics, San Diego, CA, July 1990.

General Atomics Report GA-A18182 (1985) *SP-100 Thermionic Technology Program Annual Integrated Technical Progress Report for the Period Ending September 30, 1985,* by GA Technologies, Rasor Associates Inc., Space Power, Inc. and Thermo Electron Corporation, General Atomics, San Diego, CA, November 1985.

General Atomics Report GA-18915 (1987) *Thermionic Irradiations Program Final Report, General Atomics, San Diego, CA.*

Gverdtsiteli, I. G., N. E. Menabde, V. K. Tshakaya and L. M. Tsakadze, (1978) "Low Temperature Thermionic Converter with an Expanded Area Collector," Sov. Phys. Tech. Phys., 24:8:990.

Gverdtsiteli, I. G., A. G. Kalandarishvili, and P. D. Chilingarishvii (1979) "Cesium Vapor Source with a Gas-Controlled Heat Pipe for Thermionic Energy Converters," *Sov. Phys. Tech. Phys.*, 24:8:1764-1765.

Hansen, L. K. (1967) "Ion-Current and Schottley Effects in Thermionic Diodes," *J. Appl. Phys.*, 38:11:4345-4354.

Hatch, G. L. (1988) *Thermionic Technology Program: Thermionic Converter Performance Final Report*, NSR-25-25, E-533-003-B-053188 (DOE Contract No. DE-ACO3-86SF15954).

Hatsopoulos, G. N. and J. Kaye (1950) "Measured Thermal Efficiencies of a Diode Configuration of a Thermo Electron Engine," *J. Applied Physics,* 29 (7).

Howard, R. C. and J. B. Dunlay (1968) "Converter SP-4 Design and Summary of Test Results, "*Second International Conference on Thermionic Electrical Power Generation*, Stressa, Italy.

Jacobson, D. L. and A. E. Campbell (1969) "The Characterization of Bare and Cesiated CVD 75% Tungsten/25% Rhenium Electrodes," *Proc. Thermionic Conversion Specialist Conference,* (NAS7-100), held in Carmel, CA, October 1969, IEEE, New York, 1969:26-33.

Levine, J. D. et al. (1965) "Oxygen as a Controllable, Reversible and Beneficial Additive on the Cesium Converter," *Thermionic Conversion Specialists Conference*, San Diego, CA.

Lieb, D. and S. Kitrilakis (1966) "Oxygen as a Steady-State Electronegative Additive in a Cesium Thermionic Converter," in *Proceedings of the 1966 Thermionic Conversion Specialists Conference*, held in Houston, TX, IEEE, New York, 1967: 348-354.

Lieb, D. and F. Rufeh (1970), "Thermionic Performance of CVD Tungsten Emitter with Several Collector Materials," *Thermionic Conversion Specialists Conference*, held in Miami Beach, FL, IEEE, New York, 1970:471-480.

Lieb, D. and F. Rufeh (1972) "The Utilization of Tungsten Oxides for Additive Thermionic Converters," in Proceedings of the *Third International Conference on Thermionic Power Generation*, Julich, Germany, F46:12.

Liu, D. S. and A. P. Majidi (1988) "High Temperature Creep of SiC Whisker-Reinforced Alumina Composites," *NASA Conference Publications*.

McVey, J. B. (1984) "Close Spaced High Temperature Knudsen Flow," *Annual Report for Air Force Office of Scientific Research,* NSR-22-1, Rasor Associates, Inc., Sunnyvale, CA.

Nikolaev, Yu. V. (1990) "Refractory Metals for Thermionic Fuel Elements of NEP," *Transactions Anniversary Specialists Conference on Nuclear Power Engineering in Space, Part 1,* Obninsk, Kaluga.

Nikolaev, Yu. V., R. Ya. Kurcherov, et al. (1990) "Close Spaced Thermionic Converter," *Proc. Seventh Symposium on Space Nuclear Power Systems,* Albuquerque, NM.

Norcross, D. W. (1962) "Departure from LTE in the Thermionic Converter," *27th Annual Conference on Physical Electronics,* MIT, Cambridge, MA.

NyRen, T., M. Korringa, et al. (1984) "Design and Testing of a Combustion Heated Nineteen Converter SAVTEC Array," *Proc. Nineteenth Intersociety Energy Conversion Engineering Conference,* San Diego, CA, 4:2300, American Nuclear Society, San Francisco, CA.

Pigford, T. H. and B. E. Thinger (1968) "Performance Characteristics of a 0001 Rhenium Thermionic Converter," in *Conference Records of 1968 Thermionic Conversion Specialist Conference,* Carmel, CA, IEEE, New York.

Poirier, V. L. (1972) "Irradiation of Metal/Ceramic Sheath Insulators," Topical Report, Thermo Electron Report No. TE4122-42-72., U.S. Atomic Energy Commission Contract No. AT(111)3056, Thermo Electron Corporation, Waltham, MA., October 1971.

Rasor, N. S. (1982) "Thermionic Energy Conversion," *Applied Atomic Collision Physics,* 5:170-199, Massey, McDaniel, and Bederson, eds., Academic Press, New York.

Rasor, N. S. and C. Warner (1964) "Correlation of Emission Processes for Adsorbed Alkali Films on Metal Surfaces," *J. Appl. Phys.,* 35(9), pp. 2589-2600.

Rice, R. W. (1985) "Mechanisms of Toughening in Ceramic Matrix Composites," Naval Research Laboratory, Washington, DC.

Rufeh, F., D. Lieb and F. Fraim (1967) "Recent Experimental Results on Electronegative Additives," *in Proceedings of the 1967 Thermionic Conversion Specialists Conference*, pp. 25-58, IEEE, New York.

Shimada, K. (1977) "Low Arc Drop Hybrid Mode Thermionic Converter," in *Proceedings of the 12th IECEC Conference*, 2:1568-1574, American Nuclear Society, La Grange Park, IL.

Shimada, K. (1979) "Recent Progress in Hybrid Mode Thermionic Converter Development," in *Proceedings of the 14th IECEC Conference*, American Chemical Society, Washington, DC, 2:1890, Boston, MA.

Thermo Electron Technologies Corporation (1990) *Final Technical Report Thermionic Technology Program Ceramic Composites and Structured Electrodes*, Report No. TE4430-190, Thermo Electron Corporation, Waltham, MA.

Acknowledgments

The work on graphite-cesium reservoirs reported herein was performed under the Thermionic Fuel Element Verification Program, General Atomics (1985-1990).

Nomenclature

d	Interelectrode gap (units)
J	Output current density (A/cm^2)
J_C	Collector electron back emission = $J_R(\phi_C, T_C)$ (A/cm^2)
J_E	Electron emission from the virtual emitter = $J_R(\phi_E + \Delta V, T_E)$ (A/cm^2)
J_R	Current from the Richardson-Dushman equation: $J_R = 120T^2\exp(-\phi/kT)$ (A/cm^2)
J_S	Schottky-effect modified electron emission from the emitter = $J_R(\phi_S, T_E)$
J_S'	Zero-field electron emission from the emitter = $J_R(\phi_E, T_E)$ (A/cm^2)
k	Boltzmann's constant, $(11605)^{-1}$ (eV/K)
T_a	Average neutral atom temperature = average of T_E and T_C (K)
T_C	Collector temperature (K)
T_E	Emitter temperature (K)
T_{eC}	Electron temperature at the plasma-collector sheath interface (K)
T_{eE}	Electron temperature at the plasma-emitter sheath interface (K)
T_R	Liquid cesium reservoir temperature (K)
T_s	Equilibrium electron temperature (K)
V_C	Collector sheath height (units)
V_d	Arc drop (V)
V_E	Emitter sheath height (units)
Vi	Ionization potential for cesium = 3.89 (eV)
V_{out}	Converter output voltage (V)
ΔV	Emitter sheath barrier height (units)
ΔV_{rad}	Portion of arc drop associated with photon loss from the plasma (V)
ϵ_λ	Spectral emissivity of the electrodes at the cesium resonance wavelength
λ_e	Electron mean free path (units)
λ_{ea}	Electron-neutral mean free path (units)
λ_{ei}	Electron-ion mean free path (units)
ϕ_C	Collector work function (V)
ϕ_E	Emitter work function (V)
ϕ_n	Equilibrium plasma potential (V)
ϕ_S	Schottky-effect modified emitter work function (V).

TABLE 1. Scope of Thermionics Programs in the 1980s.

1983 - 1985	SP-100 Phase I Design Effort
1984 - 1985	SP-100 Technology Program
1984 - 1987	Thermionic Technology Program
1986 - Present	TFE Verification Program
1986 - Present	Multiple SBIR Contracts
1987 - 1989	Multimegawatt Design Effort
1989 - Present	Advanced Thermionic Initiative
1990 - Present	USAF Design and Technology Programs
1991 - Ongoing	TSET Program
1992 - Ongoing	DOE/SDIO/USAF Systems Programs

TABLE 2. Fueled Emitter Test Parameters.

Capsule Number	Cell Number	Converter Position	Design Emitter Temperature (K)	Emitter Wall Thickness (in)	Emitter i.d.(in)
1	1	Top	1750	0.100	0.900
	2	Middle	1850	0.100	0.900
	3	Bottom	1750	0.100	0.900
2	1	Top	1650	0.070	0.960
	2	Middle	1850	0.070	0.960
	3	Bottom	1750	0.070	0.960
3	1	Top	1750	0.040	1.020
	2	Middle	1750	0.070	0.960
	3	Bottom	1750	0.070	0.960

TABLE 3. Irradiation History: Fueled Emitters.

Cell Number	TRIGA Insertion Date	Total Irradiation Time, Hours	Fast Fluence 10^{22} nvt	Burnup a/o
1.1	Feb 21, 1985	19,400	.15	.33
1.2			.22	.49
1.3			.17	.39
2.1	Jan 13, 1985	39,000	.26	.51
2.2			.47	.92
2.3			.38	.75
3.1	July 19, 1985	23,900	.28	.50
3.2			.31	.60
3.3			.27	.53

TABLE 4. Peak Emitter Distortion, Mils.

	Time of Neutron Radiographs															
	Jan '85	Feb '85	May '85	Jul '85	Sep '85	Dec '85	Mar '86	May '86	Jul '86	Oct '86	Feb '87	Jun '87	Sep '87	Sep '88	Jul '89	Aug '90
Capsule 1: Hours Irradiated		0	2100		4560	6342	7862	9489	10861	12250	14578	17090	19425			
Cell 1					0		1	1		2	2	2				
Cell 2					0		1	1		2	2	2				
Cell 3					0		2	2		3	3	3				
Capsule 2: Hours Irradiated	0	800	2900	3972	5388	7173	8693	10319	11691	13080	15408	17920	20255	27844	34595	39000
Cell 1					0	0	1	2	2	3	4	4	6			
Cell 2				0			2	2	4	5	6	6	9			
Cell 3				0	0	2	1	3	3	4	4	5				
Capsule 3: Hours Irradiated				0	1413	3198	4719	6314	7716	9105	11433	13945	16281	23871		
Cell 1					0	0	3	5	8	11	10	14	17			
Cell 2					0		1	2	2	3	3	4	5			
Cell 3					0		1	2	2	3	3	4	4			

TABLE 5. Emitter Distortion Summary.

Cell No.	Total Hours Irrad.	Emitter Temperature		Failure Mode	Valid Distortion Data	
		Design K	Actual K		Distortion Mils	Hours Operation
1.1	19,400	1750	-	External short	-	-
1.2	19,400	1850	1852	Seal	2	10,900
1.3	19,400	1750	-	Seal	<1	6,000
2.1	39,000	1650	-	Seal	-	-
2.2	39,000	1850	1791	External short	8	20,100
2.3	39,000	1750	1759	Seal	4	18,000
3.1	23,900	1750	1781	External short	14	14,200
3.2	23,900	1750	1769	Seal	3	13,600
3.3	23,900	1750	1773	External short	3	16,200

TABLE 6. TRIGA Irradiation Experiment Insulators Test Matrix.

Sample	Material	Length (in.)	Design Temperature °C	Applied Voltage V	Current (mA)	nvt x 10^{-13}	Resistivity Mohms-cm
1	Al_2O_3	0.31	761.0	9.98	0.021	2.59	177.5
2	Y_2O_3	0.18	758.0	9.96	1.566	2.59	1.4
3	Al_2O_3	0.68	763.0	9.96	1.220	2.61	6.6
4	Y_2O_3	0.75	751.0	1.01	3.49	2.57	0.26
5	Y_2O_3	1.30	716.0	9.94	1.97	2.45	7.5

TABLE 7. Composite and Cermet Materials.

Material	Type	Method	Testing			Test Result
			Resistivity ohms-cm	Test Hours	Test Temp K	
Alumina-YAG	Powder mix composite	Filter press	10^9	>1000	1100	Cracked as trilayer
Alumina-SiAlON	Powder composite		10^6	Degraded		Many cracked
Boron Nitride Alumina	Commercial composite	Pressure bonded	10^9	Degraded		Weak, cracked
YAG-Boron nitride	Commercial composite	Pressure bonded				Weak, cracked
Alumina Silicon Carbide	Fiber composite	Pressure bonded	Low			
Alumina Chromium Oxide	Powder composite	Hot pressed	10^7	700	1100-1300	Good
YAG Chromium Oxide	Powder composite	Filter pressed				Cracked
Aluminum Oxynitride	Commerical composite	Pressure bonded	10^7	500		Some good Some cracked
Alumina Silicon Nitride	Powder fiber composite	Pressure bonded	10^9	500	1250	Good bond
Alumina-Niobium	Powder sphere cermet	Slip cast		UCA-1		
YAG-Niobium	Powder sphere cermet	Filter pressed		UCA-1		

TABLE 8. Test Diode Summary.

Planar Diodes	Emitter	Collector	Comments
PD-1	EE FL W	Nb (4000 ppm O_2)	Similar to PD-2
PD-2	EE FL W	Nb	Fluoride W(100) reference
PD-3	EP FL W	Nb (15000)	Additive behavior
PD-4	EP FL W	Nb (3000)	Additive behavior
PD-5	EP FL W	Nb (2000)	Additive behavior
PD-6	DUPLEX W	Nb	Chloride W(110) reference
PD-7	DUPLEX W	Nb (12000)	Additive behavior
PD-8	DUPLEX W	Nb (1000)	Additive behavior
Cylindrical Diodes	**Emitter**	**Collector**	**Comments**
CD-1	DP FL W	Nb (3000)	Oxide behavior
CD-2	DUPLEX W	Nb	Chloride W(110) reference

NOTES:
EE.......... Electro-Etch
EP.......... Electro-Polish
DUPLEX W.... Chloride CVD tungsten or fluoride CVD tungsten substrate
FL W........ Fluoride CVD tungsten
(XXXX)...... PPM oxygen in molybdenum oxide.

TABLE 9. Performance Mapping Parameters.

Cesium families for:	
Emitter temperature (T_E):	1600, 1700, 1800, 1900, 2000 K
Collector temperature (T_c):	600, 800, 1000, 1100 K
Interelectrode Spacing (d):	5, 10, 20 mils (.125, .25, 5 mm)

TABLE 10. Barrier Index Results of Planar Converters.

Emitter Temperature (K)	Spacing (Inch)	Emitter	Collector	Barrier Index (Ev)	Collector Temperature (K)
1700	0.010	Smooth	Smooth	209	979
		Grooved	Smooth	206	903
		Smooth	Grooved	206	953
		Grooved	Grooved	209	953

TABLE 11. Cesium Loading for the Cs-Graphite Compounds.

Stage	Formula	Cesium Loading g Cs/g C
1	C_8Cs	1.384
1	$C_{10}Cs$	1.108
2	$C_{24}Cs$	0.461
3	$C_{36}Cs$	0.308
4	$C_{48}Cs$	0.231
5	$C_{60}Cs$	0.185
6	$C_{72}Cs$	0.154

TABLE 12. Selected Results From SAVTEC Experimental Program.

Test Converter	Temperatures (K)			Electrode Emitter Lead		
	Emitter	Collector	Cesium	Spacing (μm)	Current (A/cm^2)	Efficiency (%)
SAVTEC 10	1100	737	502	10	3.3	3.3
SAVTEC 12A	1200	743	461	10	0.7	7.2
SAVTEC 15	1750		591	13	4.2	18
SVT-1B	1170	800	507	6.5	1.0	7.5
HOTSHELL B	1250	960	565	10	0.5	5.6
HOTSHELL N	1300	965	550	10	0.75	7.5

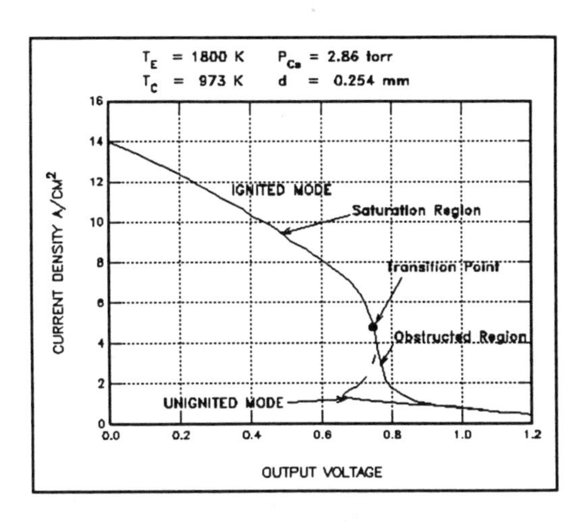

FIGURE 1. Thermionic Converter I-V Curve with Operating Regions.

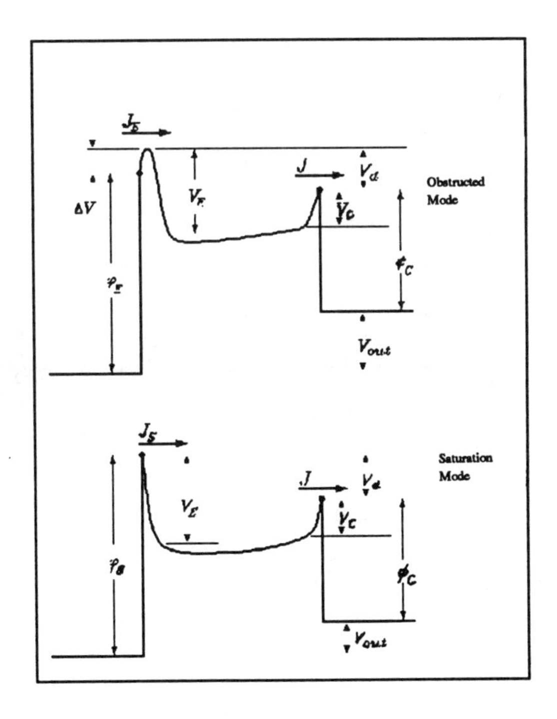

FIGURE 2. Motive Diagrams for the Obstructed and Saturation Regions.

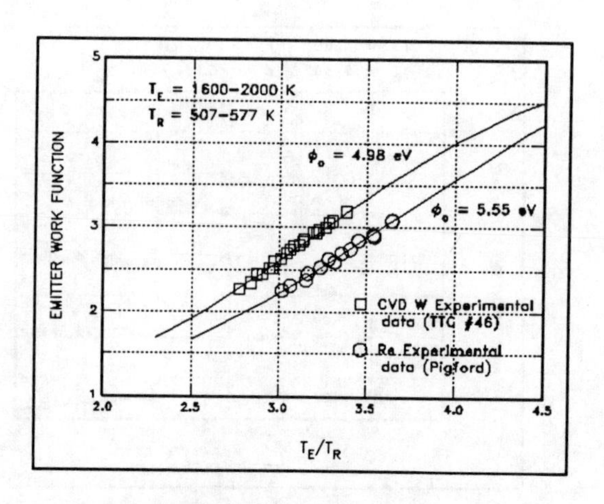

FIGURE 3. Comparison of Experimental and Calculated Emitter Work Functions.

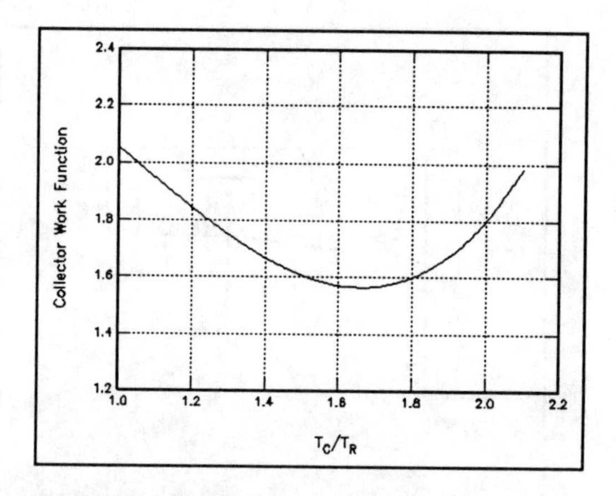

FIGURE 4. Empirical Curve Used to Calculate Collector Work Functions.

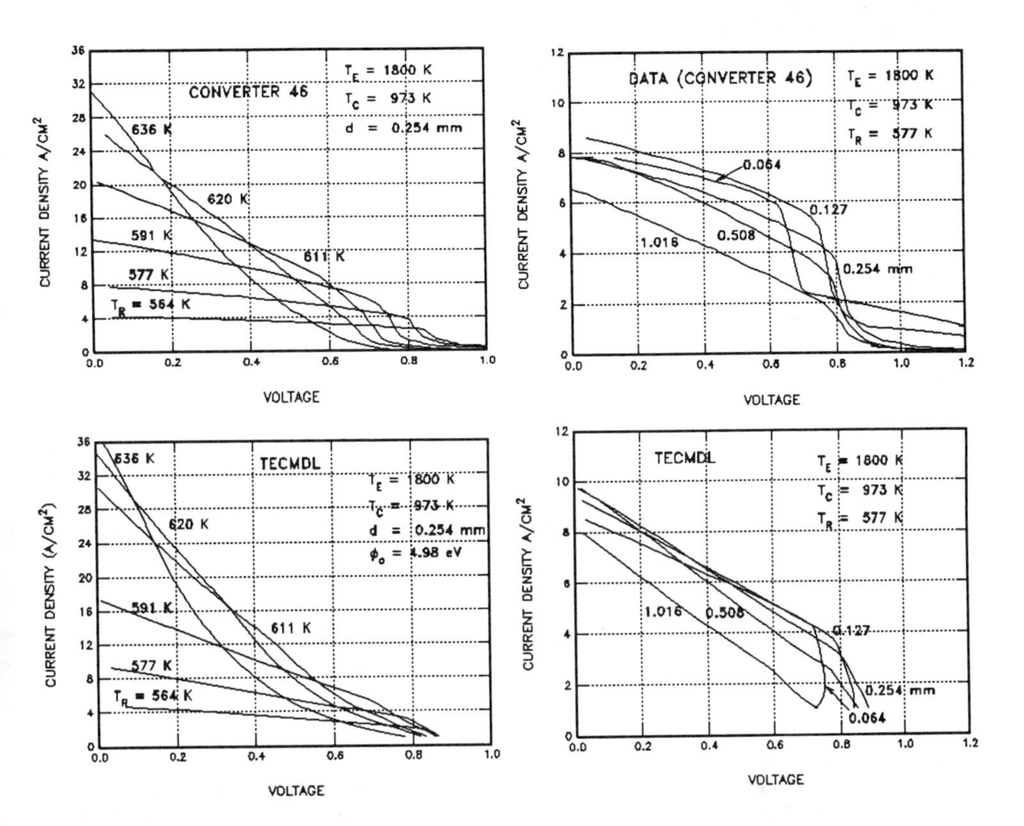

FIGURE 5. Comparison of Calculated and Experimental Cesium Pressure Families of I-V Curves.

FIGURE 6. Comparison of Calculated and Experimental Spacing Families of I-V Curves.

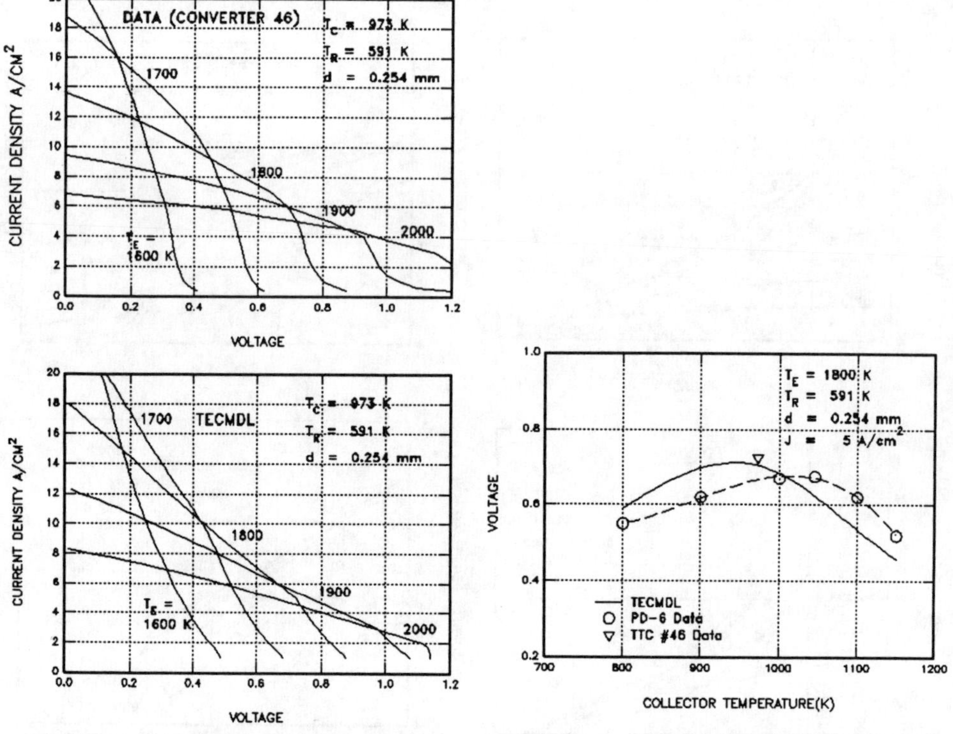

FIGURE 7. Comparison of Calculated and Experimental Emitter Temperature. Families of I-V Curves.

FIGURE 8. Comparison of Experimental and Calculated Output Voltage as a Function of Collector Temperature.

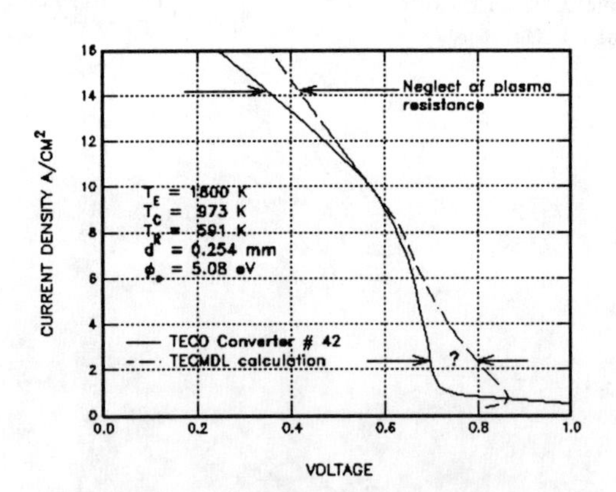

FIGURE 9. Comparison of Experimental and Calculated Output Voltage as a Function of Collector Temperature.

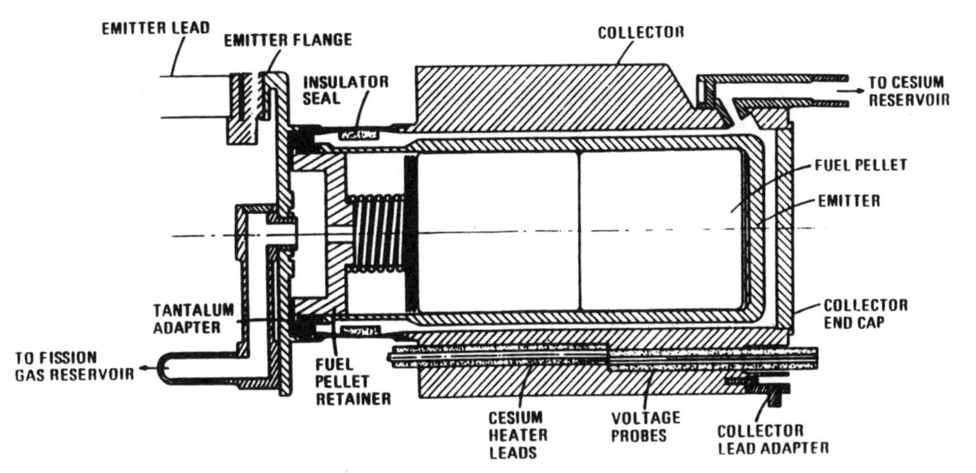

FIGURE 10. Thermionic Technology Program Fueled Emitter Cell.

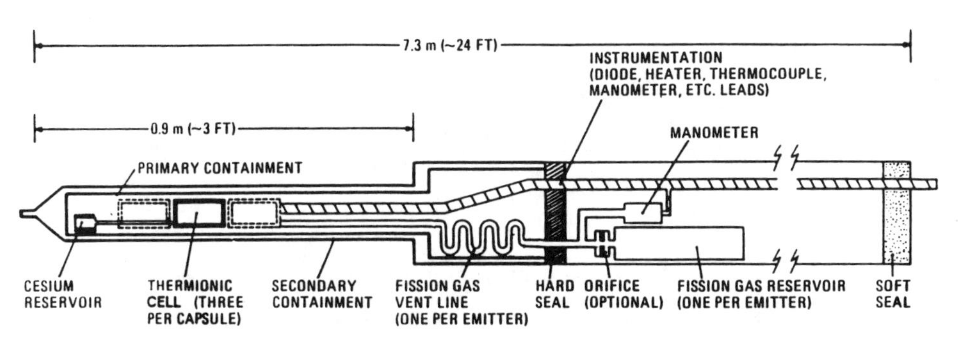

FIGURE 11. Test Capsule Containing Three Fueled Elements.

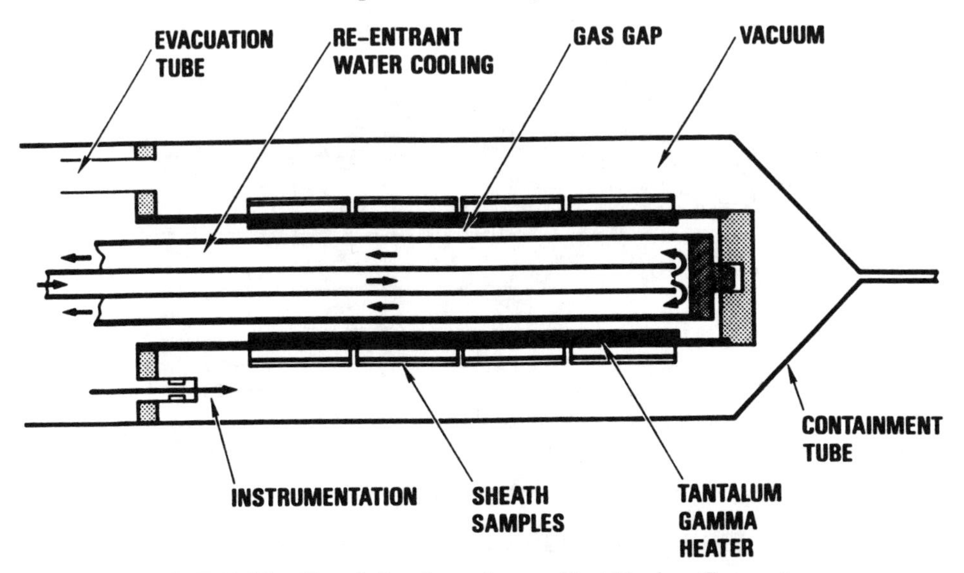

FIGURE 12. Sheath Insulator Incore Test Design Concept.

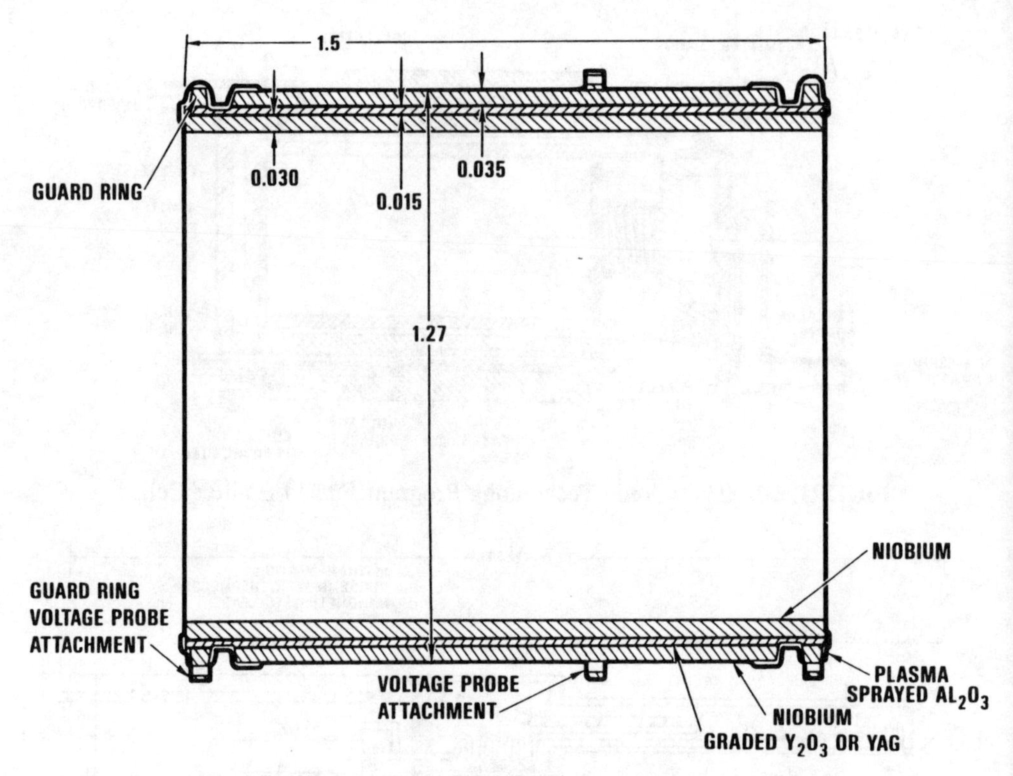

FIGURE 13. TRIGA Insulator Capsule Irradiation Test Specimen.

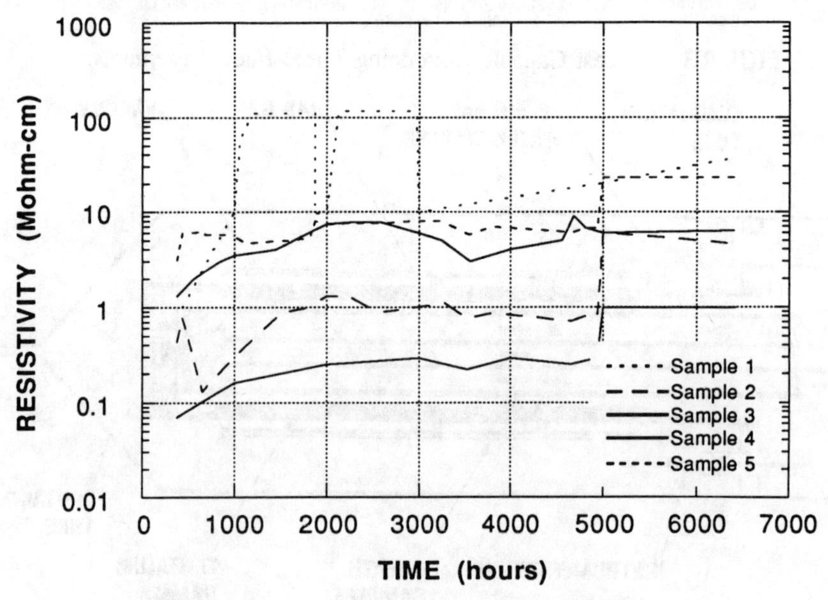

FIGURE 14. TRIGA Insulator Resistivity.

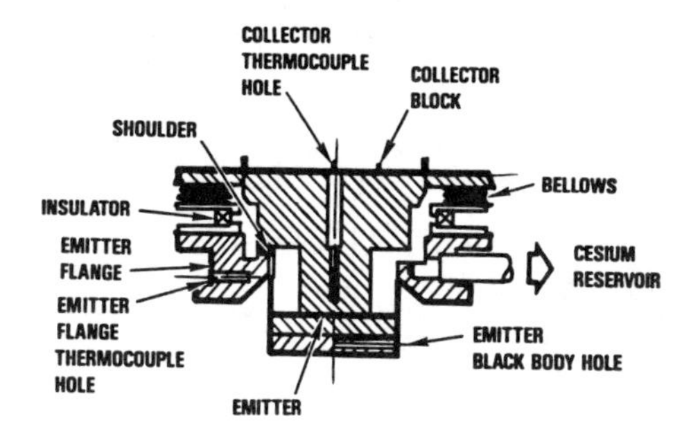

FIGURE 15. Drawing of a Planar Test Diode.

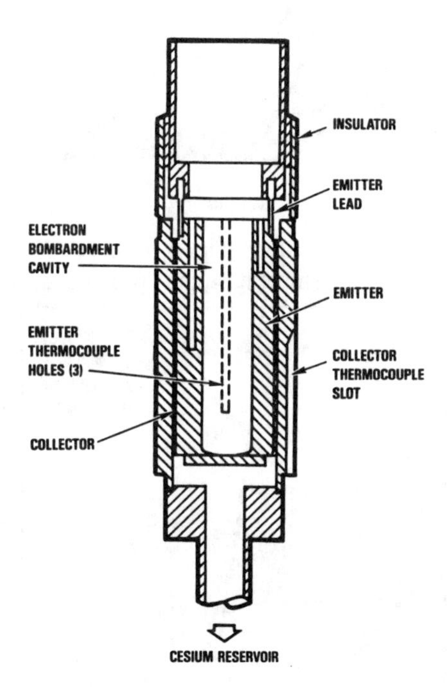

FIGURE 16. Drawing of a Cylindrical Test Diode.

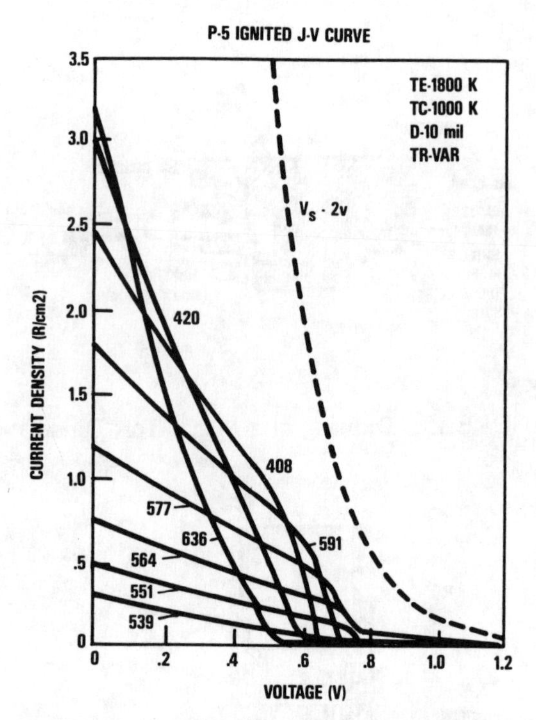

FIGURE 17. Cesium Vapor Family for the Temperature and Spacing Indicated.

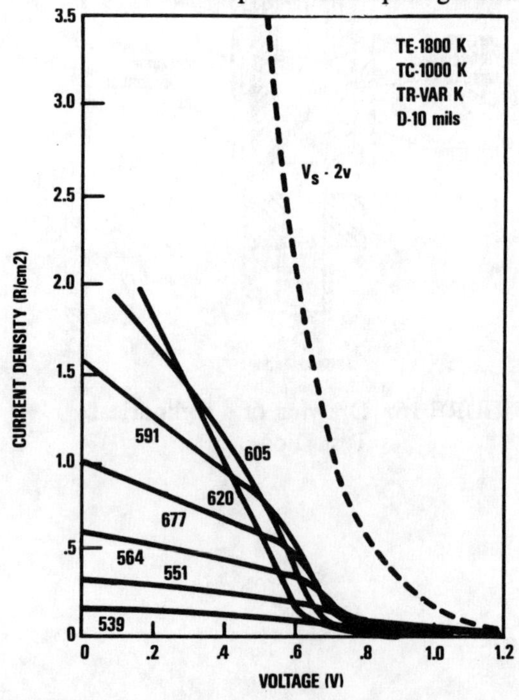

FIGURE 18. Cesium Vapor Family for the Temperature and Spacing Indicated.

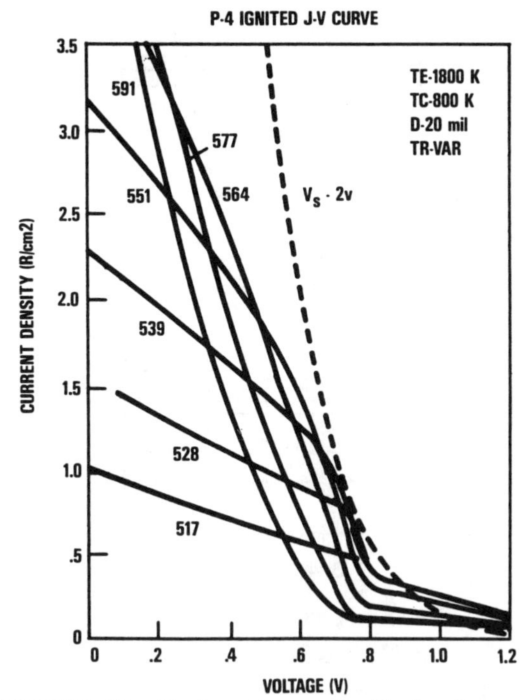

FIGURE 19. Cesium Vapor Family for the Temperature and Spacing Indicated.

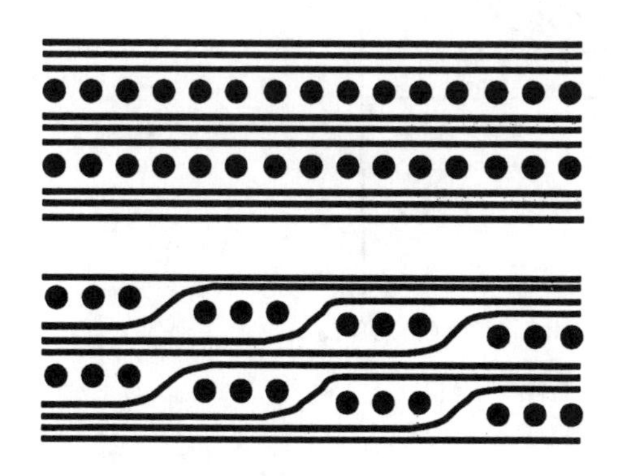

Figure 20. Classical (Top) and Island Model of 3_{rd} Stage Cs-Graphite.

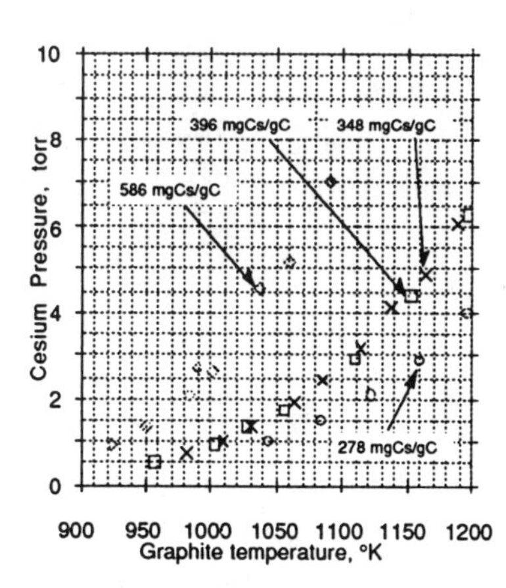

Figure 21. Cs Vapor Pressure Over Cs-Intercalated POCO.

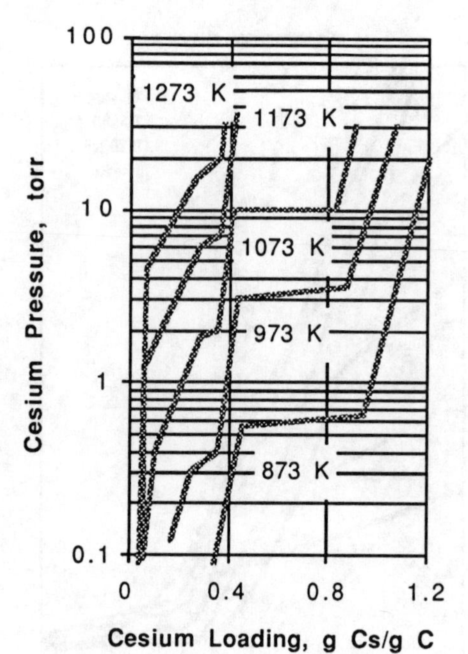

Figure 22. Cs-Graphite Isotherms.

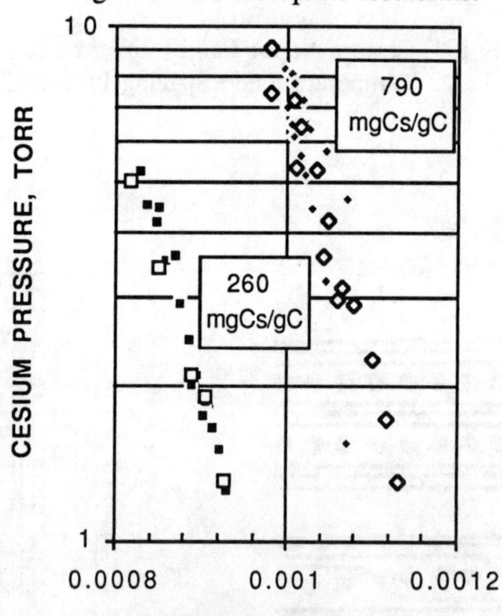

Figure 23. Cs-POCO Isosteres. Open Symbols Are Preirradiation Data and Closed Systems Were Obtained in Postirradiation Measurements.

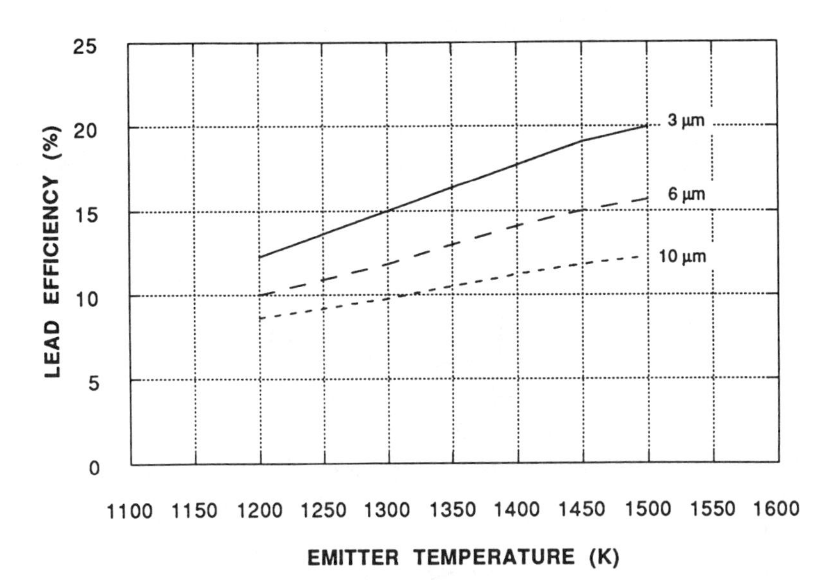

FIGURE 24. Efficiency as a Function of Emitter Temperature and Interelectrode Spacing.

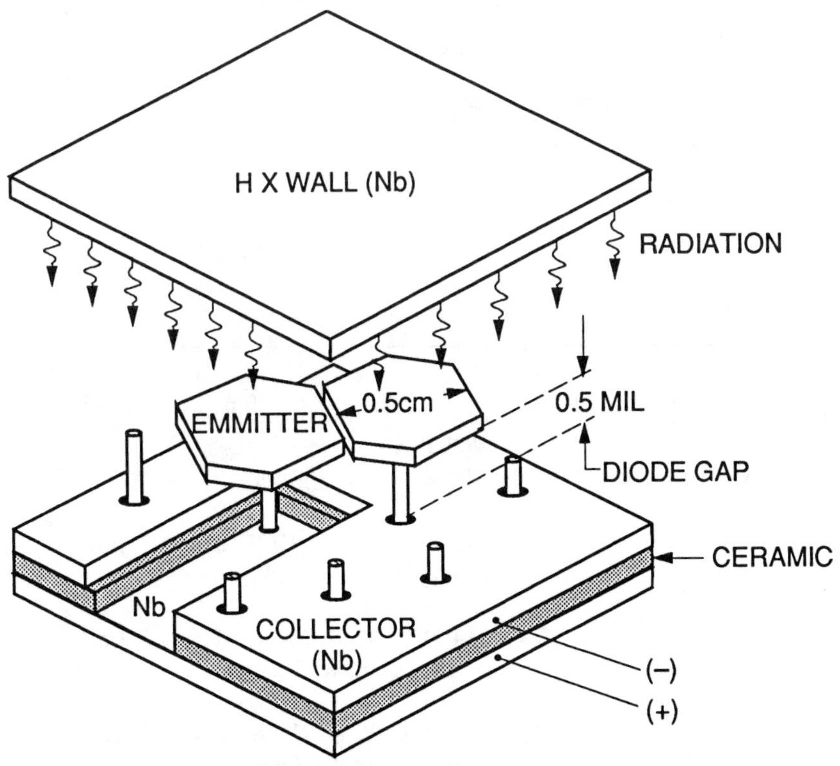

FIGURE 25. SAVTEC Power Conversion System Schematic.

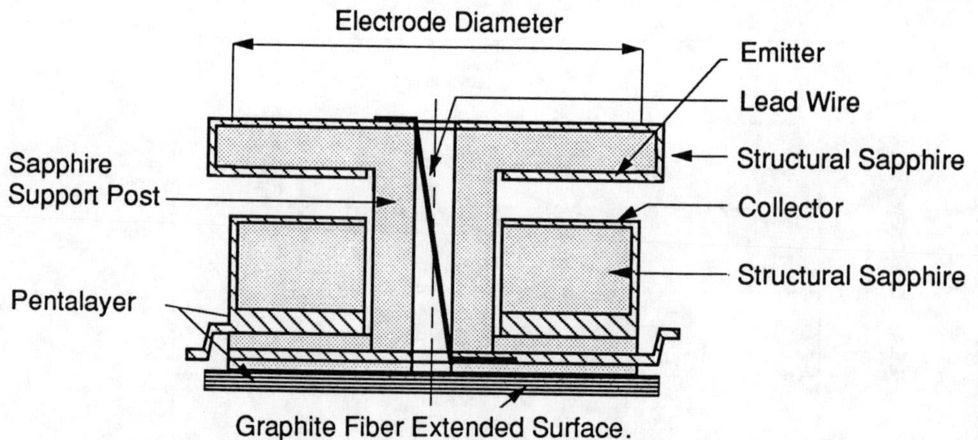

FIGURE 26. NETCON Converter

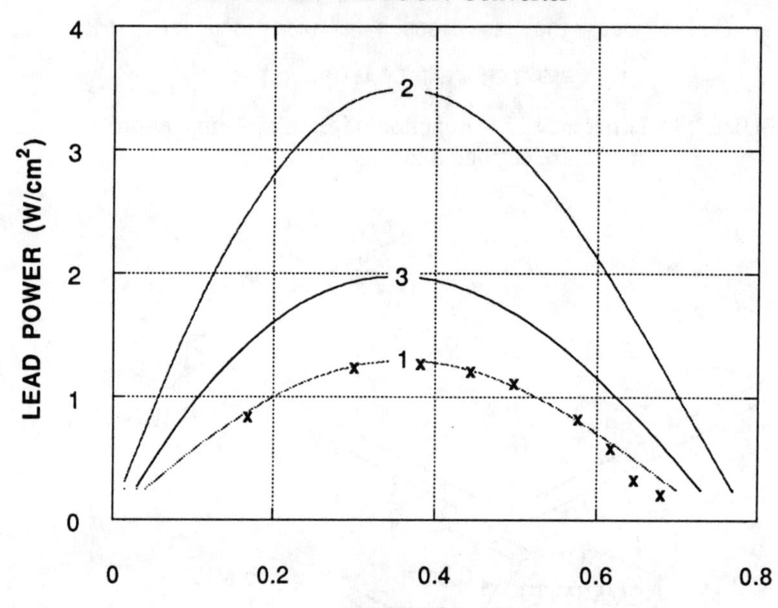

Test: x T_E = 1000 K, T_c = 810 K, P_{Cs} = 133 Pa

Calculations:

1 - T_E = 1330 K, T_c = 910 K, d = 4.9 μm
X_E = 1.94 eV, X_c = 1.57 eV

2 - T_E = 1400 K, T_c = 950 K, d = 3.0 μm
X_E = 1.95 eV, X_c = 1.55 eV

3 - T_E = 1330 K, T_c = 910 K, d = 3.0 μm
X_E = 1.94 eV, X_c = 1.57 eV

FIGURE 27. TEC Lead Power vs Output Voltage.

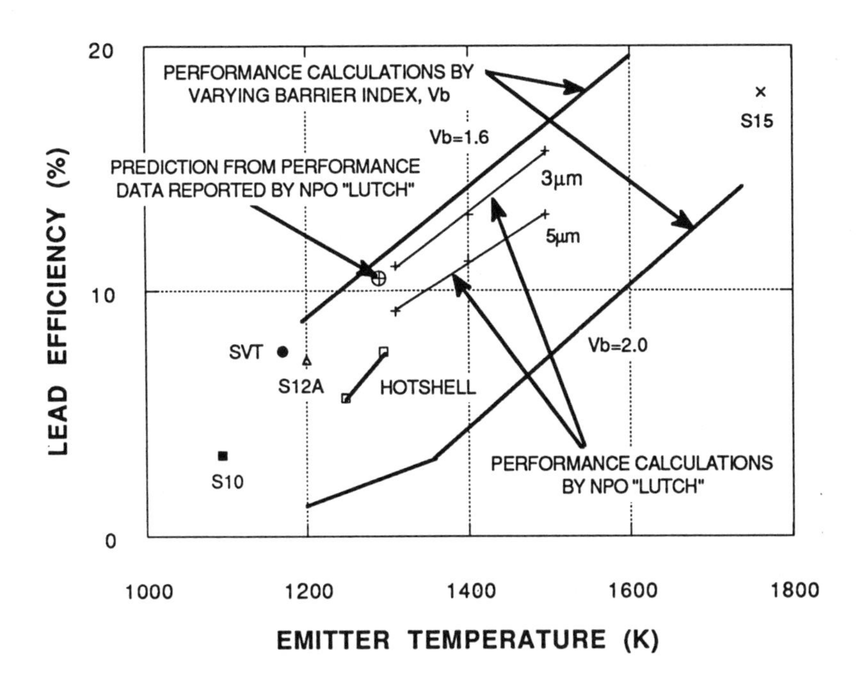

FIGURE 28. Recent Data From NPO LUTCH Along With SAVTEC Experimental Results and Calculated Performance Lines.

SPACE NUCLEAR POWER
HEAT PIPE TECHNOLOGY ISSUES REVISITED

Michael A. Merrigan
Los Alamos National Laboratory
P.O. Box 1663, MS J576
Los Alamos, NM 87544
(505) 667-6466

Abstract

Potential applications of high temperature heat pipes in space power systems are categorized in terms of operating temperatures, radial and axial heat fluxes, transport distances, and operating lifetimes. The current state-of-the-art of heat pipe development is reviewed in terms of these requirements with emphasis on experimentally demonstrated performance. The state of knowledge of transient heat pipe behavior is discussed and current modeling efforts reviewed. Lifetime test data for high temperature heat pipes is summarized with some discussion of compatible envelope material/working fluid combinations. Ongoing development of light-weight material combinations such as carbon-carbon is reviewed. Current plans for microgravity testing of liquid metal heat pipes are summarized. It is concluded that there have been only minor changes in heat pipe requirements and demonstrated capability for space power applications over the last 10 years.

INTRODUCTION

In 1985, state-of-the-art of high temperature heat pipes for space nuclear power systems was summarized in Merrigan (1985). That technical paper discussed potential applications of high temperature heat pipes in space power systems and categorized those applications in terms of their demands on heat pipe technology (for example, operating temperatures, radial and axial heat fluxes, transport distances, and operating lifetimes). The paper considered the heat pipe technology available to meet those requirements in terms of wick and enclosure materials, working fluids, fabrication, and processing methods. The state of knowledge of transient heat pipe behavior was discussed and lifetime test data summarized. When we review these requirements and their solution in hardware in terms of today's requirements, we find that while the limits have generally grown there has been no fundamental change in either the requirements or the technology available to meet the requirements. With the expansion of proposed space nuclear power system designs to include thermionic systems and active conversion systems such as the Stirling or Brayton cycles in addition to thermoelectrics, the overall range of heat pipe operating temperatures of interest has widened from the 700 K to 1400 K range considered in 1982 to the 500 K to 2000 K range. These temperature ranges are considered in terms of heat rejection system applications covering the range from about 500 K to 1000 K and primary heat transport applications, covering the range from about 1000 K to 2000 K. Table 1 summarizes the current range of design values. The bulk of the research efforts conducted in the last 10 years has been focused on the lower of these ranges with applications in heat rejection radiators predominating. Design values for axial and radial power densities have remained nearly constant, with the boundaries established by the experimentally demonstrated limits. These demonstrated limits have increased

TABLE 1. Space Power Heat Pipe Performance Requirements.

	Primary Ht. Trans.	Radiator Ht. Trans.
Temperature:	1000 to 2000 K	500 - 1000 K
Axial Power Density:	10 to 15 kW/cm^2	0.5 to 2.0 kW/cm^2
Radial Power Density:	50 to 150 W/cm^2	2 to 50 W/cm^2
Transport Distance:	1 to 5 m	1 to 10 m
Temperature Difference:	< 10 K	< 20 K

fractionally over the intervening 10 years. For example, radial flux performance of potassium heat pipes has been demonstrated to values of greater than 100 W/cm^2, a factor of 3 or 4 greater than had been demonstrated in 1983. (Woloshun et al. 1990). Similarly, the absolute value for axial heat flux has been raised from 19 kW/cm^2 to about 24 kW/cm^2 (Merrigan et al. 1986). However, these increased values have not significantly affected the designs proposed for space power systems.

Experimentally demonstrated operating lifetimes of alkali metal/refractory metal heat pipes have increased significantly with sodium-molybdenum tests having been run to more than 50,000 hours at 1400 K, better than double the demonstrated life of 10 years ago. (Lundberg 1987 and Ernst 1985) More significantly, some container material/working fluid combinations of interest for space radiator heat rejection applications, such as titanium-potassium, for which essentially no data existed in times past, have been verified in tests for periods of greater than 10,000 hours. (Sena and Merrigan 1990) Thermochemical modeling techniques, for use in predicting corrosion limited heat pipe lifetimes, have been developed and used for heat pipe operating life predictions. (Feber and Merrigan 1987). There has, however, been little correlation of the thermochemical predictions of mass transport with experimental data (Alger 1992).

TRANSIENT HEAT PIPE MODELING

Much of the heat pipe research that has been conducted over the past 10 years has been directed to improving the understanding of heat pipe behavior, rather than to raising the demonstrated performance limits. In the field of transient performance prediction, in particular, a large number of attempts to model heat pipes have been undertaken. These efforts have had mixed success, with no single model available at present that meets the general requirements for rapid, accurate, general purpose and performance prediction. Work on the development of transient models continues as an understanding of the heat pipe start-up, operational transient, and shutdown behavior is necessary for their implementation in space reactors. Liquid metal heat pipe performance limiting mechanisms during power transitions in space nuclear power systems include evaporator entrainment, freeze out of the working fluid inventory in the condenser, evaporator capillary limits, nucleate boiling departure in the evaporator, thermochemical changes inhibiting re-wetting of the wick, and radial sonic limits in the evaporator. Recently, a number of investigators have developed computational models of various aspects of liquid metal heat pipe transient behavior. To date, most models have characterized basic transient heat pipe physics. Little work has been done to characterize actual heat pipe transient performance limits. Generally, the models employ different sets of assumptions and are useful in different circumstances. A wide variety of approaches to numerical solution have been attempted. Most of these numerical solutions have been either explicit or implicit finite difference formulations.

Their results are documented in numerous reports and journal papers. A brief summary of works related to liquid metal heat pipe transient behavior follows. Journal cites are for the most recent or the most extensively documented papers readily available in the literature.

Costello et al. (1987) used the Kachina numerical technique in a one dimensional, transient, compressible vapor space solution. The SIMPLER numerical method was used for the wick region in a circumferential-axial incompressible liquid flow solution. The Kachina method is an iterative semi-implicit scheme while the SIMPLER method is an explicit numerical scheme. The vapor space was linked with the liquid region with a radial submodel. The transition in the vapor region from free molecule to continuum flow was handled with a Knudsen number based correction to the laminar flow friction factor. The resulting code was very slow in execution due to its overall explicit numerical formulation.

Jang and Colwell (1987) developed a numerical model of heat pipe start-up from the frozen state related to hypersonic vehicle leading edge cooling. This modeled was designed for integration into a finite element structural analysis code of airfoil-shaped leading edge. An explicit finite element technique was used to solve the conduction in the leading edge structure. Conduction and melting effects were modeled in the wick. In the vapor space solution, separate one-dimensional, quasi-steady governing equations were used for regions in free molecule, choked, and compressible continuum flow. A Runge-Kutta technique solved the governing equations. Mass coupling between the wick and the vapor space was computed with kinetic theory.

Seo and El-Genk (1989) developed a model of an annular wick heat pipe. The model assumed a molten liquid metal working fluid. An implicit finite difference formulation was used to solve the governing equations. The spatial variables were discretized by a central difference method. The time variables were discretized by a backward difference method. A steady state, axially one-dimensional approximation was used in solving the energy, mass, and momentum equations in the vapor region. The liquid region was modeled in the axial and radial directions. The interphasic pressure difference was computed using the vapor void fraction value in the wick pore in a similar manner to Hall (1988). This code calculated the capillary limit, the sonic limit, the boiling limit, and the entrainment limit. The results of this study were compared with data from Merrigan (1986).

Hall and Doster (1990) developed an implicit first order accurate scheme which is one dimensional and compressible in the vapor space. The vapor space and the liquid region were coupled with a radial sub model. A block tri-diagonal solution solved for blocks of 15 unknowns with 15 equations at each axial location. The unknowns included all transport variables. The interphasic pressure difference was computed using the vapor void fraction value in the wick pore. The pressure solution was linearized with a truncated Taylor series expansion. The mass interchange between the liquid and vapor regions was modeled using kinetic theory. The transition between the free molecule flow regime and continuum flow was handled with a Boltzmann equation based diffusion relation in the continuity and energy equations. Accommodation coefficients in the condenser and evaporator regions were adjusted during validation. The resulting code was compared with data from Merrigan (1986).

Issacci et al. (1991) developed a model of a heat pipe vapor space with a rectangular cross section during startup. A detailed treatment of compressibility effects included a treatment of the shock front. The vapor model was two-dimensional using a cartesian coordinate system. Discretization of the convective terms was calculated with a center-differenced finite difference scheme with a nonlinear filtering technique. Multiple wave reflections were reported in the evaporator and adiabatic regions. The vapor space time response constant was found to be on the order of seconds.

Cao and Faghri (1992) used the SIMPLE explicit numerical scheme to model the startup of a heat pipe from the frozen state. The heat pipe region in the free molecule flow regime was modeled with a self-diffusion relation in the continuity and energy equations. The continuum regime was modeled in the axial and radial directions with two-dimensional continuity, momentum, and energy equations. The results of this study were validated using a sodium-stainless steel heat pipe. No transient performance limits were tested with this heat pipe data.

MATERIALS

Another area of heat pipe research that has received much attention in the last decade is that of mass reduction, particularly for heat pipes in the refractory metal temperature range. Refractory metals have been considered for radiator applications in the <1000K range as well as for primary heat transport because of the concern for survival of space power systems under hostile thermal threats. The search for lightweight high temperature heat pipe systems has been focused on two approaches. The first considered is reduction in enclosure thickness without change in the enclosure material used. This approach has led to membrane heat pipes in niobium and molybdenum materials, Trujillo et al. (1990) and Woloshun et al. (1986) (Figure 1). The limiting characteristic of this approach (other than fabrication difficulties) is the danger of penetration of the enclosure envelope by hypervelocity impact of debris or natural micro-meteoroids. As the enclosure material thickness is decreased, the area of an individual heat pipe element must be decreased to minimize the probability of impact and the resultant loss in radiator area per impact. This approach to radiator heat pipe survival may not result in a minimum weight radiator and consequently other means of providing protection to the heat pipes have been considered. These methods have included use of thermally transparent cover materials as bumpers and the use of high strength woven ceramics as armor (Schuller et al. 1991). The use of such cover materials in space radiator heat pipe applications entails some performance penalty because of the reduced effective thermal conductivity of the heat pipe wall material. Figure 2 illustrates the performance penalty measured for a sodium heat pipe operated with and without a woven ceramic cover. A second approach that has received considerable attention is the use of a lightweight, non-metallic heat pipe envelope material such as silicon carbide or carbon-carbon with some thin metallic liner provided to ensure compatibility with the working fluid. The liner is necessary because the alkali metal working fluids will generally attack the non-metallic enclosure material leading to rapid failure of the heat pipe. This approach has been actively pursued by NASA for the SP-100 Advanced Radiator Program, (Rovang et al. 1992). In this effort, CVD metal deposition has been attempted as a method of applying the protective inner liner. More recently, the effort has been redirected to investigation of brazed-in-place metal foil liners used in carbon-carbon tubes. Much work remains to be done in the development of techniques for attaching end closures to the tubular elements, ensuring that carbon diffusion through the thin metallic liner is controllable, and demonstrating that the resulting assemblies will survive extended operation and thermal cycling. The state-of-the-art of these lightweight heat pipe developments is such that system designers generally feel confident in predicting ultimate radiator designs with specific performance in the 5 kg/m^2 range, however such performance has not yet been demonstrated.

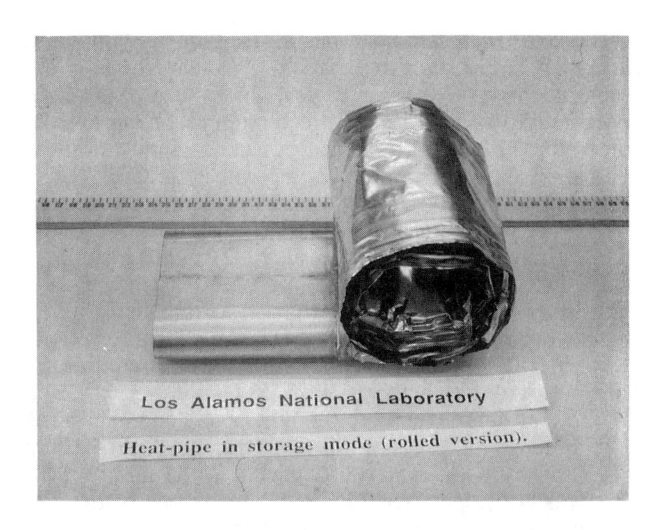

FIGURE 1. A Deployable Membrane Heat Pipe using Potassium Working Fluid. Shown in Storage Mode, it is Self Deploying Under Load.

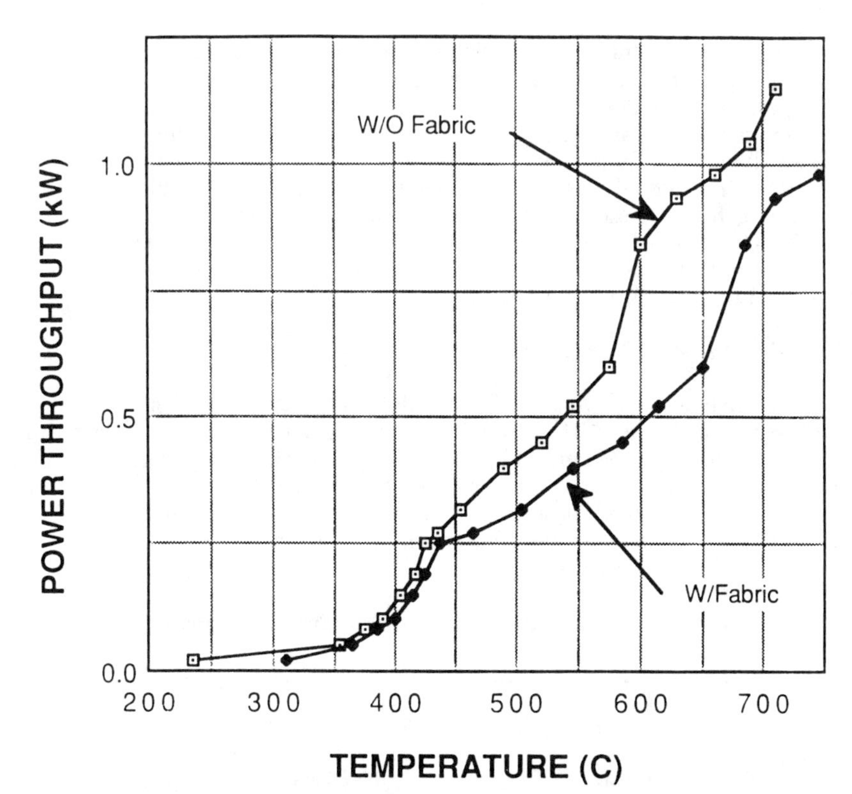

FIGURE 2. Sodium Heat Pipe with High Emissivity Coating Performance With and Without Carbon Fabric Cover.

Enclosure Materials

The traditional choice of enclosure materials for the range of temperatures of interest for primary heat transport has been the refractory metals Nb, Mo, W, Ta, and their alloys. New developments over the last 10 years have been primarily directed to the evaluation of specific alloys for increased creep strength or improved properties. In the range of temperatures of concern for heat rejection applications, there has been more interest and more development of composite structures for reduced mass. Various approaches have been investigated, including ceramic fabric reinforced metal foils, (Kiestler et al. 1992) layered metal-oxide composites, (Rosenfeld et al. 1991) metal lined carbon-carbon structures (Hunt et al. 1992) and bi-metallic structures such as the niobium tube-beryllium fin assemblies proposed for the SP-100 radiator heat pipes. In general, the approach has been to use a thin metal inner layer that is compatible with the working fluid (typically potassium at heat rejection temperatures) and to reinforce the outside of the metal liner with a high strength/high conductivity material. Problems involved in the use of these composite assemblies are in maintenance of good thermal conduct between the layers and in controlling the inter diffusion of the elements of the system through the thin layers. Perhaps the most straight-forward of these mixed material concepts is the niobium lined beryllium configuration that was considered for SP-100. This is one of the lightest possible material combinations and is limited in application primarily by the difficulties of fabricating and joining the beryllium material.

Working Fluids

In the higher temperature portion of the operating range for space nuclear power system heat pipes, there has been no significant change in heat pipe operating fluids over the past 10 years despite some work on alternatives, such as lead, for special purpose applications. (Merrigan 1990). The primary working fluids of interest are still the alkali metals, lithium, sodium, potassium, and cesium. In the range from 500 K to about 700 K, normally reserved for mercury, some alternative fluids are currently under investigation. These include sulfur, iodine, water, diphenyls, and fluorinated biphenyls. Thermal stability, operating capacity, and heat pipe internal pressure is a concern with all of these non-metallic fluids (Grzyll 1991). Water is limited by thermal stability concerns to about 550K.

Wick Materials and Structures

Wick requirements for high performance, high temperature liquid metal heat pipes are summarized in Table 2. These requirements have not changed significantly in the past 10 years, although the current interest in thermionic power systems has lead to a decrease in minimum pore sizes of interest. Some special configurations, such as that used for the SCEPTRE design and in capillary pumped electro-chemical systems such as AMTEC, generate unique requirements for reduced pore size at the cost of lower permeability (Anderson 1992).

TABLE 2. Wick Structure Requirements for Space Power Heat Pipes.

Pore Diameter:	10 to 50/mm
Permeability:	6×10^{-12} m^2 artery
	6×10^{-14} m^2 annular
Other:	1. Bendable/Flexible
	2. Arteries incorporated
	3. Complex shapes

Prior to 1983, a number of methods of producing wick structures, having an effective pore size in the 20 to 50 micron range, had been investigated. These investigations were triggered by the fact that there were, at that time, no commercially available materials, such as screens or felts, in the required fine pore sizes. In the early SP-100 effort, Los Alamos investigated screen compaction, CVD overcoating, sintered powder metal, and sintered microsphere wicks before developing 400 mesh Mo-41 Re screen wick material (Merrigan et al. 1982). Since that time, work has continued on most of these wicking concepts with the notable addition of etched pore foil wick structures developed by General Electric for the SP-100 program. Thermacore Corp. has concentrated on sintered metal powder wick applications (Ernst et al. 1985) with screen separators used at artery interfaces in some cases to suppress boiling in the arteries. The sintered metal fabrication technique has advantages in adaptability to irregular surfaces and in the ability to achieve pore sizes in the <10 micron range. Limitations of this wick configuration are in the difficulty of achieving a narrow range of pore sizes and in the low permeability of the resulting structures when the fine powders necessary to produce small pore sizes are used. An additional restriction, in cases where weight minimization is essential, is the need to maintain minimum section thicknesses of approximately 10 times the particle size to insure consistent pore size. The thick wick and artery configuration resulting from this requirement is illustrated in Fig. 3, which shows a niobium microsphere wick developed by Wah Chang for the leading edge cooling of the National Aerospace Plane (Wojcik et al. 1991). These thick wick sections provide a secondary benefit in structural integrity, providing a wick structure that will generally retain its configuration through repeated melt freeze transients.

Minimum weight is the primary attraction of the etched foil wick configuration chosen by GE for the SP-100 radiator heat pipes (Kirpich 1986). These wicks were developed with both single and double layers of 1 mil foil that was etched with a grid of identical pores. Figure 4 shows a typical GE produced SP-100 potassium radiator heat pipe wick. The heat pipes, using this wick, are generally built in an annular

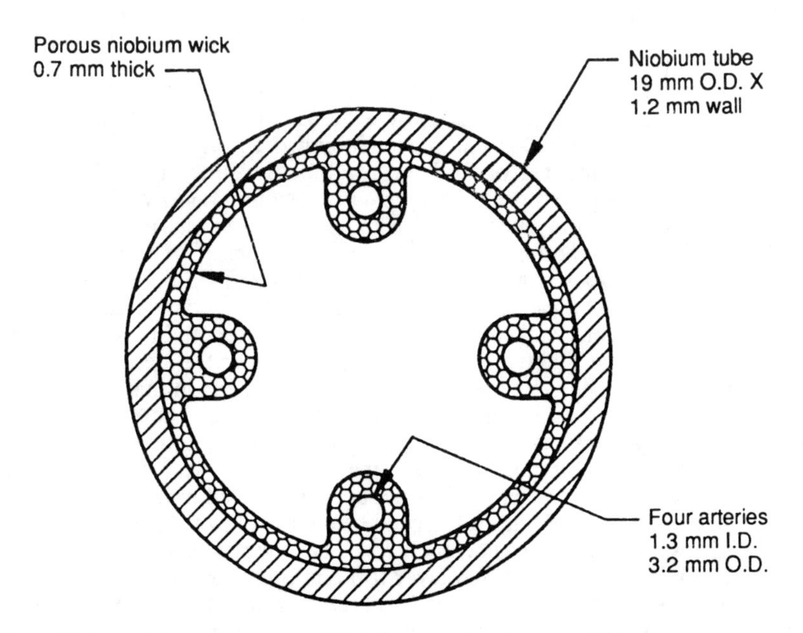

FIGURE 3. Niobium Microsphere Artery Wick Structure Developed by Wah Chang.

HEAT REJECTION SUBSYSTEM

SINGLE FOIL HEAT PIPE CONDENSER END

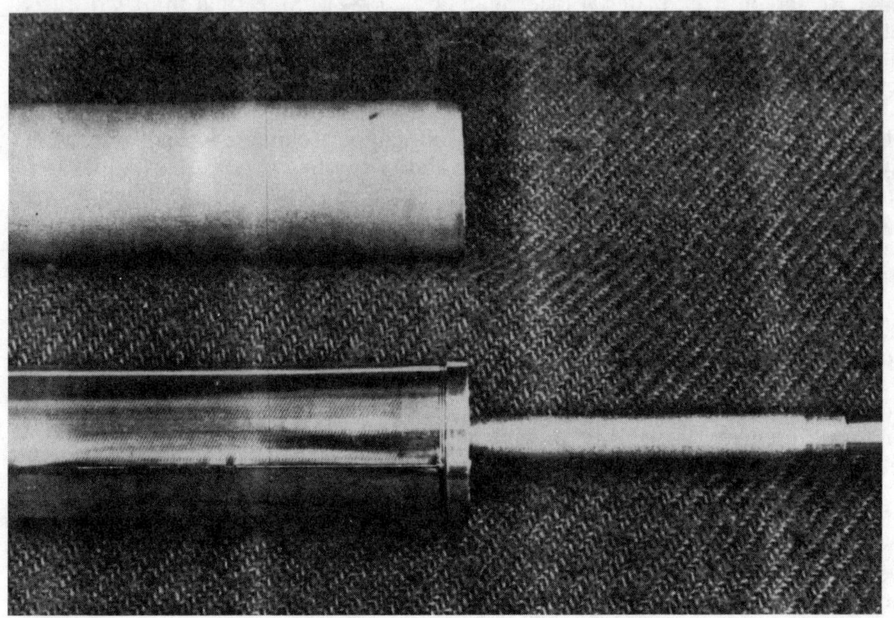

TP0617/A123/P-465

FIGURE 4. Etched Foil Wick Developed by GE for the SP-100 Radiator Heat Pipes.

artery configuration with the foil layer supported at the ends of the heat pipe. This wick provides what is essentially a single size of pore, within manufacturing tolerances. The wick material is produced by photo-resist etching techniques and has been fabricated in stainless steel, titanium and niobium-1 Zr. Limitations of this wick structure are the result of the lack of in-plane (axial) permeability, the reduced area of liquid-vapor interface (approximately 25%) and the extreme fragility of the thin foils. The lack of longitudinal permeability is potentially of concern in start-up and initial wet-in of the heat pipes, although no problems have been observed under 1-G start-up conditions with annular wick configurations. The reduced liquid-vapor interface area seems to impose a limit on radial heat flux below that demonstrated for similar configurations with screen wicks. The lack of physical strength in this type of wick is a concern in applications that require repeated shut-down and re-start because of the high local forces that can be developed due to liquid expansion in melt-freeze cycles. As used in the annular configuration for SP-100, there may be concern for the possibility of the wick buckling into the vapor space with consequent changes in the fluid inventory required for normal heat pipe operation. As with the issue of longitudinal permeability, this concern has not been evidenced in laboratory tests. In threat response testing of GE built, foil wicked heat pipes at Los Alamos, no physical disruption of the wick structure has been observed despite repeated dry out of the test heat pipes. The wick structure, a two layer foil configuration, appears undistorted, despite this heat pipe having been started from the frozen state and taken to a radial flux induced dryout repeatedly during the test program (Sena and Merrigan 1992).

The state-of-the-art of screen wicks has not changed significantly during the ten year period of this update. In 1983 the finest refractory metal wicks available were those developed for the SP-100 program. These 400 mesh, Mo-41Re screens were woven in Europe from wire drawn by Plansee Corporation of Austria. A total of about 150 ft^2 of the material was delivered and was of uniformly high quality although expensive (more than $100/$ft^2$). As far as is known, no finer refractory screen wick material has been produced to date. At present, NASA is in the process of duplicating these materials for use on the NASP leading edge cooling heat pipes. These fine screen wicks offer many advantages as heat pipe wicking material. They can be inspected and their performance verified prior to installation in the heat pipe, they have generally very consistent fluid properties. Their performance is predictable, and in multiple layer, sintered structures and have excellent structural characteristics. Their primary limitations are availability and the difficulty of forming them into complex shapes.

Lifetime Demonstration

The accumulation of life test data for high temperature heat pipes has continued with the total time at temperature for single heat pipes exceeding 50,000 hours for Mo/Na (Lundberg 1987) and the total exposure hours for some material combinations exceeding 100,000 hours. (Sena 1990) A significant period of exposure has now been accumulated for Ti/K and Nb-1Zr/K, material combinations of interest for heat rejection applications in space power systems. Table 3 presents a summary of radiator heat pipe life tests conducted at Los Alamos for the SP-100 program. It is worth noting that, despite the comparatively high evaporator radial heat flux values used in these tests, no failures have been observed. Where total hours are indicated in the table, the tests were terminated as scheduled without failure. The Ti/K heat pipes used for these tests were fabricated by GE as part of the SP-100 heat rejection system development program and are prototypic in configuration with etched foil wicks and thin wall containment.

The low failure rate in these tests may be interpreted as evidence of low contaminant levels in the heat pipes at fabrication. Thermo-chemical models of the corrosion processes in high temperature heat pipes indicate that concentration of contaminants in the evaporator region of the heat pipes is the most common cause of failure. (Lundberg et al. 1984) In superalloy heat pipes, using alkali metal working fluid, condenser solubility of the wall materials, primarily nickel, in the working fluid has been identified as an alternative failure mode.

With current interest in mass reduction of space power heat rejection systems, through the use of composite enclosure materials, such as metal lined carbon-carbon or titanium-alumina, the concern for lifetime demonstration should shift to concerns of diffusion limited material transfer to the working fluid and the evaluation of barrier coatings for control of carbon and oxygen transfer into the heat pipes. These heat pipe concepts are generally still in the fabrication development stage. However, as the fabrication issues are resolved, lifetime demonstration programs should be initiated.

MICROGRAVITY PERFORMANCE TESTING

Low temperature heat pipes have been repeatedly tested and used in microgravity environment since the first water heat pipe was demonstrated in space in 1967 (Deverall 1967). However, in spite of almost 30 years of heat pipe research efforts, no liquid metal heat pipes have operated in a micro-gravity environment. Steps are now underway to fill this gap in the technology of space nuclear power. Three liquid metal heat pipes have been designed, fabricated, and tested for a planned Space Shuttle

TABLE 3. Niobium-1% Zr/Potassium Heat Pipes (1.52 cm Diameter).

HEAT PIPE NUMBER	LENGTH (cm)	INPUT HEAT FLUX (kW/m^2)	TEMPERATURE (K)	HOURS OF OPERATION
LHPK-1	46	189	838	14,071 (total)
LHPK-3	73	181	873	13,885 (total)
LHPK-4	30	152	870	10,042
LHPK-5	38	138	830	10,042
LHPK-6	48	152	805	10,042
LHPK-7	61	253	873	13,213
LHPK-8	76	290	833	13,213
LHPK-9	97	300	795	13,213
TITANIUM/POTASSIUM HEAT PIPES (1.9 cm diameter)				
MC-2	47.6	187	885	11,198
MC-3	93.31	226	820	11,198
MC-5	93.06	183	805	11,198

experiment under a program conducted jointly by Los Alamos Laboratory and the USAF Phillips Laboratory. (Woloshun et al. 1993). The heat pipes to be flown use potassium working fluid with stainless steel wick and envelope structures. Three wick structures are being tested; homogeneous, annular gap, and arterial. A summary of the heat pipe designs is given in Table 4. In the planned shuttle experiment, the heat pipes will be contained in an HHG canister as shown in Fig. 5.

The principal objective of the high temperature microgravity heat pipe experiments is to demonstrate and characterize heat pipe start-up from a frozen working fluid state to the fully operating heat pipe at a steady-state, isothermal condition. This will be followed by cooldown to the frozen state, and a repeated frozen start. Power throughput for the heat pipes will be approximately 350 W.

Acknowledgments

This research was sponsored and performed under the auspices of the U.S. Department of Energy and conducted at Los Alamos National Laboratory. The U.S. Government retains a nonexclusive, royalty-free license to publish or reproduce the published form of this contribution, or to allows others to do so, for U.S. Government purposes.

References

Alger, D. L. (1992) "Heat Pipe Heat Transport System for Stirling Space Power Converter," Sverdrup Tech., Inc., *Presented at the 24th IECEC, Report No. 929399* San Diego, CA, August 1992.

Anderson, W.G. (1992) "Sodium Wide-Pumping Experiments for a Vapor-Fed AMTEC System," *Presented at the 24th IECEC, Report No. 929146* San Diego, CA, August 1992.

TABLE 4. Microgravity Heat Pipe Design Summary.

ALL:	Potassium working fluid		
	SST tube and wick		
	2.30 cm OD		
	.089 cm wall thickness		
	2.12 cm ID		
	61 cm long:	10 cm evaporator	
		5 cm adiabatic	
		46 cm condenser	
HOMOGENEOUS WICK:			
	Wick thickness		0.226 cm
	Pore Radius:		0.013 cm
	Permeability:		0.0002 cm^2
ARTERIAL WICK:			
	2 arteries (one redundant)		
	1.5 mm diameter arteries		
	Wick thickness:		0.03 cm
	Pore Radius:		0.005 cm
ANNULAR GAP WICK:			
	1 mm annulus		
	Wick thickness:		0.03 cm
	Pore Radius:		0.005 cm

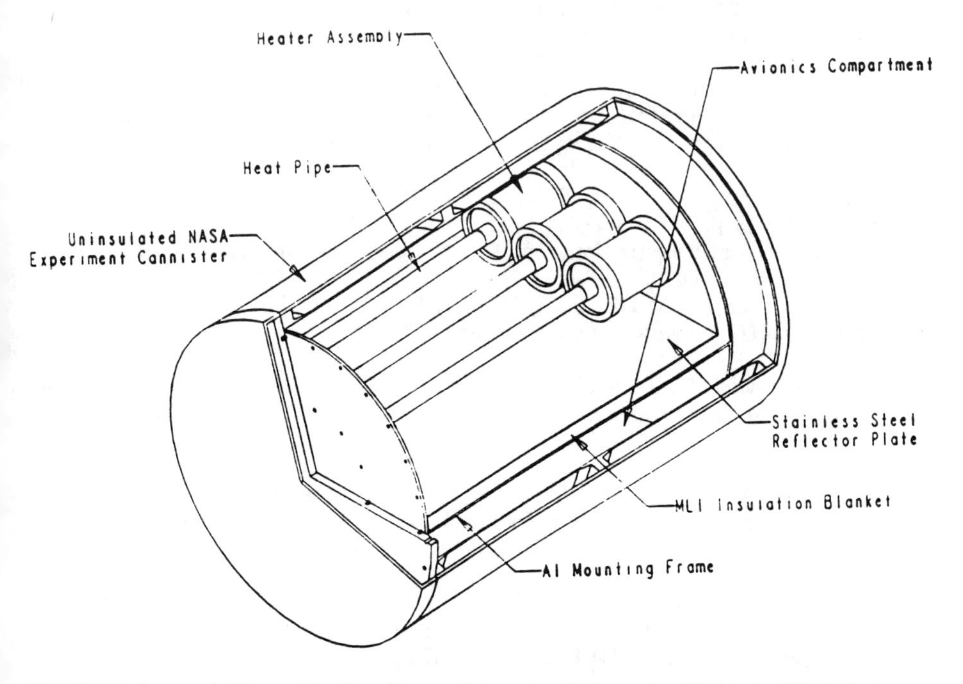

FIGURE 5. Isometric View of the Heat Pipes and Associated Electronics Within the HH-G Canister.

Cao, Y. and Faghri, A. (1992) "Analyses of High Temperature Heat Pipe Startup from the Frozen State - Part 1: Numerical Simulation," *Proceedings of the ASME National Heat Transfer Conference*, San Diego, CA, August 1992.

Costello, Frederick A., Montague, Allen F., Merrigan, Michael A., and Reid, Robert S. (1987) "Detailed Transient Model of a Liquid-Metal Heat Pipe," *Presented at the 4th Symposium on Space Nuclear Power Systems 1987.* Published in the Transactions of the Fourth Symposium on Space Nuclear Power Systems, edited by M.S. El-Genk and M.D. Hoover, Orbit Book Company, Malabar, FL, 1988, I: 197-204.

Deverall, J.E., and Kemme, J.E., "High Thermal Conductance Devices Utilizing the Boiling of Lithium or Silver," Los Alamos National Laboratory, Report No. LA-3714, June 5, 1967.

Ernst, D.M. and Eastman, G.Y. (1985) "High Temperature Heat Pipe Technology at Thermacore - An Overview," *AIAA 20th Thermophysics Conference*, Williamsburg, Virginia 1985.

Feber, Roy C. and Merrigan, M.A. (1987) "Thermochemical Modeling of Mass Transport in High-Temperature Heat Pipes, *VI International Heat Pipe Conference*, Grenoble, France (May 25-28, 1987).

Grzyll, (1991) "Investigation of Heat Pipe Working Fluids for Thermal Control of the Sodium/Sulfur Battery," *26th IECEC*, Boston, Massachusetts, August 1991.

Hall, M.L. and Doster, J.M. (1989) "THROHPUT Computer Code (Thermal Hydraulic Response of Heat Pipes Under Transients)," North Carolina State University, Nuclear Engineering Department.

Hall, M.L., (1988) "Numerical Modeling of the Transient Thermohydraulic Behavior of High Temperature Heat Pipes for Space Reactor Applications," North Carolina State University for the U.S. Department of Energy (Ph.D thesis), 1988.

Hunt, M.E., Rovang, R.E. and Palamides, T.R. (1992) "An Innovative Carbon-Carbon Heat Pipe Radiator for Space Nuclear Power Applications," Nuclear Technologies for Space Exploration - 92, Jackson, Wyoming, August 1992.

Issacci, F., Catton, I. and Ghoniem, N.M. (1991) "Vapor Dynamics of Heat Pipe Startup," *ASME Journal of Heat Transfer*, 113, 985-994.

Jang, J. H., Colwell, G. T., and Camarda, C.J. (1987), "Modeling of Startup from the Frozen State," *proceedings of the International Heat Pipe Conference*, Grenoble, France, pp. 165-170.

Kiestler, W.C., Marks, T.S. and Klein, A.C. (1992) "Design and Testing of Fabric Composite Heat Pipes for Space Nuclear Power Systems," Nuclear Technologies for Space Exploration - 92, Jackson, Wyoming, August 1992.

Kirpich, A., Biddiscombe, R., Chan, J., and McNamara, E., (1986) "Comparison of Concepts for a 300 kWe Nuclear Power System," presented at the *IECEC 86 Conference*, Paper No. 869484 held in San Diego, CA (August 25, 1986).

FUELS FOR SPACE NUCLEAR POWER AND PROPULSION: 1983-1993

R. Bruce Matthews, Ralph E. Baars, H. Thomas Blair, Darryl P. Butt, Richard E. Mason,
Walter A. Stark, Edmund K. Storms, Terry C. Wallace
Los Alamos National Laboratory
PO Box 1663
Los Alamos, NM 87545
(505) 667-2556

Abstract

Ten year's progress in the fabrication and performance of nuclear reactor fuels for space power and propulsion applications is reviewed. Thermochemical and engineering properties for fuel materials in the uranium-oxygen, uranium-carbon, and uranium-nitrogen systems are examined, and the known phase diagrams of these systems are presented. The analyses are extended to the uranium-refractory metal ternary (U,Zr)C and (U,Nb)C, and the quaternary (U, Nb, Zr)C systems. Historical fabrication processes are reviewed for these fuel forms, and recent advances are highlighted.

Known ex-pile and in-pile performance testing of pin type uranium oxide fuel for thermionic fuel elements and uranium nitride for the SP-100 space power reactor is presented. Swelling and fission gas release trends with temperature and burnup are discussed, as is the compatibility of the fuel and cladding with fission gas release; changes in chemical and physical properties caused by fission product metals or radiation are of less concern because of the short operating times. The major performance limiting phenomena are melting of the fuel, mass loss caused by vaporization and chemical interaction with the propellant, and thermal-stress-induced failure. Progress in modeling these phenomena is discussed.

The developmental requirements for future space nuclear power and propulsion in the next decade are proposed, and the challenge is to combine the best attributes of the different known fuel forms into those types having the optimum combination of safety and performance characteristics is issued.

INTRODUCTION

Uranium oxides, carbides, and nitrides have been developed and tested for space nuclear propulsion and power reactors. Desirable fuel characteristics include high density, high thermal conductivity, high melting point, high temperature stability, chemical compatibility, predictable irradiation performance, and ease of fabrication. In general, oxides have predictable irradiation performance and are the easiest to fabricate; however, they have relatively low thermal conductivity and density, and they react with liquid metals. During irradiation, oxides release more fission gas but swell less than carbides or nitrides. Carbides tend to swell more than nitrides, while nitrides are less stable than carbides and are more difficult to fabricate. Table 1 summarizes properties and characteristics of oxide, nitride, and carbide fuels. The combination of high uranium density and high thermal conductivity make the overall performance capability of carbides and nitrides relatively more attractive than the performance of uranium oxide.

Uranium oxide fuel has been used extensively in Russian thermionic reactors (Bennett 1989) and for the US thermionic program (Brown 1992). The thermionic fuel element verification program was initiated in 1986 to demonstrate technology readiness of a thermionic fuel element suitable for use in a thermionic space reactor with a full-power lifetime of seven years. Technical issues include high temperature fuel swelling with subsequent emitter distortion and compatibility between the fuel, metallic fission products, and tungsten emitter at 1800 K.

Performance feasibility of UO_2 in thermionic fuel elements is well established; however, operation has been limited to one year and 4 atomic % burnup at the high operating temperatures required for efficient energy conversion (Lawrence 1991).

TABLE 1. Characteristics of Space Reactor Fuels.

Characteristic	UO_2	UN	UC	UC_2	$(U_{0.2}Zr_{0.8})C_{0.99}$
U Density,g/cc	9.66	13.52	12.97	10.60	2.88
Melt Point, K	3100	3035	2775	2710	3350
Thermal Cond. W/mK	3.5	25	23	18	30
Relative Stability	Moderate	Low	High	High	High
Relative Swelling	Low	Mid	Mid	Low	High
Fission Gas Release	High	Low	Mid	Low	Mid
Fabricability	Easy	Moderate	Easy	Difficult	Difficult

Uranium nitride fuel is to be used in the SP-100 liquid-lithium cooled, refractory metal, fast flux reactor. The SP-100 Program is aimed at developing, testing, and demonstrating a 100 kWe reactor for space-based power (Mondt 1989). The main thrust of the project has been design, construction, and testing of nuclear components; qualification and fabrication of UN fuel and fuel pins are virtually complete (Matthews 1992).

A renewed interest in manned exploration of space and the Space Exploration Initiative revitalized interest in the nuclear rocket technology developed during the 1960's and the potential for advancing that technology (Bennett 1991). Nuclear thermal propulsion reactors offer higher specific impulse than conventional chemical rockets because of higher exhaust temperature and the low molecular weight of the hydrogen propellant. Carbide fuel performance, melting point, stability, fabricability, and compatibility are recognized as key technology issues (Clark 1991). Fuel development for the multimegawatt reactors and particle bed reactor will not be discussed here because of the sensitive nature of these reactor programs and the lack of published data.

HISTORY

This paper reviews recent applications of oxides, carbides, and nitrides for space nuclear power and propulsion systems. H. Matzke's (1986) monograph on the "Science of Advanced Liquid-Metal-Fast-Breeder-Reactor Fuels" is an outstanding book that covers all aspects of carbide and nitride fuels.

Uranium Oxide

Uranium oxide, in the form of tungsten metal matrix cermets, was also developed for nuclear rocket fuels during the early 1960's (Kruger 1991). Although cermet fuel is robust, compatible with hot hydrogen, and resistant to fission product release, further development was discontinued in favor of carbide fuels for the Rover Program. Oxide cermets will not be reviewed here because no work has been done during the past ten years.

Uranium Nitride

Uranium nitride (UN) fuels were originally developed for the Systems for Nuclear Auxiliary Power Program;

Bauer (1972) has published a comprehensive review of the properties and performance of UN. Several laboratory-scale techniques to synthesize sinterable UN powders were developed. The most commonly used process was hydriding of uranium metal followed by nitriding (Anselin 1963). A carbothermic-reduction/nitriding process was developed (McLaren, 1968) and demonstrated on a small scale in Japan (Muromura 1977 and 1980). Arc-melting and casting, hot isostatic pressing, and cold-pressing and sintering have been used to fabricate UN fuel pellets. Cold pressing and sintering studies (McLaren 1965 and Tennery 1971) indicated that UN was difficult to sinter to high density.

Irradiation performance of UN fuels clad in Nb-1% Zr, tantalum alloys, and tungsten was examined in thermal reactor testing during the Systems for Nuclear Auxiliary Power Program (DeCrescente 1965 and Weaver 1969). Over 100 helium-bonded pins were irradiated in thermal reactors with various parameters, including: tungsten, tantalum, molybdenum, and niobium alloy cladding; cladding temperatures ranging from 1100 - 2200 K; burnups to 4.5 at.%; fuel densities ranging from 80% - 96% TD; and various UN grain sizes and stoichiometries. The results from these irradiations varied greatly, but general trends can be summarized as follows:

- Niobium alloys were the most viable cladding material for operation below 1400 K; however, tungsten liners were required to prevent chemical interactions between the fuel and cladding.
- Fission gas release rates were low for stoichiometric UN.
- The high creep strength of UN resulted in swelling forces that are not constrained by niobium cladding at high temperatures.
- Lower amounts of fission gas were released from large grain size UN.
- Swelling and fission gas release were very high from hypostoichiometric UN.

Based on this early experience, UN was selected as the reference fuel for the SP-100 space nuclear power reactor. Uranium oxide was not selected for SP-100 because of its low density, low thermal conductivity, and reaction with liquid lithium. Carbide was rejected because of perceived compatibility and swelling problems.

Uranium Carbide

Demonstration of stable high-temperature carbide fuels was the key to the success of the Rover and the Nuclear Engine for Rocket Vehicle Applications Programs (Taub 1975 and Koening 1986). Twenty reactor tests were completed, and performance improved with evolving fuel designs, materials improvements, and fabrication advances. The effect of increasing hydrogen exit temperature on specific impulse is shown in Fig. 1. Specific impulse was calculated with the classical relationship

$$I_{sp} = f(T_c/M)^{1/2}:$$

where

I_{sp} = specific impulse

T_c = coolant exit temperature, and

M = molecular weight of the coolant.

The data points are ideal specific impulse values calculated from hydrogen exit temperatures measured for Rover reactor tests. The I_{sp} values are not specific to a particular nozzle design, but rather show the relative potential for increasing nuclear rocket performance with advanced carbide fuels. These data indicate that a propellant exit temperature as high as 3400 K would theoretically result in a specific impulse greater than 1000 seconds, a significant increase over the ~850 second range achieved for the Rover reactors. However, as indicated by the shaded area, the operating uncertainty increases with increasing temperature thereby reducing operating margins.

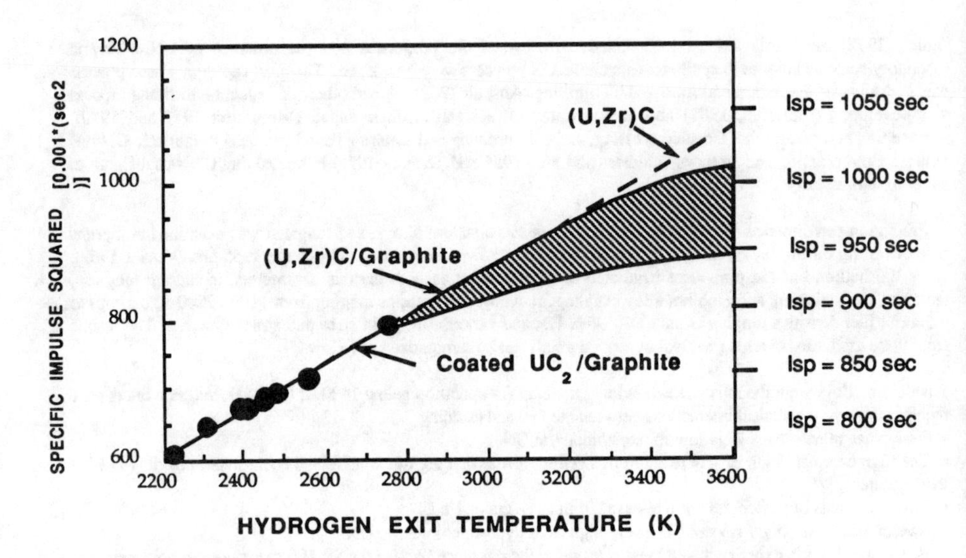

FIGURE 1. Potential Specific Impulse as a Function of Hydrogen Exit Temperature.

The potential operating temperature capabilities of Rover fuel types are indicated with arrows on Fig. 1. Three types of fuel were developed with the goal of increasing operating time and temperature: (1) pyrolytic carbon-coated UC_2 spheres, dispersed in a graphite matrix and coated with NbC; (2) composite graphite/uranium, zirconium carbide fuel coated with NbC or ZrC; and (3) single-phase uranium, zirconium carbide ($(U,Zr)C_x$) fuel. The relevant information about the Rover fuels is summarized in Fig. 2, including a schematic representation of the macrostructures of the three Rover fuel types. During the Rover and Nuclear Engine for Rocket Vehicle Applications Fuel Element Development Programs at Los Alamos, Westinghouse, and Oak Ridge, extrusion was used to produce tens of thousands of fuel rods with both hexagonal and cylindrical cross-sections. Dispersed fuel was the most highly developed type and will be the benchmark for comparing performance of the new generation of nuclear rocket fuels. Early dispersed fuels were UO_2 particles mixed with graphite and a binder. The mix was homogenized and extruded into the typical Rover nineteen hole fuel rods. The fuel rods were baked to remove volatiles and heat treated to convert the oxide to carbide fuel and complete graphitization.

Attempts to eliminate mid-band corrosion led to the development of (U,Zr)C + graphite composite fuel. Because both components were continuous and interconnected, the composite fuel rods were more robust and chemically stable than the bead-loaded fuel rods. Single phase (U,Zr)C was developed to eliminate the carbide/carbon eutectic limitation and thereby permit higher operating temperature (up to 2800 K). Only seven carbide rods were fabricated and tested in the nuclear furnace (Lyon 1973), and insufficient data were generated to fully evaluate performance potential. The solid solution fuel was difficult to fabricate, tended to crack, and showed some corrosion in flowing hydrogen (MacMillan, 1991). Nevertheless, single-phase carbide fuel has the greatest potential for improving the performance of the next generation of nuclear rockets.

Type	Bead Loaded Graphite	Carbide-Graphite Composite	Carbide
Development Period	1958 -1967	1967 - 1972	1970 - 1972
Fuel Element			
Macrostructure			
Scale reference	20 vol. % particles; dia. = 125 ± 25 μm	35 vol. % carbide; carbide filament cross section 5-20 μm.	Mean grain size, 15 μm.
Fuel Composition	Pyrographite Coated UC_2 Beads	$(U_{0.085}Zr_{0.915})C_{0.98}$ + C Composite	$(U_{0.025}Zr_{0.975})C_{0.958}$ Carbide
Density, g/cm^3	2.30	3.50	5.50
U Loading, g/cm^3	0.400	0.400	0.300
Melting Temperature, K	2725	2900	3200
CTE (300 to 2400 K), μm-/m•K	5.8	6.7	7.8
Thermal Conductivity (300 K), W/m•K	110	80	7
Flexure strength (300K), MPa	35	55	76
Peak Fuel Temperature, K	2600	2525	2525
Exit Gas Temperature, K	2550	2450	2450
Carbon Loss Rate by Hydrogen Corrosion, g/cm^2-s			
NbC Protective Coating	$11.2 \cdot 10^{-6}$	--	--
ZrC Protective Coating	$5.5 \cdot 10^{-6}$	$2.8 \cdot 10^{-6}$	--
Carbide Fuel Form	--	--	$0.6 \cdot 10^{-6}$ [a]

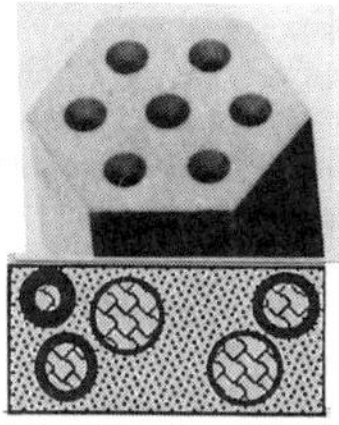

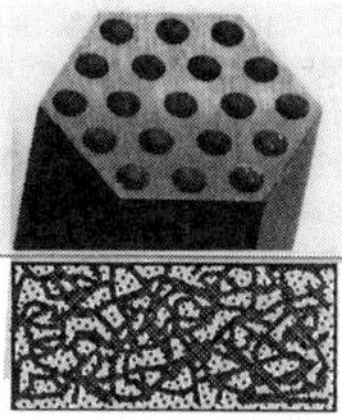

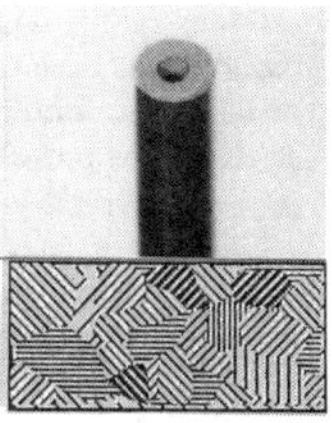

[a] Estimated for fuel temperature of 2525 K from data of McMillan (1991).

FIGURE 2. Evolution of Rover Fuel Types.

Chemical-vapor-deposited NbC and ZrC coatings were developed to protect the graphite fuel rod surfaces and coolant channels from hot hydrogen corrosion (Wallace, 1991). Fabrication and irradiation performance have been well documented (Davidson 1991, Taub 1975, Koening 1986, Kirk 1990, and Lyon 1973), and in general the fuel behavior exceeded expectations. Pyrocarbon coats were used to protect the UC_2 particles from air oxidation; the mixed carbide ternary fuel compounds are stable in air, and need no such protection. Later, NbC and ZrC coatings were used to prevent hydrogen reaction with the graphite matrix; the chemical-vapor-deposited coatings were not intended to be a fission product barrier but did act as a fission product diffusion barrier (Bokor 1990). Pyrolytic carbon coated UC_2 particles dispersed in ZrC-coated graphite operated at 2555 K for 40 minutes in the PEWEE-1 nuclear rocket test (Koening 1986). Two operating limits were defined: the carbon/UC_2 eutectic limiting operating temperature to 2725 K, and "mid-band" corrosion cracking (Lyon 1973). The "mid-band" corrosion was caused by thermal expansion mismatch between the graphite matrix and ZrC coating, subsequent cracking in the ZrC coating, reaction between hot H_2 and graphite, and measurable mass loss from the fuel element.

Carbides have also been tested for fuel-pin-type space power reactors, but development was discontinued because of apparent swelling and compatibility issues (Rankin 1987). In addition, the Russians used UC_2 for the Romashka out-of-core thermionic reactor and some recent work was done for the STAR-C reactor in the U.S. (Chidester 1990). This application of carbides will not be discussed further. Based on the successful performance of mixed (U,Pu) carbides for the liquid metal fast breeder reactor (Matthews 1983), the early rejection of carbides for space power reactors may have been premature. Uranium carbide is relatively easy to fabricate, is stable at high temperatures, can be alloyed to form high melting compounds, and is relatively ductile at operating temperatures. The apparent high swelling was probably caused by formation of hypostoichiometric UC_{1-x}, and the compatibility problems could be controlled with liners, gap size, and carburization resistant alloys.

THERMOCHEMICAL PROPERTIES

Thermodynamic and crystallographic properties data for five uranium-bearing nuclear fuel materials are listed in Table 2. These fuels are discussed in the following sections.

TABLE 2. SelectedThermodynamic and Crystallographic Properties* of Uranium Compounds.

Property	UO_2	UC	UC_2	UN	$U_{.2}Zr_{.8}C_{.99}$
Theoretical Density, g/cc	10.96	13.63	11.68	14.32	8.01
Theoretical Uranium Density, g/cc	9.66	12.97	10.60	13.52	2.88
Lattice Parameter, Å	5.4690	4.9605	5.488**	4.8893	4.758
Crystal Structure	fcc(CaF_2)	fcc(NaCl)	fcc(KCN)†	fcc(NaCl)	fcc(NaCl)
Cation:Anion Coordination Number	8:4	6:6	6:6††	6:6	6:6
Melting Point, K	3100	2775	2710	3035§	3350§§
Free Energy of Formation, kJ/mol (1500 K)	-827	-108	-103	-173	-186
Enthalpy of Formation, kJ/mol (1500 K)	-1092	-110	-95	-307	-206
Heat Capacity, J/mol•°C (1500 K)	71.1	66.5	98.8	65.3	57.8
Debye Temperature, K	160	328	350¶	366	192
Uranium Vapor Pressure, Pa (2000 K)	$4 \cdot 10^{-3}$	$1 \cdot 10^{-5}$	$2 \cdot 10^{-5}$	$2 \cdot 10^{-2}$	$5 \cdot 10^{-6}$
Thermal Expansion Coefficient, µm/m•K, (1273 K)	10.1	11.2	12	8.9	7.6
Thermal Conductivity, W/cm•°C (1273 K, theoretical density)	0.035	0.23	0.1	0.25	0.3

* All properties are room temperature values at 1 atm unless otherwise noted. The data were based on 19 references denoted by asterisks in the Reference section at the end of this chapter.

** At 2173 K.

† Structure is face centered cubic above and face centered tetragonal below ~1765°C.

†† There are two carbon atoms (C_2) located on the anion site of the UC_2 rock salt structure. Thus, the 6:6 cation:anion ratio for UC_2 considers each C_2 "dumbell" to be a single anion.

§ UN dissociates at this temperature, $T_m(K) = 3035(P_{N_2})^{.02832}$.

§§ Solidus Temperature.

¶ Estimate for tetragonal phase of UC_2.

The Uranium-Oxygen System

The most important solid oxides of uranium are UO_2, U_4O_9, U_3O_8, and UO_3, which are shown in the uranium-oxygen binary phase diagram of Fig. 3.

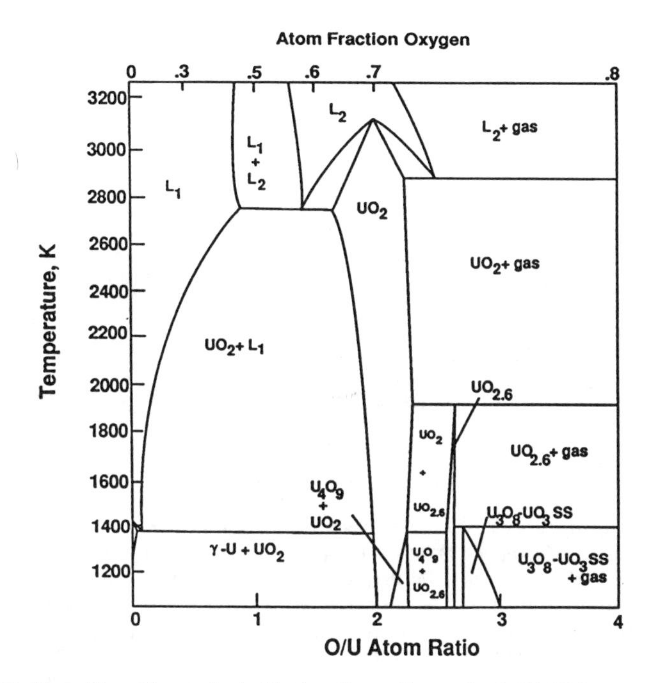

FIGURE 3. Binary Phase Diagram for the Uranium-Oxygen System (Levinskii 1974).

Uranium dioxide exists over a range of composition, from $UO_{1.64}$ to $UO_{2.27}$, because of the occurrence of oxygen vacancies and/or interstitial oxygen atoms in the oxygen sublattice. Uranium dioxide is face centered cubic with a fluorite crystal structure and a lattice parameter of 5.4690 Å. Because of its relatively large negative free energy of formation (-827 kJ/mol at 1500 K), UO_2 is a very stable compound in oxidizing environments. A comparison of the free energy of formation of UO_2 with those for the carbide and nitride materials listed in Table 2 reveals that, in oxidizing atmospheres, the carbides and nitrides are thermodynamically unstable and will therefore be converted to the oxide form or will grow oxide surface films.

The Uranium-Nitrogen System

In the uranium-nitrogen binary phase diagram (Fig. 4), it can be seen that there are three solid compounds which comprise the uranium-nitrogen system: UN, U_2N_3, and UN_2. Recent results (Katsura 1993) suggest that the existence of the UN_2 phase is suspect.

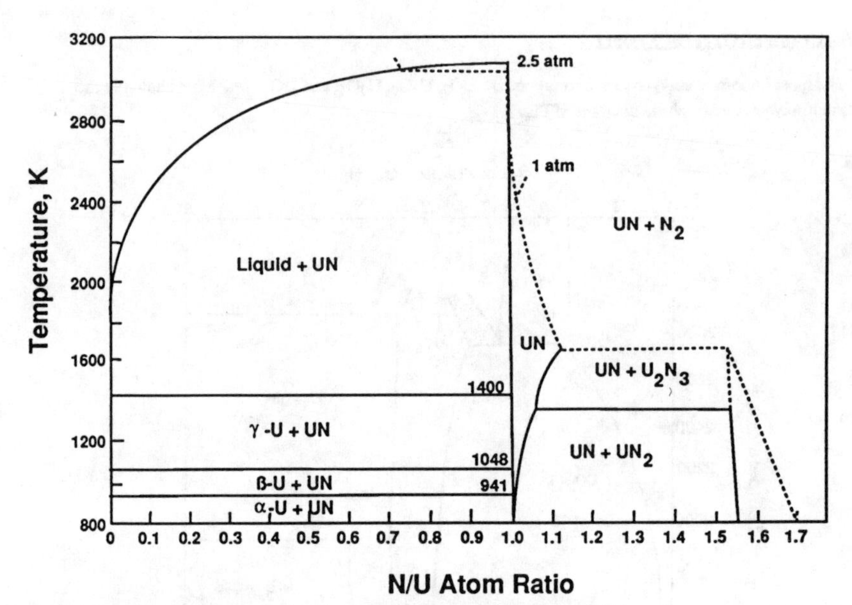

FIGURE 4. Binary Phase Diagram for the Uranium-Nitrogen System (Storms 1988).

Uranium mononitride is a face centered cubic material having a rock salt structure with a lattice parameter of 4.8893 Å. Uranium nitride has a single-phase composition range from $UN_{1.00}$ to a maximum nitrogen content of approximately $UN_{1.14}$ at 1650 K (Storms 1988). The range of stoichiometry narrows rapidly as the temperature is increased above 1650 K and approaches $UN_{1.00}$ at the dissociation temperature of 3050 K. The relationship between nitrogen pressure and UN composition at 1650 K is shown in Fig. 5. Within the single-phase, mononitride phase boundaries, the N_2 overpressure changes nearly one order of magnitude. The pressure remains high and relatively independent of composition in Region II; when Region I is entered, the N_2 pressure drops sharply as nitrogen is removed from the UN lattice. Because of its relatively high uranium and nitrogen vapor pressures, UN has a lower usable temperature compared than the other compounds described in Table 2.

The Uranium-Carbon System

As shown in the uranium-carbon phase diagram (Fig. 6), the uranium-carbon system is a complex system, consisting of three compounds: UC, U_2C_3 and UC_2. Uranium monocarbide and uranium dicarbide are immiscible in one another at temperatures between approximately 2100 and 2350 K, depending on the C/U ratio. At temperatures above 2350 K, UC_2 and UC form a solid solution with stoichiometries ranging between $UC_{0.92}$ and $UC_{1.95}$. At approximately $UC_{1.85}$ and 2673 K, there is a minimum congruent melting point. The UC-UC_2 solid solution is a face centered cubic material having a rock salt structure with lattice parameters ranging from approximately 4.96 to 5.49 Å. The solid solution forms between the two compounds because, rather uniquely, either one or two carbon atoms can be located on the anion site of the UC_x rock salt structure. Uranium monocarbide is slightly more refractory than uranium dicarbide and exhibits somewhat superior thermal expansion and thermal conductivity behavior.

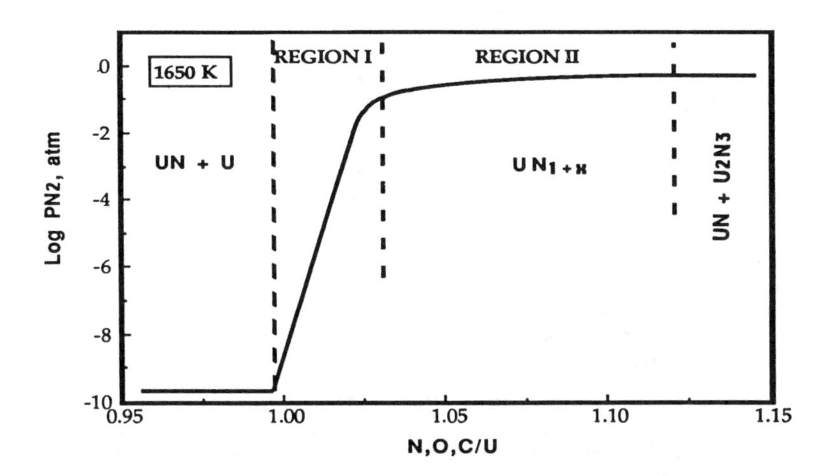

FIGURE 5. Relationship between Nitrogen Pressure and UN Composition and 1650 K. N,O,C is the Sum of Oxygen, Carbon, and Nitrogen; N,O,C/U is the Effective Stoichiometry of UN.

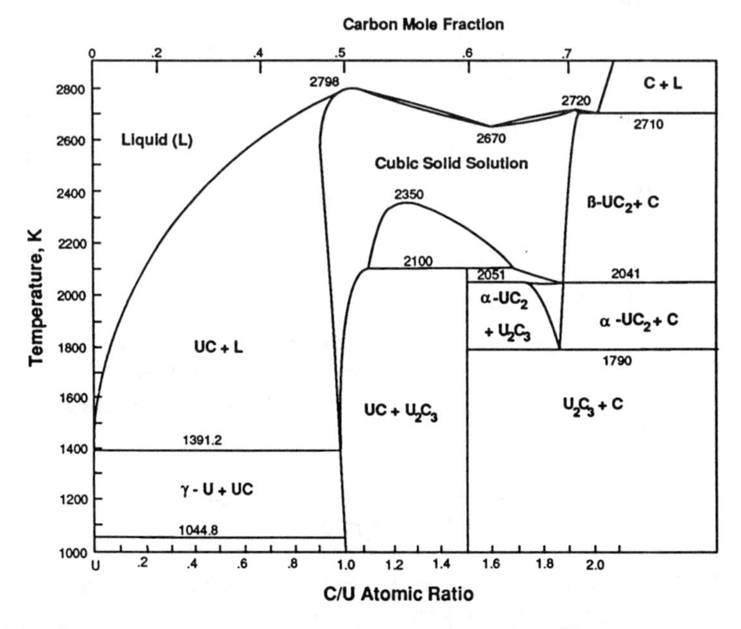

FIGURE 6. Binary Phase Diagram for the Uranium-Carbon System (Holley 1984).

The Uranium-Zirconium-Carbon System

Table 2 also provides thermochemical data on the compound $U_{0.2}Zr_{0.8}C_{0.99}$. This specific composition was included because it illustrates how the melting and vaporization behavior of a uranium fuel can be improved by incorporating uranium into a more refractory compound. Uranium carbide and zirconium carbide (ZrC), which has a melting point of 3693 K, form a nearly ideal solid solution. The extent of the solid solution can be seen in the UC-ZrC pseudobinary solidus-liquidus phase diagram (Fig. 7). The pseudobinary diagram shown is a projection across the ternary phase diagram from $UC_{1.0}$ to the maximum melting point of ZrC_x, and thereby represents the maximum solidus and liquidus temperatures. The specific values between the two end points will change with different end point compositions.

$U_{0.2}Zr_{0.8}C_{0.99}$ has a face centered cubic rock salt structure with a lattice parameter of 4.758 Å, which is smaller than that of UC. In addition to having a relatively high melting point and low equilibrium vapor pressure, the mixed carbide has a lower thermal expansion coefficient and higher thermal conductivity compared with the other compounds shown in Table 2. $U_{0.2}Zr_{0.8}C_{0.99}$ is therefore more refractory and may have superior thermal shock resistance compared with the other uranium compounds in the table.

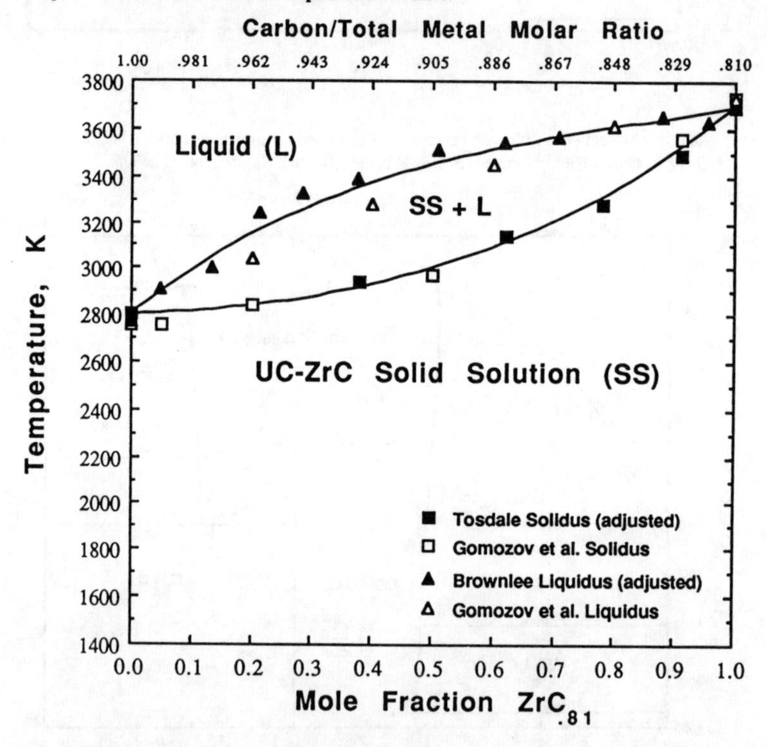

FIGURE 7. The UC-$ZrC_{0.31}$ Pseudobinary Phase Diagram Showing Solidus and Liquidus Temperature Data for the $(U_yZr_{1-y})C_x$ Solid Solution (Butt 1993).

Vaporization Behavior of Uranium Compounds

As discussed above, nitride, oxide, and carbide uranium compounds do not have to be stoichiometric. At elevated temperatures, vaporization of these materials generally occurs incongruently, and the stoichiometry of the material is subsequently altered, particularly at the surface. During vaporization, the composition will shift in such a way as to minimize the total system free energy. Given sufficient time, the stoichiometry will shift toward a stable composition referred to as the congruently vaporizing composition. The congruently vaporizing composition for a particular compound is in effect a balance of the rate at which species can be removed from the surface during vaporization, the rate at which species can diffuse to the surface through solid state diffusion, and the configurational energy associated with changes in composition. The congruently vaporizing compositions for UC, UC_2, and UO_2 are shown in Fig. 8. Because UN experiences incongruent vaporization, it is not particularly useful to refer to a congruently vaporizing composition because, as nitrogen is evolved quite rapidly from the surface, the material tends toward U(l) + N_2(g) rather than a more stable compound of UN.

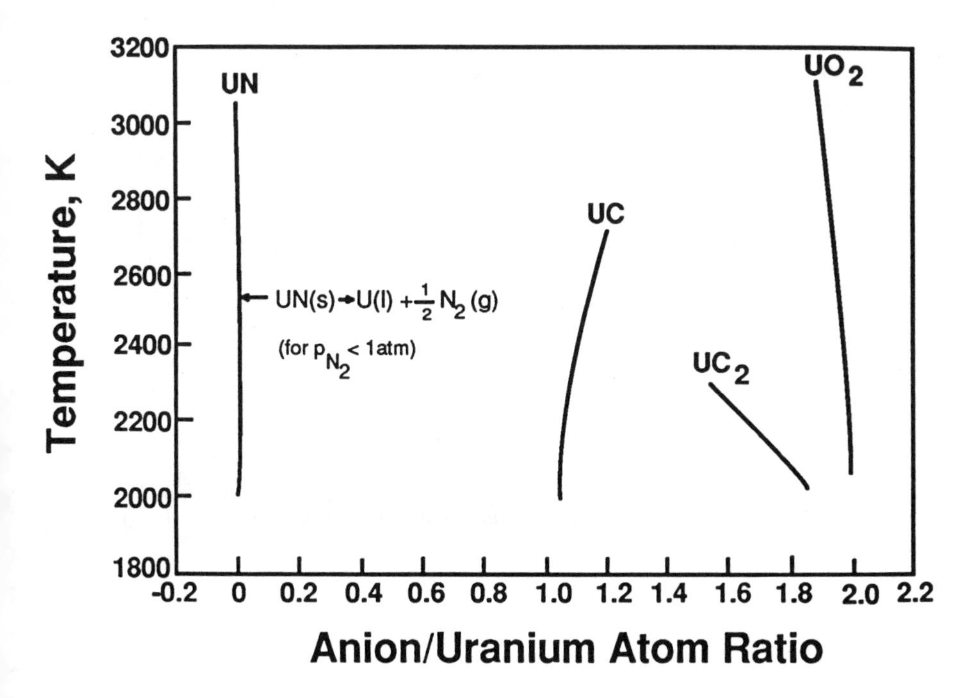

FIGURE 8. Congruently Vaporizing Surface Compositions for UC, UC_2, and UO_2 Vaporizing into Vacuum. The Curves were Calculated Using the Methods of Storms (1991) and Wallace (1994).

As the congruently vaporizing composition develops, the rate of vaporization decreases until steady state conditions are reached. Fig. 9 shows the steady state recession or vaporization rates (into vacuum) for the uranium compounds compared to several of the most thermally stable refractory carbide materials. As the figure indicates, UN experiences the most rapid vaporization. Uranium monocarbide and dicarbide lose material at rates approximately an order of magnitude less than that of UO_2. Uranium, zirconium carbide ($U_{0.1}Zr_{0.9}C$) loses material at a rate substantially less than the other uranium compounds.

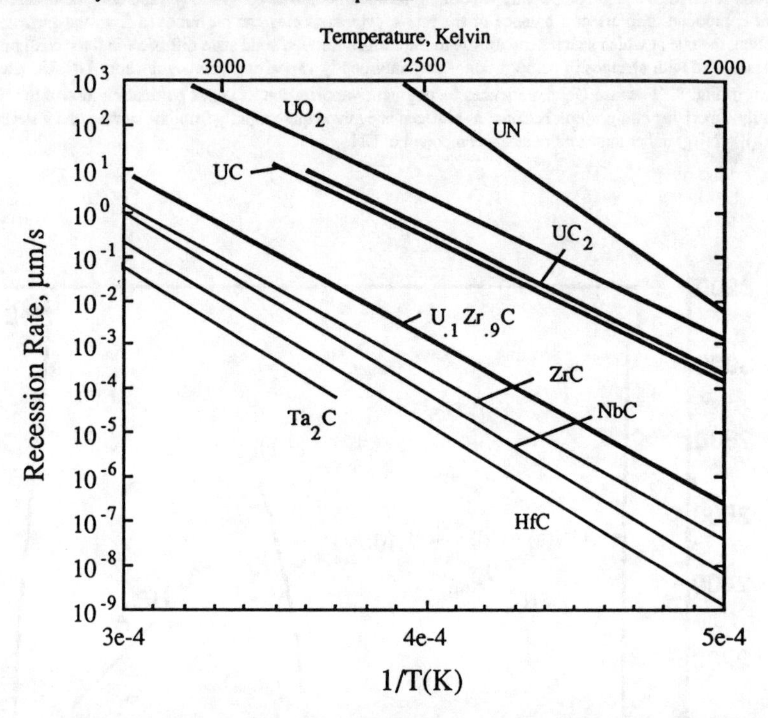

FIGURE 9. Arrhenius Plots of Recession Rates for Uranium Compounds Compared to More Stable Refractory Carbides. The Curves were Calculated Using the Methods of Storms (1991) and Wallace (1994).

FABRICATION

Uranium Dioxide

The fabrication of UO_2 by conventional cold pressing and sintering techniques is well established throughout the nuclear industry and will not be discussed. The main issue with thermionic fuel performance is fuel swelling and emitter distortion; various techniques to modify fuel structure and reduce swelling have been developed. Cored fuel pellets, "soft" fuel, additives, fuel with a dense rind, and porous cores have been fabricated and tested, but details of fabrication techniques are usually proprietary.

Uranium Nitride

The development of the process for producing UN fuel, carbothermic reduction of a UO_2 and graphite mixture followed by synthesis to UN in a nitrogen-6% hydrogen atmosphere at high temperatures, has been well documented (Mason 1985 and 1987, Matthews 1988, Blair 1989, and Matthews 1992). Significant progress has been made in the fabrication of UN fuel for the SP-100 Space Nuclear Power Program since this project began in 1984. Over 55,000 UN pellets (0.25 inch in diameter by 0.34 inch in length) were fabricated for the Nuclear Assembly Test core. In addition, pellets having different enrichments, densities, stoichiometries, oxygen and carbon contents, and microstructures were produced for irradiation tests in the Experimental Breeder Reactor (EBR-II) or the Fast Flux Test Facility (FFTF). The first step in UN fuel fabrication is the conversion of enriched uranium feed stocks to ceramic grade UO_2 followed by the reaction of UO_2 plus carbon to UN by a carbothermic reduction/nitriding process. The conversion and synthesis processes were developed, qualified, and demonstrated by converting over ninety 2.5-kg batches of oxide to nitride during the fabrication campaign. Figure 10 shows a schematic of the synthesis steps; basically, UO_2 and graphite are blended and pressed into compacts; the compacts are heated in vacuum to initiate the reduction reaction to UC, and nitrogen is introduced to replace carbon. Carbon is removed with hydrogen, and finally, the stoichiometry is adjusted to monatomic UN. The final stoichiometry adjustment is required for effective sintering of UN powders into high density pellets.

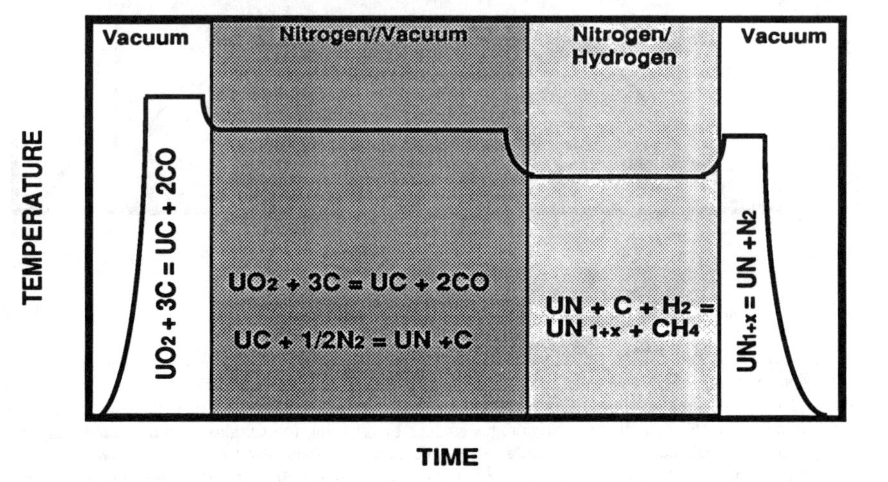

FIGURE 10. Uranium Nitride Synthesis Steps.

The synthesized UN powder is then milled into a fine powder in a high-energy, vibratory ball mill using UN pellets as the milling media (Blair 1989). The self-milling process was developed, and it can eliminate free iron contamination. The vibratory ball mill was required to reduce particle size in a reasonable time. Sintered UN density and grain size could be increased to approximately 95% of theoretical density (TD) and 30 µm by increasing milling time. After milling, the UN powder is slugged, granulated, pressed into green pellets, and sintered.

The key to sintering UN into high density, stoichiometric fuel pellets is understanding and controlling phase relationships at high temperature. A schematic of the UN sintering cycle is shown in Fig. 11. Green pellets are heated in vacuum to reduce sesquinitride and to remove residual oxygen. Nitrogen is introduced and the pellets heated to the time and temperature required to attain specified density and grain size. Sintering in N_2 prevents dissociation of UN and the subsequent formation of free uranium. A vacuum cool-down prevents formation of U_2N_3. Hypostoichiometric UN has large grains and high density but contains free uranium which will melt and swell during irradiation. Hyperstoichiometric UN pellets have a rind of U_2N_3 which dissociates to N_2 during high temperature operation and pressurizes the fuel pin. Near stoichiometric UN has a finite single phase region (see Fig. 4), and a homogeneous microstructure can be attained with careful control of sintering time and atmosphere. No UN sintering aids have been reported; however, the free uranium present in hypostoichiometric UN probably enhances diffusion rates at sintering temperatures and thereby acts as a self-sintering aid.

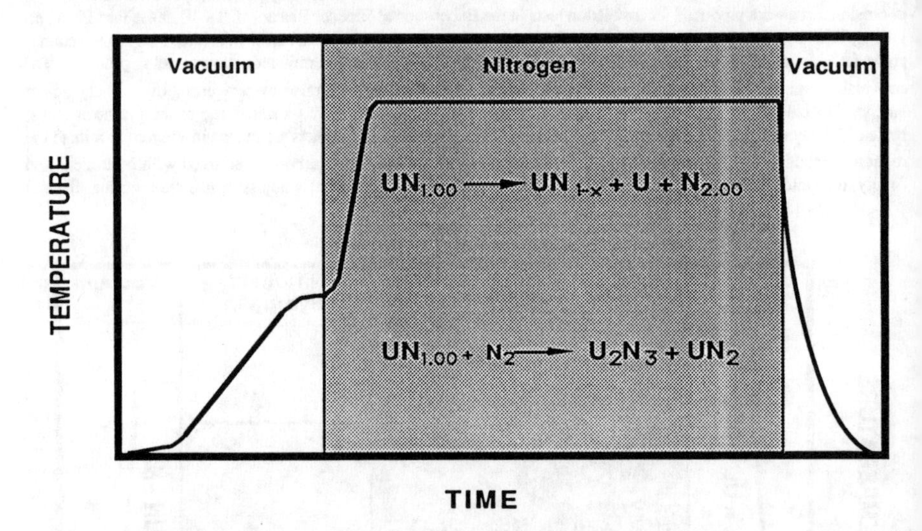

FIGURE 11. Uranium Nitride Sintering Cycle.

Batch qualification data for the pellets produced to date are summarized in Table 3, and all parameters are in specification. The original UN fuel pellet specification had no grain size requirements, and the average UN grain size of the production batches was 14 µm. However, uranium nitride fission gas release correlations were based on fuel having grain size greater than 50 µm. Fission gas escapes by diffusing from grains to grain boundaries where fission gas bubbles coalesce and form free paths to the pellet surface. Fission gas bubbles that accumulated at UN grain boundaries after irradiation can be seen in the photomicrograph in Fig. 16. Because fission gas diffuses to the grain boundaries inversely with grain size, it was decided to increase the UN grain size to match the existing UN irradiation data base and help reduce fission gas release. The results of grain growth studies are illustrated in Fig. 12; by increasing sintering time and temperature, grains greater than 100 μm were grown in UN pellets. A grain size requirement was added to the UN specification as shown in Table 3. Resintering process parameters to increase UN grain size to 40 μm were developed and qualified, and resintering of the original batches increased the mean grain size from 14 μm to 42 μm.

TABLE 3. UN Fuel Pellet Production Results

Batch	C (ppm)	O (ppm)	Fe (ppm)	Density(%TD)	Grain Size(μm)
Specification	<3000	<1000	<300	94.5±1.5	NR*
Demonstration	1930	1020	290	94.6	78
Qualification	2280	600	170	94.3	67
Insulators	1430	680	45	94.2	59
Production	1300	170	<5	94.6	14
New Spec.	<3000	<1000	<100	96.5±1.5	30
Resintered	1200	130	NA*	97.2	42

* NA (not measured); NR (not required)

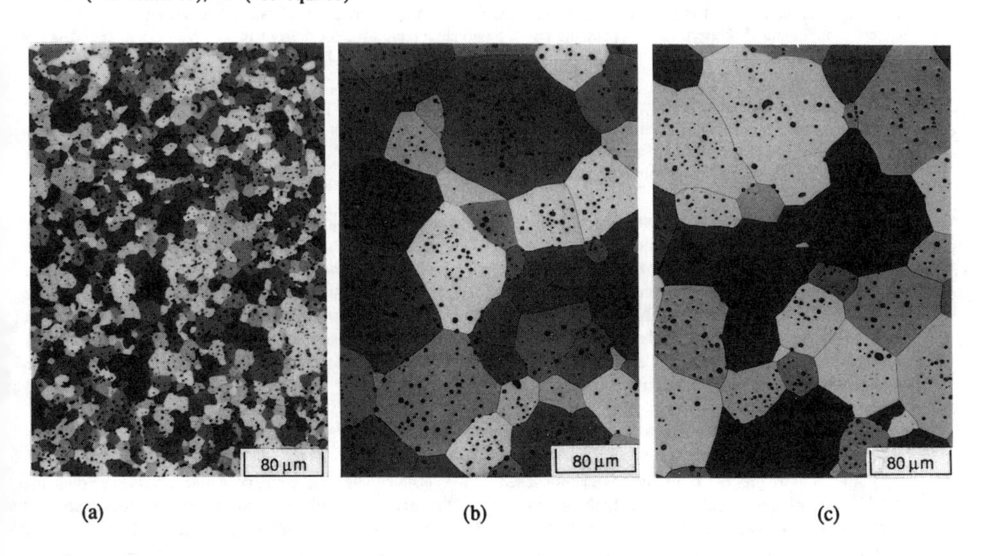

FIGURE 12. Microstructures of UN Fuel (a) as Originally Sintered, (b) after Resintering at Higher Temperature, and (c) after Resintering at Higher Temperature and Longer Time.

Resintering also increased the mean density of the fuel to 97.2%TD. In summary three process parameters are available to control UN density, grain size, and stoichiometry: (1) milling conditions to reduce particle size and increase sinterability, (2) sintering time and temperature to increase density and grain size, and (3) atmosphere control to maintain UN stoichiometry.

Uranium Carbides

Recent activities in carbide fuel fabrication development have focused on the high temperature mixed carbides for nuclear rockets. The very properties which make the ternary carbide fuel compositions so desirable as a reactor fuel also make their fabrication difficult. There are a number of methods for producing refractory carbide nuclear fuels including extrusion, hot pressing, arc melting, cold pressing and sintering, sol gel, combustion synthesis, and freeze drying.

Extrusion has been used to produce fuel rods. Extrusion has the advantage that large quantities of material can be rapidly processed, and is viewed as a relatively low risk, highly proven method for producing refractory carbide fuel rods of various dimensions. Near the end of the Rover program, 1/4-inch diameter rods of solid solution (U,Zr)C were extruded (Taub 1975). The lack of significant quantities of graphite in the mix required high extrusion pressures.

Arc melting has been used for fabrication of carbide test samples, for fabrication of metal fuel rods, and for fabrication of carbides for the Canadian organic-cooled-reactor (Jones 1973). Hot pressing is another proven technique for producing carbide fuel materials, primarily for making laboratory test specimens. Hot pressing has the disadvantage of being a relatively slow processing technique, but can be made more efficient through the use of complex dies that allow for pressing of multiple parts. It has the advantage of producing very dense parts with excellent mechanical properties. Hot pressing could be a viable technique for designs in which stacks of fuel disks or short rods are used.

Cold pressing and sintering are reliable methods that could be used to produce variable fuel geometries. Cold pressing is used a great deal in the ceramics industry for producing spheres of varying diameters (≈0.1 to 5 cm). Cold pressing and sintering are also processes of choice for fabricating pellets and short rods, and they are relatively low risk methods for producing refractory carbide materials.

Combustion synthesis is a promising way of synthesizing mixed carbide solid solutions. Because of the high stability of the mixed carbide solid solutions, the formation reactions of these mixed carbides from their elemental components are highly exothermic and self propagating. Generally, the combustion products are porous and therefore would have to be pulverized to form powder feed stock. However, because of the high melting point, the product can retain its original shape and may be suitable for forming porous preforms. The combustion synthesis process has several important advantages. It requires relatively simple equipment setup, it is quick, it is energy efficient, and it can produce a relatively pure final product. The latter results from the high reaction temperature which effectively expels the volatile impurities.

Gel-sphere techniques have been extensively developed and used for fabricating carbide spheres for gas-cooled power and propulsion reactors. UC_2 gel-spheres were used in the dispersed Rover fuel rods and for particle bed reactor fuels. A relatively new process utilizing cryochemical techniques (Blair 1991) has some potential advantages over the established gel-sphere techniques for making mixed-carbide spheres. Basically, the process quickly freezes sprayed droplets of aqueous suspensions of solids to form frozen spheres. The frozen spheres, containing UO_2, ZrC, and carbon, are freeze dried to remove water, then converted to the carbide, and sintered to density. The process has the advantages of having only a few simple steps, being applicable to a variety of uranium compounds, and being capable of various sizes and densities. In addition, the liquid waste streams and rejected spheres are readily recycled. A photograph of mixed-carbide spheres made by the cryochemical process is shown in Fig. 13.

FIGURE 13. Microspheres of $(U_{0.1},Zr_{0.9})C_x$ Made by Cryochemical Processing.

Carbide Coating Materials

Chemical-vapor-deposited, refractory metal carbide coatings are applied to graphite matrix fuel rods and spheres to protect the graphite from reacting with the hydrogen propellant. In addition, graphite matrix fuel rods are coated with ZrC. The chemical-vapor-deposition techniques for the refractory carbide coatings are generally well known (Wallace 1991), and used currently to prepare high quality inert carbide coatings. Careful control of temperatures, gas flow, and feed gas composition is required. These techniques are usually chlorine based, which creates some manageable leaching problems. Newer thermal decomposition techniques that use organometallic precursors for coating (Blair 1991) show promise for applying refractory metal carbide coatings, but are somewhat slow, and need development to provide very dense coatings.

PERFORMANCE

Space Nuclear Power Reactors

The performance of fuel-pin-type reactors is strongly influenced by fuel and cladding interactions under prototypical temperature and burnup conditions. Fuel interacts with metallic cladding by (1) swelling out against the cladding, causing radial or axial deformation of the cladding (fuel-cladding mechanical interaction); (2) release of fission gas, causing the cladding to deform from internal pressure; and (3) interacting chemically with the cladding, causing corrosion, thinning, and/or embrittlement of the cladding (fuel-cladding chemical interaction). In addition, cladding alloys may creep more rapidly during irradiation, may swell because of the formation of voids, or may become brittle. For the most part, temperatures in space power reactor applications are high enough to anneal out irradiation-induced defect damage in the cladding so that thermal phenomena dominate.

The primary source of mechanical strain on cladding is from fuel swelling caused by fission products generated during fission (two or three fission product atoms are produced for every uranium atom fissioned). Fuel swelling is caused by a combination of solid fission products and fission gas bubbles that have not escaped the fuel matrix. Swelling caused by solid fission products is generally accepted to be approximately 0.6 volume percent ($\Delta V/V\%$) per atomic percent (at.%) burnup. Very little can be done to mitigate swelling from solid fission products because they dissolve into the fuel lattice or precipitate from the fuel lattice. Control and understanding of the behavior of noble fission gases, krypton and xenon, constitute the major share of fuel performance evaluations.

Previous irradiation data (Hilbert 1971 and Chubb 1973) suggest that high-density, large-grained-size UN fuels release less fission gas because fission-gas bubbles are less mobile in a fully dense, high-strength fuel structure. The movement of fission-gas atoms to grain boundaries can be limited by increasing grain size, and therefore fission-gas release can be delayed. Fission gases appear to be highly mobile in hypostoichiometric UN_{1-x}, which releases more fission gas than stoichiometric UN because of the presence of free uranium (Hilbert 1970). Bubble nucleation and coalescence produce fuel swelling. Fission gases usually collect in bubbles and along grain boundaries, and exert pressure on the surrounding matrix thereby increasing fuel dimensions. Release of fission gases occurs by a combination of grain boundary diffusion and connection of the bubbles to the fuel surface. Breakaway swelling and fission gas release are of particular concern during long irradiation times or high temperatures. At low temperatures and/or low burnups, the fission gas atoms are not mobile enough or have not had sufficient time for the atoms to diffuse to a pore or grain boundary, so the swelling and fission gas release remain low. Space power reactors operate at high temperatures (1400 - 1800 K) and moderate burnups (>6 at.%), in which case fission gases become mobile enough to concentrate in bubbles in the grains and at the grain boundaries. Therefore, both fission gas release and swelling can become life-limiting phenomena. A large temperature gradient favors movement of the bubbles up the gradient with an accompanying movement of fuel down the gradient. It is not surprising, then, that large gas releases are typical of uranium dioxide with its low conductivity, whereas rather small releases are typical of high-conductivity uranium nitride and carbide fuels.

Chemical interaction between fuel and cladding can lead to performance-limiting phenomena. For example, the nitrogen over UN reacts with niobium-based cladding, leading to a slow decomposition of the fuel and potential nitriding of the cladding. If long life is desired, it is necessary to physically separate the fuel and cladding. Additional fuel-cladding chemical interaction phenomena is the formation and migration of metallic fission products. Metallic fission products tend to migrate down the thermal gradient from the fuel center to the cladding where they can react and decrease cladding strength.

All of these phenomena -- fuel-cladding mechanical and chemical interactions, and fission gas release -- can interact with each other to further complicate matters. Both mechanistic and empirical models have been developed to predict fuel pin performance. Mechanistic models frequently must be adjusted to satisfactorily fit the data; consequently, empirical correlations should not be used outside the range of the data conditions. For a more comprehensive discussion of fuel models, the reader is referred to Matzke (1986).

Uranium Oxide for Thermionic Fuel Elements

Irradiation Performance

Considerable effort has been devoted to modeling the performance of uranium dioxide fuels for terrestrial power reactor conditions. This modeling is not generally applicable to the higher operating temperatures experienced by the thermionic fuel elements, which have the highest temperature application and appreciable burnup. Most modeling work has been done using the LIFE code (Roth 1982) which was developed for liquid-metal, fast-breeder reactor applications. The LIFE code has been selected to predict thermionic fuel element performance because of its universal UO_2 mechanistic formulations.

Temperature predictions are relatively straightforward, using standard heat transfer procedures. Predicting temperature drops across the gap between fuel and cladding is more complex, and the Ross-Stoute (1962) approach is usually used. However, thermionic elements are sealed in a vacuum so that as-manufactured, heat transfer between the fuel and cladding (or emitter) is very poor. The fuel is believed to vaporize on startup and recondense on the inner surface of the emitter, and relatively good heat transfer between the fuel and emitter continues from that point on. Direct verification of temperature predictions is not available.

The fission gas release from thermionic fuel appears to be very high, as might be expected from the high operating temperature. However, the release is even greater than one would expect from the LIFE model extrapolations. There appears to be no incubation time, as is the case with lower temperature fuel. Furthermore, the release is not complete, as appreciable emitter deformation is observed, thereby implying fuel swelling beyond that caused by solid fission products. Some small amount of gas must be retained to account for the swelling; retention of gas in the central void has been proposed as a mechanism. Fuel pellet ratcheting during reactor startup and shutdown may also contribute to cladding deformation. The code has been calibrated to release enough gas so that the amount of gas retained agrees with Zimmerman's (1978) measurements. Cladding (emitter) strain in thermionic fuel elements is not caused by released fission gas pressure because fission gas is vented.

Swelling is predicted mechanistically in LIFE, depending on the gas retained, the fuel creep strength, surface tension, temperature, and constraining forces. It appears that extrapolation of the various expressions for these phenomena from formulations used for liquid-metal fast-breeder reactor fuel is not satisfactory. A calibration of the code to available data has been obtained for thermionic fuel elements by changing adjustable constants in expressions for the models (Baars 1991). Predictions of fuel swelling made by the current calibration are shown in Fig. 14. Considerably more data are needed, particularly for detailed behavior, before this calibration can be validated.

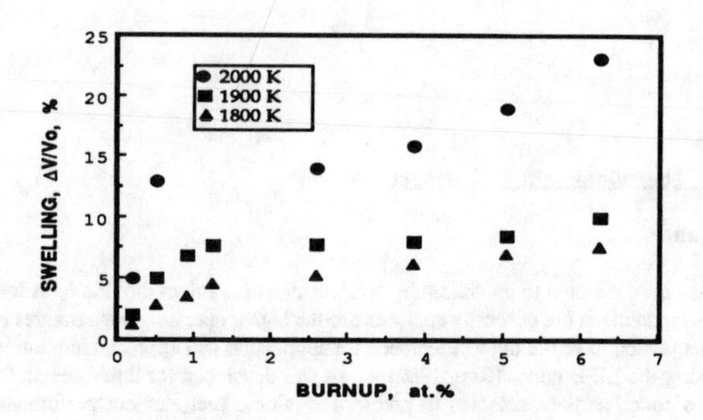

FIGURE 14. Unrestrained UO_2 Swelling as Predicted by the LIFE-Space Code.

Uranium Nitride for SP-100

Approximately 90 SP-100 UN fuel pins were tested in the Experimental Breeder Reactor (EBR-II) and the Fast Flux Test Facility (FFTF) fast spectrum reactors with the following variables:

Design Features
Cladding: 5.84 - 7.62 mm-o.d./Nb-1% Zr or PWC-11 (Nb-1% Zr + 0.1% C)
Liner: Tungsten, rhenium, and bonded-rhenium
UN density: 87% - 96%TD
Irradiation Conditions
Temperature: 1200, 1400, & 1500 K cladding
Burnup: 1 - 6 at.%

Many details of these irradiation tests are not currently available; however some performance trends can be found in the literature. First, even though the SP-100 pins were tested at relatively high temperatures - up to 1900 K fuel centerline - no failures occurred; therefore the fuel pin design is robust. Second, metallic fission products, especially ruthenium (Matthews 1988), were found to migrate down the thermal gradient to the fuel pellet surface and into the cladding. Third, no breakaway swelling or fission gas release was observed, and high density fuel released less fission gas than low density fuel (Makenas 1991). Fourth, large fission gas pores formed in the center of the low-density UN fuel pellets, while the high density fuel showed no evidence of restructuring or cracking. Finally, chemical-vapor-deposited tungsten liners cracked during irradiation but a thin liner of wrought rhenium metallurgically bonded to the inside diameter of Nb-1Zr cladding was found to be an effective barrier to fuel/cladding/chemical and mechanical interactions (Truscello 1992). Mechanistic performance models, predicting the performance of UN fuels for space reactors have been developed (Vaidyanathan 1993).

Microstructural analysis of irradiated UN showed that low density fuel tended to restructure and form central voids. Transport of UN from the fuel centerline to the cladding walls was seen, suggesting dissociation of UN from hot regions and reformation in cold regions. Chemical interactions between fuel, fission products, liners, and cladding were also found in the low density fuel pins. Metallographic cross sections from an irradiated UN/PWC-11 fuel pin with a free standing tungsten liner are shown in Fig. 15. The fuel was low density (~85% TD) and operated at approximately 1900 K centerline temperature; the high magnification photomicro-graphs show ruthenium deposits on the liner and cracks in the tungsten liner. The high density UN displayed rather innocuous changes in microstructure. The microstructures of irradiated high-density UN in Figure 16 show that fission gas bubbles nucleate and coalesce at grain boundaries with increasing burnup.

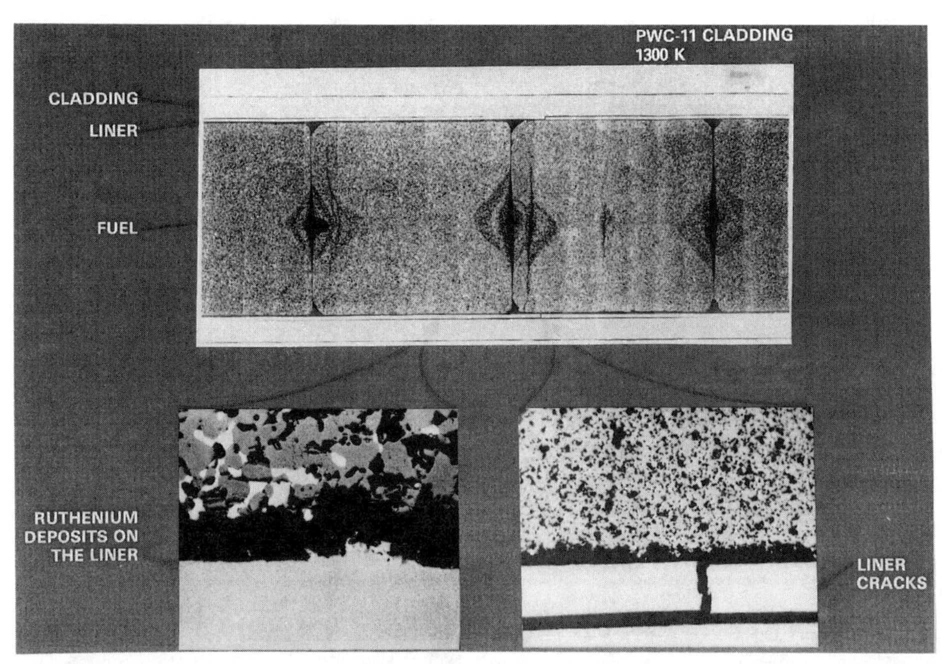

FIGURE 15. Longitudinal Cross Section of an Irradiated UN/PWC-11 Fuel Pin with Low Density UN (Makenas).

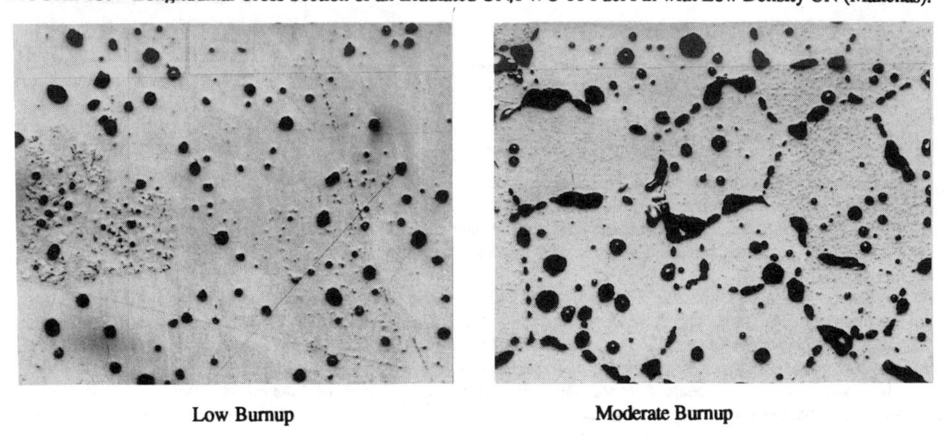

FIGURE 16. Microstructures of Irradiated, High Density UN Fuel (Makenas).

Fig. 17 graphically displays the hypothetical performance of UN fuel pellets based on the observations from the U.S. space reactor fuel development program. The black marks represent metallic fission products migrating to the pellet surface and into the cladding. The voids represent the formation of center line fission gas pores. High density UN (>95 % TD) operating at moderate temperatures (fuel center line <1650 K, cladding <1400 K) has low swelling, low fission gas release, no fission product interaction, and the potential of operating to high burnups. Low density, hypostoichiometric UN, and high operating temperatures are undesirable characteristics that will limit the useful lifetime of UN fuel. As a consequence of these observations, UN fuel pellets for the SP-100 reactor were fabricated to the following specification:

Density:	96.5 ±1.5% TD
Stoichiometry:	1.00 -1.05 (N+C+O)/U
Grain Size:	>30 μm
Carbon: <3000 ppm	
Oxygen: <1000 ppm	

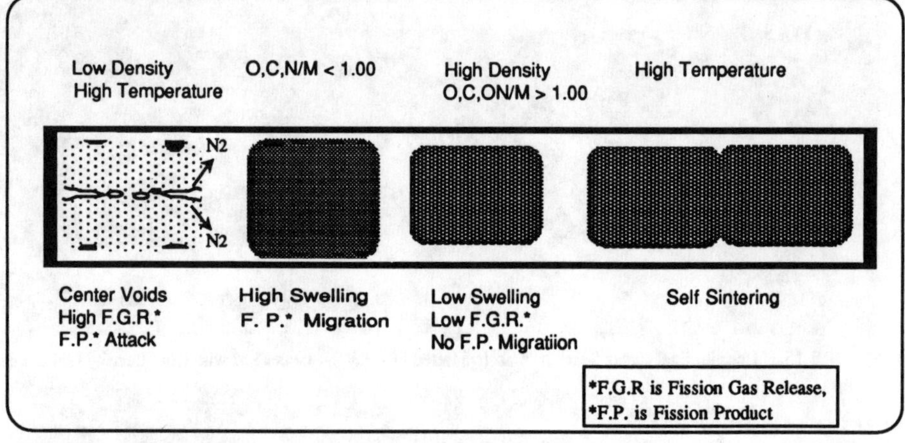

FIGURE 17. Schematic Representation of UN Fuel Pellet Performance.

UN Fission Gas Release

Variables influencing irradiation induced fission gas release include temperature, burnup, fuel density, temperature gradient, fuel stoichiometry, impurities, and grain size. Unfortunately, the only reliably reported variables are burnup and starting density; operating temperatiires are estimated, and the other variables are generally not reported in the literature. However, sufficient information is available to determine trends.

Fission gas release from the SP-100 UN irradiation tests is plotted as a function of burnup in Fig. 18-a and temperature in Fig. 18-b. The gases were determined by puncturing fuel pins after irradiation, measuring volume of gas, and analyzing for Xe and Kr. The high density UN released considerably less fission gas than low density fuel as would be expected because the greater open porosity in low density fuel allows better access to the plenum gap. The fission gas release appears to increase linearly with burnup to 5 at.%, and the release from low density fuel is approximately three times that for high density UN. The temperature trends in Fig. 18-b suggest that low density UN begins to release fission gas at temperatures greater that 1600 K, while high density UN appears to retain most of the generated fission gas up to 1800 K.

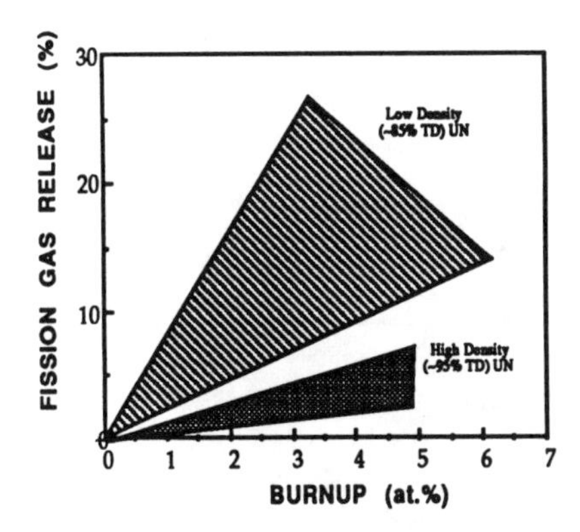

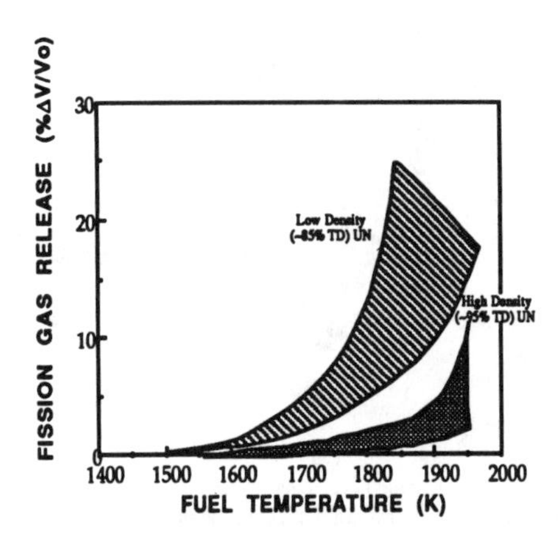

FIGURE 18. SP-100 UN Fission Gas Release; (a) as a Function of Burnup, and (b) as a Function of Temperature.

Storms (1988) proposed an empirical fission gas release equation of the form:

$$FGR = 100/[\exp(0.0025\{90TD^{0.77}/BU^{0.09} - T\}) + 1],$$

where FGR is the percent of fission gas released, TD is the percent of theoretical fuel density, BU is burnup in atomic percent of uranium, and T is the average fuel temperature in Kelvin. This correlation is strictly an empirical fit to the available fission gas release data as a function of burnup, temperature, and fuel density. The correlation suggests that release declines with increasing burnup, accelerates rapidly with increasing temperature, and is sensitive to fuel density.

A comparison of available UN and (U,Pu)N fission gas release data is plotted as a function of burnup and temperature in Fig.s 19 and 20. The very high fission gas releases are probably from either high burnup tests or hypostoichiometric UN_{1-x} . The very low fission gas releases are probably from gas escape during measurement. Nevertheless, results that suggest fission gas release from UN can be kept very low at temperatures up to 1600 K; higher temperatures clearly result in increasing release rates. Although the data contain many unannotated variables — for example, power density, burnup rate, smear density, temperature estimates, stoichiometry, grain size, and impurity characteristics — there is a relatively good consistency, and the figures match the correlation suggested by Storms (1988).

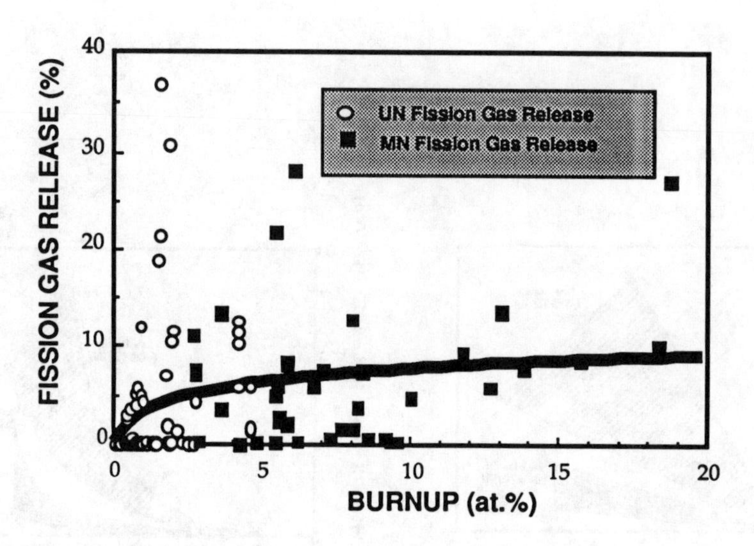

FIGURE 19. Fission Gas Release from Nitride Fuels as a Function of Burnup. Differences in Temperature, Density, and Fuel Characteristics are not Considered, and the Heavy Curve Suggests the General Trend of the Data.

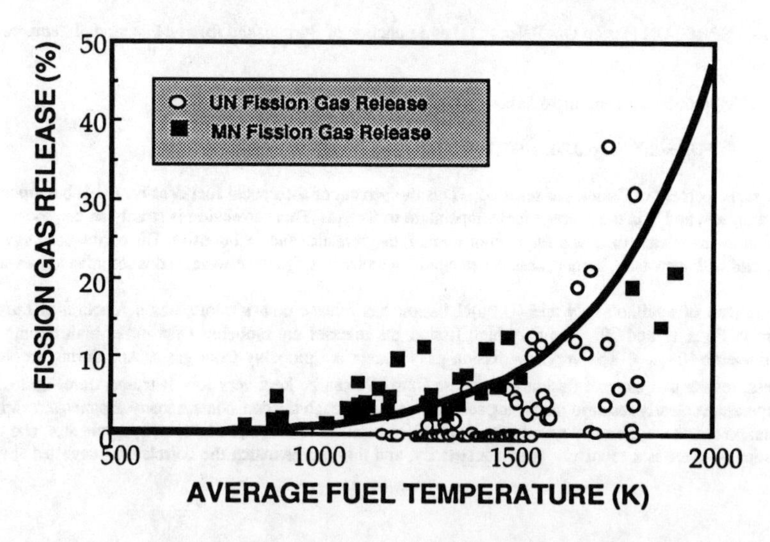

FIGURE 20. Fission Gas Release from Nitride Fuels as a Function of Temperature. Differences in Burnup, Density, and Fuel Characteristics are not Considered, and the Heavy Curve Suggests the General Trend of the Data.

UN Swelling

Two UN swelling correlations have been published (Thomas, 1988 and El Genk, 1987) of the form:

$\Delta V/V_0 = a \times 10^{-b}.(T)^c.(BU)^d.(TD)^d$
where a,b,c, & d are fit constants.

These correlations are not particularly good because of the number of variables in the irradiation data base, the uncertainty of temperature measurements, and various swelling measurement techniques. Uranium nitride swelling has been measured on 45 fuel pins irradiated during the SP-100 program. Fuel pellet dimensional changes were measured by digitizing densitometry traces of neutron radiographs of the irradiated pins. The swelling data of the SP-100 UN plotted in Fig. 21 includes cladding temperatures from 1250 to 1550 K and high and low density UN. The decrease in swelling rate at higher burnup is probably caused by resistance of the cladding, and the initial, high swelling represents unrestrained UN swelling.

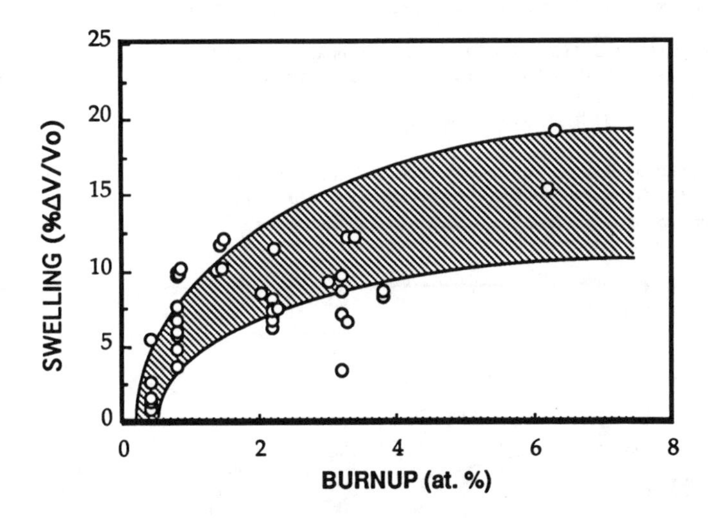

FIGURE 21. Swelling Rate of SP-100 UN Fuel Irradiated in FFTF and EBR-II. Differences in Operating Temperature and Fuels Density are not Discriminated.

Probably the greatest uncertainty in these fission gas release and swelling correlations lies in the actual operating fuel temperature because the test data only report cladding temperatures estimated from reactor coolant temperatures. The average fuel temperatures are then calculated based on uncertain fuel/cladding gap size, gap conductivity, and fuel thermal conductivity information. A review of detailed test characteristics followed by consistent temperature calculations might help narrow down the scatter in the data and permit more accurate predictions of uranium nitride fission gas release and swelling. Nevertheless, by applying the performance trends described in this review of nitride fuel irradiation data, a reliable nitride fuel pin with high burnup capability can be postulated based on the following constraints and recommendations:

- Fuel density greater than 95% TD to reduce restructuring and fission gas release,

- Peak fuel operating temperature less than 1800 K to reduce fission gas release, swelling, and fission product migration,
- Hyperstoichiometric UN_{1+x} to eliminate metallic fission product formation,
- Pin diameter, cladding thickness, fuel/cladding gap thickness, plenum volume based on performance requirements but constrained by Storms' fission gas release correlation and swelling at 1.5% $\Delta V/V_0$ per at.% burnup.

UN Compatibility

Stoichiometry is the most important nitride fuel characteristic to regulate because UN dissociates at high temperatures (Storms 1988). The resulting free nitrogen pressurizes the fuel pin plenum and contributes to cladding creep. The free uranium can react with cladding components to reduce the creep strength of the cladding. Fig. 5, shown earlier, is a plot of nitrogen pressure as a function of UN composition at 1650 K within the single-phase, mononitride phase boundaries. Nitrogen pressure remains relatively constant in Region II; in this region N_2 overpressure contributes to cladding creep. Nitrogen pressure changes sharply with composition in Region I; UN fuel in this region uranium can react with fission products and cladding components.

The SP-100 fuel pin cladding is lined with rhenium to protect the Nb-1Zr from reacting with UN fuel. Ex-pile compatibility testing on the UN/Re system shows that UN and rhenium will react at operating temperatures according to:

$$UN_{1+x} + Re = URe_2 + N_2$$

if x is sufficiently small. Reaction thermodynamics and kinetics were measured, (Storms 1989) and Fig. 22 plots the equilibrium reaction conditions between UN and rhenium to form URe_2; nitrogen pressures above the line prevented the reaction, while those below the line allowed the reaction to occur.

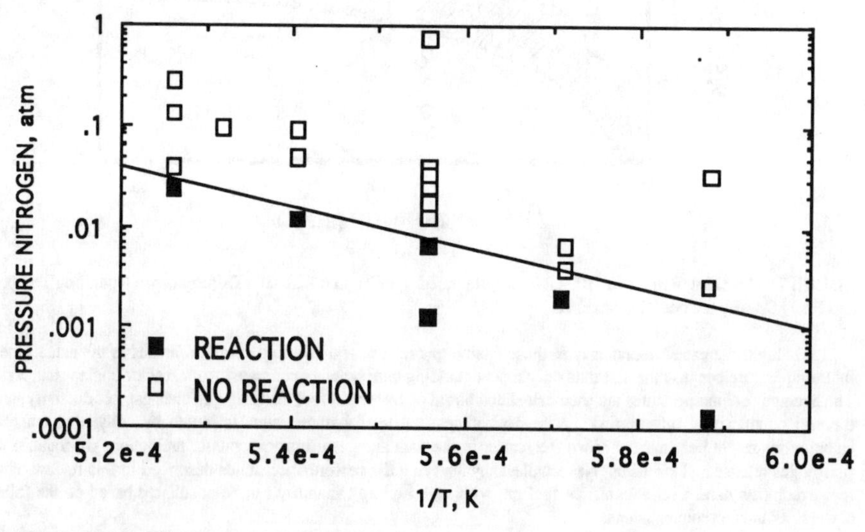

FIGURE 22. Effect of Nitrogen Pressure on URe_2 Formation as a Function of Temperature.

Postirradiation examinations of test pins show the formation of molten fission product uranium/ruthenium alloys near the pellet/cladding interface as shown schematically in Fig. 17. Below some critical stoichiometry (Region I in Fig. 6) UN will react with metallic fission products to form low melting alloys that increase atomic mobility in the fuel, thereby greatly increasing fission gas release, swelling, and fuel/cladding chemical interactions. Gradual loss of nitrogen from nitride can produce the following hypothetical reaction:

$$UN_{1+x} + (Ru,Pd,Mo,Rh) = U(Ru,Pd,Mo,Rh)liquid + N_2 + UN_{1-y}$$

The molten fission product alloys will react with cladding components and present a threat to the fuel pin. Ruthenium is the predominant metal fission product seen in irradiated UN. The uranium/ruthenium phase diagram (Mason 1993) shows that the solidus and liquidus temperatures on the ruthenium rich portion are approximately 300 K lower than previously thought. This low melting point explains fission product effects observed in irradiated UN pins. Calculations predict that the reaction will not occur during the seven-year lifetime of the SP-100 reactor if stoichiometry, temperature, and liner integrity are controlled.

Carbides for Nuclear Rockets

Although the fuels developed for Rover demonstrated considerable performance advantages for the nuclear rocket over chemical systems, modern requirements have demanded additional improvements. Recent carbide fuels activities have focused on increasing operating temperature, lifetime, and margins to failure. Because the manned Mars program never "took-off", very little experimental work was done; however, some effort was put into developing and understanding carbide fuel performance limits (Storms 1991 and Butt 1993). Three major performance limiting phenomena have been identified: (1) melting point of the fuel, (2) mass loss caused by chemical interactions and vaporization, and (3) brittle failure caused by thermal stress. The combination of preferential loss of primary constituents from the fuel and the resulting decrease in melting point is probably the primary time/temperature limiting mechanism. Compatibility, fission gas release, and changes in chemical and physical properties caused by fission products metals and radiation are of less concern because of the short operating times of nuclear rockets.

Melting

Solid solutions of uranium carbide with carbides of the refractory metals zirconium, niobium, hafnium, and tantalum are the most promising high temperature fuels. These ternary compositions take advantage of very high melting points of metal carbides. Small additions of uranium carbide will only have a modest effect in depressing the melting point of the refractory carbide. The individual carbide systems themselves are characterized by compounds having a wide range of single phase composition, and solid solutions between the two carbides have useful ranges of miscibility. The nominal melting points for the important refractory carbides are presented in Table 4.

TABLE 4. Binary Carbide Melting Point Maxima.

Compound	Melting Temperature, K
$ZrC_{0.81}$	3693 ± 20
$NbC_{0.85}$	3871 ± 50
$HfC_{0.88}$	4200
$TaC_{0.88}$	4273
$UC_{0.98}$	2798 ± 30
$UC_{1.84}$	2720

Combinations of these binary carbides yield the pseudobinary (UC/MC) compositions. Some phase diagram work, especially melting point determination, has been carried out on the pseudo-binary systems having uranium carbide as one component. While many of these results have been inconsistent, Butt and Wallace (1993 a and 1993 b) have optimized the phase diagrams of U-Zr-C and U-Nb-C by rationalizing the published solidus-liquidus data, and using thermodynamics-based calculational techniques to map out the two phase (liquid + solid) regions of the phase diagrams.

The solidus-liquidus phase relationships for the U-Nb-C system are displayed in Fig. 23 (the UC-ZrC pseudo-binary phase diagram is shown in Fig. 7. The solidus - liquidus temperature-composition curves are crucial to the development of a phase diagram, but more information is needed to clarify the sensitivity of the melting point on composition.

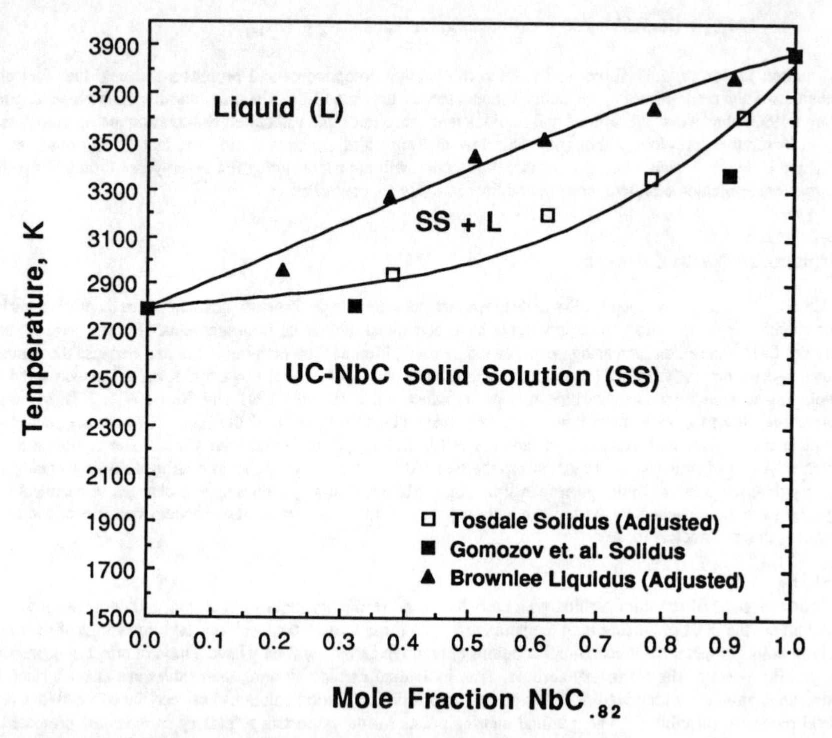

FIGURE 23. The UC-$NbC_{0.82}$ Pseudobinary Phase Diagram (Butt 1993).

Four isothermal cuts across the composition space of the U-Zr-C system are shown in Fig. 24. As the temperature is increased from 2873 K to 3693 K, the region of solid phase, solid solution is seen to dramatically decrease in area, and finally disappear by 3693 K. The U-Nb-C system behaves similarly. These phase diagram sections illustrate two important facts: 1) the differential loss of components from the solid solution mixture, by vaporization, diffusion, or chemical reaction, can have a strong effect on melting point; and 2) the amount of uranium in the solid solution must be kept small to achieve the highest melting points.

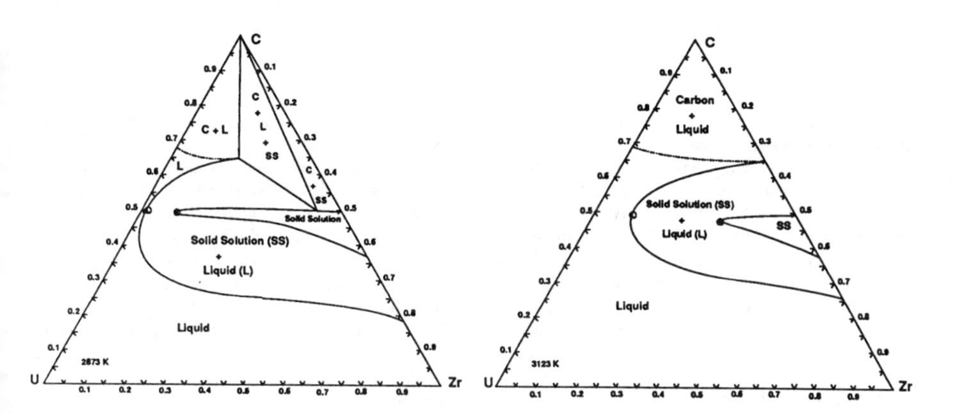

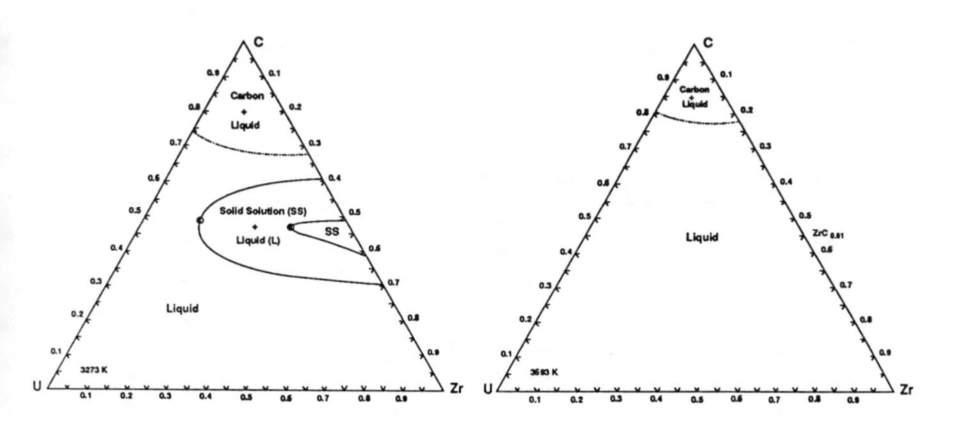

FIGURE 24. Calculated Isothermal Sections of the U-Zr-C Ternary Phase Diagrams (Butt 1993).

Diffusion and vaporization of carbon and uranium from the surface of the fuel will change the composition and hence the melting point. Carbon loss by vaporization from ZrC or NbC is very rapid at high temperatures. Carbon loss rate drops with time and quickly approaches a relatively constant value once diffusion and vaporization rates are matched. The surface composition of this material is called the steady-state congruently vaporizing composition. Because of the rapid loss of carbon, carbide fuel on the carbon-rich side of the solid solution is preferable for achieving longer lifetimes The presence of <20 at.% uranium has only a small effect on the congruently vaporizing composition because metal diffusion rates are slow. However, the equilibrium pressure for uranium is sufficiently high that surface uranium concentration will be low, but with a steep gradient. Because the uranium content of the surface is low, the surface of uncoated fuel will have a melting point that is higher than the bulk material.

Mass-Loss

The hydrogen gas exiting from a nuclear rocket core will contain significant carbon and some uranium as the respective concentration gradients are formed in the fuel (MacMillan 1991). As the reactor continues to operate, three regions having different behavior will form as illustrated in Fig. 25. At the lowest temperature (Region I), below ~1500 K, an insignificant reaction is expected. As temperature increases along the fuel element, hydrogen corrosion will occur in Region II. When hydrocarbons are added to the propellant gas, or are acquired by hydrogen corrosion upstream, the mass loss rate is reduced. Hydrogen reaction will peak and become negligible at higher temperatures. In spite of the corrosion reaction, mass-loss and physical changes in Region II are expected to be small. Added carbon has a smaller and smaller inhibition effect as temperature is increased, and attempts to reduce the vaporization rate in the high temperature region by addition of carbon to the propellant will be relatively ineffective. If surface temperatures near the exit end exceed 2900 K (Region III), mass-loss caused by simple vaporization of the component elements in the fuel will become significant. The mass loss will depend on the operating time, the maximum surface temperature, and the surface area of the fuel exposed to the high temperature. High fuel density is crucial to very high temperature nuclear rocket applications because porosity enhances vaporization and surface diffusion, which in turn limits lifetime. Fig. 26 compares the calculated loss rates from a particle bed fuel element with a prismatic-type fuel element; particle fuel is predicted to have a higher mass lose rate at the high temperatures because of the higher exposed surface area of fuel (Storms 1991).

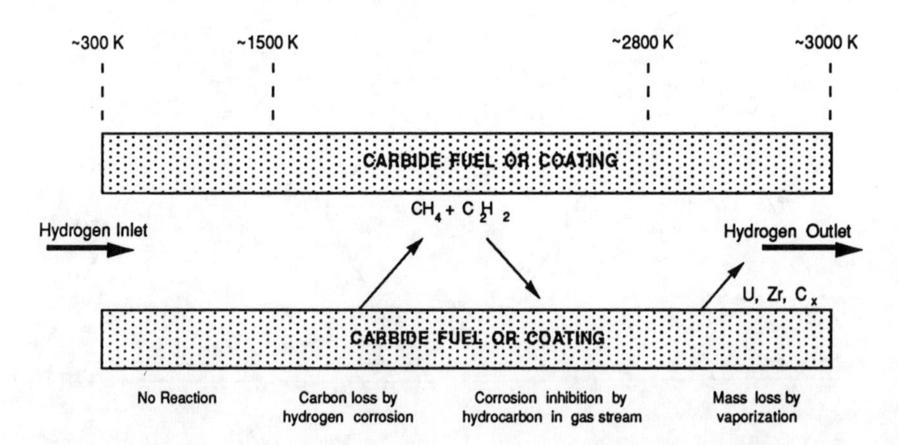

FIGURE 25. Schematic Representation of the Reactions Occurring in Carbide Fueled Nuclear Rocket Core.

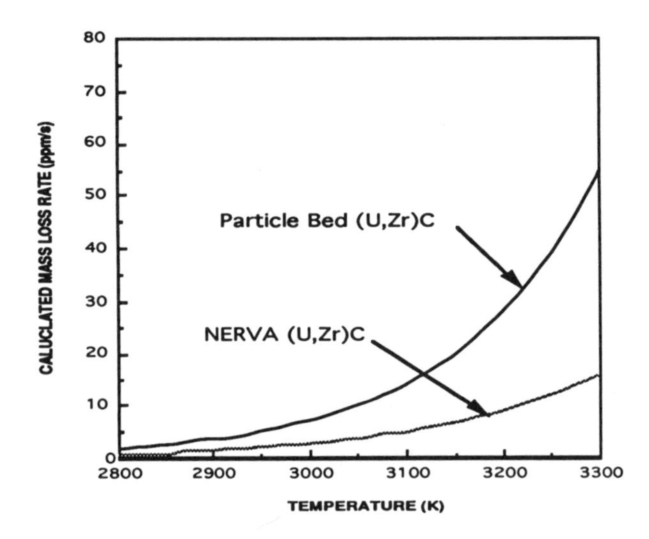

FIGURE 26. Comparison of Predicted Mass Loss Rates from Particle Bed and Prismatic Nuclear Rocket Core Configurations (Storms 1991).

Because of these composition changes, the highest practical operating time and temperature of carbide fuel is not defined by the maximum melting point. The practical operating conditions are determined by either (1) the time and temperature that causes enough fuel mass loss or structural deterioration to cause reactor shutdown, or (2) the time and temperature at which the fuel surface composition has changed to the composition of the solidus. At this point, liquid will form on the evaporating surface and destructive processes will be accelerated.

The maximum temperature for both ZrC and NbC based UC fuels is expected to be in the 3100 -3200 K range for reactors operating for <30 minutes. Although modest temperature increases might be achievable, very careful design and a minimum uranium content will be required. Bulk carbon content of the fuel should be as high as possible to achieve the highest surface melting temperatures and lowest mass loss rates. However, free carbon should be avoided because its rapid removal by hydrogen will produce pores that will increase uranium loss. Addition of carbon to the hydrogen propellant either by upstream reaction or as added methane is expected to reduce low temperature corrosion but have no effect on high temperature mass loss or melting. Addition of niobium to zirconium-based fuel is expected to lower the total metal loss rate while raising the melting point of the surface, thereby increasing the effective operating temperature. The above discussion is based on extrapolation of available experiments and modeling based on known thermochemical data. Considerable fundamental data generation and relevant testing will be required to verify these predictions.

Brittle Failure

Mechanical failure of carbide fuels and coatings can be caused by a combination of thermal shock and thermal expansion. The inherent thermal shock resistance of brittle materials is related to the temperature gradient thermal transient and the mechanical properties of the fuel. All he refractory carbides have relatively similar thermal conductivities and mechanical properties. Therefore, relatively little difference in the thermal shock resistance of mixed carbide fuels is anticipated. However, mechanical failure will be significantly affected by fuel element design.

For example, the coated Rover fuel element suffered from thermal expansion mismatch between the graphite and ZrC coating, causing cracking of the protective coating. Thermal expansion could affect the particle bed design by causing the bed to expand and either fracture coating materials or distort the containment. Thermal shock and brittle fracture could be the most important limit for the Russian fuel element strip design (Bulman 1993).

FUTURE

Background

Before forecasting the state of space reactor fuels development in 2003, the status in 1984, the year of the first Space Nuclear Reactor Symposium, should be remembered. Nearly fifteen years had passed since the Space Nuclear Auxiliary Power program and associated nitride fuels development was terminated, ten years since work on oxide fuels for thermionic reactors ended, and over ten years since the Rover nuclear rocket program was terminated along with research and development on carbide fuels. Although Soviet programs continued to develop space nuclear fuels, not much information was published in the open literature. The SP-100 program, with a requirement for UN fuel, stepped in to fill the void; a nuclear ground test was predicted to start by 1986 and a flight phase was projected for 1991 (Wright 1984). Therefore, a 1984 prediction of today's status of space reactor fuels would estimate that UN fuel would be space qualified for seven years operation, processes for fabricating large quantities of fuel and cladding would be in place, the core for the ground test would be tested, and fabrication of the first flight system would have started.

In fact, during the past ten years UN fuel development evolved from a bench-top process to pilot plant production, 90 fuel pins were irradiated in the Experimental Breeder and Fast Flux Test Facility, high-temperature irradiation performance is well characterized, two-year lifetime is verified, seven-year lifetime is achievable, processes to fabricate rhenium-bonded, niobium-1% zirconium cladding have been developed, and a full core loading of UN fuel pellets has been fabricated to tight specifications. In addition, progress has been made on developing and testing oxide and carbide fuels for space reactors. The thermionic fuel element verification program started in 1985 to demonstrate performance of thermionic reactor fuels. Several thermionic fuel elements have been fabricated and irradiation tested in Experimental Breeder Reactor and Training, Research, and Production General Atomic (TRIGA) reactor, and one-year lifetime has been demonstrated with the potential of operating for five years. A process for fabricating high density UC_2 was developed for the out-of-core thermionic reactor, and irradiation testing started. Renewed interest in nuclear thermal propulsion has renewed interest in the carbide fuel developed during the Rover program. High melting point (U,Zr)C fuel has been processed, new techniques for fabricating spherical fuels have been developed, thermochemical properties have been characterized, particle bed fuels have been tested for the Space Nuclear Thermal Propulsion Program, and steps to recapture the extrusion processes developed for Rover fuels have started. While the postulated 1984 predictions on fuel development were not completely fulfilled, considerable progress was achieved in several unanticipated areas.

TABLE 5. Future Space Nuclear Power Fuels and Cladding Technology Needs.

DESIRED ATTRIBUTES	BACKGROUND & ISSUES
• High Burnup • Low Fission Gas Release & Swelling • Fuel/Cladding/Fission Product Compatibility • Fuel Pin Integrity • Thermionic Fuel Element Integrity • Benign Off-Normal Behavior	SP-100 • Demonstrated UN Operation to 6 at.% at 1400 K • Operation to 10 at.% Burnup Plausible • Issues: UN Dissociation at High Temp. & Burnup THERMIONIC FUEL ELEMENT • Demonstrated UO_2 Operation at 1800 K for 2 Yrs • Potential Operation for 5 Yrs at 2000 K • Issues: Fuel Swelling, Emitter Distortion, UO_2 Vaporization

Criteria for future fuels performance can be derived from perceived mission requirements. When this paper was initially written, those new missions potentially included: (1) application of space power reactors (SP-100 and thermionic systems) for orbital transfer, surveillance, and lunar and planetary surface power missions; (2) utilization of space power reactors for nuclear electric propulsion for deep space scientific missions, and Mars cargo transfer systems; and (3) development and testing of an advanced nuclear thermal rocket engine for piloted Mars missions and a high specific impulse, high thrust-to-weight nuclear thermal rocket for military missions. Desired fuel attributes and issues for space nuclear power and propulsion reactors are summarized in Tables 5 and 6.

TABLE 6. Future Nuclear Thermal Propulsion Fuels Technology Needs.

DESIRED ATTRIBUTES	BACKGROUND & ISSUES
• High Fission Product Retention • Thermal Stability (Low Mass Loss) • High Melting Point (>3200 K) • High Fuel Density ([U] >10%) • Thermal Shock Resistance • Slow Degradation Mechanisms • Chemical Compatibility with Coating • High Surface Area to Volume Ratio • Fabricability	Prismatic Carbide Fuel (Most Experience) • Proven Operating Experience to 2550 K for 2 Hr • Subject to Thermal Shock, and Cracking • Plausible Designs up to 3000 K Gas Exit Temp Cermet Refractory Fuel (Safest) • Robust Design, Compatible with H_2 • High Fission Product Retention • Low Specific Impulse and Thrust/Weight Particle Bed Carbide Fuel (Best Performance) • High Thrust/Weight, High Operating Temp • Excessive Fuel Loss and Fission Product Release • Minimal Long Life Experience Ribbon Fuel (Unknown) • Potential for Long Life at High Temperature • Subject to Cracking and Fission Gas Release • Difficult to Fabricate

Recent political, fiscal, and international events have dramatically changed the requirements for space nuclear power and propulsion systems. However, without requirements, the future discussion becomes moot; therefore, the hypothetical mission set and subsequent fuels requirements will be retained.

Safety and performance are two generic requirements for space reactors that will be influenced by fuel type. Safety requirements will narrow fuel options to those with as-low-as-reasonably-achievable fission gas release, coolant and cladding compatibility, thermal shock resistance, high temperature stability, and high melting point. Ideally these fuel characteristics should lead to robust reactors with high reliability, long lifetimes, restart capability, and large margins to failure. Performance requirements emphasize fuels with high uranium inventory, high burnup, low swelling and fission gas release, high operating temperature, high thermal conductivity, and high temperature stability. These fuel characteristics should lead to propulsion reactors with high specific impulse and low mass and power reactors with low specific mass and long lifetimes. High operating temperature and long lifetime are the crux requirements: the higher the operating temperature, the better the performance but shorter the lifetime and the greater the risk. This tradeoff is schematically illustrated in Fig. 27, where increasing temperatures lead to more rapid degradation of core materials. Swelling and fission gas release can cause fuel/cladding mechanical failures, accelerated diffusion and chemical reactions can cause fuel/cladding/coating chemical interactions, vaporization of the fuel or coating materials lead to mass loss, and fuel melting is the ultimate the operating limit.

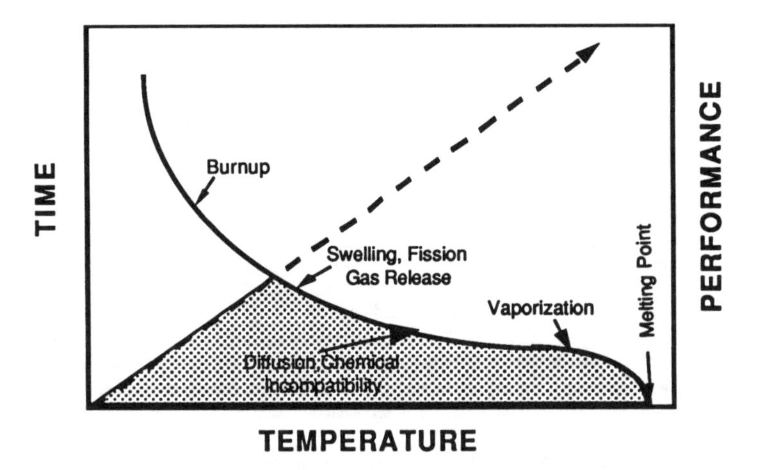

FIGURE 27. Trade-offs between Efficiency and System Lifetime as a Function of Temperature.

The generic question for future fuels development is: What is the optimum combination of high temperature, long life, and low risk? During the next ten years, a significant amount of effort should be spent on developing high temperature fuels and characterizing their time and temperature limitations, principally for nuclear propulsion reactors.

Fortunately, for the fuel developer the problem narrows down to developing the highest temperature fuel allowed by nature, and only three general fuel compounds are available: oxide, nitride, and carbide fuels. In the future, well thought out fuels programs will include the following steps: (1) preparation of an approved fuel development plan that quickly narrows the number of fuel choices, (2) establishment of an out-of-reactor data base of fundamental properties such as phase diagrams, mechanical properties, stability, compatibility, and melting points to screen candidate high temperature uranium/metal compounds and define margins to failure; (3) development of environmentally benign and quality assured synthesis, fabrication, and disposal processes; (4) capsule tests to identify irradiation effects on properties and failure mechanisms; (5) single element and bundle irradiation tests under operation and transient conditions to provide a performance data base of swelling, fission gas release, fission product migration, failure mechanisms, and operating margins; (6) safety testing to establish safe launch conditions and crew radiation safety; (7) reactor system testing and qualification, and (8) flight testing and/or use.

What specifically will and should be done to develop and qualify space reactor fuels during the next ten years?

General Trends

First, lab-scale testing to screen candidate fuels and materials and to define general performance trends and operating margins should be expanded. Non-nuclear testing is relatively inexpensive and technical feasibility issues, operating envelopes, margins to failure, and reliability data can be established with data developed from out-of-reactor testing and small capsule tests. Predictive models that can be verified in a few single element and bundle irradiation tests should be developed. Second, as launch and manned operation of space nuclear reactor systems becomes a reality, greater emphasis would be placed on safety testing for shipping, launch approval, manned-operations, and transient behavior. Because of the lack of approved missions, fuel safety has been inadequately tested in current space reactor fuels development programs; future applications would require significant verification of safety models. Third, because of recent emphasis on environmental and quality issues, advanced fabrication processes to reduce wastes, avoid mixed hazardous wastes, and increase scrap recycle would be developed for the production of space nuclear reactor cores.

Space Nuclear Power

An advancement remaining for UN fuels is to increase burnup and raise operating temperature and specific power for nuclear electric propulsion. Subtle changes to UN characteristics, such as increased grain size to decrease fission gas release and additives to stabilize stoichiometry and immobilize metallic fission products, should be developed and tested. Formation and migration of metallic fission product inclusions have been observed in UN (Matthews 1988) and (U,Pu)N (Giacchetti 1971) at high powers and temperatures. Therefore, it is important to maintain a low temperature gradient and control stoichiometry. Excess carbon in the form of M_2C_3 has been found to be effective in maintaining hyperstoichiometry during (U,Pu)C irradiations (Matthews 1983). Similarly, excess nitrogen in the form of U_2N_3 might help stabilize nitride fuels during irradiation. The effects of excess N_2 on gap conductivity and internal pin pressure would have to be considered in the fuel pin design. In addition, new fabrication techniques, such as high pressure sintering, would have to be developed to fabricate nitride fuels with excess U_2N_3.

Advanced cladding materials such as Mo/Re or monocrystalline alloys to reduce creep and enhance temperature capabilities would also offer improvements. If SP-100 is used for manned missions, reliability, maintainability, and man-rating will be priorities that would require statistical irradiation testing to verify predictive codes and reliability analyses. By the year 2002, utilization of UN fuel pins could become routine.

Assuming that a TOPAZ flight program continues, investigation and fabrication of UO_2 for thermionic reactors would become a major emphasis during the next few years. Analysis of Russian data on the performance of thermionic fuel elements will lead to the development of empirical codes to predict the performance of TOPAZ-I & II type reactors. The TOPAZ fuel program would also require some in reactor performance verification of US fabricated single cell thermionic fuel elements. A production campaign to fabricate UO_2 fuel pellets to load into a Russian TOPAZ-II core could also be completed.

The US thermionic program needs a breakthrough in fuels and emitter properties in order to develop thermionics for long life and high power operations. Collaboration on the TOPAZ program could lead to utilization of the thermionic fuels and materials advances made in the former Soviet Union to reduce life limiting-phenomena (fuel swelling, emitter creep, fission product interactions) that have restricted consideration of thermionic reactors for some long life missions. A more aggressive thermionics technology program should be completed during the next ten years that characterizes failure mechanisms and investigates Russian fuel technologies (low swelling UO_2 and mixed carbides) and emitter materials (single crystal tungsten and rhenium) that may enhance the performance of thermionic reactors.

Nuclear Thermal Propulsion

The next generation of nuclear rockets would start where Rover technology ended, but with a more rigorous set of operating requirements including operating lifetime of minutes to several hours, specific impulse up to 1000 seconds, hydrogen exhaust temperatures up to 3000 K, low fission product release, thrust-to-weight up to 40, and enhanced compatibility with hydrogen. A propellant exit temperature as high as 3400 K theoretically would result in a specific impulse greater than 1000 seconds, a significant increase over the 825-second range achieved for the Rover reactors. The desired high specific impulse leads to concepts with very high temperatures that exceed currently known fuels properties and performance capabilities. The next ten years should see the recapture of Rover fuels and development of high temperature (U,metal)carbide compounds including in-reactor testing under simulated nuclear rocket conditions.

Four fuel forms have been suggested by nuclear thermal propulsion reactor designers —prismatic, cermet, particle bed, and the Russian ribbon fuel— and all have relative advantages and disadvantages. These fuel forms are schematically represented in Fig. 28. Prismatic fuel is basically uranium carbide or (U,Zr)C dispersed in graphite with H_2 coolant channels. Carbide fuels have proven operating experience up to 2550 K for two hours; however, prismatic fuels are subject to thermal shock, cracking, and H_2 corrosion. Cermet fuel contains UO_2 or UN dispersed in tungsten. Cermets are compatible with H_2, resist thermal shock, and minimize fission gas release. However, cermet fuels are heavy and can result in low thrust/mass nuclear rockets. Particle bed reactors have a bed of coated (U,Zr)C spheres inside concentric porous frit tubes with H_2 flowing through the bed. Particle bed reactors have the highest performance capability with high thrust/mass, high specific energy, and high specific impulse. However, particle fuels are subject to high fuel loss and fission gas release and are sensitive to flow blockage. The Russians apparently have developed and tested several nuclear rocket fuel types including ribbons of (U,Zr)C stacked in segmented columns. Although not much has been published on this fuel form (Goldin 1991), the concept is intriguing because it may represent a good compromise between the low surface area to volume ratio prismatic and high surface area particle fuels. However, the fuel appears to be difficult to fabricate, susceptible to brittle failure, and prone to high fission gas release.

The challenge over the next decade will be to combine the best attributes of these fuel forms to match the combination of safety and performance characteristics required for specific missions. During the next ten years, these four fuel types should be tested on a laboratory scale so that a selection of the best type can be made. New fuel forms with the highest potential combination of operating temperature and lifetime, similar to the ribbon fuel or a composite fuel (Matthews 1991), should also be investigated and developed as appropriate.

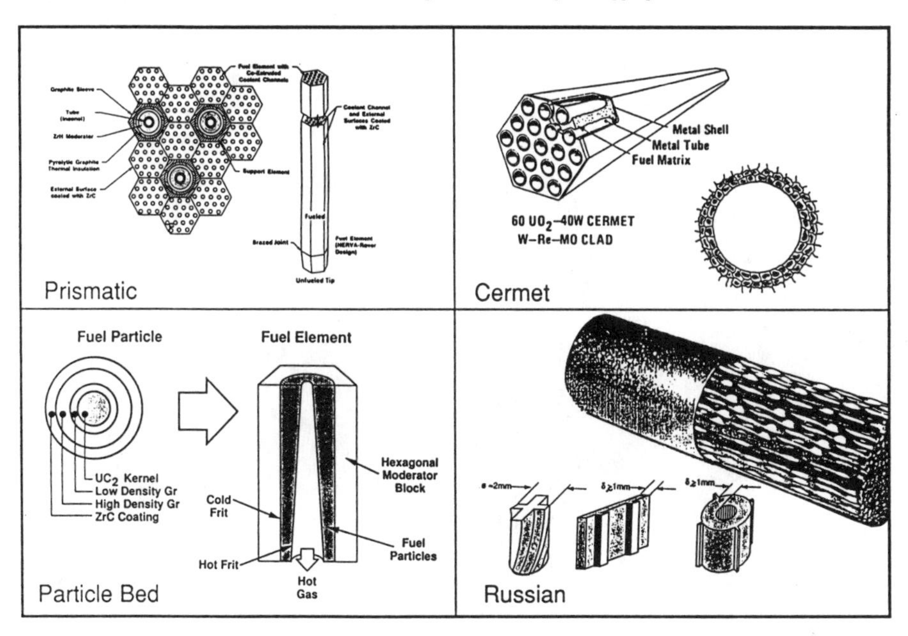

FIGURE 28. Nuclear Thermal Propulsion Reactor Fuel Types.

ACKNOWLEDGEMENTS

The information in this report was supported by projects funded by DOE, NASA, and SDIO, represents collaboration of many organizations including ANL, GA, GE, HEDL, INEL, ORNL, and WHC, and summarizes the efforts of individuals far too numerous to name.

References

*Alexander, C. A. (1967) "Vapor Properties," in *Uranium Dioxide; Properties and Nuclear Applications*, J. Belle ed., (Naval Reactors, Division of Reactor Development, Atomic Energy Commission.

Anselin, F. (1963) "Etude des Nitures d'Uranium de Plutonium et de Leurs Solutions Solides," *Nucl. Mater.*, 10: 301.

Baars, R. E. (1991) "Modeling Emitter Deformation for Thermionic Fuel Elements Using the LIFE Code: Interim Report," Los Alamos National Laboratory report LA-CP-91-2.

*Bauer, A. A. (1972) "Nitride Fuels: Properties and Potentials," *Reactor Tech.* 15 (2): 87-104.

*Belle, J. (1961) "Uranium Dioxide: Properties and Nuclear Applications," U.S. Atomic Energy Commission, Washington, D.C.: 173-304.

Bennett, G. L. (1989) "A Look at the Soviet Space Nuclear Power Program," in *Trans. 24th Intersociety Energy Conversion Engineering Conference*, Washington, D.C.

Bennett , G. L. and T. J. Miller (1991) "The NASA Program Plan for Nuclear Propulsion," in *Trans. 8th Symposium on Space Nuclear Power Systems* Albuquerque, NM: 524.

Blair, H. T., D. W. Carroll, and R. B. Matthews (1991) "Cryochemical and CVD Processing of Spherical Carbide Fuels for Propulsion Reactors," in *Trans. 8th Symposium on Space Nuclear Power Systems*, CONF-910116, Albuquerque, NM: 1194-1206.

Blair, H. T., B. J. Beer, K. M. Chidester, and R. B. Matthews (1989) "NAT Fuel Production Accomplishments," in *Trans. 6th Symposium on Space Nuclear Power Systems*, CONF-890103-Summs., Albuquerque, NM: 128-131.

Borkor, P. C., W. C. Kirk, R. J. Buhl (1991) "The Behavior of Fission Products During Nuclear Rocket Tests," in *Trans. 8th Symposium on Space Nuclear Power Systems*, CONF-910116, Albuquerque, NM: 1194-1206.

Brown, C. and D. Mulder (1992) "An Overview of the Thermionic Space Nuclear Power Program," in *Trans. 9th Symposium on Space Nuclear Power Systems*, Albuquerque, NM.

Bulman, M. J., et. al (1993) "US/CIS Integrated NTRE Concept," in *Trans. 10th Symposium on Space Nuclear Power and Propulsion*, Albuquerque, NM:591.

Butt, D. P. and T. C. Wallace (1993) "The U-Zr-C Ternary Phase Diagram, *J. Am. Ceram. Soc.*, 76 (6): 1409-1419.

Chidester, K. M. and R. B. Matthews (1990) "The Role of Carbon Diffusion in Densification of Uranium Dicarbide," in *Trans. 8th Symposium on Space Nuclear Power Systems*, Albuquerque, NM: 902-909.

Chubb, W., V. W. Storhok, and D. L. Keller (1973) "Factors Affecting the Swelling of Nuclear Fuels at High Temperatures," *Nucl. Technology*, 18: 231.

Clark, J. S. (1991) "A Comparison of Nuclear Thermal Propulsion Concepts: Results of Workshop," in *Trans. 8th Symposium on Space Nuclear Power Systems*, Albuquerque, NM: 740.

*Cordfunke, E.H.P. and R.J.M. Konings (1990) *Thermochemical Data for Reactor Materials and Fission Products*, Elsevier Science Publishers, New York, NY.

Davidson, K.V. (1991) "A Brief Review of Rover Fuel Development at Los Alamos," in *Trans. 8th Symposium on Space Nuclear Power Systems*, Albuquerque, NM: 1015-1023.

DeCrescente, M. A., M. S. Freed, and S. D. Caplow (1965) "Uranium Nitride Fuel Development - SNAP-50 PWAC-488", Pratt and Whitney Aircraft.

El-Genk, M. S. (1987) "Uranium Nitride Fuel Swelling and Thermal Conductivity Correlation," in *Trans. 4th Symposium on Space Nuclear Power Systems*, Albuquerque, NM.

Giacchetti, G. (1971) "Actinides and Fission Products Distribution in Fast Breeder Nitride Fuels," *Nucl. Tech.*, 28.

Goldin, A.Y., A. A. Koroteev, A. D. Konopatov, V. A. Pawhook, N. N. Ponomarev-Stepnoy, V. F. Sernyonov (1991) "Development of Nuclear Rocket Engines in the USSR," AIAA Propulsion Meeting, Sacramento, CA.

*Gomozov, L. I., I. Dedjurin, S. G. Titov, O. S. Ivanov, and A. A. Baikov (1976) "Physical and Mechanical Properties of Some Uranium-Containing Carbides," H. Blank and R. Lindner, eds., in *Plutonium and Other Actinides*, North-Holland Publishing Co., Amsterdam, Netherlands: 915-934.

*Hayes, S. L., J. K. Thomas, and K. L. Peddicord, "Material Property Correlations for Uranium Mononitride, I) Physical Property, II) Mechanical Properties, III) Transport Properties and, IV) Thermodynamic Properties," *J. Nucl. Mater.*, 171: 262-318.

Hilbert, R. F., V. W. Storhok, and W. Chubb (1970) "High Temperature Irradiation Behavior of UN, UC, UO2 Fuels Compared," in *Trans. Am. Nucl Soc.*, 13: 102.

Hilbert, R. F., V. W. Storhok, W. Chubb, and D. L. Keller (1971) "Swelling of UC and UN at High Temperatures," *Proc. Conf. Fast Reactor Fuel Element Technology*, R. Farmakes, ed. American Nuclear Society, Illinois, ANS Conference Proceeding No. : 753.

*Holley, Jr. C. E., M. H. Rand, and E. K. Storms (1984) "Part 6. The Actinide Carbides," *The Chemical Thermodynamics of Actinide Elements and Compounds*, International Atomic Energy Agency, Vienna, Austria.

Jones, R. W. and J. L. Crosthwarte (1973) "Uranium Carbide Fuels for Organic Cooled Reactors," AECL-4443.

Katsura, M. and M. Miyake (1993) "Some Problems in Nonstoichiometry of l-Uranium Sesquinitride," *J. Alloys and Compounds* 193: 101-103.

*Katz, J. J. and E. Rabinowitch (1951) *The Chemistry of Uranium, Part I. The Element, Its Binary and Related Compounds*, McGraw-Hill, New York, NY: 244-395.

*Kingery, W. D. (1959) "Thrmal Conductivity: XIV, Conductivity of Multicomponent Systems," *J. Am. Ceram. Soc.*, 42: 617-627.

*Kingery, W. D., J. Francl, R. L. Coble, and T. Vasilos (1954) "Thermal Conductivity: X, Data for Several Pure Oxide Materials Corrected to Zero Porosity," *J. Am. Ceram. Soc.*, 37: 80-84.

Kirk, W. L. (1990) "Nuclear Rocket Performance Based on Rover/Nerva Technology," AIAA Space Programs and Technologies Conference and Exhibit.

Koenig, D. R. (1986) "Experience Gained from the Space Nuclear Rocket Program (Rover)" Los Alamos National Laboratory report LA-10062-H, Los Alamos, NM.

Kruger, G. B. (1991) "A Comet Fuel Reactor for Nuclear Thermal Propulsion," presented to the Nuclear Thermal Propulsion Workshop, Cleveland, OH.

Lawrence, L. A., B. J. Makenas, and L. L. Begg (1992) "Performance of Fast Reactor Irradiated Fueled Emitters for Thermionic Reactors," in *Trans. 9th Symposium on Space Nuclear Power Systems*, Albuquerque, NM.

*Levinskii, Y. V. (1974) *At. Energ.*, 37 (4): 339.

Lyon, L. L. (1973) "Performance of (U,Zr)C-Graphite (Composite) and of (U,Zr)C (Carbide) Fuel Elements in the Nuclear Furnace 1 Test Reactor," Los Alamos National Laboratory report LA-5398-MS, Los Alamos, NM.

MacMillian, D. P. (1991)"Mass Loss Rates of Uranium-Zirconium Carbide in Flowing Hydrogen and Hydrogen-Hydrocarbon Mixtures," in *8th Symposium on Space Nuclear Power Systems*, Part Three, AIP CONF-910116: 103.

Makenas, B. J., J. W. Hales, and A. L. Ward (1991) "Fuels Irradiation Testing for the SP-100 Program," in *Trans. 8th Symposium on Space Nuclear Power Systems*: 886.

Mason, R. E., K. M. Chidester, C. W. Hoth, and R. B. Matthews (1985) "Uranium Nitride Fuel Fabrication for SP-100 Reactors," in *Space Nuclear Power Systems*, Albuquerque, NM.

Mason, R. E., K. M. Chidester, and R. B. Matthews (1987) "UN Pellet Production with Stoichiometry Control," in *Trans. 4th Symposium on Space Nuclear Power Systems*, CONF-870118-Summs., Albuquerque, NM: 135-141.

Mason, R. E. and M. S. El-Genk (1993) "Experimental Investigation of the Uranium-Ruthenium and Uranium-Rhenium Binary Systems," *J. Nucl. Mater.*

*Matsui, T. and R. W. Ohse (1987) "Thermodynamic Properties of Uranium Nitride, Plutonium Nitride and Uranium-Plutonium Mixed Nitride," *High Temperatures-High Pressures*, 19: 1-17.

Matthews, R. B., et. al. (1991) "Carbide Fuels for Nuclear Thermal Propulsion," AIAA/NASA/OAI Conference on Advanced SEI Technologies, Cleveland, OH.

Matthews, R. B. (1992) "Ceramic Fuel Development for Space Reactors," *Ceramic Bulletin*, 71: 96-101.

Matthews, R. B., K. M. Chidester, C. W. Hoth, R. E. Mason, and R. L. Petty (1988) "Fabrication and Testing of Uranium Nitride Fuel For Space Power Reactors," *J. Nucl. Mater.*, 151: 334-344.

Matthews, R. B. (1983) "Uranium-Plutonium Carbide Fuel LMFBR," *Nucl. Tech*, 163: 9-22.

*Matzke, H. (1986) "Science of Advanced LMFBR Fuels," North-Holland Physics Publishing, Amsterdam, Netherlands.

McLaren, J. R. and P.W.M. Atkinson (1965) "The Sintering of Uranium Nitride," *J. Nucl. Mater.*, 17: 142.

McLaren, J. R., R. J. Dicker, J. D. L. Harrison, and L. E. Russel (1968) "The Preparation and Sintering Behaviour of Uranium Mononitride Made by the Carbothermic Reduction of Uranium Oxides in Nitrogen," in *Proc. 2nd European Sym. Powder Metallurgy, Vol. 2* .

Mondt, J. F. (1989) "Development Status of the SP-100 Power System," in *AIAA/ASME/SAE/ ASEE 25th Joint Propulsion Conference*, AIAA-89-2591, Monterey, CA.

Muromura, T. and H. Tagawa (1977) "Formation of Uranium Mononitride by the Reaction of Uranium Dioxide with Carbon in Ammonia and a Mixture of Hydrogen and Nitrogen - I Synthesis of High Purity UN," *J. Nucl. Mater.*, 71: 65.

Muromura, T. and H. Tagawa (1980) "Formation of Uranium Mononitride by the Reaction of Uranium Dioxide with Carbon in Ammonia and a Mixture of Hydrogen and Nitrogen - II Reaction Rates," *J. Nucl. Mater.*, 81: 330.

*Nikol'skii, S. S. and I. N. Levina (1969) "High-Temperature Thermodynamics of Uranium Carbides. II. Calculations of Uranium and Carbon Activities on the Basis of Storms' Data," *Teplofiz. Vys. Temp.*, 7 (6): 1014-1020.

*Rama Rao, G. A., S. K. Mukejee, V. N. Vaidy, V. Venugopal, and D. D. Sood (1991) "Oxidation and Hydrolysis Kinetic Studies on UN," *J. Nucl. Mater.*, 185: 231-241.

Rankin, R. A. (1988) "Swelling of UC-ZnC Fuels Irradiated at High Temperatures, *Space Nuclear Power Systems 1988*, M. S. El-Genk and M. D. Hoover eds., Orbit Book Company, Inc., Malabar, FL, 8: 231-242.

Roth, T., Ed. (1982) "LIFE-4 Programmer's Manual," Ward-94000-12, Vol. 2, Rev. 0.

Ross, A. M. and R. L. Stoute (1962) "Heat Transfer Coefficient Between UO_2 and Zircalloy-2," AECL-1552 (CRFD)-1075.

Storms, E. K. (1988) "An Equation Which Describes Fission Gas Release from UN Reactor Fuel," *J. Nucl. Mater.*, 158.

Storms, E. K. (1989)"Compatibility Issues in the SP-100 Fuel System," Los Alamos National Laboratory report LA-11468-MS, Los Alamos, NM.

Storms, E. K., D. Hanson, W. Kirk, and P. Goldman (1991) "Effect of Fuel Geometry on the Life-Time Temperature Performance of Advanced Nuclear Propulsion Reactors," in *AIAA/NASA/OAI Conference on Advanced SEI Technologies*, AIAA 91-3454, Cleveland, OH.

*Storms, E. K. (1967), *The Refractory Carbides*, eds. Academic Press, New York, NY: 171-213.

*Storms, E. K. and J. Griffin (1973) "Thermodynamic and Phase Relationships of the Zirconium-Uranium-Carbon System," *High Temp. Sci.*, 5: 423-437.

*Storms, E. K. (1972) "Phase Relationships and Electrical Properties of Refractory Carbides and Nitrides," *Solid State Chemistry*, 10: 37-78.

*Taub, J. M. (1975) "A Review of Fuel Element Development for Nuclear Rocket Engines," Los Alamos National Laboratory report, LA-5931, Los Alamos, NM.

Tennery, V. J., T. G. Godfrey, and R. A. Potter (1971) "Sintering of UN as a Function of Temperature and N_2 Pressure," *J. Am. Ceram. Soc.*, 54: 327.

Thomas, J. K. (1988) "Material Property and Irradiation Performance Correlations for Nitride Fuels," in *5th Symposium on Space Nuclear Power Systems*, Albuquerque, NM.

Truscello, V. C. and L. L. Rutger (1992) "SP-100 Power System," in *Trans. 9th Symposium on Space Nuclear Power Systems:* 1-23.

Vaidyanathan, S. (1993) "Uranium Nitride Fuel Pin Performance Model," *10th Symposium on Space Nuclear Power Systems*, Albuquerque, NM.

*Venugopal, V., S. G. Kulkarni, C. S. Subbanna, and D. D. Sood (1992) "Vapour Pressures of Uranium and Uranium Nitride over UN(s)" *J. Nucl. Mater.*, 186: 259-268.

Wallace, T. C. (1991) "Review of Rover Fuel Element Protective Coating Development at Los Alamos," in *Trans. 8th Symposium on Space Nuclear Power Systems* .

Wallace, T. C., Sr. and D. P. Butt (1994) "Diffusion and Vaporization of Transition Metal Carbides as it Relates to Diffusion Coupled Vaporization Processes: A Review," Los Alamos National Laboratory report LA-UR-93-2903, to be published in the Proc. of the Hydrogen Corrosion Workshop, *10th Symposium on Space Nuclear Power and Propulsion*, Albuquerque, NM.

Weaver, S. C., R. L. Senn, J. L. Scott, and B. H. Montgomery, "Effects of Irradiation on Uranium Nitride Under Space Reactor Conditions," Oak Ridge National Laboratory report ORNL-4461, Oak Ridge, TN.

Wright, W. E. (1984) "Accomplishments and Plans of the SP-100 Program," *Space Nuclear Power Systems:* 37.

Zimmerman , H. (1978) (KfK) "Investigations on Swelling and Fission Gas Behavior in Uranium Dioxide," *J. Nucl. Mater.*, 75: 154-161.

PRELUDE TO THE FUTURE: A BRIEF HISTORY OF NUCLEAR THERMAL PROPULSION IN THE UNITED STATES

Gary L. Bennett
NASA HQ
Code CT
Washington, DC 20546
(202) 358-4676

Harold B. Finger
7837 Laurel Leaf Dr
Potomac, MD 20854
(301) 983-9343

Thomas J. Miller
Lewis Research Center
21000 Brookpark Road
Cleveland, OH 44135
(216) 891-2199

William H. Robbins
Analytical Engineering Corp.
2511 County Club Blvd
North Olmsted, OH 44070
(216) 779-0181

Milton Klein
48 Politzer Drive
Menlo Park, CA 94025
(415) 329-9261

Abstract

The national space policy objective of expanding human presence into the solar system has rekindled interest in advanced propulsion concepts such as nuclear thermal propulsion (NTP). NTP has an exciting history in the U.S., beginning with the first detailed studies after World War II which indicated the feasibility and benefits of nuclear rockets followed by the Rover/NERVA program which demonstrated that nuclear rockets could be built and successfully operated. Some 20 reactors were built and tested in the Rover program showing that long-duration operation sufficient for a piloted mission to Mars was possible. When the U.S. effort was terminated by budgetary considerations in 1973, a tremendous technological legacy was left for future generations to build on in humanity's eventual voyage to Mars and beyond.

INTRODUCTION

On 2 November 1989, President Bush approved a national space policy that reaffirmed that a long-range goal of the civil space program is to "expand human presence and activity beyond Earth orbit into the solar system" [White House 1989]. The long-term focus of the original Space Exploration Initiative (SEI) was to place humans on Mars by 2019. Studies such as those conducted by NASA, the National Research Council (NRC) Committee on Human Exploration of Space, and The Synthesis Group have confirmed studies dating to the 1950s in identifying nuclear propulsion (nuclear thermal propulsion and/or nuclear electric propulsion) as greatly enhancing and realistically enabling piloted missions to Mars [NASA 1989a, NASA 1989b, NRC 1990, and Synthesis Group 1991]. In particular a number of recent studies have shown that nuclear propulsion for Mars missions offers several major advantages over all-chemical propulsion systems [Adams et al. 1990, Bennett and Miller 1991, Bennett et al. 1991, Borowski et al. 1989 and 1990, Braun and Blersch 1989, Frisbee et al. 1990, Hack et al. 1990, Holdridge et al. 1990, and Palaszewski 1989]:

- Reduces the transit time for long stay-time missions for the same initial mass into low-Earth orbit (IMLEO)

  - Minimizes crew exposure to zero gravity, solar flares, and galactic cosmic rays

  - Increases the percentage of mission time spent at Mars

- Reduces round trip times for short stay-time missions for the same IMLEO

and/or

- Reduces the IMLEO (propellant mass) required for the same mission duration

  - Reduces number of Earth-to-orbit (ETO) launches and/or the ETO vehicle lift requirement

- Allows greater mission design flexibility

  - Allows accomplishment of both long-duration and short-duration mission classes with a common vehicle design

  - Increases the Earth and Mars departure windows

  - Increases the propulsion margin available for mission aborts

These benefits are consistent with an earlier exhaustive study of the various uses of nuclear propulsion [TRW 1965]. To varying degrees the nuclear propulsion systems offer similar advantages over chemical propulsion systems using aerobrakes. Aerobraking forces the crew into a highly elliptical orbit about Mars which limits the choice of landing sites. Moreover, the development of aerobrakes for piloted missions to Mars is at least as technically challenging and probably as expensive as developing nuclear propulsion. Unlike nuclear propulsion, aerobrakes will most likely require experimental verification flights to Mars.

Three basic facts about nuclear rockets make it a better performing propulsion system than chemical rockets [Corliss and Schwenk 1971 and Shepherd and Cleaver 1948-1949]:

- Fission-generated heat, which comes from a source (the nucleus) that is inherently more energetic than the chemical source (atomic electrons), is converted into the kinetic energy of the rocket propellant;

- Chemical combustion is not needed thereby eliminating the need to carry an oxidizer and allowing utilization of propellants with low molecular weights to attain high exhaust velocities; and

- Reactor fuel has more energy packed in it than chemical fuels and it is not limited by chemical heats of combustion.

Given the foregoing advantages of nuclear propulsion and the interest that was shown in nuclear propulsion for the Space Exploration Initiative, it is appropriate to review the original nuclear rocket program which was code-named Project Rover and which enjoyed early presidential support. President Kennedy delivered a special message on 25 May 1961 to a joint session of Congress on urgent national needs. Included in this message were the commitment to go to the Moon and "Secondly, an additional 23 million dollars, together with 7 million dollars already available, will accelerate development of the Rover nuclear rocket. This gives promise of some day providing a means of even more exciting and ambitious exploration of space, perhaps beyond the moon, perhaps to the very end of the solar system itself" [Kennedy 1962]. The Rover/NERVA [Nuclear Engine for Rocket Vehicle Applications] program consisted of five major elements: (1) the Kiwi non-flyable nuclear test reactors developed by Los Alamos with support from Rocketdyne; (2) the NERVA project operated by Aerojet and Westinghouse; (3) the Phoebus and Pewee advanced reactors developed by Los Alamos; (4) the RIFT [Reactor-in-Flight Test] project operated by Lockheed; and (5) the Nuclear Furnace fuel-testing project operated by Los Alamos. An overall chronological diagram of the program is shown in Figure 1 [Gabriel 1974 and Gunn 1989].

The following sections provide background information on the historical development and testing of nuclear rockets (specifically the Rover/NERVA program) in the United States. A recent overview of other nuclear propulsion concepts considered by the U.S. may be found in JPL 1992 and NASA 1991.

THE BEGINNINGS

The history of nuclear propulsion can be traced to the writings of Dr. Robert H. Goddard and others before World War II [Goddard and Pendray 1961 and Reupke 1992]. As noted by Bussard and DeLauer [1965]: "The possibility of using a fission 'pile' to heat a working fluid to high temperatures for use as rocket propellant was discussed briefly as early as 1944 at both the Los Alamos Scientific Laboratory and the University of Chicago Metallurgical Laboratory". Following World War II there were a number of studies which led to separate nuclear propulsion and nuclear power programs for space applications [Dewar 1974]. The following paragraphs provide an overview of the major early studies.

Early Studies

Much of the early work had a military bent because the world had just come out of World War II and then entered a long cold war. World War II had involved the first ballistic missile (V-2) and the first nuclear weapons so there was interest in developing the technology to propel ballistically a nuclear weapon. Because it did not appear that chemical propulsion systems had the power to send a nuclear weapon over intercontinental distances there was a continuing interest in making use of the new-found power of the atomic nucleus. A 1945 study of nuclear rocket propulsion undertaken by the U.S. Air Force (USAF) Scientific Advisory Board (SAB) at the request of General H. H. Arnold, under the direction of Theodore von Karman with H. S. Tsien, recommended no action because of the technical difficulties

[Dewar 1974 and Koenig 1986]. Similar pessimistic conclusions were drawn by early studies at what was then the Lewis Engine Laboratory of the National Advisory Committee for Aeronautics (NACA) and the Johns Hopkins University Applied Physics Laboratory (APL), largely because of limitations in the existing materials technology [Dewar 1974 and Ruark et. al. 1947]. The APL report did conclude that hydrogen was the best propellant for nuclear rockets and that graphite and metallic carbides were the most likely materials to be used in the construction of nuclear rocket reactors. A key aspect of the negative conclusions can be traced to the fact that "the necessary data do not exist, and their accumulation is the *sine qua non* for further progress in this field" [Bussard and DeLauer 1965 and Ruark et al. 1947].

Project Rand, which began as a separate department of the Douglas Aircraft Company, published a short memo in 1946 by Richard Serber in which a heat exchanger rocket was judged to be the most practical approach to nuclear propulsion [Dewar 1974]. A much more comprehensive study was completed in 1947 by the staff of the Aerophysics Laboratory of the North American Aviation Corporation following an initial survey by Hubert P. Yockey and Thomas F. Dixon [Dixon and Yockey 1946 and North American Aviation 1947]. The North American Aviation team concluded that a nuclear rocket having a reactor composed of uranium bearing fuel elements made of graphite and using hydrogen as the propellant was feasible and the staff conducted experiments which led to the recommendation of a carbide film to protect the graphite from erosion by hot hydrogen [Dewar 1974]. Nuclear rockets were considered briefly in the 1948 Massachusetts Institute of Technology (MIT) Lexington critical study on NEPA [Dewar 1974]. The best unclassified study of nuclear propulsion produced in the 1940s was done by two members of the British Interplanetary Society. This study contained the following as one of its six conclusions: "The hope of achieving interplanetary flight must therefore be centred largely on some application of atomic energy, though the radiation difficulties involved in such a development indicate that chemical motors may still play an important subsidiary role" [Shepherd and Cleaver 1948-1949].

Nuclear Airplane Heritage

The next phase of nuclear rocketry is loosely associated with the two nuclear airplane programs: NEPA (Nuclear Energy for the Propulsion of Aircraft), which ran from 1946 to 1951, and its successor the ANP (Aircraft Nuclear Propulsion) program. While working on the ANP program at the Oak Ridge National Laboratory (ORNL), R. W. Bussard did some of the pioneering analyses which sparked the interest of John von Neumann, chairman of the Air Force Strategic Missiles Evaluation Committee [Dewar 1974]. As Bussard noted upon his arrival at ORNL in 1952: "It seemed, from the immediate-post-WW II work, that nuclear rocketry had been effectively written off as a dead end by most of the missile and rocket people (who didn't really understand nuclear energy and liked chemical energy better anyway) and most of the reactor people (who thought the whole idea of nuclear flight of any sort was generally loony)" [Bussard 1962]. Bussard's work in turn sparked new efforts at Lewis, North American Aviation, General Electric, and ORNL [Dewar 1974].

In his seminal 1953 study R. W. Bussard concluded ". . . that nuclear-powered rocket vehicles will be lighter than comparable chemically powered vehicles for vehicle velocities greater than 15,000 ft/sec with payload weights of 1,000 to 10,000 lb, or for vehicle velocities greater than 7,000 ft/sec with payload weights greater than 10,000 pounds" [Bussard 1953]. Bussard was also heavily involved in the critical 1955 Los Alamos study which concluded that "A combination of a chemical booster and a nuclear rocket is proposed to accomplish the Atlas and/or more exacting missions" [Aamodt et al. 1955]. This was the key report upon which positive recommendations came from the Air Force Scientific Advisory Board. Corliss and Schwenk [1971] have observed that "It was Bussard, in fact, who did much to resurrect nuclear rocketry. Working in the nuclear aircraft development program at Oak Ridge National Laboratory (ORNL) in Tennessee in the early 1950s, he was able to show that the earlier nuclear rocket studies had been too negative and too conservative. He was convinced that nuclear rockets *could successfully compete* with chemical rockets on long flights with heavy payloads. Bussard's studies and personal salesmanship were decisive. The Air Force in early 1955 decided to reexamine nuclear rockets as ICBM thrusters" [emphasis in original]. It should also be noted that von Neumann "was a prime mover in establishing the program" through his involvement in the various studies that took place following the publication of Bussard's work [Dewar 1974]. As an historical note it must be mentioned that even in the perceived "faster, cheaper, better" days of the 1950s, studies, reviews, and more studies were required in order to start a program. And money was tight -- less than one million dollars in exploratory research money was made available for nuclear rocket research in mid 1955.

Rover Starts

The nuclear rocket program officially began on 2 November 1955 at two Atomic Energy Commission (AEC) laboratories: Los Alamos Scientific Laboratory (now Los Alamos National Laboratory) and Lawrence Radiation Laboratory (now Lawrence Livermore National Laboratory), although it had unofficially been under way since April 1955 when both laboratories formed divisions to work on nuclear propulsion following the positive recommendations in March 1955 of an Ad Hoc Committee of the Air Force Scientific Advisory Board. At Los Alamos the newly formed division, known as N-Division, was headed by Raemer E. Schreiber with Roderick W. Spence as its alternate leader. At Livermore the nuclear propulsion section was headed by Haywood Gordon and Ted Merkle [Dewar 1974].

The problem facing these pioneering researchers was summed up in Spence [1965]: "If you've ever done much work with hot tungsten filaments, you know that a 2000°C filament is mighty bright--too bright to look at without smoked glasses. And 2500°C it is really dazzling. Now, instead of a tiny filament, imagine a mass of material about the size and shape of an oil drum, and imagine that some internal energy source only slightly less powerful than Hoover Dam is heating the drum to an average temperature close to 2000°C. Imagine further that cold hydrogen is being pumped into one end of this incandescent mass at a rate of 70 lb/sec, and hot hydrogen at a temperature of 2000°C is scorching out the other end. You are now imagining a nuclear rocket engine in operation".

Five years later Dr. Glenn T. Seaborg, Chairman of the USAEC would echo Spence's remarks: "Lest you get the impression that the development of such a nuclear rocket is as simple as its principle sounds, let me point out what is involved in it. What we must do is build a flyable reactor, little larger than an office desk, that will produce the 1500 megawatt power level of Hoover Dam and achieve this power in a matter of minutes from a cold start. During every minute of its operation, high-speed pumps must force nearly three tons of hydrogen, which has been stored in liquid form at 420°F below zero, past the reactor's white-hot fuel elements which reach a temperature of 4,000°F. And this entire system must be capable of operating for hours and of being turned off and restarted with great reliability" [Seaborg 1970].

At Los Alamos one of the basic decisions was selection of the fuel element and the core support materials because "they would influence if not determine reactor design and fabrication. Three candidate materials were examined, graphite, metals, and ceramics" [Dewar 1974]. As recorded in Dewar [1974]: "Ceramics were considered to be of the future as little was known about them in 1955, leaving refractory metals and graphite as the leading contenders. Each had positive and negative qualities. Among refractory or heat resisting metals, tungsten was the leading candidate, remaining strong and integral at very high temperatures. Also tungsten could be fueled with uranium oxide compounds diffused through the metal or could be used as cladding on a tungsten/uranium oxide core. And the metal did not react with the different candidate working fluids, hydrogen, ammonia, or methane. Negatively, tungsten was very expensive, difficult to fabricate, and a relatively strong neutron absorber, thereby directly influencing reactor design.

"On the other hand, graphite was an excellent neutron moderator and therefore posed few restrictions on reactor design. Weaker than tungsten at normal room temperatures, graphite had the unique quality of gaining strength as the temperature increased; at very high temperature, graphite's strength equalled [sic] if not excelled that of tungsten's, because tungsten loaded with uranium lost some of its high temperature qualities. Graphite did not when so loaded. Furthermore, graphite was very inexpensive and easy to fabricate. The material's most striking limitation, however, was its vigorous reaction with hot hydrogen, by itself or in methane or ammonia. The unprotected graphite was eaten away almost at centimeters per minute by the hydrogen eroding the carbon out of the graphite to form acetylene or methane, depending on the temperature. Thus an unprotected graphite reactor having hot hydrogen pass through it would be likely to collapse within a few minutes. But a few minutes of reactor lifetime were all that were necessary in 1955 because the ICBM mission was considered paramount; a running time longer than five minutes was not needed to complete the mission. Thus, the reasons of cost, fabricability, neutronic design, and operational lifetime for the mission determined the selection of graphite . . . N-Division felt the erosion problem could be solved either by protecting the graphite with a metallic carbide coating or adding a carbon-base compound to the working fluid" [Dewar 1974].

After surveying several options ranging from 600 MWt to 2000 MWt, N-Division decided to design a 1500-MWt reactor using ammonia as the propellant because it allowed the design of a smaller booster for an ICBM and because it was easier to handle than the desired hydrogen (although Los Alamos had almost immediately conducted electrically heated tests of graphite fuel elements in hot hydrogen). In contrast, the Livermore team decided that a 10,000-MWt graphite reactor with liquid hydrogen as the propellant operating as a single stage offered the best way to overcome in a short time the many obstacles to nuclear propulsion rather than the building block approach of Los Alamos. Some envisioned this approach as a way to space flight using the Eniwetok atoll. Initially both laboratories looked at the ANP test complex in Idaho as the test site for ground testing nuclear rockets but they concluded that the ANP test site was not feasible and, instead, selected the Nevada Test Site (NTS) ". . . because of its remoteness from heavily populated areas, its general proximity to the two laboratories, its potential for expansion, and its generally favorable climate -- permitting testing eleven months a year. Both laboratories then surveyed the arid desert, selected two different locations for their test complex, started development studies for the facility, and initiated some preliminary construction. The Los Alamos area was called Jackass Flats, the Livermore Cain Springs" [Dewar 1974]. Figure 2 is a schematic of the Nuclear Rocket Development Station (NRDS) and Figure 3 is an overview of the hydrogen dewars and reactor test cell at NRDS.

However, the move toward hardware development entailed increased costs with the two-laboratory approach. Subsequent military studies, most particularly the Loper committee report in 1956, coupled with the improvements in chemical propulsion systems for ICBM use and the lack of funding led the AEC to concentrate the nuclear rocket work at Los Alamos which was judged to have the superior facilities for materials research [Dewar 1974]. Major contractors in the Rover program were (then) North American Aviation's Rocketdyne Division (liquid hydrogen turbopump and nozzle) and Aerojet General (flow control system), ACF-Erco (pressure shell), and Edgerton, Germeshausen and Greer [EG&G] (instrumentation) plus a dozen or more small contractors [Nucleonics 1958]. Livermore went on to work on Project Pluto, the nuclear ramjet (note the "Tory program" mentioned in Figure 1) [Dewar 1974]. The two programs overlapped in the sense that both involved the development of extremely high-temperature reactor systems [Nucleonics 1958]. As noted in Dewar [1974]: "Ironically, although Livermore was taken out of the program, their division nickname, Rover, became the code word for the project at Los Alamos. In the meantime, N-Division began preparations to carry out the two newly defined and approved objectives of the Rover program: to demonstrate feasibility and to carry on research leading to advanced nuclear systems". Table 1 summarizes the nuclear tests conducted in the Rover/NERVA program.

KIWI - THE FLIGHTLESS BIRDS

The Los Alamos team began the Rover program by adopting a dual course between two extremes: "a limited research program into advanced nuclear propulsion systems and materials, and a specific basic reactor testing effort called the KIWI program. Named by Thomas Gittings after the flightless New Zealand bird, KIWI was a two-phase sequence, moving from the relatively easy to the more difficult steps in establishing feasibility. KIWI-A, the easier, was to determine the basic data both in reactor and testing procedures; KIWI-B, the more difficult, was to use liquid hydrogen as the working fluid" [Dewar 1974]. One reason the Kiwi program began with gaseous hydrogen was the unavailability of liquid hydrogen pumps. In designing the Kiwi reactors the Los Alamos researchers had to consider [Dewar 1974]:

- Heating the working fluid uniformly to as high a temperature as the materials allowed

  - Hot and cold spots had to be avoided because they could degrade reactor performance and control

  - Neutron peaking in the core edges near the reflectors had to be calculated accurately to avoid problems with thermal expansion, thermal stress, and heat transfer

  - Metal parts heated by nuclear radiation from the core had to be cooled and allowances made for possible changes in their physical properties

- Thermal expansion of the core

  - The lateral support had to constrain the core without being so rigid that the fuel elements cracked yet also without being so open that the core "wobbled"

  - The fuel elements had to withstand without buckling or breaking the rapid rise to a high internal temperature which caused different rates of expansion from the interior to the exterior

- Stresses induced by the flow and pressure of the propellant

  - The fuel elements had to withstand flow-induced vibrations or fluctuations

  - The core had to be supported so that the pressure drop across it did not force the core out the nozzle

- Reactor controllability

  - The flow of gas into the reactor and the operation of the reactor had to be coordinated and controlled to reach high temperatures quickly but stably

  - The reactor had to operate in a stable steady state mode (and later to deal with throttling)

As an aside it is worth noting the analyses done by the Reaction Motors Division of Thiokol Chemical Corporation: "Developing a satisfactory fuel element is first on the list of problems. One starts with one of the high-temperature materials such as graphite as combination moderator and matrix for the uranium fuel. Then it is necessary to fabricate this mixture into shapes rugged enough to withstand violent thermal and mechanical stresses and yet thin enough to minimize thermal gradients and provide adequate surface area. . . With other things equal the design with the smallest pressure drop is preferred" [Levoy and Newgard 1958]. The Thiokol researchers studied the relative pressure drops of fuel spheres, wire screens, Raschig rings, staggered wires with cross flow, staggered wires with parallel flow, and flat plates finding flat plates and staggered wires with parallel flow to be superior. The flat plate concept allowed two-phase flow relief for hot spots. Spherical fuel elements and Raschig rings had the highest pressure drops [Levoy and Newgard 1958].

Kiwi-A

Los Alamos began the Kiwi series with a simple low power (100 MWt) design with these objectives [Dewar 1974]:

- To demonstrate that a high power density reactor could heat a propellant quickly and stably to high temperature
- To establish basic testing procedures
- To determine the basics of the graphite-hydrogen interaction

Kiwi-A, which is shown schematically in Figure 4, consisted of an annular stack of flat-plate, graphite fuel elements loaded with highly enriched urania particles using graphite cylinders for the inlet and radial reflectors (separated from the core by a carbon wool), all surrounding a central hole of D_2O to moderate the neutrons (reduced critical mass) and to cool the control rods. (The central hole or island was not envisioned for a flight design.) Flat plates were selected because they were judged to be easier to make although they were not coated. Power flattening was achieved by varying the fuel loading. On 1 July 1959, Kiwi-A was operated in Test Cell A for 5 minutes at 70 MWt, which was less than the design of 100 MWt; however, because of a loss of the graphite closure plate and graphite wool a compensatory test approach led to much higher fuel temperatures (up to 2900 K) than anticipated. Temperatures were monitored upstream and no calculations had been performed on the gamma heating on the closure plate; in effect the whole bottom plate broke under the thermal stress. The high fuel temperatures led to melting of the UC_2 fuel and high erosion of the graphite fuel plates but no plates were ejected [Dewar 1974 and Koenig 1986].

"Even so, the test was an eminent success: KIWI-A had achieved its test objectives, demonstrating that a high power reactor could operate stably and controllably at high temperatures for a period of time. Too, invaluable data and experience had been gained on testing procedures. On the other hand, the failure of the graphite part, even though it caused the reactor core to run much higher in temperature than expected, (in fact, much hotter than any KIWI) was not serious -- the KIWI-A core design was not to be used again. The erosion of the graphite was expected, but the extent, the rapidity, and the exact location of the corrosion was what was sought to be learned. The data would be useful in developing techniques to overcome corrosion in the KIWI-B fuel elements. In essence, KIWI-A convinced many that the nuclear rocket was feasible" [Dewar 1974].

Kiwi-A' and Kiwi-A3

Los Alamos was already pushing ahead with a 1000-MWt reactor called Kiwi-B that contained a completely redesigned core. In order to test some of the Kiwi-B design features during the time when the new, larger reactor, test cell (Test Cell C), turbopump and nozzle were being built, Los Alamos decided to test certain of the new reactor-core features in the Kiwi-A geometry. For the new Kiwi-A reactors, known as Kiwi-A' and Kiwi-A3, the Kiwi-A core design was changed from whims and fuel plates to UO_2-loaded cylindrical fuel elements contained in graphite modules as shown in Figure 5. The graphite modules were supported from a flat graphite plate screwed into a metal disk at the inlet end. Each cylinder had four axial coolant channel holes that were coated by a chemical vapor deposition (CVD) process with NbC to reduce hydrogen corrosion of the graphite. The cylindrical fuel elements were comprised of stacked segments about 23 cm long which enabled testing different NbC coating thicknesses and techniques (carbide and CVD) [Dewar 1974, Koenig 1986, and Spence 1968].

Kiwi-A' was tested for nearly 6 minutes at 85 MWt on 8 July 1960 and Kiwi-A3 was operated in excess of 5 minutes at 100 MWt on 10 October 1960. As described in Dewar [1974]: "Both hot tests indicated essentially the same results. With A', the fuel element erosion was moderate, but in the improved A3, erosion was minimal compared to that suffered in the A test. Thus, both carbiding and vapor deposition appeared promising solutions to one major developmental problem and long reactor lifetimes suitable for space missions seemed feasible. On the other hand, fuel elements in both reactors were cracked and ejected out the nozzle, indicating that the B core design had a serious weakness. Until the core, the most important part of a nuclear rocket, could be made structurally sound, the program could not continue".

The Los Alamos team now focused on three new core support concepts in 1960: Kiwi-B-1, Kiwi-B-2, and Kiwi-B-4. During this same period NASA and the AEC formed the joint Space Nuclear Propulsion Office (SNPO) under the direction of Harold Finger with Milton Klein as his deputy to manage the nuclear rocket program. The newly established NASA took over the USAF responsibilities for nuclear propulsion research. (USAF had moved on to other propulsion research because it had no requirement for a nuclear rocket.) In consideration of flying a nuclear rocket engine the RIFT program was initiated by NASA at the Lockheed Corporation in 1960. RIFT is discussed in a separate section of this paper. By the summer of 1961 SNPO had selected the team of Aerojet-General (Aerojet Nuclear Systems Company or ANSC) and Westinghouse Electric Corporation (Westinghouse Astronuclear Laboratory or WANL) to perform the development phase (NERVA) of the flight nuclear rocket. [Dewar 1974, Koenig 1986 and Robbins and Finger 1991].

Kiwi-B1A

The 1000-MWt-class Kiwi-B series of reactors were based on hexagonal fuel elements with six of these elements clustered around a single tie-rod support element. In order to maintain the size of the reactor at this higher power level the number of fuel elements and coolant holes were increased. As part of the sizing, the internal D_2O moderator and control rods were replaced with 12 boron carbide control drums mounted external to the core in the cooler reflector region. Two support arrangements were then being considered: the B-1 cold end support and the B-2 hot end support. Initially, the 1000-MWt-class Kiwi-B reactors were envisioned as a stepping stone toward the eventual 10,000-MWt-class reactors desired but soon the 1000-MWt-class became an end in itself [Dewar 1974, Koenig 1986 and Walton 1991].

In order to meet schedules within the constraints of existing facilities (Test Cell A), Los Alamos decided to operate the Kiwi-B1A reactor at only 300 MWt with gaseous hydrogen. The test began on 7 December 1961 but a "leak in the seal between the nozzle and pressure vessel resulted in a very large hydrogen fire, with the highly pressurized gas acting like a welder's torch, cutting through the fittings on the nozzle and pressure vessel and threatening to destroy the reactor itself" [Dewar 1974]. As a result Kiwi-B1A was only operated for 30 seconds at 300 MWt but it did partially fulfill its test objectives. Dewar (1974) has written: "Confidence was substantiated in the control drums, beryllium reflector, and pressure vessel -- all new design features. Even the nozzle, which could be tested adequately only in an actual reactor operation, performed flawlessly; some indentations were noticed in the cooling passages, but they were not considered serious". However, the early shutdown limited the data collected on fuel element and support structure performance which in turn left unanswered questions regarding future Kiwi-B tests and also affected the ongoing NERVA design work (in particular the support system) [Dewar 1974].

Kiwi-B1B

By now it was clear that the Kiwi-B4 design with its superior fuel element support plan (see discussion on Kiwi-B4) was the preferred option but delays in its fabrication led Los Alamos to test Kiwi-B1B in order to obtain some data on operation of a reactor with liquid hydrogen even though the core concept was considered marginal. Los Alamos engineers tested a nonnuclear Kiwi-B1B mockup with liquid hydrogen and found no problems, so they reassembled it with uranium-bearing fuel elements and began nuclear testing on 1 September 1962. The reactor startup began at low power and achieved proper chilling. Then the reactor was taken to 600 MWt without any power excursion caused by the presence of liquid hydrogen. However, "the core then failed, and spectacularly and regularly ejected fuel elements from the nozzle" [Dewar 1974]. The reactor was ultimately taken to about 900 MWt where it was allowed to operate for a few seconds with the continuing loss of fuel elements and then shut down following a leak in the nozzle instrument which caused a hydrogen fire around the nozzle [Dewar 1974 and Koenig 1986].

As Dewar (1974) observed: "Notwithstanding technical failure in one area, B-1B was quite successful in another. The worries and doubts about liquid hydrogen startup, about two-phase flow, proved groundless; in fact, far from hindering reactor handling, hydrogen in the core tended to regulate reactor operation. Stability was maintained. Too, non-nuclear components performed well -- the turbopump, the drive system, and the nozzle. The nozzle hydrogen fire resulted from a sensor instrument failure creating an opening, not from any deficiency in the nozzle itself. Confidence was gained also in test facility operation" [Dewar 1974].

Kiwi-B4A

"Different from the modular types of cores of B-1 and B-2 in which the fuel elements were inserted into a graphite module support structure, the B-4 featured an all-loaded core with a unique, radical, but superior fuel element support plan. Each fuel element was an extruded, hexagonal graphite rod about three quarters of an inch in width, having nineteen systemmatically [sic] arranged holes to allow for gas passage, one in the center surrounded by six surrounded by twelve. Each six-sided fuel element was combined with six others to form a cluster, one in the center surrounded by six. Bundled together, each cluster was combined with many others to form the core which, from a distance, resembled a honey comb. Filling in the edges along the core's periphery, to make the core cylindrical, were unloaded half-hexagonal fuel elements" [Dewar 1974]. Figure 6 shows the general layout of the Kiwi-B4A fuel element cluster which became the basis for the NERVA fuel element design [Gunn 1989]. Core support was achieved by using metal tie rods held by an aluminum plate at the inlet or cold end and fastened to a heat-resistant molybdenum core which, in turn supported a graphite block [Dewar 1974].

On 30 November 1962 Kiwi-B4A, the first design intended as a prototype flight reactor, was tested. The liquid hydrogen startup was successful [Dewar 1974 and Koenig 1986]. "But paralleling the rapid increase in power was a rapid increase in the frequency of flashes of light from the nozzle; on reaching 500 MW, the flashes were so spectacular and so frequent that the test was terminated and shut down procedures began. Initial disassembly confirmed that the flashes of light were reactor parts being ejected from the nozzle; further disassembly and analysis revealed that over 90% of the reactor parts had been broken, mostly at the core's hot end" [Dewar 1974].

The Kiwi-B4A test had far-reaching consequences. President Kennedy had been scheduled to visit Los Alamos and the Nevada Test Site in December 1962. President Kennedy's visit to NRDS following the Kiwi-B4A failure led to a presidential decision to slow down flight testing which meant limited funding for paper studies in the RIFT program pending successful completion of Kiwi-B4B hot testing. Meetings and technical debates in January 1963 led to SNPO insisting on a more systematic development program with further hot flow testing. The program was put on hold until a determination and a solution had been made of the cause of the Kiwi-B4A failure. A special task group was established and WANL engineers began working with Los Alamos engineers to develop a better hot end design [Dewar 1974].

Kiwi-B4D

Cold flow tests using gaseous nitrogen, helium, and hydrogen were run on Kiwi-B-type reactors having fuel elements without fissionable material. The first test showed that flow gas caused severe vibrations that in turn cracked the fuel elements. Adding a gas seal and strengthening the girdle curtailed the vibration and led to a successful cold flow test. SNPO then approved the continuation of the hot testing which began with Kiwi-B4D a reactor containing the hot end seal. The Kiwi-B4B core "consisted of full-length, UO_2-loaded, 19-hole, hexagonal fuel elements with bores NbC coated by the tube-cladding process" [Koenig 1986]. An initial cold flow test showed no vibration problem with Kiwi-B4D. The nonnuclear fuel elements were replaced with nuclear fuel elements and hot testing started on 13 May 1964. "Starting quickly and completely automatically, B-4D reached and maintained full power of 1000 MW for about a minute until a leak in the nozzle forced termination of the test. Small amounts of liquid air trapped in the nozzle tubes were thought to have caused small explosions when mixed with the liquid hydrogen, resulting in a rupture of the nozzle. Other than the nozzle failure, disassembly confirmed that the test was a complete success; the core was intact, no vibration had occurred, the core design was good" [Dewar 1974]. S. V. Gunn has commented that the explosions resulted from the accumulation of liquid air between the pressure shell and the outer surfaces of the nozzle tubes. The intense radiation environment led to the formation of ozone in the liquid air. The ozone started to detonate [Gunn 1993].

Kiwi-B4E

The Kiwi-B4E reactor, which was the last of the Kiwi reactors to be tested, had the same core design as Kiwi-B4D but used an improved fuel element consisting of beaded or coated particles in a graphite matrix [Dewar 1974]. "The core consisted of full-length, 19-hole, hexagonal fuel elements, loaded for the first time with UC_2 particles. The bores were NbC coated by the tube-cladding process" [Koenig 1986].

The first test of Kiwi-B4E was conducted on 28 August 1964. "Running for eight minutes at 900 MW, the duration of the test was limited by storage capacity of the liquid hydrogen dewars. Startup and control were smooth and stable; the core performed well with no flashes; the exit gas temperature was 2000 degrees C, slightly lower than B-4D" [Dewar 1974]. In order to obtain more information on fuel element lifetime and reactor reliability Kiwi-B4E was tested again on 10 September 1964 at nearly full power for 2.5 minutes [Dewar 1974 and Koenig 1986]. "No problems were encountered. Subsequent disassembly and analysis revealed that the beaded fuel elements suffered only minimal corrosion and that the core remained intact. The reactor could have run much longer" [Dewar 1974].

One final note on the Kiwi reactor program: In September 1964 "Measurements, at zero power, of the neutronic interaction of two Kiwi reactors positioned adjacent to each other verifi[ed] that there is little interaction and that, from a nuclear standpoint, nuclear rocket engines may be operated in clusters similar to chemical engines" [Koenig 1986; also cited in Finseth 1991].

Following the Kiwi tests Los Alamos focused on advanced nuclear rocket materials and concepts while ANSC and WANL converted the Kiwi-B4 core design into the NERVA core design. But by now the focus had shifted from developing an early flight engine to conducting basic research and technology [Dewar 1974].

THE KIWI-TNT NUCLEAR SAFETY TEST

Safety had long been a part of the nuclear rocket program [Decker 1962 and Dix 1967]. As part of the Rover Flight Safety program Los Alamos ran a full-scale destructive excursion experiment using the Kiwi type of reactor. The experiment was named Kiwi-TNT which was an acronym for Transient Nuclear Test [King et al. 1966]. The primary objective of the experiment was "To supply experimental information on the total energy produced, the kinetic or explosive energy release, and the fission product dispersal from a maximum type of accidental reactor excursion. This information is of great importance to the Rover Flight Safety program since these results can be directly compared with theoretical predictions. Suitable parameters in such calculations can then be adjusted so that they match the experimental results. The prediction of any other type of accident can then be made with confidence since extrapolations and uncertain assumptions are eliminated" [King et al. 1966]. Second objectives were [King et al. 1966]:

- To supply experimental information on core fragmentation which is of interest to the Self Destruct Concept of reactor disposal in space;

- To supply information on decontamination problems, potential missile damage, and reactor component dispersal from an accidental excursion; and

- To supply a large short burst of neutrons to external sample experiments. This was of particular importance to power reactor and Rover type fuel studies since experimental results could be extended into a transient flux region hitherto unattainable by any other means.

As described by King et al. 1966, "The Kiwi-TNT was a nuclear transient reactor having the same basic configuration and nuclear characteristics as the Kiwi-B-4E-301. The primary function of this reactor was to generate a nuclear transient of an appropriate order-of-magnitude for evaluation and study. No propellant was used". The reactor, which was mounted on a specially constructed railroad car, was deliberately destroyed on 21 January 1965 at NRDS. The destruction was accomplished by rotating the control drums as fast as possible (about 100 times the normal rate) [Finseth 1991 and King et al. 1966]. For the control drums all the way out Kiwi-TNT was 7.3$ above prompt critical [King et al. 1966]. Basically what occurred was a thermally induced mechanical explosion of the reactor core.

As summarized by Finseth 1991, "Some of the results were: 1) Core temperature measurements indicated a temperature of about 2167 K (3900 R); 2) Within a 7620 m (25000 ft) radius, only about 50% of the core material could be accounted for. The remainder presumably either burned in the air or was so fine as to be carried further downwind in the cloud; 3) It was estimated on the basis of the total energy which was produced by the excursion that only 5-15% of the core could have been vaporized; 4) The heaviest piece of debris found was a portion of the pressure vessel approximately 0.91 m (3 ft) square and weighing 67 kg (148 lb). It was located 229 m (750 ft) from the reactor. Another piece of the pressure vessel weighing 44 kg (98 lb) was found 457-533 m (1500-1750 ft) from the reactor; 5) The total number of fissions was determined to be approximately 3.1×10^{20}."

Finseth (1991) concluded that "The experimental results from the KIWI TNT excursion provided the basic experimental information required for the general analysis of potential accidents of interest to the ROVER Flight Safety Program".

NERVA - THE FLIGHT ENGINE

In June 1961, NASA and the AEC awarded the NERVA contract to Aerojet and Westinghouse to develop the RIFT engine which was to be based on the Kiwi-B design. At that time the NERVA engine was to produce a thrust of 245 kN and a specific impulse of 7450 m/s at a power of 1100 MWt. As noted in the preceding section, structural problems with the Kiwi-B design in 1962 led to successful design changes (such as the improved core support) to overcome the vibration induced by the liquid hydrogen. Separately the rising cost of the space program led to the cancellation of the RIFT program in 1963 ". . . and with it the loss of a well defined mission [technically RIFT was a "use" for the NERVA engine and not a "mission" per se] for the NERVA engine . . . The loss of a defined mission did imply that a more valuable product of any future effort be the improvement of general nuclear rocket technologies rather than the qualification of a specific engine system. The program was redirected along this line in early 1964. THE NRX series of tests began in April of 1964 and culminated with the successful one hour full power run of NRX-A6 in December of 1967. Integration of the reactor, pressure vessel, nozzle, turbopump and valves which had been tested on NRX/EST [Engine System Test], was reaching the final stages for prototype system testing as the XE' engine. Full power tests of this configuration achieved all of the objectives of the NERVA program while operating under simulated altitude conditions in the engine test stand with 28 startups and 228 minutes of total operation." [WANL 1972].

Objectives and Program Logic

The primary objectives to be demonstrated by completion of the program as revised in 1964 were [WANL 1972]:

- Capability to operate for 60 minutes at full power;
- Restart after full power from a fully cooled condition anywhere in the life cycle;
- Startup and shutdown temperature ramp rates up to 83 K/s;
- Cool down using only liquid hydrogen (LH_2);
- Bootstrap startup of the engine; and
- Determination of system operational margins, limits, and reliability.

The distribution of the major test objectives to the individual full-scale tests actually performed is shown in Table 2. The overall NERVA program logic involved a step-by-step process of defining the critical objectives and comparing them with the unknowns which needed to be determined to achieve those objectives. The systems were modeled and tests were run followed by comparisons between predicted behavior and experimental results. As noted in WANL (1972): "Although the logic sounds so simple that one would be inclined to consider it to be only common sense; the NERVA test program is one of the few in which this type of logical implementation was used consistently. The management attention to the rigorous use of analysis to predict results before performing the tests, followed by a comparison of measured results versus the prediction, was a powerful factor in completing a successful program".

Reactor Description

This section, which is essentially verbatim from WANL (1972) presents a description of those aspects of the NRX-A2, NRX-A3, NRX-A4/EST, NRX-A5, NRX-A6 and XE-Prime reactors which bear significantly on the nuclear and radiation and shielding design and analysis technology. The basic reactor configurations of all of the above systems except the NRX-A6 were similar and they were designed for a nominal power of 1100 MWt. NRX-A6 used a thicker beryllium reflector than did the others and did not employ a graphite inner reflector. The design features of the NRX-A6 reflector system were of a type that would exist in reactors up to 5000 MWt. The assembled NRX reactor is shown in Figure 7 and the flow path is shown in Figure 8. The reactor included the core assembly, the outer beryllium reflector, and the lateral support. Clusters of graphite-uranium fuel elements (shown in Figure 9) which constituted the core assembly were supported by metal tie rods from the core support plate where the temperatures and pressure were lower. The core exhibited a 60° symmetry. Immediately surrounding the core was the graphite lateral support region which acted primarily as a neutron reflector and moderator as well as a barrier between the core coolant flow and the reflector coolant flow, as a thermal and pressure barrier between the core and the outer reflector, as an absorber and transmitter for lateral loads, and as a retainer for filler strips. A thin-wall aluminum barrel was slipped over the graphite barrel to provide a means of handling the core during assembly and to retain lateral support system parts in place in the event of their failure and it also provided a convenient method of controlling and distributing coolant flow. The beryllium reflector consisted of twelve segments, each covering 30° of the periphery and containing one rotatable control drum. Because a flight shield was not required in the NRX-A reactors, but would be needed in subsequent flight reactors, a simulated shield was included in the NRX-A designs, and was required to simulate the flight version with respect to flow path and impedance, and also the overall envelope size. The overall core dimensions were 1.32 m in length and approximately 0.89 m in diameter. The pressure vessel was approximately 21 mm thick, 1.9 m in length and 1.3 m in outer diameter. (Occasionally in the literature there are references to NERVA I and NERVA II -- the former was the 1120-MW design and the latter was a 5000-MW design. The NERVA II concept came about as a means of reducing the running time but was abandoned when the lower power NERVA tests showed that lifetimes on the order of one hour could be achieved.)

As noted in Koenig (1986): "The flow of hydrogen coolant through the reactor was as follows [see Figure 8]: liquid hydrogen entered the aft end of the nozzle to cool the nozzle wall before entering the reflector plenum. From this plenum the hydrogen traveled forward through the reflector and control drums, also cooling the pressure vessel. It entered a plenum again before flowing forward through the outer region of the simulated shield. The flow discharged from the shield and entered the plenum region between the shield and the dome of the pressure vessel. Here the flow reversed, and the gas flowed aft through the inner region of the shield, then through a fine mesh screen and the core support plate. Most of the coolant then flowed through the channels in the fuel elements where it was heated to a high temperature. A small part of the flow cooled the periphery region between the core and the beryllium reflector, and some coolant also flowed past the tie rods in the core. These coolant flows were mixed in the nozzle chamber at the reactor exit before expulsion through the nozzle". While the "hot bleed cycle" was considered (see the description of the NRX/EST in a later section) the NERVA engine was to be a "full-flow cycle" or "topping cycle".

Koenig (1986) goes on to note that "One aim of the developmental series of tests conducted by Westinghouse Electric was to reduce the fraction of coolant flow that did not pass through the fuel in order to obtain the highest-possible gas temperature in the nozzle chamber. This aim was achieved by applying design modifications described below for the Phoebus reactors. The duration of full-power runs was gradually increased with each NRX reactor until the test in December 1967 in which the NRX-A6 ran continuously for 60 min at 1125 MW with an exit coolant temperature at or above 2280 K, corresponding to a vacuum specific impulse of 730 s . . . The test duration and power level exceeded the NERVA design goals at that time".

The standard NRX-A fuel element (see illustrative example in Figure 9) was a long hexagonal rod, with 19 full-length coolant channels arranged in a modified triangular pitch. NRX-A2 and NRX-A3 contained 1626 fuel elements. The NRX/EST and subsequent reactors contained 1584 fuel elements. The fuel elements were assembled in clusters of six supported by a tie rod in the central location as illustrated by Figure 9. Each coolant channel was coated internally with niobium carbide (NbC) as protection against the corrosive effect of the high temperature hydrogen coolant. The coolant flow distribution was controlled by orifices in the inlet end of each coolant channel which were sized to provide approximately the same exit gas temperature for all channels. The reactor used 182 kg of uranium enriched to 93.1% in uranium-235 and had a lattice which was strongly undermoderated. Power flattening was achieved by varying the fuel loading. The central fueled region was nuclearly characterized as an epithermal neutron region with neutrons having a median fission energy of 400 eV to 500 eV with many neutrons having energies lying within the resonance energy range of ^{235}U and ^{238}U. The spectrum of the core region adjacent to the beryllium reflector was relatively thermal because of the essentially thermal neutrons returning from the reflector. In addition to this position dependence of nuclear properties resulting from the varying neutron energies there was an extreme temperature gradient between the core and reflector with part of the fueled region operating at temperatures exceeding 2500 K while the reflector was at cryogenic temperatures. In addition to nuclear effects this temperature difference also affected the thermal and radiation induced stresses.

NRX-A2

NRX-A2 was the first NERVA reactor tested at full power by WANL (NRX-A1 having been a cold-flow test item). The major objectives of the power test were [Finseth 1991]:

- To provide significant information for verifying the steady-state design analysis for power operation

- To provide significant information which will aid in assessing the suitability of the reactor to operate at the steady-state power level and temperatures required for the reactor to be a component of an experimental engine system

As noted by Finseth (1991): "The power test was conducted on September 24, 1964 and the reactor operated in the range of half to full power (1096 MW) for 6 minutes, with full power operation lasting 40 seconds. The reactor was restarted on October 15, 1964 to investigate the margin of control in the low flow, low power regime over a broad range of hydrogen density inlet conditions". The test was limited by the available hydrogen supply but it was judged successful and demonstrated an equivalent vacuum specific impulse of 7500 m/s (760 s) [Koenig 1986]. Incipient corrosion of the fuel elements, particularly along the fuel element flats at the core periphery toward the hot end of the core, was found in the post-mortem inspection [Finseth 1991].

NRX-A3

On 23 April 1965 testing of the NRX-A3 reactor began with these objectives [Finseth 1991]:

- To operate at full power for 15 minutes with margin for operation at full power for a period of 5 minutes after restart
- To shut down and cool down on liquid hydrogen
- To start up from a low-power, low-flow, steady-state operating condition and to shut down from a medium power level on liquid hydrogen flow control only.
- To check the stability of certain new control concepts
- To determine the acceptability of design changes and modifications to this test article
- To verify the limits of the predicted steady-state power flow operating map up to medium power

The NRX-A3 reactor was operated for 8 minutes, with 3.5 minutes at full power (1093 MWt). The test was terminated early because of a spurious overspeed trip of the turbopump believed to have been caused by a loose electrical connection in the turbine overspeed circuit. This unplanned automatic shutdown caused overheating of the core tie rod assembly; however, a comprehensive review of the test data indicated that the reactor was not damaged. The reactor was restarted on 20 May 1965 and was run for 16 minutes, with 13 minutes at full power (approximately 1072 MWt) with the run time being dictated by the available supply of hydrogen. Based on the chamber temperature (not the higher, but unmeasured, fuel exit temperature) the equivalent ideal vacuum specific impulse achieved at the full-power hold was at least 7870 m/s (803 s) while the calculated thrust was over 237 kN. A third and final test was made on 28 May when NRX-A3 was operated for 46 minutes in the low- to medium-power range to explore the limits of the reactor operating map [Finseth 1991]. As Finseth (1991) has observed: "This test showed definitely that the reactor was inherently stable on liquid hydrogen flow control only. Once the reactor was at stable low power, the reactor power could be controlled at the desired core exit temperature up to high power by increasing the turbopump speed, with the control drums used only as a fine trim of the core exit temperature".

Post-mortem examination of NRX-A3 found improved corrosion resistance for the peripheral fuel elements showing the benefit of the NbC coating. A total of 3301 pinholes were observed on 928 fuel elements indicating that pinhole formation was probably related to the formation of corrosion pockets [Finseth 1991]. The reactor achieved a total operating time of 66 minutes with over 16.5 minutes at full power [Koenig 1986].

NRX/EST (ENGINE SYSTEM TEST)

The NRX/EST, which is shown schematically in Figure 10, was the first NERVA "breadboard" power plant in which the major engine components were connected in their flight functional relationship. The NRX/EST engine system was comprised of a basic NRX-A reactor (essentially the NRX-A4) subsystem, an engine propellant feed system, and a hot-bleed port nozzle all of which were installed on an NRX-A-type test car. The test car also included piping for the necessary auxiliary systems such as normal and emergency cool down, diagnostic and control instrumentation, and the necessary lines and valves for purging and venting the engine system. Among the significant objectives and major milestones achieved during the test series were [Finseth 1991]:

- Demonstration of the bootstrap startup capability of the engine system
- Evaluation of the effects of the test conditions on the structural integrity of the entire system
- Evaluation of engine system stability under transient and steady-state conditions

The NRX/EST was run at intermediate power levels on 3 February and 11 February 1966. A full power (1055 MWt) run was performed on 3 March 1966 using a bootstrap startup (at 483 kPa Dewar pressure) and engine duration tests were performed on both 16 March and 25 March 1966. In all, 11 startups and shutdowns (including 3 aborts) were performed with a total of 1 h 56 m of power of which 29 m were at powers in excess of 1000 MWt and 30.3 m at chamber temperatures of 2056 K or greater [Finseth 1991 and Koenig 1986]. As Koenig (1986) has observed: "These times were by far the greatest achieved by a single nuclear rocket reactor as of that date".

Fluid oscillations were observed during the NRX/EST tests but these oscillations caused no operational difficulties. Peripheral fuel elements and elements located at the core center sustained damage. Two types of damage were observed: (1) aft end corrosion and (2) mid-element internal bore corrosion. A total of 528 fuel elements were broken because of high localized pinhole density and formation of gross corrosion pockets that caused a general weakening of the elements [Finseth 1991]. Overall, "The NRX/EST test series was a significant milestone in the development of a nuclear rocket engine. The hot bleed bootstrap principle of nuclear rocket engine operation was demonstrated for the first time, system stability under a number of control modes and over a wide operating range of pressure and temperature was demonstrated, the multiple restart capability of the engine system was demonstrated, and significant reactor engine operating endurance at rated conditions was demonstrated" [Finseth 1991].

NRX-A5

The general objective of the NRX-A5 test series was to operate at design conditions for a total of 40 minutes. Secondary objectives were [Finseth 1991]:

- To evaluate reactor control concepts

- To start from a low-power, subcritical condition to near full power with liquid hydrogen flow control and constant drum position

The NRX-A5 was operated on 8 June 1966 at full power (approximately 1120 MWt) for 15.5 minutes with chamber temperature above 2056 K. The second run on 23 June 1966 went for 14.5 minutes at a power of about 1050 MWt with a chamber temperature over 2222 K [Finseth 1991]. The liquid hydrogen capacity was not sufficient to permit 30 minutes of continuous operation at design power [Koenig 1986]. Power oscillations occurred in the first run probably because of a noisy thermocouple which affected the temperature control system. The second test was terminated when the control drum reached its limit as a result of a loss of reactivity [Finseth 1991]. The post-test examination showed a 13% higher fraction of broken elements than in the NRX/EST; however, a limited examination of molybdenum-bore-coated fuel elements indicated that molybdenum coating may have been beneficial in reducing midband corrosion [Finseth 1991]. The most significant operations and accomplishments of the NRX-A5 test series were [Finseth 1991]:

- The test assembly was operated for 29.6 minutes at, or above, chamber temperatures of 2111 K and for 22.4 minutes at, or above, chamber temperatures of 2222 K

- Operation of a new eight-decade neutronic system was demonstrated

- The reactor was checked out and operated at rated conditions using a temperature and control system without the neutronics power control as an inner loop

- The acceptability of a start up from low power to near rated conditions using programmed liquid hydrogen flow with drums in a fixed position was demonstrated

- The initial criticality of the reactor was performed after all poison wires were removed

<u>NRX-A6</u>

The primary operational objective of the NRX-A6 test was to achieve a full-power (1120 MWt) run to a predetermined loss of reactivity or for a time of 60 minutes. As noted by Finseth (1991): "The principal differences in the NRX A6 design from previous reactors was [sic] the elimination of the graphite inner reflector and consequent modifications in core periphery and core lateral support systems, and that the reactor was supported from the aft (nozzle) end. These design differences resulted from two of the NRX A6 design objectives:

1. That the reactor have a basic structure applicable to reactors of greater power density and size.

2. That the core periphery and lateral support system design lead to the reduction of core corrosion."

In addition to other changes NRX-A6 included a molybdenum overcoating on the fuel element channel bores to reduce midband corrosion.

After an aborted run on 7 December 1967 caused by electrical transients, a full power run was successfully completed on 15 December 1967. This run lasted 60 minutes at or above 2278 K chamber temperature, 4 MPa chamber pressure, and 1125 MWt. Post-test examination showed axial cracks in the reflector assembly which were attributed to a temperature spike at the end of the test and irradiation effects on the beryllium [Finseth 1991]. As observed by Finseth (1991): "The performance of the NRX A6 fuel elements was characterized by inter-element bonding, mild surface corrosion, low pinhole densities, lower midband weight losses, and higher hot end weight losses relative to the NRX A5".

Finseth (1991) has concluded that ". . . the total run time of 62 minutes above a nozzle chamber temperature of 2278 K (4100 R) more than doubled the full power and temperature endurance of previous reactors with a reduction of 75-80% in the fuel element time rate of corrosion compared with that observed in the NRX/EST and NRX A5 reactors. The increase in core corrosion performance is attributed to the combination of improved fuel element coating techniques, across flats dimensional control, attention to coefficient of thermal expansion, flattened core power distribution, and changes in core interstitial pressure distribution. All NRX A6 test objectives were achieved".

XE-PRIME

XE-Prime (or XE'), which was the first down-firing prototype nuclear rocket engine (see Figure 11), had among its major objectives [Finseth 1991]:

- Operate the hot-bleed-cycle engine in a flight-type configuration (that is, a flight-type, close-coupled propellant feed system) at rated conditions
- Conduct engine start-ups without the use of nuclear instrumentation
- Conduct engine start-ups using different control logic sequences
- Start and restart the engine from a variety of different initial conditions including: (1) different core, reflector, and pump material temperatures; (2) different source power levels; and (3) different pump-inlet fluid conditions
- Demonstrate liquid hydrogen pulse cooling using run-tank flow
- Demonstrate the Engine Test Stand 1 (ETS-1) design concept
- Remotely remove a "hot" engine from the test stand and perform remote disassembly at the Engine Maintenance, Assembly and Disassembly (E-MAD) facility.

The XE-Prime engine was designed to produce a nominal thrust of almost 247 kN with the reactor operating at a power of approximately 1140 MWt, chamber temperature of 2272 K, chamber pressure of over 3.8 MPa, nozzle flow rate of 31.8 kg/s, and total flow rate of 35.8 kg/s. The XE-Prime engine had an overall specific impulse of 7 km/s (710 s) at rated conditions. [Note: As in the other tests the Aerojet nozzle had an exhaust expansion area ratio of 10:1.] The length of the XE-Prime test article was 6.9 m, the diameter was 2.59 m, and the mass was approximately 18 144 kg [Finseth 1991].

As summarized by Koenig (1986): "The reactor was operated at various power levels on different days [from 4 December 1968 through 11 September 1969] for a total of 115 min of power operation that included 28 restarts. Individual test times were limited by the facility's water storage system, which could not support operations longer than about 10 min at full reactor power. This test series was significant milestone in the nuclear rocket program and demonstrated the feasibility of the NERVA concept".

PHOEBUS AND PEWEE - THE ADVANCED REACTORS

With the Apollo mission firmly based on chemical propulsion (in particular the use of the lunar orbit rendezvous technique in place of the direct ascent approach thereby eliminating the need for the Nova vehicle with its nuclear upper stage) and NERVA focused on future lunar and Earth orbital applications studies showed that having a nuclear rocket with a thrust of about 890 kN would be ideal for piloted Mars missions [Dawson 1991, Gunn 1989, and Spence 1968]. In addition a nuclear rocket of such thrust would span the range of other missions being considered. Accordingly, SNPO agreed that, in parallel with the NERVA engine development and technology program, LANL should design and build a 5000-MWt reactor which would serve as the prototype of the reactor in an engine of 1.02 MN thrust [Spence 1968; other sources state a 1.11-MN thrust]. As noted by Spence and Durham (1965): "The name 'Phoebus,' which originally was used to denote a very high power density reactor of the Kiwi-B size, has gradually come to represent the entire LASL advanced graphite-based reactor program. The broad aims are, as they have always been, to achieve, over a considerable range of powers, higher temperatures, longer running times, and lower specific weights". The design parameters for Phoebus-2 were [Spence and Durham 1965]:

Reactor power	5000 MWt
Hydrogen flow rate	129 kg/s
Chamber pressure	4.3 MPa
Reactor-core pressure drop	1.3 MPa
Reactor-inlet pressure	6.2 MPa
Fuel-element exit gas temperature	2500 K

The goal was to achieve a specific impulse of over 8 km/s. Figure 12 shows the evolution and relative sizes of the LANL reactors. To obtain early data on the performance improvements expected with new fuel element designs LANL proposed to begin the Phoebus series with two additional 1500-MWt-class test reactors, Phoebus 1A and Phoebus 1B, which were to be derived from the Kiwi-B4E configuration.

Phoebus 1A

The Phoebus 1A core consisted of 1534 full-length (1.3 m), hexagonal fuel elements (1.91 cm across the flats) loaded with pyrocoated UC_2 particles. (Note: The term "pyrolytic-graphite-coated" has often been used in the literature. Shepherd 1993 points out that this is an incorrect description and that it is really pyrocarbon.) Each fuel element contained the standard 19 coolant holes although with larger diameters (2.79 mm instead of 2.54 mm) to achieve a higher power density by reducing thermal stress and core pressure drop [Finseth 1991]. The test objectives of the Phoebus 1A test were [Finseth 1991]:

- To operate the reactor at design-point conditions of mass flow rate, temperature, and power for the maximum times allowed by the liquid hydrogen supply in order to evaluate (by postmortem examination) the relative merits of various design changes aimed at reducing corrosion
- To obtain data which could be used to predict the operation of the first full power restart
- To obtain data to be used in determining temperature, power, and pressure transfer functions in the reactor system

Phoebus 1A was tested for 10.5 minutes at 1090 MWt (flow rate 31.4 kg/s) on 25 June 1965 reaching a chamber temperature of 2278 K and a fuel temperature of 2444 K. The radiation environment caused the instruments measuring the liquid hydrogen tank to give erroneous readings which in turn led to operation after the supply was exhausted. The reactor, which overheated, was shut down by an automatic scram triggered by turbopump overspeed [Finseth 1991]. As summarized in Finseth (1991): "During reactor disassembly it was found that the reactor damage was confined almost exclusively to the core. The entire hot end of the core was fused together (except for portions of the badly disrupted central part) by a metallic-appearing melt which was probably melted stainless steel liners from center elements. In spite of the wide-spread damage, peripheral corrosion on the NbC coated sector was practically non-existent".

Phoebus 1B

The Phoebus 1B reactor consisted of 1498 fuel elements with coolant holes having diameters of 2.54 mm (49 elements had coolant holes with diameters of 2.79 mm) and coated with about 70 g of NbC overcoated with 2 g of molybdenum to reduce the midband corrosion in the central one-third of the core. In order to avoid the hydrogen depletion problem that occurred during the Phoebus 1A test an emergency supply of hydrogen was added. The objectives of the Phoebus 1B test were [Finseth 1991]:

- Operate the reactor at a power of 1500 MWt with an average fuel element exit gas temperature of 2500 K to obtain fuel element corrosion and thermal stress data in a reactor environment at fuel element power densities approaching those planned for the Phoebus 2A reactor

- Operate at full power for 30 minutes or until the control drums had turned 20 degrees

- Obtain information on bore corrosion and on external corrosion of the fuel elements

The reactor was run at an intermediate power of 588 MWt for 2.5 minutes on 10 February 1967; although a power spike of about 3500 MWt occurred during shutdown from this test. On 23 February 1967 the Phoebus 1B reactor was run for 46 minutes of which 30 minutes were above 1250 MWt with a maximum power of 1450 MWt and gas temperature of 2444 K being achieved [Finseth 1991]. As summarized by Finseth (1991): "The PHOEBUS 1B reactor increased the average fuel element power density to 1 MW/element and the fuel elements demonstrated improved corrosion resistance. Additionally, the core exit pressure and hydrogen flow rate was [sic] increased over the PHOEBUS 1A".

Phoebus 2A

The Phoebus 2A reactor, which was the most powerful nuclear rocket reactor ever built, contained 4068 fuel elements similar to those in the earlier Phoebus tests plus 721 regeneratively cooled support elements [Koenig 1986]. As noted by Finseth (1991): ". . . the single pass cooling of the metal core support structure of earlier reactor designs was changed to two pass regenerative cooling by diverting about 10% of the liquid hydrogen to the core support and returning this coolant to the main flow through the core at the inlet of the fuel elements. This new coolant path eliminated the performance degradation associated with the single pass tie rod system by preventing the mixing of the lower temperature core support coolant with the core exit gas in the nozzle chamber". The 203-mm-thick beryllium-reflector assembly contained 18 control drums rather than the 12 drums used on the earlier smaller reactors. The overall active core dimensions were 1.39 m in diameter and 1.32 m in length. The 2.54-mm-thick aluminum pressure vessel had a diameter of 2.07 m and an approximate length (excluding the nozzle) of 2.5 m. The reactor mass (including the pressure vessel) was 9.3 Mg [Koenig 1986]. The overall test objectives of the Phoebus 2A test were [Finseth 1991]:

- Demonstrate the capability of the reactor and test system to operate at 5000 MWt and at a chamber temperature of 2500 K

- Operate at the design point for a maximum of 20 minutes or until 15 degrees of control drum motion had taken place, whichever occurred first, to obtain endurance information on the fuel elements and the structural components

- Evaluate the structural and thermal flow performance of the new regeneratively cooled tie-tube core support setup

Phoebus 2A was tested four times [Finseth 1991 and Koenig 1986]:

- 8 June 1968 - 2000 MWt to evaluate the operation of the reactor and the facility systems

- 26 June 1968 - 32-minute run with 12.5 minutes above 4000 MWt, reaching a peak power of 4082 MWt (time limited by available hydrogen)

- 18 July 1968 - 1280 MWt and at 3500 MWt (about 30 minutes of operation) to test controls

As noted by Koenig (1986): "The reactor could not be operated up to the design power level of 5000 MW because part of the aluminum pressure vessel assembly was overheating prematurely as a result of unexpected poor thermal contact with an LH_2-cooled clamp ring. The maximum fuel-element exit-gas temperature attained was 2310 K, and the maximum nozzle chamber temperature, nearly as high, was 2260 K. This small temperature difference is an indication of the effectiveness of the measures taken to reduce mixing of cold coolant with the core exit gas. At design power, the core power density would have been nearly twice that of the Kiwi-B reactors".

Koenig (1968) has concluded that "The successful conclusion of the Phoebus-2A tests was a milestone in nuclear rocket technology because of the high-power capability that the test demonstrated. Some problems remained, particularly in the area of fuel longevity and temperature capability, but the feasibility of practical nuclear space propulsion had been convincingly demonstrated by this stage of the Rover program".

Pewee

The Pewee reactor was a small test bed designed to test full-size Phoebus and NRX fuel elements and other components. (A second reactor, Pewee-2, was not built.) As noted by Koenig (1986): "The general design was directed toward providing a realistic nuclear, thermal, and structural environment for the fuel elements in a core containing one-fourth the number of elements in these reactors, and one-tenth the number of elements in Phoebus-2A". The core contained 402 fuel elements and 132 support elements. Of the 402 fuel elements, 390 elements had 19 coolant holes and 12 had 12 coolant holes. The Pewee fuel elements contained 27 different combinations of graphite matrix, coating process, hot-end tips, etc. While most of the coolant channels were coated with NbC several were coated with ZrC. The overall core diameter was 533 mm. The necessary reactivity was provided by adding sleeves of zirconium hydride around the tie rods in the support elements [Finseth 1991 and Koenig 1986]. While the reactor had a number of objectives, the principal one was to demonstrate the capability of the reactor as a fuel element test bed [Koenig 1986].

The Pewee reactor was tested on three occasions, beginning with a checkout run on 15 November 1968 and then a short-duration, near-full-power run on 21 November 1968. The short-duration run achieved 472 MWt (flow rate 18.1 kg/s) at an average fuel exit temperature of 2450 K [Finseth 1991]. The full-power test consisted of two 20-minute holds at 503 MWt and an average fuel-element exit-gas temperature of 2550 K [Koenig 1986]. As summarized by Koenig (1986): "This temperature was the highest achieved in the Rover program. It corresponds to a vacuum specific impulse of 845 s, a level in excess of the design goal set for the NERVA. The peak fuel temperature also reached a record level of 2750 K. The average power density in the core was 2340 MW/m^3, also a record high and greater than that required for the NERVA. The peak power density in the fuel was 5200 MW/m^3. . . The ZrC-coated fuel elements performed significantly better [than the NbC-coated elements]".

RIFT - THE FLIGHT PROJECT

The RIFT project originally consisted ". . . of the design of the nuclear stage, the associated research and development necessary to qualify materials and components which will adequately withstand the added environment of nuclear radiation, the fabrication and assembly of the stages for static testing and launching, integration of the Nerva engine and other systems into a complete stage and the stage into a complete vehicle, and, finally, static testing and flight demonstration" [Fellows 1962]. RIFT was planned to be used as the third stage on what was then the Saturn C-5 or as a second or third stage on the planned Nova launch vehicle [Fellows 1962]. The RIFT program began in August 1960 when NASA let four cost sharing contracts with Convair, Douglas, Lockheed, and Martin to investigate the major aspects of a reactor flight test program. This work followed on earlier studies dating back to the 1950s conducted by various aerospace contractors. As described by Dewar (1974): ". . . the program would study missions, preliminary vehicle design, funding requirements, facility requirements, and flight test methods. How to test a nuclear rocket, even though its operational use was intended to be an upper-stage, received particular attention because of political, economic, and technical consequences". It was recognized that RIFT could be tested in an orbital start, an atmospheric start or through a ground launch. The four contractors preferred the atmospheric start because it involved a suborbital flight ending in the Atlantic Ocean where the recording instruments could be retrieved. In May 1962 the Nuclear Vehicle Project Office (NVPO) of NASA's Marshall Space Flight Center (MSFC) awarded the RIFT contract to Lockheed with the goal of a flight in 1967. RIFT became the integrating third member of a triad of nuclear propulsion projects, the other two being the Kiwi technology project and the NERVA engine project [Dewar 1974 and Fellows 1962]. RIFT was envisioned to provide both a backup to the chemical upper stages on the Saturn V and a means to conduct post-Apollo manned missions to the Moon (for example, lunar bases) and beyond [Dewar 1974, Fellows 1962, McLaughlin 1964, and Nucleonics 1963].

The initial phase (Phase 1) of the RIFT contract, which was established to incorporate information from the Kiwi project, was to extend for 10 months until May 1963 at a cost of $6.6 million. During Phase 1, Lockheed was to develop a preliminary design for the RIFT stage, develop a manufacturing and tooling plan and determine the test facility requirements. It was envisioned that RIFT ground tests would be conducted at NRDS using special facilities to be built for RIFT such as a Stage Assembly and Maintenance (SAM) Building, three test stands, and a Components Test Area. RIFT engines were to be examined in the (then) planned NRDS Engine Maintenance and Disassembly Building (E-MAD) [Fellows 1962]. Other facilities used in the RIFT project included a hangar at the Navy's Moffett Field to be used for hydrogen tests; the Lockheed-operated Georgia Nuclear Laboratories (GNL) (formerly an ANP facility); and the USAF Nuclear Aerospace Research Facility (NARF)(another ANP facility) in Texas [Dewar 1974 and Fellows 1962].

Ten RIFT stages were planned for fabrication [Fellows 1962]:

Designation	Type	Destination	Purpose
S-N-TA 1	"Battleship"	NRDS	RIFT testing/facility checkout
S-N-TA 2	"Battleship"	NRDS	RIFT testing/facility checkout
S-N-T 1	Preliminary Flight Configured	NRDS	RIFT testing
S-N-T 2	Flight Configured	NRDS	RIFT testing
S-N-T 3	Flight Configured	NRDS	RIFT testing
S-N-D	Flight Model	MSFC	Dynamic testing
S-N-1	Flight Configured	Atlantic Missile Range	Flight Mission 1
S-N-2	Flight Configured	AMR	Flight Mission 2
S-N-3	Flight Configured	AMR	Flight Mission 3
S-N-4	Flight Configured	AMR	Flight Mission 4

The RIFT launch-vehicle system was to consist of a Saturn S-IC first stage powered by five F-1 engines; a Saturn S-II stage powered by five J-2 engines (this stage would be a dummy stage in the RIFT flights); the S-N nuclear stage; the instrument unit; the nose cone; and the automated-check-out launch system [Fellows 1962].

The nuclear rocket program began to be questioned by President Kennedy and then by President Johnson leading to a decision in December 1963 to reorient the Rover program to a research and technology effort with NERVA being redirected to a ground testing reactor program and RIFT being terminated. This decision presaged the beginning of the end of plans for a vigorous post-Apollo manned space program [Dewar 1974].

NF-1 - THE NUCLEAR FURNACE

The Nuclear Furnace, which was designed to provide an inexpensive means of testing full-size nuclear rocket reactor fuel elements, was the last reactor test of the Rover program. The reactor, which is illustrated in Figures 13 and 14, consisted of two parts: a permanent, reusable portion that included the reflector and external structure; and a temporary, removable portion that consisted of the core assembly and associated components [Finseth 1991, Kirk 1973, and Koenig 1986]. The Nuclear Furnace was designed to permit removal and disassembly of the core section leaving the permanent structure to be reused with a new core. The first (and only) Nuclear Furnace was designated NF-1 and it was tested during the summer of 1972 at NRDS [Kirk 1973]. The Nuclear Furnace test had two major objectives [Finseth 1991 and Koenig 1986]:

- To check out the operating characteristics of both the Nuclear Furnace and the effluent cleanup facility

- To operate the reactor at an average fuel element exit-gas temperature of 2444 K for at least 90 minutes

As described by Koenig (1986) "The NF-1 core was a 34-cm-diameter by 146-cm-long aluminum can that contained 49 fuel elements as compared to 402 in Pewee. This core was surrounded by a 27-cm-thick beryllium radial reflector that accommodated six rotating control drums. The fuel inventory was about 5 kg of uranium (93% enriched). Sufficient reactivity for critical configuration with such a small fuel inventory was obtained by designing the core as a heterogeneous water-moderated thermal reactor. Each fuel cell contained a standard 19-hole, hexagonal fuel element encased in an aluminum tube as described in Fig. [15]. The cell tubes were inserted inside aluminum sleeves, and water flowed through the core in two passes, first between the sleeves and the cell tubes to the aft end of the core, where the flow turned around and went back between the elements. The hydrogen coolant, after making several passes in the reflector assembly, made a single pass through the core within the fuel coolant channels.

"The hydrogen exhaust gas was handled differently than [sic] in previous reactors. Instead of being exhausted through a convergent-divergent nozzle directly to the atmosphere, the hot hydrogen was first cooled by injecting water directly into the exhaust gas stream as shown in Fig. [16]. The resulting mixture of steam and hydrogen gas was then ducted to an effluent cleanup system to remove fission products before release of the cleaned gas to the atmosphere."

Six experimental plans were carried out in NF-1 involving two new types of fuel elements. Overall the NF-1 reactor was operated at the design power of 44 MW for 109 minutes at a fuel-element exit-gas temperature of approximately 2440 K and for 121 minutes with temperatures at or above 2220 K. A maximum exit temperature of about 2550 K was reached [Kirk 1973 and Koenig 1986]. Of the 49 fuel cells, 47 contained (U,Zr)C graphite (composite) elements and 2 contained (U,Zr)C (carbide) elements. As reported by Koenig (1986): "The carbide elements withstood peak power densities of 4500 MW/m^3 but experienced severe cracking. These elements were small (5.5 mm across the flats), hexagonal elements with a single 3-mm-diameter coolant hole. Redesign, by reducing the web thickness by 25%, would substantially decrease the temperature gradients and reduce the cracking. The composite elements withstood peak power densities in the fuel of 4500-5000 MW/m^3 and achieved better corrosion performance than was observed previously in the standard, graphite-matrix, Phoebus-type fuel element".

Figure 17 summarizes the performance improvements achieved during the NRX and Pewee series of tests (where corrosion measured is in terms of mass loss normalized to unity for the NRX-A2 and NRX-A3 tests)[Koenig 1986]. Projections of future endurance limits have been qualitatively estimated in Figure 18 [Koenig 1986]. These figures are, at best, general relationships. As noted by Shepherd 1993, these relationships depend upon the details of the core and fuel element design and the operating conditions, including temperature gradients. As stated by Koenig (1986): "These projections indicate that the composite fuel should be good for 2-6 h in the temperature range of 2500-2800 K. Similar performance can be expected at 3000-3200 K for the carbide fuels, assuming that the cracking problem can be reduced through improved design. For 10 h of operation, the graphite-matrix fuel would be limited to a coolant exit temperature of 2200-2300 K, the composite fuel could go to nearly 2400 K, and the pure carbide to about 3000 K".

SUMMARY AND CONCLUSIONS

The Rover program, which ended in 1973 because of budget problems and a lack of post-Apollo missions, was a very successful technical program. Seventeen reactors, one nuclear safety reactor, and two ground experimental engines were tested with these summary record performances [Gunn 1989]:

Power (Phoebus 2A)	4,100 MW
Thrust (Phoebus 2A)	~930,000 N
Hydrogen flow rate (Phoebus 2A)	120 kg/s
Equivalent specific impulse (Pewee)	~8,300 m/s (~848 s)
Minimum reactor specific mass (Phoebus 2A)	2.3 kg/MW
Average coolant exit temperature (Pewee)	2,550 K
Peak fuel temperature (Pewee)	2,750 K
Core average power density (Pewee)	2,340 MW/m^3
Peak fuel power density (Pewee)	5,200 MW/m^3
Accumulated time at full power (NF-1)	109 minutes
Greatest number of restarts (XE)	28

On the basis of the results achieved in the Rover program the practicality of solid graphite reactor/nuclear rocket engines has been established and the technology has been demonstrated to support future space propulsion requirements, using liquid hydrogen as the propellant, for thrust requirements ranging from 111 kN to 1.1 MN with vacuum specific impulses of at least 8300 m/s and with full engine throttle capability [Gunn 1989]. In looking toward future plans for nuclear propulsion, Robbins and Finger (1991) have concluded: "Reactor and engine system tests were conducted at temperature, pressure, power levels, and durations commensurate with today's propulsion system requirements Feasibility of the nuclear rocket has been clearly established so that future nuclear propulsion development associated with new space exploration initiatives can be directed to incremental performance, reliability and lifetime improvements. In addition, a model for an effective management approach involving two government agencies in a major technology development has been demonstrated. The real future development challenge will be associated with engine and reactor system ground testing in an environmentally acceptable fashion".

Acknowledgments

The authors would like to acknowledge with a deep debt of gratitude the dedication and tireless work of the hundreds of nuclear rocket pioneers at Los Alamos National Laboratory, Westinghouse, Aerojet, Rocketdyne, Lockheed, NRDS, and the joint AEC/NASA Space Nuclear Propulsion Office. Mr. Finger created the joint office and managed it. Mr. Klein was the deputy manager and then manager of the office. Mr. Robbins was in charge of the systems branch at SNPO-C. Mr. Miller was Chief of the Nuclear Propulsion Office at LeRC. Dr. Bennett was a reactor physicist at SNPO-C and is program manager at NASA Headquarters for advanced propulsion.

References

Aamodt, R. L. et al. (1955) *The Feasibility of Nuclear-Powered Long Range Ballistic Missiles*, Los Alamos Scientific Laboratory report LAMS-1870 (rev), March 1955.

Adams, A., C. C. Priest, P. Sumrall, and G. Woodcock (1990) "Overview of Mars Transportation Options and Issues", AIAA paper 90-3795 prepared for AIAA Space Programs and Technologies Conference held in Huntsville, Alabama, 25-28 September 1990.

Bennett, G. L. and T. J. Miller (1991) "Planning for the Space Exploration Initiative: The Nuclear Propulsion Option", *Proceedings of the Eighth Symposium on Space Nuclear Power Systems*, CONF-910116, held in Albuquerque, New Mexico, 6-10 January 1991.

Bennett, G. L., S. R. Graham, and K. F. Harer (1991) "Back to the Future: Using Nuclear Propulsion to Go to Mars", AIAA paper 91-1888, presented at the AIAA/SAE/ASME 27th Joint Propulsion Conference held in Sacramento, California, 24-26 June 1991.

Borowski, S. K., M. W. Mulac, and O. F. Spurlock (1989) "Performance Comparisons of Nuclear Thermal Rocket and Chemical Propulsion Systems for Piloted Missions to Phobos/Mars", IAF paper 89-027 presented at the 40th Congress of the International Astronautical Federation, held in Malaga, Spain, 7- 13 October 1989.

Borowski, S. K., T. J. Wickenheiser, and J. J. Andrews (1990) "Nuclear Thermal Rocket Technology and Stage Options for Lunar/Mars Transportation Systems", AIAA paper 90-3787 presented at the AIAA Space Programs and Technologies Conference, held in Huntsville, Alabama, 25-28 September 1990.

Braun, R. D. and D. J. Blersch (1989) "Propulsive Options for a Manned Mars Transportation System", AIAA paper 89-2950, presented at AIAA/ASME/SAE/ASEE 25th Joint Propulsion Conference, held in Monterey, California, 10-12 July 1989.

Bussard, R. W. (1953) *Nuclear Energy for Rocket Propulsion*, ORNL Central Files Number 53-6-6, Oak Ridge National Laboratory, Oak Ridge, Tennessee.

Bussard, R. W. (1962) "Nuclear Rocketry--The First Bright Hopes", *Astronautics*, Vol. 7, No. 12, pp. 32- 35.

Bussard, R. W. and R. D. DeLauer (1965) *Fundamentals of Nuclear Flight*, McGraw-Hill Book Company, New York, New York.

Corliss, W. R. and F. C. Schwenk (1971) *Nuclear Propulsion for Space,* U.S. Atomic Energy Commission booklet, Library of Congress Catalog Card Number 79-171030.

Dawson, V. P. (1991) *Engines and Innovation, Lewis Laboratory and American Propulsion Technology,* NASA SP-4306, National Aeronautics and Space Administration, Washington, D. C. (for sale by the U.S. Government Printing Office, Washington, D. C.).

Decker, R. S. (1962) "Safety and Operations with Nuclear Vehicles", *Astronautics*, Vol. 7, No. 12, pp. 63- 65 (December 1962).

Dewar, J. A. (1974) "Project Rover: A Study of the Nuclear Rocket Development Program, 1953 - 1963", unpublished manuscript, U.S. Department of Energy, Washington, D.C.

Dix, G. P. (1967) "Operational Safety of Nuclear Rockets", in *Nuclear, Thermal and Electric Rocket Propulsion*, edited by R. A. Willaume, A. Jaumotte, and R. W. Bussard, Gordon and Breach Science Publishers, New York, New York.

Dixon, T. F. and H. P. Yockey (1946) *A Preliminary Study of Use of Nuclear Power in Rocket Missiles,* North American Aviation report NA-46-574.

Fellows, W. S. (1962) "RIFT", *Astronautics*, pp. 38-47 (December 1962).

Finseth, J. L. (1991) *Rover Nuclear Rocket Engine Program, Overview of Rover Engine Tests, Final Report*, File No. 313-002-91-059, Sverdrup Corporation, Huntsville, Alabama.

Frisbee, R. H., J. J. Blandino, J. C. Sercel, M. G. Sargent, and N. Gowda (1990) "Advanced Propulsion Options for the Mars Cargo Mission", AIAA paper 90-1997, presented at the AIAA/SAE/ASME/ASEE 26th Joint Propulsion Conference held in Orlando, Florida, 16-18 July 1990.

Gabriel, D. S. (1974) "Nuclear Propulsion in The United States", *Astronautica Acta*, Vol. 18 (Supplement), pp. 3-18.

Goddard, E. C. and G. Edward Pendray (ed.) (1961) *Robert H. Goddard, Rocket Development, Liquid- Fuel Rocket Research, 1929-1941*, Prentice-Hall, INc., New York, New York.

Gunn, S. V. (1989) "Development of Nuclear Rocket Engine Technology", AIAA paper 89-2386, presented at the AIAA/ASME/SAE/ASEE 25th Joint Propulsion Conference held in Monterey, California, 10-12 July 1989.

Gunn, S. V. (1993) Personal communication to G. L. Bennett, 8 January 1993.

Hack, K. J., J. A. George, J. P. Riehl, and J. H. Gilland (1990) "Evolutionary Use of Nuclear Electric Propulsion", AIAA paper 90-3821, presented at AIAA Space Programs and Technologies Conference, held in Huntsville, Alabama, 25-28 September 1990.

Holdridge, J., K. Shepard, U. Hueter, and P. Sumrall (1990) "An Infrastructure Assessment of Alternative Mars Transfer Vehicles", AIAA paper 90-1999, presented at AIAA/SAE/ASME/ASEE 26th Joint Propulsion Conference, held in Orlando, Florida, 16-18 July 1990.

JPL (1992) *Proceedings of the Nuclear Electric Propulsion Workshop*, Jet Propulsion Laboratory report JPL D-9512 (2 volumes), May 1992.

Kennedy, J. F. (1962) "Special Message to the Congress on Urgent National Needs", *Public Papers of the Presidents of the United States, John F. Kennedy*, U. S. Government Printing Office, Washington, DC.

King, L. D. P., D. Ackworth, W. U. Geer, C. A. Fenstermacher, B. W. Washburn, and J. F. Weinbrecht (1966) *Description of the Kiwi-TNT Excursion and Related Experiments*, Los Alamos Scientific Laboratory report LA-3350-MS (August 1966).

Kirk, W. L. (1973) *Nuclear Furnace-1 Test Report*, Los Alamos Scientific Laboratory report LA-5189-MS, Los Alamos, New Mexico, March 1973.

Koenig, D. R. (1986) *Experience Gained from the Space Nuclear Rocket Program (Rover)*, LA-10062-H, Los Alamos National Laboratory, Los Alamos, New Mexico.

Levoy, M. M. and J. J. Newgard (1958) "Rocket-Reactor Design", *Nucleonics*, Vol. 16, No. 7, pp. 66-68.

McLaughlin, J. F. (1964) "Sizing nuclear orbital launch vehicles for interplanetary missions", *Astronautics & Aeronautics*, Vol. 2, No. 2, pp. 70-76.

NASA (1989a) *Report of the 90-Day Study on Human Exploration of the Moon and Mars*, National Aeronautics and Space Administration, Washington, DC.

NASA (1989b) *NASA Advisory Council Report of the Task Force on Space Transportation*, National Aeronautics and Space Administration, Washington, DC.

NASA (1991) *Nuclear Thermal Propulsion, A Joint NASA/DOE/DOE Workshop*, NASA Conference Publication 10079.

North American Aviation (1947) *Feasibility of Nuclear Powered Rockets and Ramjets*, North American Aviation Aerophysics Laboratory report NA-47-15.

NRC (1990) *Human Exploration of Space, A Review of NASA's 90-Day Study and Alternatives*, Committee on Human Exploration of Space, National Research Council, National Academy Press, Washington, DC.

Nucleonics (1958) "Nuclear-Rocket Timetable", *Nucleonics*, Vol. 16, No. 7, p. 69.

Nucleonics (1963) "How RIFT Would Establish Moon Base", *Nucleonics*, Vol. 21, No. 3, pp. 84-85.

Palaszewski, B. (1989) "Electric Propulsion for Manned Mars Exploration", paper presented at the JANNAF Propulsion Meeting held in Cleveland, Ohio, 23-25 May 1989.

Reupke, W. A. (1992) "The Rocket Pioneers and Atomic Energy", *Journal of The British Interplanetary Society*, Vol. 45, pp. 297-304.

Robbins, W. H. and H. B. Finger (1991) "An Historical Perspective of the NERVA Nuclear Rocket Engine Technology Program", AIAA paper 91-3451 (also NASA Contractor Report 187154) prepared for the AIAA/NASA/OAI Conference on Advanced SEI Technologies held in Cleveland, Ohio, 4-6 September 1991

Ruark, A. E. et al. (1947) *Nuclear Powered Flight*, Johns Hopkins University Applied Physics Laboratory report APL/JHU-TG-20, Laurel, Maryland.

Seaborg, G. T. (1970) "A Nuclear Space Odyssey", remarks to the Commonwealth Club of California, San Francisco, California, 24 July 1970, U.S. Atomic Energy Commission release number S-27-70.

Shepherd, L. R. and A. V. Cleaver (1948-1949) "The Atomic Rocket: Parts I - IV", *Journal of the British Interplanetary Society*, Vol. 7, No. 5 and 6 and Vol. 8, No. 1 and 2.

Shepherd, L. R. (1993) Personal communication to G. L. Bennett, 26 April 1993.

Spence, R. W. and F. P. Durham (1965) "The Los Alamos nuclear-rocket program", *Astronautics & Aeronautics*, pp. 42-46 (June 1965).

Spence, R. W. (1965) "Nuclear Rockets", *International Science and Technology*, July 1965, pp. 58-65.

Spence, R. W. (1968) "The Rover Nuclear Rocket Program", *Science*, Vol. 160, No. 3831, pp. 953-959.

Synthesis Group (1991) *America at the Threshold, Report of The Synthesis Group on America's Space Exploration Initiative*, available from Superintendent of Documents, U.S. Government Printing Office, Washington, DC.

TRW (1965) *Mission Oriented Advanced Nuclear System Parameters Study*, TRW Space Technology Laboratories report 8423-6005-RU000 performed for NASA Marshall Space Flight Center (March 1965).

Walton, J. T. (1991) "An Overview of Tested and Analyzed NTP Concepts", AIAA paper 91-3503 prepared for the AIAA/NASA/OAI Conference on Advanced SEI Technologies held

WANL (1972) *Technical Summary Report of NERVA Program*, Westinghouse Astronuclear Laboratory report TNR-230 (Executive Summary plus 6 volumes).

White House (1989) "Fact Sheet, U.S. National Space Policy", Office of the Press Secretary, The White House, 16 November 1989.

TABLE 1. Chronology of Rover and NERVA Reactor/Engine Tests.

Date	Test Article	NRDS Test Facility	Maximum Power	Time at Maximum Power*
1 July 1959	Kiwi-A	A	70 MW	5 minutes
8 July 1960	Kiwi-A'	A	85 MW	6 minutes
10 Oct 1960	Kiwi-A3	A	100 MW	5 minutes
7 Dec 1961	Kiwi-B1A	A	300 MW	30 seconds
1 Sep 1962	Kiwi-B1B	A	900 MW	Several seconds
30 Nov 1962	Kiwi-B4A	A	500 MW	Several seconds
13 May 1964	Kiwi-B4D	C	1,000 MW	~40 seconds
28 Aug 1964	Kiwi-B4E	C	900 MW	8 minutes
10 Sep 1964	Kiwi-B4E	C	900 MW	2.5 minutes - restart
24 Sep 1964	NRX-A2	A	1,096 MW	40 seconds
15 Oct 1964	NRX-A2	A	Restart	(performance mapping)
21 Jan 1965	Kiwi-TNT	Safety test reactor - deliberately		destroyed on power excursion
23 April 1965	NRX-A3	A	1,093 MW	3.5 minutes
20 May 1965	NRX-A3	A	1,072 MW	13 minutes
28 May 1965	NRX-A3	A	≤500 MW	46 minutes - performance maps
25 June 1965	Phoebus 1A	C	1,090 MW	10.5 minutes
3,16,25 Mar 65	NRX/EST	A	1,055 MW	1.25 min - 14.5 min - 13.7 min
8 June 1966	NRX-A5	A	1,120 MW	15.5 minutes
23 June 1966	NRX-A5	A	1,050 MW	14.5 minutes (restart)
10 Feb 1967	Phoebus 1B	C	588 MW	2.5 minutes
23 Feb 1967	Phoebus 1B	C	>1,250 MW	30 min. - low power - 10 Feb 67
15 Dec 1967	NRX-A6	C	1,125 MW	62 minutes
8 June 1968	Phoebus 2A	C	2,000 MW	~100 sec
26 June 1968	Phoebus 2A	C	4,100 MW	12 minutes
18 July 1968	Phoebus 2A	C	1,280 MW - 3,430 MW	30 minutes of total operation
3-4 Dec 1968	Pewee	C	514 MW	40 minutes
11 June 1969	XE-Prime	ETS-1	1,140 MW	3.5 minutes
NOTE: XE-Prime had 28 experimental restarts from 4 Dec 1968 to 11 Sep 1969				
29 June - 27 July 1972	Nuclear Furnace	C	44 MW	109 minutes (6 experiments)

* Note: In several cases the reactor was operated at lower powers for longer times.

TABLE 2. Distribution of Major NERVA Test Objectives

Objectives	NRX-A1	NRX-A2	NRX-A3	NRX/EST	NRX-A5	NRX-A6	XE'
Assess Mechanical Flow Stability	X						
Verify Design Analysis for Steady State Power Operation		X	X				
Full Power Endurance			X	X	X	X	
Determine Operating Map Characteristics			X	X	X		X
Start Up and Shut Down Rates				X	X		X
LH_2 Flow Rate Control (Power Control with Fixed Control Drums)					X		
Restart Capability				X			
Cooldown with LH_2					X		
Control Systems Evaluation					X	X	X
Bootstrap Start-Up				X			X
Determination of Flight-Type Operational Cycle							X

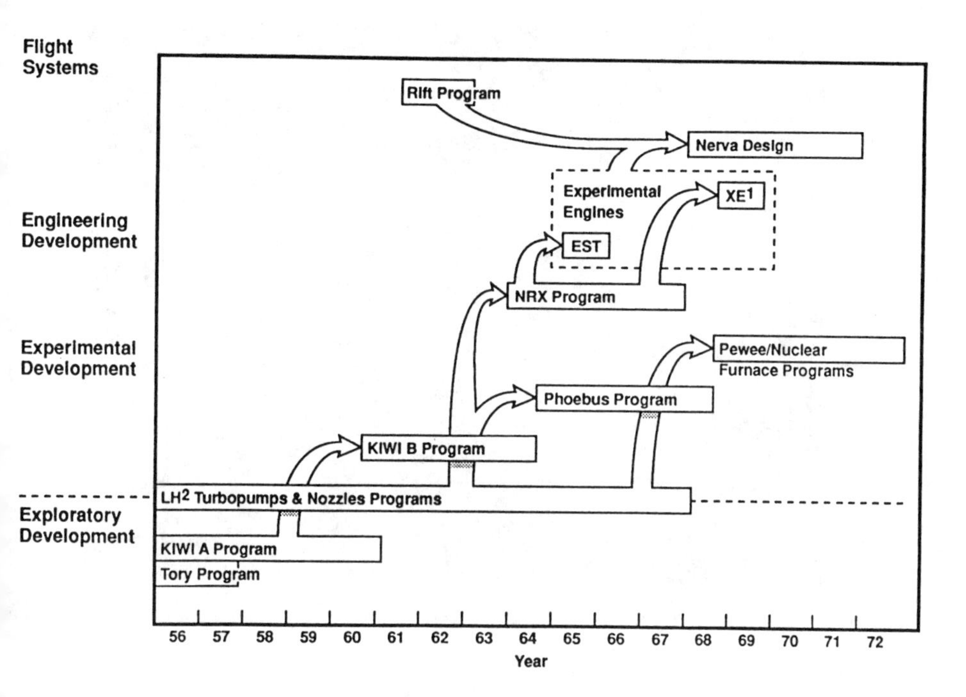

FIGURE 1. Chronological Overview of the Nuclear Rocket Program [Gunn 1989].

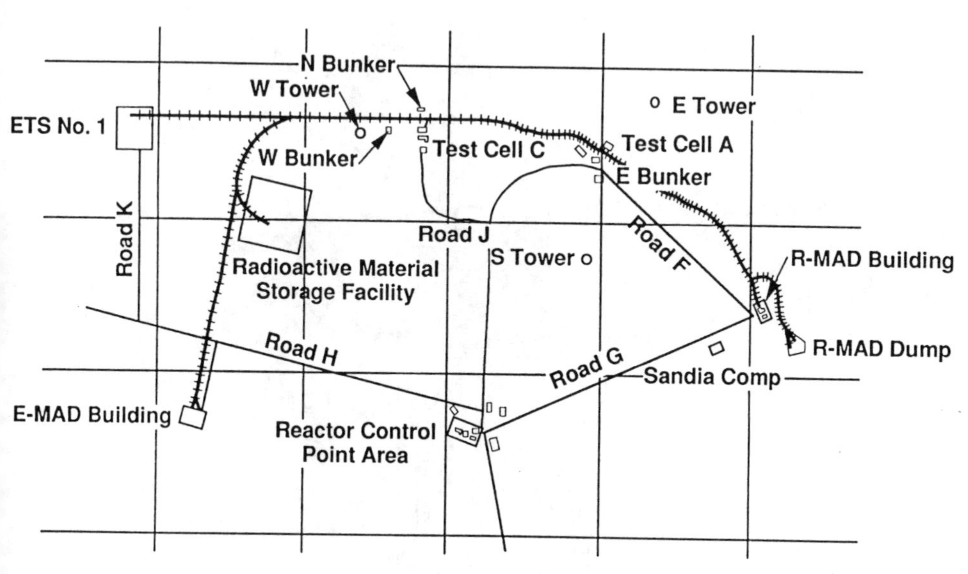

FIGURE 2. Schematic of the Nuclear Rocket Development Station (NRDS) [Gunn 1989].

FIGURE 3. Reactor Test Cell C at NRDS (Showing the Hydrogen Dewars) [Gunn 1989]

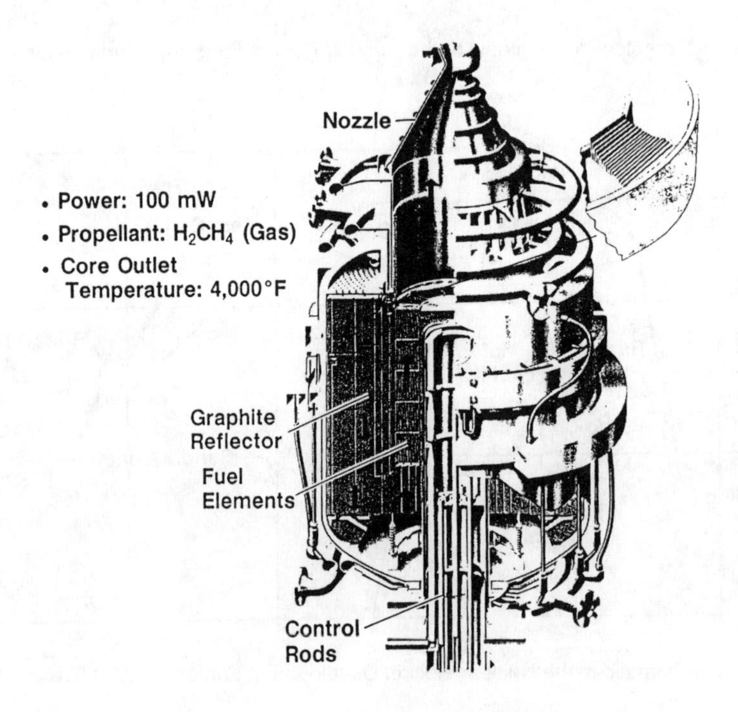

FIGURE 4. Isometric Cutaway of Kiwi-A Reactor [Gunn 1989 and Koenig 1986].

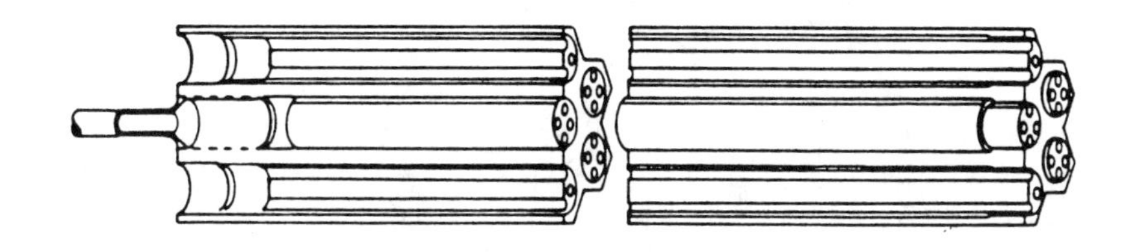

FIGURE 5. Kiwi-A' and Kiwi-A3 Graphite Module Fuel Element Assembly [Gunn 1989 and Koenig 1986].

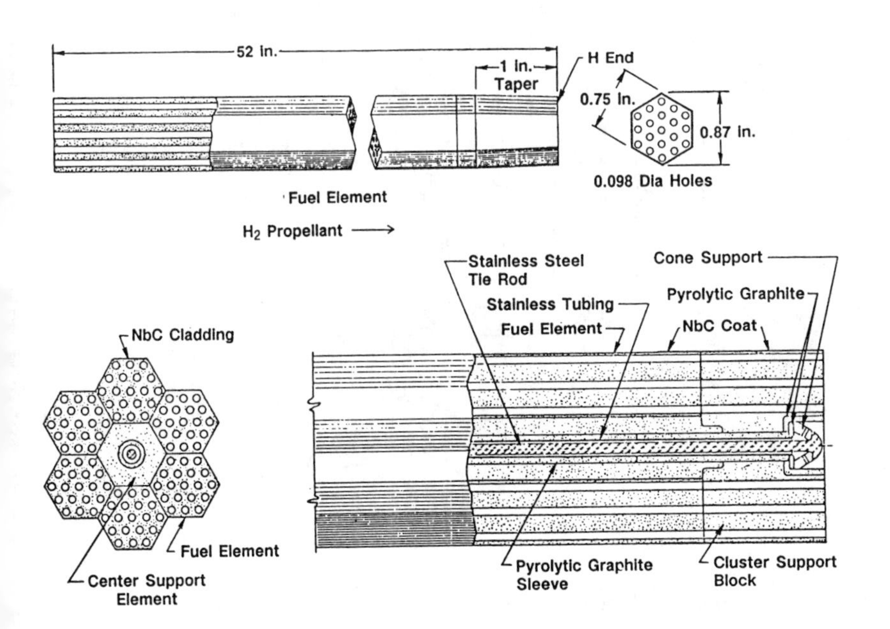

FIGURE 6. Kiwi-B4A Fuel Element Cluster [Gunn 1989].

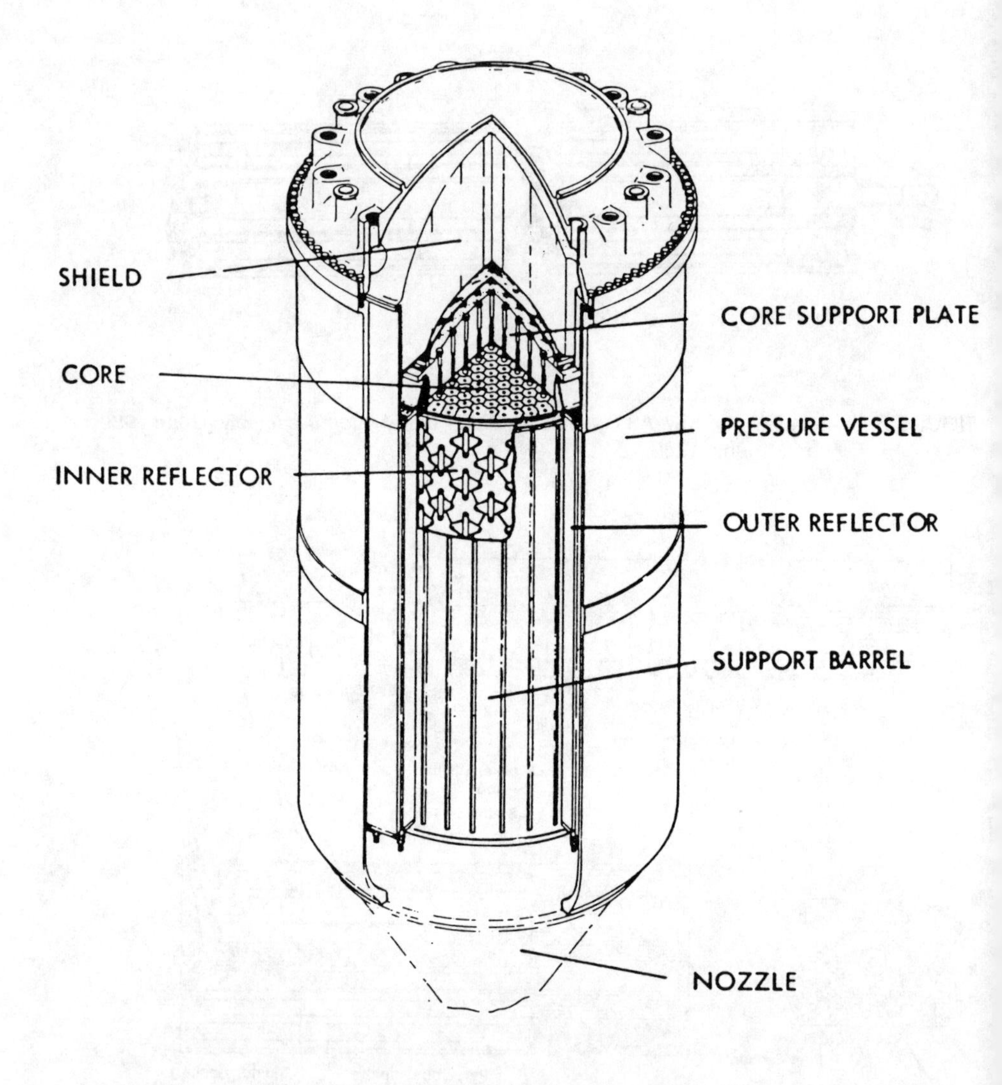

FIGURE 7. Trimetric of Assembled NRX Reactor [WANL].

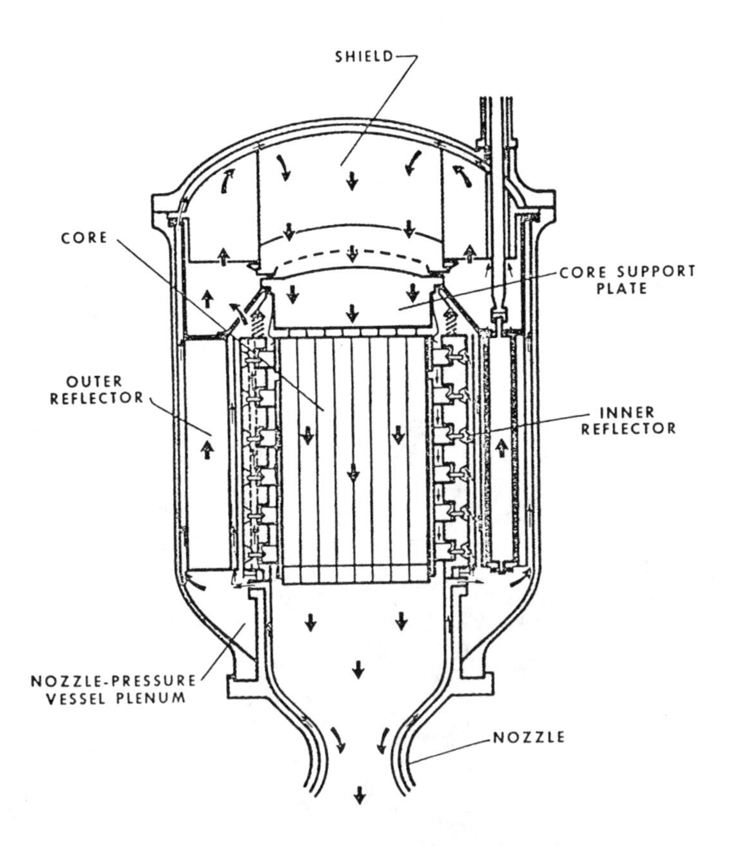

FIGURE 8. Flow Schematic Through NRX-A2 [WANL].

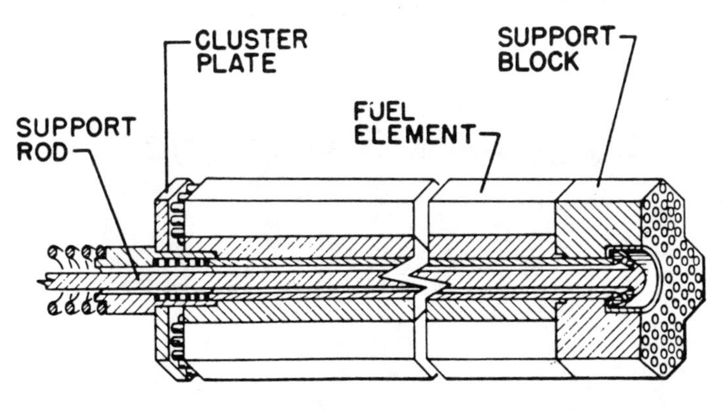

FIGURE 9. Fuel Element Cluster Employed in Most of the Later Rover Reactor Designs. It Consists of Six, Full-Length, Hexagonal Fuel Elements Supported by a Centrally Located Tie Rod. Each Extruded Graphite Fuel Element had 19 Cooling Channels [Koenig 1986].

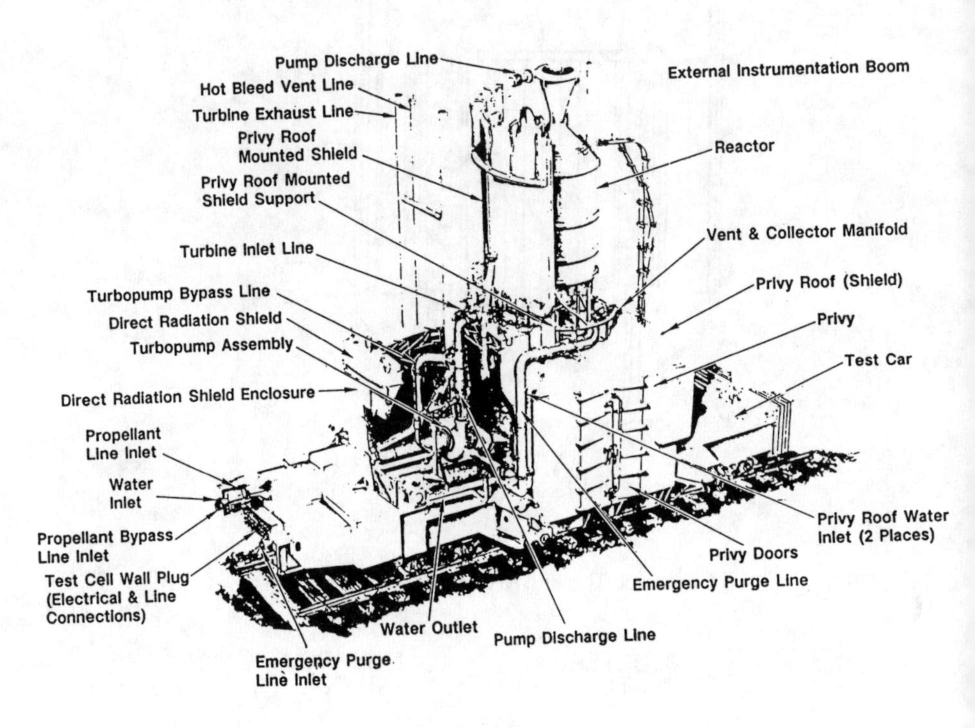

FIGURE 10. Isometric Cutaway of NRX-EST [Gunn 1989].

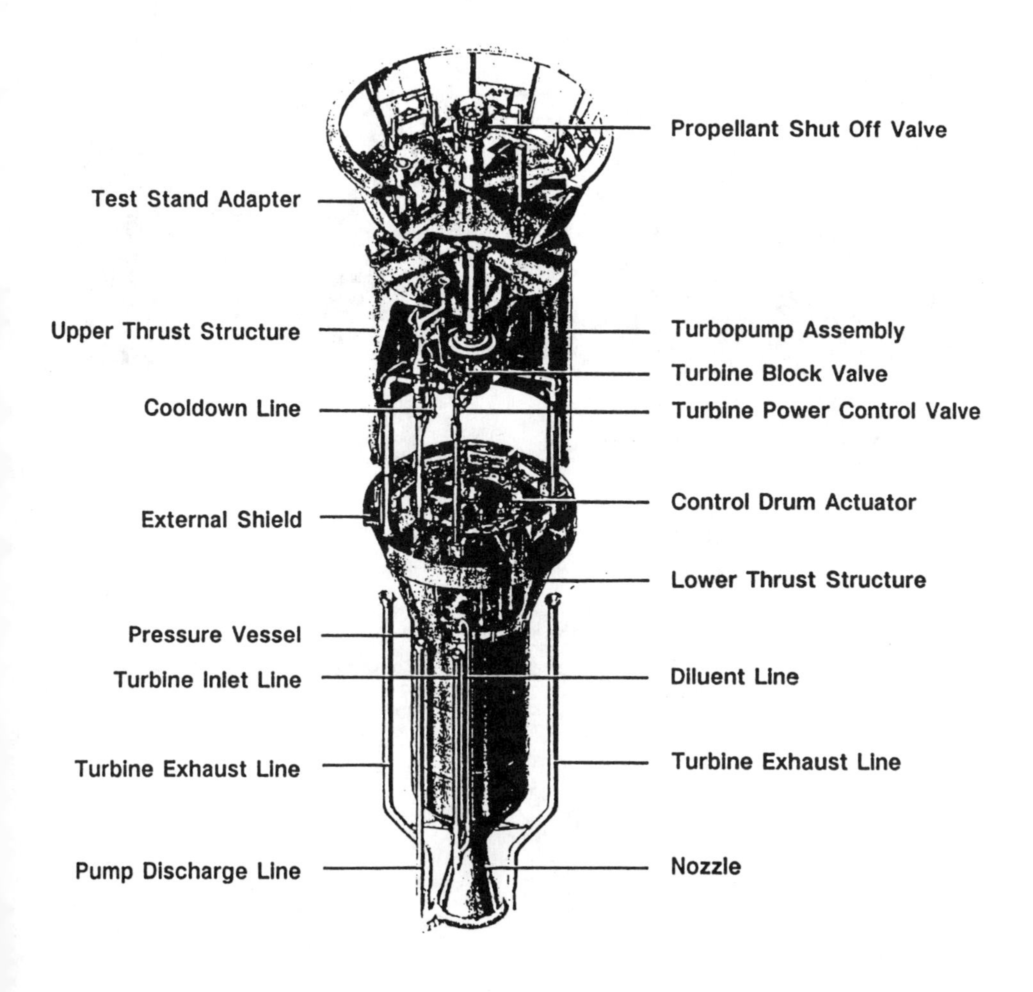

FIGURE 11. Isometric Cutaway of XE-Prime Engine [Gunn 1989 and Koenig 1986].

FIGURE 12. Evolution of the Los Alamos Rover Reactors [Gunn 1989].

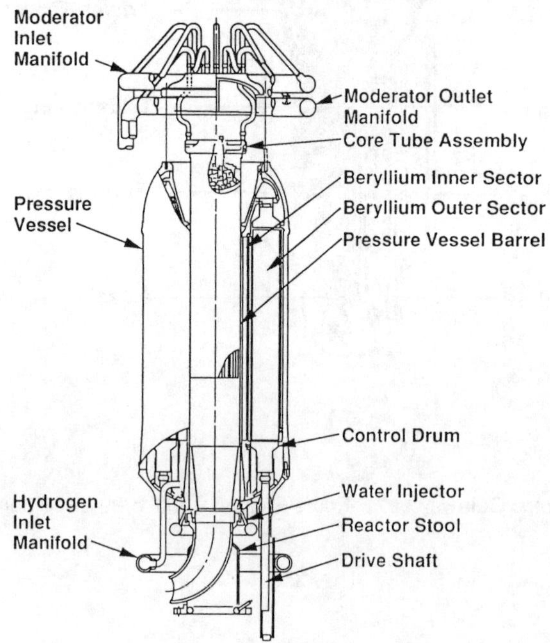

FIGURE 13. Nuclear Furnace Reactor Assembly [Gunn 1989, Kirk 1973, Koenig 1986].

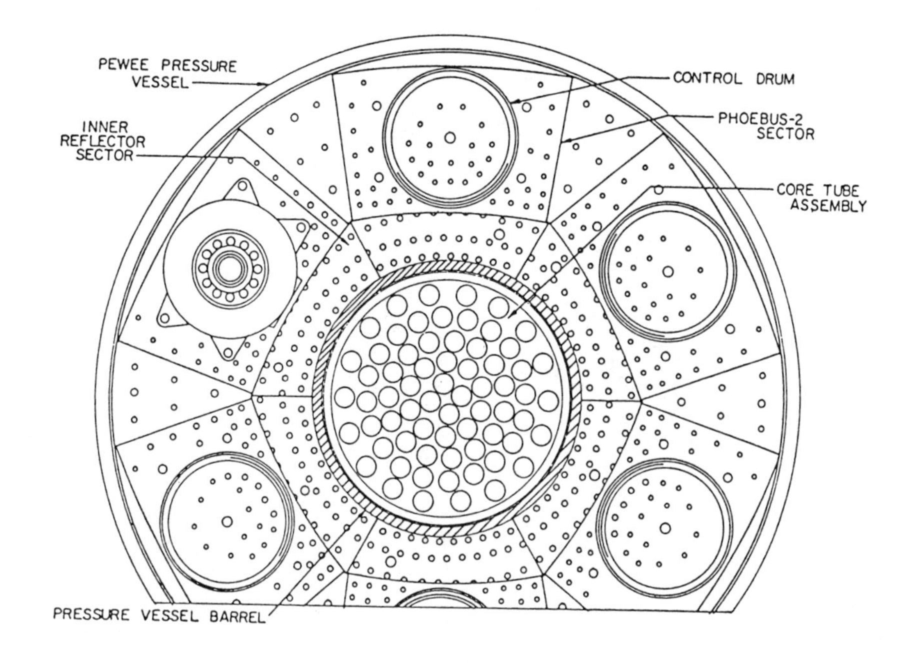

FIGURE 14. Transverse View of Nuclear Furnace-1 [Kirk 1973].

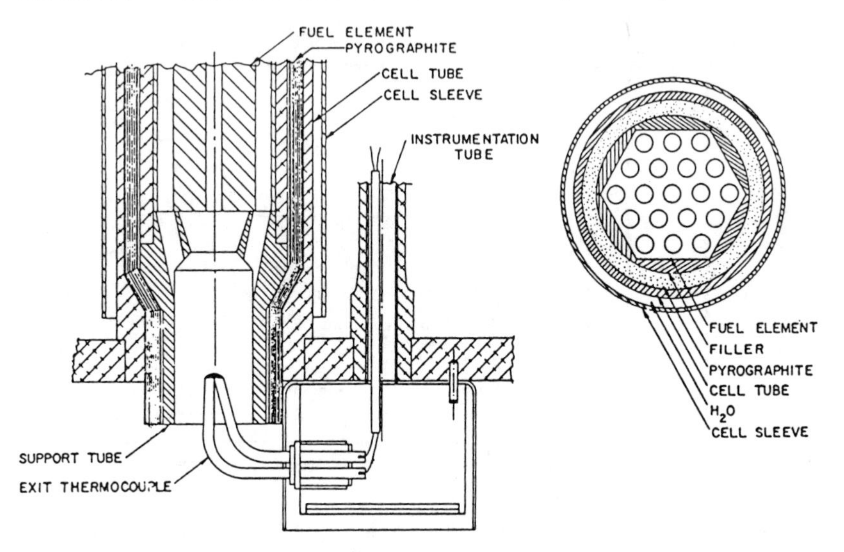

FIGURE 15. Hot End Detail of Nuclear Furnace-1 Cell [Kirk 1973].

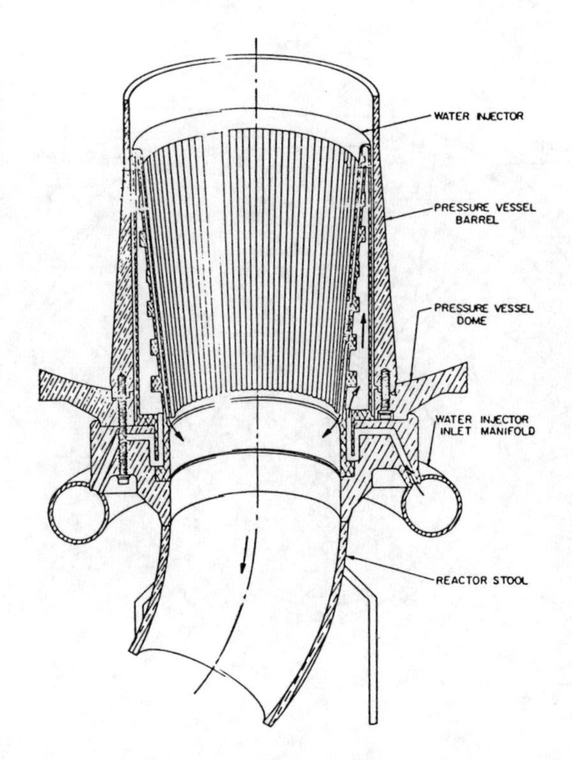

FIGURE 16. Injector and Reactor Stool for Nuclear Furnace-1 [Kirk 1973].

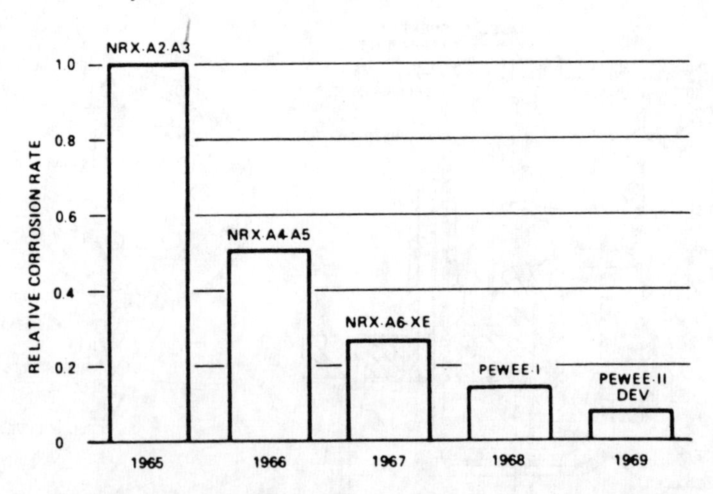

FIGURE 17. Progress Achieved in Reducing Hydrogen Corrosion of Fuel Elements Normalized to the Corrosion Rate Observed in the NRX-A2 Tests. The Corrosion Rate Shown for Pewee-2 is a Projection Because that Reactor was Never Built [Koenig 1986].

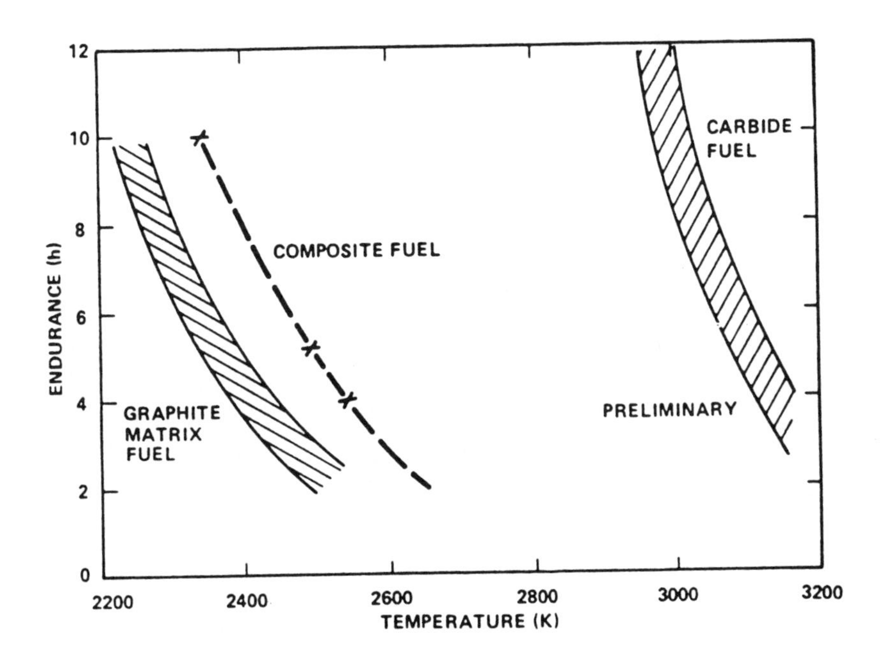

FIGURE 18. Qualitative Comparison of Projected Endurance of Several Fuels Versus Coolant Exit Temperature. The Carbide Curve is an Extrapolation. Such Relationships Depend Upon the Details of the Core and Fuel Element Design and the Operating Conditions (Including Temperature Gradients) [Koenig 1986].

U. S. SPACE NUCLEAR SAFETY: PAST, PRESENT, AND FUTURE

Joseph A. Sholtis, Jr. and
Robert O. Winchester
Air Force Safety Agency
Directorate of Nuclear Surety
2251 Maxwell, S.E.
Kirtland AFB, NM 87117-5773
(505) 846-9897

Leonard W. Connell,
Albert C. Marshall, and
William H. McCulloch
Sandia National Laboratories
Albuquerque, NM 87185
(505) 846-0495,
(505) 272-7002,
(505) 845-8696

Neil W. Brown
General Electric
6835 Via del Oro
San Jose, CA 95119-1315
(408) 365-6516

James E. Mims
Advanced Sciences, Inc.
6739 Academy Rd., NE
Albuquerque, NM 87109-3345
(505) 828-0959

Andrew Potter
NASA Johnson Space Center
Houston, TX 77058
(713) 483-5061

Abstract

Nuclear power systems are an important part of the ongoing U.S. space program and new initiatives are to make use of an even more varied array of nuclear power and propulsion systems. Each space nuclear power system must be designed with safety in mind; this is followed by extensive safety testing and analysis once the system is built. Ultimately each nuclear-powered space mission requires a thorough mission safety analysis and independent safety evaluation to obtain launch approval. These safety analyses, tests, and evaluations ensure full assessment of any potential for radiological risks. This article describes the historical evolution of space nuclear safety, including safety philosophy, safety analyses, and safety testing, and identifies opportunities to further enhance the design and operational safety of future space nuclear power and propulsion systems.

INTRODUCTION

Nuclear systems have enabled tremendous strides in our country's exploration of space. The spectacular views of the outer planets in our solar system sent back by the Voyager spacecraft simply could not have been obtained without nuclear power. Since 1961, 42 nuclear power systems have been launched by the U.S. in support of 25 space missions (Bennett 1987). One mission (SNAPSHOT, in 1965) involved a small, 500-W_e nuclear reactor (SNAP-10A); the remainder were powered by radioisotope thermoelectric generators (RTGs).

While space nuclear energy systems have contributed greatly to our knowledge of Earth and distant worlds, and offer promise for even greater advances in the future, their use poses unique safety challenges. These safety challenges must be recognized and addressed in the design of each space nuclear system. In doing so, the system's planned and potential uses must be considered, and normal as well as accident situations must be taken into account (Sholtis 1989). Safety testing and analyses must be conducted to determine the level of safety built into the system. Extensive safety analyses must establish safety adequacy for ground testing. Lastly, the risks for each nuclear-powered space mission must be assessed so an informed launch decision can be made based on risk-benefit considerations (Sholtis and Winchester 1992).

Safety in design, development, and use of nuclear technology is vital to the viability, acceptance, and continued use of nuclear energy systems in space. To reap the benefits that nuclear systems can provide in space, protection of the public, the environment, workers, property, and other resources must be assured. This can be achieved if appropriate safety objectives are established and met using proven safety philosophies, strategies, and practices.

This article provides an overview of U.S. space nuclear safety as it has evolved to its current state and where it is headed in the future.

BACKGROUND/HISTORY

A focus on safety from the outset followed by meticulous attention to safety in design, development, and planned use has permitted the U.S. to launch a variety of nuclear systems in support of civilian and military space missions (Bennett 1987 and 1991). See Table 1. Recent U.S. nuclear-powered space missions involved the October 1989 Space Shuttle launch of the Galileo spacecraft, which will be used to study Jupiter and its moons, and the October 1990 Space Shuttle launch of the Ulysses spacecraft, which will permit pioneering study of the polar regions of our Sun. These and previous nuclear-powered space missions were extensively reviewed from a system safety and mission risk perspective prior to launch. As an integral part of designing and developing U.S. space nuclear power and propulsion systems, safety tests and analyses are conducted to validate system safety.

Launch and space flight involve risks of failure. Some failures can pose severe accident environments to an on-board nuclear system. In general, the most critical periods include launch, ascent, and orbital injection, when large quantities of rocket propellants are present. Launch, ascent, and orbital injection accident environments include blast overpressure, small shrapnel and larger fragment impacts, Earth surface impacts, propellant fires, and possible atmospheric reentry. Overall, the probability of a catastrophic launch vehicle accident is in the range of several percent (NASA 1988, Martin Marietta 1992). For space reactors, the potential for internal malfunctions during operation in space must also be considered. This includes loss of coolant, or other undercooling accidents, and reactivity insertions, or other overpower accidents.

Four U.S. space missions with nuclear systems on-board have experienced failures; three were caused by the launch vehicle or transfer stage. In each case, built-in safety features performed as designed and there were no adverse consequences. A chronological discussion of these events follows.

The first failure occurred on 21 April 1964, when the TRANSIT 5BN-3 navigational satellite failed to achieve Earth orbit because a computer malfunction prematurely shut down an upper stage booster. The satellite and its RTG power supply reentered the Earth's atmos-

TABLE 1. Summary of Space Nuclear Power Systems launched by the U.S. (1961-1990)[a]

Spacecraft Designation	Mission Type	Launch Date	Power Source (# Sources/Nominal Power)	Status
TRANSIT 4A	Navigation	29 Jun 61	SNAP[b]-3B7 (1/2.7W_e)	Successfully achieved orbit.
TRANSIT 4B	Navigation	15 Nov 61	SNAP-3B8 (1/2.7W_e)	Successfully achieved orbit.
TRANSIT 5BN-1	Navigation	28 Sep 63	SNAP-9A (1/25W_e)	Successfully achieved orbit.
TRANSIT 5BN-2	Navigation	5 Dec 63	SNAP-9A (1/25W_e)	Successfully achieved orbit.
TRANSIT 5BN-3	Navigation	21 Apr 64	SNAP-9A (1/25W_e)	Failed to achieve orbit; RTG burned up on reentry as designed.
SNAPSHOT	Experimental	3 Apr 65	SNAP-10A (1/500W_e)	Successfully achieved orbit; spacecraft voltage regulator malfunction after 43 days resulted in permanent reactor shutdown as designed. Reactor in 3000+ yr orbit.
NIMBUS B-1	Meteorological	18 May 68	SNAP-19B2 (2/40W_e ea)	Vehicle destroyed during launch; RTGs retrieved intact; fuel used on later mission.
NIMBUS III	Meteorological	14 Apr 69	SNAP-19B3 (2/40W_e ea)	Successfully achieved orbit.
APOLLO 12	Lunar Exploration	14 Nov 69	SNAP-27 (1/70W_e)	Successfully placed on Moon.
APOLLO 13	Lunar Exploration	11 Apr 70	SNAP-27 (1/70W_e)	Mission aborted en route to Moon; RTG survived reentry and sank in deep ocean.
APOLLO 14	Lunar Exploration	31 Jan 71	SNAP-27 (1/70W_e)	Successfully placed on Moon.
APOLLO 15	Lunar Exploration	26 Jul 71	SNAP-27 (1/70W_e)	Successfully placed on Moon.
PIONEER 10	Outer Solar System Exploration	2 Mar 72	SNAP-19 (4/40W_e ea)	Successfully placed on inter-planetary trajectory.
APOLLO 16	Lunar Exploration	16 Mar 72	SNAP-27 (1/70W_e)	Successfully placed on Moon.
TRANSIT	Navigation	2 Sep 72	TRANSIT-RTG (1/30W_e)	Successfully achieved orbit.
APOLLO 17	Lunar Exploration	7 Dec 72	SNAP-27 (1/70W_e)	Successfully placed on Moon.
PIONEER 11	Outer Solar System Exploration	5 Apr 73	SNAP-19 (4/40W_e ea)	Successfully placed on inter-planetary trajectory.
VIKING 1	Mars Exploration	20 Aug 75	SNAP-19 (2/40W_e ea)	Successfully placed on Mars.
VIKING 2	Mars Exploration	9 Sep 75	SNAP-19 (2/40W_e ea)	Successfully placed on Mars.
LES 8	Communications	14 Mar 76	MHW (2/150W_e ea)	Successfully achieved orbit.
LES 9	Communications	14 Mar 76	MHW (2/150W_e ea)	Successfully achieved orbit.
VOYAGER 2	Outer Solar System Exploration	20 Aug 77	MHW (3/150W_e ea)	Successfully placed on inter-planetary trajectory.
VOYAGER 1	Outer Solar System Exploration	5 Sep 77	MHW (3/150W_e ea)	Successfully placed on inter-planetary trajectory.
GALILEO	Jovian Exploration	18 Oct 89	GPHS-RTG (2/275W_e ea)	En route to explore Jupiter.
ULYSSES	Solar Polar Exploration	6 Oct 90	GPHS-RTG (1/275W_e)	Successfully placed in heliocentric orbit.

a Updated from Bennett (1987 and 1991)

b SNAP stands for Systems for Nuclear Auxiliary Power; odd-numbered SNAP systems are RTGs while even-numbered SNAP systems are nuclear reactors.

MHW Multi-Hundred Watt RTG

LES Lincoln Experimental Satellite

GPHS-RTG General Purpose Heat Source RTG

phere and burned up completely, as early RTGs were designed to do, at an altitude of about 50 kilometers. Approximately 20,000 curies of plutonium-238 were released into the upper atmosphere and dispersed worldwide. Although this occurrence did not pose a threat to any member of the population, it did involve a release of radioactivity into the biosphere. Subsequently, the design requirement for RTGs under accidental reentry was changed from complete breakup and dispersal at high altitude to survival intact, that is, with fuel containment and confinement preserved through reentry.

In May 1965, 43 days after the successful launch of the SNAPSHOT experimental spacecraft, powered by a SNAP-10A reactor, a voltage regulator failure on the spacecraft caused automatic and irreversible shutdown of the reactor. There was no safety consequence from this failure. The reactor remains in a 3000+ year Earth orbit.

On 18 May 1968, approximately one minute into the launch of the RTG-powered NIMBUS B-1 meteorological satellite, the range safety officer destroyed the launch vehicle by command destruct action so its errant flight trajectory would not put the public in danger. Although the launch vehicle and satellite were completely destroyed, the two on-board RTGs survived intact, with no release of radioactive fuel, and were retrieved from the Santa Barbara channel. The nuclear fuel recovered from these RTGs was used on a subsequent mission.

Lastly, in April 1970, the Apollo 13 mission was aborted on the way to the Moon because of an explosion of an oxygen tank in the service module. An RTG was on the lunar lander to power a lunar surface experiment package. Because the lunar lander returned to Earth with the crew reentry module, it and the RTG experienced atmospheric reentry. The RTG survived reentry intact, with no release of radioactive material, and sank to a depth in excess of 7000 feet at the bottom of the Tonga trench in the South Pacific, where it remains.

Although failures have occurred, they were anticipated by prior analyses and specifically accounted for in the design of the on-board nuclear systems to prevent harmful radiological consequences. The historical record illustrated in Table 1, including the failures discussed above, indicates that the U.S. space nuclear safety program has worked extremely well.

THE FOUNDATIONS OF SPACE NUCLEAR SAFETY

Purpose, Objectives, Approach, and Philosophy of Space Nuclear Safety

The primary purpose of space nuclear safety is to protect the public, the environment, workers, property, and resources from undue risk of injury or harm. To achieve this purpose, three objectives must be met: (1) create a safe product, (2) demonstrate safety—convincingly, and (3) obtain ground test and launch approvals. The approach used to meet these objectives is to build safety into the design, to conduct safety tests and analyses to determine how well safety has been integrated, and to perform safety/risk analyses and evaluations to assess the level of risk involved in the specific application planned. To guide these efforts, an important nuclear safety philosophy has been adopted, namely, to reduce risks to levels that are as low as reasonably achievable, considering the system's planned and potential use, by minimizing the potential interaction of radiation and radioactive materials with the Earth's population and environment. This philosophy is important in setting the proper tone for safety in design; for implementation, it is indispensable for ensuring safety objectives will be met. Moreover, it provides confidence that launch approval can be obtained for a wide spectrum of space missions because it reflects the basic tenet upon which launch decisions for nuclear-powered space missions are made: mission risks must be weighed against the benefits to be accrued.

Nuclear Safety Policy

A number of safety policy statements have been formulated and applied during the design and development of U.S. space nuclear systems. The most recent statement was developed by a joint National Aeronautics and Space Administration/U.S. Department of Defense/U.S. Department of Energy (NASA/DOD/DOE) nuclear safety policy working group formed to recommend nuclear safety policy for nuclear propulsion under the President's Space Exploration Initiative (Marshall et al. 1992). This policy statement, which follows, is broad enough to encompass all types of space nuclear systems. It also represents a synthesis of the best attributes of previous U.S. space nuclear system safety policy statements.

> *Ensuring safety is a paramount objective of the . . . program; all program activities shall be conducted in such a manner as to achieve this objective. The fundamental program safety philosophy shall be to reduce risks to levels as low as reasonably achievable. In conjunction with this philosophy, stringent design and operational safety requirements shall be established and met for all program activities to ensure the protection of individuals and the environment. These requirements shall be based on applicable regulations, standards, and research.*
>
> *A comprehensive safety program shall be established. It shall include continual monitoring and evaluation of safety performance and shall provide for independent safety oversight. Clear lines of authority, responsibility, and communication shall be established and maintained. Furthermore, program management shall foster a safety consciousness among all program participants and throughout all aspects of the . . . program.*

The following elaboration expands on and discusses the implications of each of the principles embodied in this safety policy statement.

- The first sentence of the policy statement reflects a commitment to safety at the highest levels of program management.

- Safety is defined as a condition judged to be of sufficiently low risk. In the context of this policy, safety includes not only the health and safety of the public and program personnel, but also protection of terrestrial and nonterrestrial environments. Safety also includes safeguarding nuclear systems and special nuclear materials against unauthorized use or diversion and protecting facilities and equipment from damage or loss. Safety must be a primary consideration in all phases of a program and should be integrated into the design from conception. It must be a key consideration in all design, operational, and programmatic decisions.

- To ensure the protection of individuals and the environment, the program safety philosophy shall be to reduce risk to as low as is reasonably achievable. Economic and social factors and technology maturity must be taken into account in making judgments on what is reasonably achievable. Risk is a measure of the potential for harm or damage, incorporating probabilities of undesirable consequences and magnitude of consequences. Reduction of risk means reducing the collective probability and/or consequences of potentially adverse events.

- Stringent design and operational safety requirements are essential to reduce risk to as low as is reasonably achievable. Safety requirements must also address severe accidents of extremely low probability. These safety requirements must be established and met for all program activities. They must be based on applicable regulations and standards and must incorporate relevant developments from research activities. These requirements ensure compliance with all applicable national and international regulations.

- A comprehensive safety program must cover all program phases. This safety program must be directed toward making a thorough search for hazards, establishing requirements for eliminating and controlling hazards, executing experimental and analytical evaluations of the associated risks, and providing reasonable assurance that all safety issues have been identified and adequately treated. The program must include continual monitoring and evaluation of activities important to safety and must provide for periodic, competent, and independent safety oversight. The safety program should include a clear delineation of the safety assurance function and should be coordinated with the quality assurance and reliability functions.

- Effective safety administration is essential and must be guided at the policy level to ensure adequate implementation. Clear lines of authority and communication for safety must be established and maintained to ensure safety responsibility and accountability. Expected and potential environments for systems interfacing with nuclear systems must be communicated among all system developers to ensure appropriate mission level safety evaluations. The definition of an organizational structure for safety and the assignment of responsibilities within that structure are vital to an effective safety program.

- It is important to foster open communication with program participants and the public regarding both the benefits and risks associated with the program. The specific emphasis placed by the program on safety and the progress toward meeting the safety objectives should also be communicated. Early planning should develop a process that
  - addresses the need for meaningful public involvement by providing a forum for two-way communication between the public and the program;
  - allows public concerns to be addressed;
  - improves public understanding of space nuclear safety; and
  - serves space nuclear program goals.

- Safety is essential to the success of any nuclear program. Management should take specific actions to develop and maintain safety consciousness among program participants and throughout all aspects of the program. This requires that management be committed to the health and safety of its workers and the public and to the protection of the environment, that this commitment be communicated to all program participants, and that safety be given explicit consideration in all technical, operational, and programmatic decisions. Safety should so pervade the program that all participants recognize their safety responsibilities and automatically consider the safety implications of their work.

One of the most important aspects of this safety policy statement is that it involves a hierarchical approach. It establishes a high-level yet rigorous framework for the formulation of mission-specific safety rules that can (1) be developed without overly constraining nuclear system developers and mission planners and (2) still provide adequate assurance to evaluators, decision makers, and the general public that sufficient measures are being taken to ensure the safe design and application of space nuclear systems.

Although nuclear safety criteria and requirements are necessary to allow space nuclear system design and development to proceed, they cannot be formulated without knowledge of the type of nuclear system in question and its planned or potential space mission use. These choices and factors essentially determine the range and types of safety issues that must be addressed via design and operational safety measures; the safety issues then lead to design and operational safety requirements.

The safety policy outlined above has a number of advantages and has been recommended as a general safety policy for U.S. space nuclear systems (Marshall 1992). It provides guiding principles that are generally applicable to any space nuclear program. Once the mission is defined, the guiding principles provide the basis for developing appropriate safety requirements applicable to the specific mission.

MISSION PHASES, SAFETY ISSUES, AND SAFETY STRATEGIES

To implement a comprehensive and thorough space nuclear safety program that can meet the stated safety objectives, safety issues must first be recognized. These safety issues are tied to the type of nuclear system and are scoped by considering situations that could arise in each of the mission phases from prelaunch through ultimate disposal or disposition of the nuclear system. Once the safety issues are identified, appropriate design safety criteria and requirements can be formulated to ensure each issue is specifically addressed in the design and operation of the space nuclear system. These topics are discussed in the following subsections.

Mission Phases

Actual phases vary depending on the specifics of the mission, such as the launch vehicle and the upper stage to be used. Nevertheless, generic mission phases are useful for scoping safety issues. Any mission can be segmented into seven generic phases: prelaunch, launch, ascent, orbital injection, in-space operations, mission termination, and ultimate nuclear system disposal or disposition. These generic mission phases are briefly described below.

- Prelaunch: The prelaunch phase is defined here to begin with arrival of nuclear system hardware at the launch site and it ends with rocket engine ignition for launch. It encompasses nuclear system handling, storage, transportation, mating/integration, and checkout at the launch site as well as any activity not accomplished prior to shipment, such as assembly, fueling, and coolant fill. The prelaunch phase could be expanded to include earlier activities, including nuclear system development and ground testing.

- Launch: The launch phase begins with rocket engine ignition and typically ends with first stage burnout and separation.

- Ascent: The ascent phase begins with first stage separation and ends with achievement of a temporary low Earth parking orbit.

- Orbital Injection: The orbital injection phase begins with achievement of a temporary low Earth parking orbit and ends with injection of the space platform in its desired stable operational orbit or flight trajectory via action of the upper stage space booster.

- In-Space Operations: The in-space operations phase begins with attainment of the intended operational orbit or flight trajectory and ends with mission termination. This phase includes all actions taken to deploy, start up and operate the spacecraft and nuclear system over their full operational life in space, including any shutdowns and restarts.

- Mission Termination: The mission termination phase begins with initiation of a final spacecraft and nuclear system shutdown signal and ends with attainment of final, permanent shutdown of the space platform.

- Ultimate Nuclear System Disposal/Disposition: This phase begins with verification of final, permanent space platform shutdown and ends with ultimate disposal or disposition of the nuclear system. As a variation, it is possible for nuclear system operation to be used to propulsively dispose of the system. For this special case, the last action in the on-orbit operations phase would coincide with initiation of the disposal phase, which is then followed by the mission termination phase.

Safety Issues and Strategies

Space nuclear systems are designed so that no single credible event will result in a significant radiological exposure to workers or the public or the release of radioactive materials into the environment. This requires identifying components with safety functions subject to failure, considering common mode/cause failures and effects associated with those failures, and incorporating design features to ensure that no single, credible event leads to adverse radiological consequences. Generally, inherent or passive safety features are preferred over active engineered safety features. Designs that require some system response to preclude a hazardous effect are sometimes necessary. It has become standard practice for all nuclear systems, including those destined for space applications, to incorporate explicit, comprehensive consideration of human factors engineering to minimize both the probability and potential impact of human error.

Space nuclear safety issues depend on whether the system is a reactor or an isotopic system. In either case, safety issues are scoped by considering each mission phase, in sequence, until the entire mission life cycle has been taken into account.

Issues and Strategies for Isotopic Power Systems

For isotopic power systems, the key safety issue is release of radioactive fuel material into the biosphere. The strategy of providing several rugged containment barriers to protect against fuel dispersal is the fundamental design safety principle. The multiple barrier approach begins by forming nuclear fuel into physical forms stable in the environments that they are likely to encounter. This usually involves ceramic fuel forms such as oxides, carbides, or nitrides. The fuel is then placed inside a series of physical containers to isolate it from the environment. The fuel may be suspended in a matrix to protect its integrity and improve its performance in the given application. The first barrier is often a coating or cladding on the fuel itself, which retards chemical attack from the outside and the release of radioactive species from the inside. The fuel form is then encapsulated in one or more metallic or carbon containers.

There are two classes of isotopic space power systems: those that require active cooling, such as dynamic isotope power systems (DIPS), and those that do not, such as RTGs. Because RTGs inherently are passively coolable, require no active control, and are constructed with multiple containment barriers, there is no credible potential for radioactive fuel release into the biosphere as a result of internal system faults or failures. The only way a radioactive material release could occur is as a result of externally imposed severe environments, such as launch vehicle explosion or inadvertent reentry of the spacecraft. Explosions create overpressure and fragments. Reentry is accompanied by heating, ablation, aerodynamic forces, and oxidation, and is followed by terminal velocity Earth surface impact. In general, isotopic space power systems are built to withstand these environments and prevent radioactive material release. In addition, isotopic systems are designed to enhance immobilization of any material that might escape. In the U.S., the preferred response for isotopic systems to inadvertent reentry has evolved from high-altitude dispersion to surviving atmospheric reentry intact. Intact reentry requires that the fuel remain in its configuration as a contiguous or contained mass through atmospheric reentry, but some disassembly (and release) at impact on hard surfaces is permissible, provided the radiological hazard can be readily removed from the impact area via cleanup. Virtual intact reentry prevents distribution of radiological materials and provides more predictability concerning system response for a wide range of reentry accident conditions.

For isotopic systems that require active cooling, like DIPS, the prime safety concern is also the release of radioactive fuel material into the biosphere. In addition to externally imposed environments, internal malfunctions that can lead to severe undercooling (loss of coolant, loss of heat sink, loss of flow, and others) must also be considered as a potential cause of radioactive material release for these systems. Consequently, above some threshold thermal power level for this class of systems, highly reliable cooling is required in addition to a rugged system design with multiple containment barriers.

Issues and Strategies for Reactor Systems

Unlike isotopic systems, nuclear fuel release into the biosphere is not the principal safety concern for reactors. Space reactors, typically fueled with uranium-235, are maintained radioactively "cold" (with no appreciable inventory of radioactive fission or activation products) until their intended startup and power operation in space. This is accomplished by limiting reactor prelaunch power operations during ground testing. A space reactor can be ground-tested either at zero power or at full power for a short period and still have no appreciable inventory of fission and activation products at the time of launch so long as sufficient time for radioactive decay is provided prior to launch. The potential release of radioactively cold space reactor fuel is a concern, but primarily from a cleanup perspective—not from a radiological perspective—because virtually no radiological health risk is posed by the dispersal of unirradiated uranium-235 since it is radioactively "cold." Reactor systems pose little risk during the launch, ascent, and orbital injection phases as long as no appreciable inventory of radioactive fission and activation products exists. The practice of precluding reactor operations extends until the system is successfully placed in its planned operational orbit or trajectory. In general, this orbit or trajectory must have enough stability and altitude to ensure that sufficient decay time after power operation will elapse before the system could potentially reenter the atmosphere and return to Earth.

The prime safety concern for space reactors during prelaunch, launch, ascent, and the orbital injection phases of a mission is inadvertent criticality. Inadvertent criticality can result from internal malfunctions as well as externally imposed accident environments. Should inadvertent criticality occur, its effects are manifold. Not only does it generate a fission and activation product inventory, it produces a substantial, usually short duration neutron and gamma radiation field. Under certain circumstances, it can also cause thermal or mechanical disassembly of the reactor. An inadvertent criticality therefore has the potential to create and disperse radioactive materials and generate a short-duration, high-intensity radiation field. To prevent inadvertent criticality, space reactors typically incorporate sufficient neutron absorber materials into the core so that the reactor will remain subcritical for all credible accident situations and configurations that could result.

It is possible for an inadvertent criticality to occur if the moderation and/or reflection of the reactor is drastically altered. This could result from reactor core breach together with burial in soil or immersion and flooding in water or other fluids like liquid hydrogen. Criticality due to core compaction resulting from reactor impact must also be considered. These threats can also be resolved by incorporating thermal neutron absorbers into the reactor core and by constructing to limit core deformation. Such design features can effectively make inadvertent criticality as a result of accidents noncredible.

With inadvertent criticality precluded, a release of fission products into the biosphere can only occur after reactor power operation in space, generally as a result of (1) accidents during reactor operation in Earth orbit, or (2) inadvertent reentry of a radioactively "hot" reactor. Highly reliable control and cooling are provided for space reactors and typically they are only operated in orbits or flight trajectories that provide sufficient decay time before atmospheric reentry could potentially occur.

Because space reactors are fueled with a significant quantity of special nuclear materials, safeguarding the reactor system and its fuel from potential loss and diversion is also important. Land or shallow water impact of a space reactor is of particular concern. The safeguards issue for space reactors during and following launch is typically addressed by using rugged, coded telemetry transmitters or beacons to provide a positive means of locating and, if necessary, recovering the nuclear materials and reactor hardware.

U.S. FLIGHT SAFETY REVIEW AND LAUNCH APPROVAL PROCESS

For any U.S. space mission involving the use of nuclear energy systems with significant quantities of radioactive or fissile material, launch approval must be obtained from the Office of the President (White House 1977). The approval decision is based on a consideration of the projected benefits and risks of the mission. It is also based on an established and proven review process that includes an independent assessment of mission safety/risk by an Interagency Nuclear Safety Review Panel (INSRP).

Chartered by Presidential directive (White House 1977), INSRP is chaired by three members, one each from DOD, NASA, and DOE. The chairs, who are appointed by high-level management from within each agency's oversight safety office, are independent of mission as well as nuclear system involvement, and have the freedom and authority to candidly raise and confront issues at any level.

Experts from academia, private industry, national laboratories, and government comprise five INSRP subpanels, which provide technical and analytical support to the INSRP chairs. These subpanels are responsible for technical evaluation in five sequential analysis areas:

- Launch Abort Subpanel—evaluates prelaunch, launch, and ascent accidents along with their probabilities and associated environments;
- Reentry Subpanel—evaluates reentry accidents, their probabilities, and their effects on the nuclear system, including characterization of any atmospheric radioactive material releases (source terms) postulated;

- Power System Subpanel—evaluates the nuclear system response to prelaunch, launch, ascent, and post-reentry Earth impact accidents. The evaluation includes assessments of the conditional release probability and characterization of any postulated source terms;

- Meteorology Subpanel—evaluates the atmospheric transport of postulated source terms; and

- Biomedical and Environmental Effects Subpanel—evaluates the environmental and health effects of postulated source terms.

This structure ensures breadth and depth of coverage. More importantly, it is geared to quantitative mission radiological risk characterization based on probabilistic methods (Sholtis et al. 1991).

The U.S. flight safety review and launch approval process for nuclear-powered space missions is illustrated in Figure 1. It begins with a preliminary safety analysis report (PSAR) for the space mission. In all, three mission safety analysis reports (SARs) are typically produced—preliminary, updated, and final, termed the PSAR, USAR, and FSAR, respectively. These documents are developed by the mission-sponsoring agency and the project office responsible for developing the nuclear power or propulsion system within the DOE; thus, they represent a project assessment of mission risk.

The INSRP conducts its independent mission safety/risk evaluation in three steps—sequentially following the PSAR, USAR, and FSAR. Although the SARs are prime inputs, the INSRP gathers other pertinent information and conducts its own analyses and tests, as needed. The results of the INSRP evaluation are documented in a nuclear Safety Evaluation Report (SER). The SER contains an independent evaluation of the mission radiological risk and is formally sent to two government agencies: (1) the agency sponsoring the mission, either DOD or NASA; and (2) the Office of Science and Technology Policy within the Office of the President.

The mission-sponsoring agency distributes the SER to other cognizant government agencies (such as DOE, NASA, and DOD) and solicits their concurrence in its request for launch approval. The request for launch approval subsequently is sent to the Office of Science and Technology Policy. The request for launch approval includes a discussion of the mission and its benefits. Because the SER is attached, the request for launch approval also includes an independent characterization of the mission risk. Ultimately, a launch decision is made, based on risk-benefit considerations, within the Office of the President. A commitment by the space nuclear system developer to keep risks as low as reasonably achievable helps ensure a judgment that mission benefits will outweigh its risk.

Historically, preliminary, updated, and final SARs have been issued approximately three years, two years, and one year respectively before the scheduled launch date. Although this schedule has emerged as a convention, it is not a requirement. In fact, there are incentives to begin the process earlier and complete it sooner, if possible. These incentives include keeping the process from becoming a critical path activity that could impact the mission schedule and enhancing safety issue identification and resolution through early design feedback. When very similar missions are involved, the process is conducted rigorously for the mission set prior to the first mission, but then is streamlined thereafter. The Apollo missions to the moon were treated in this manner.

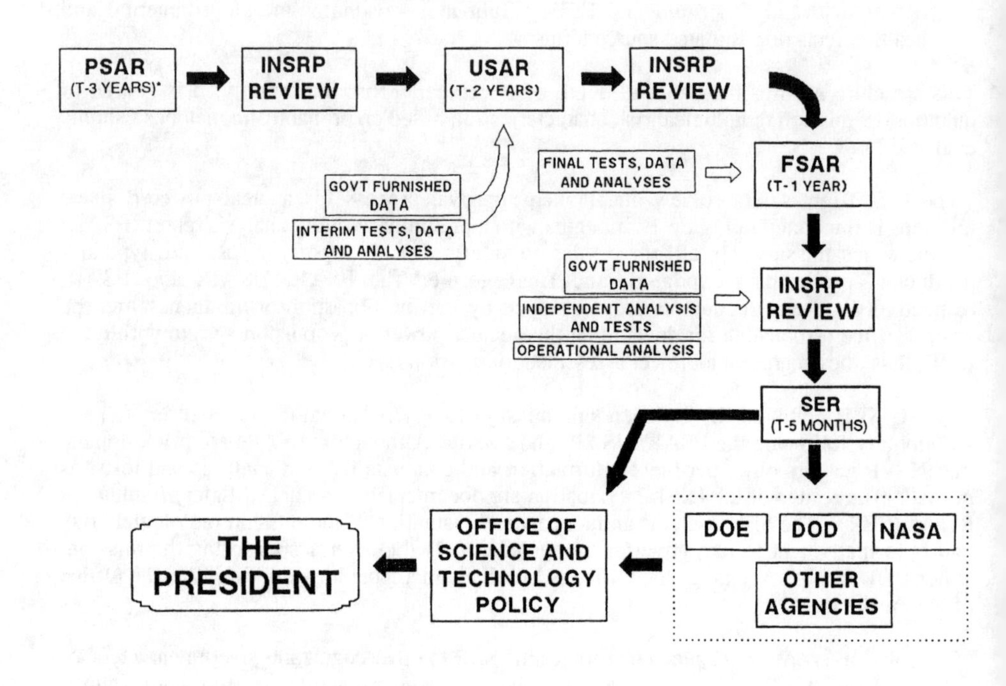

* "T-" notation specifies time before scheduled launch date. For example, "T-3 YEARS" means three years before scheduled launch date.

FIGURE 1. U.S. Space Nuclear Safety Review/Approval Process.

SAFETY TESTING

Safety testing is conducted to validate the level of safety built into the design of the system. It also serves as the basis for extensive system response modeling and safety analyses used to assess risk. Safety testing therefore serves to establish benchmarks for design and analysis.

Isotopic Space Power System Safety Testing

The U.S. isotopic space power system currently in use is the general purpose heat source radioisotope thermoelectric generator (GPHS-RTG), which uses multiple containment barriers to prevent fuel release to the environment under credible accident situations. See Figures 2 and 3. The safety test program conducted for GPHS-RTG hardware is representative of safety test programs for isotopic space power systems (GE 1990b).

The GPHS development test program was initiated in the late 1970s. After preliminary design, response of the GPHS modules to postulated accident environments was verified. The previous Multi-Hundred Watt (MHW) RTG program demonstrated the ability of heat source components to survive potential liquid and solid propellant fires. Nevertheless, solid propellant fire tests were repeated for GPHS hardware; no releases occurred (GE 1990b).

Before the Challenger accident, GPHS hardware was subjected to an extensive test program based on the environments and other conditions attendant to Space Shuttle/Centaur launch vehicle accidents considered to be predominant. The test program was structured to complement the development of the GPHS-RTG and to contribute to the safety evaluation for the Galileo and Ulysses space missions. The test program subsequently was expanded and modified as more data and information concerning the definition of the accident environments and their effects on the GPHS-RTG emerged.

Although the test program prior to Challenger was extensive, many tests were specific to Galileo and Ulysses launch vehicles and to those mission configurations (Space Shuttle, Centaur). The Challenger accident and the attendant investigative findings, deletion of the Centaur as the upper stage, and reconfiguration of the Galileo and Ulysses space mission profiles necessitated additional safety tests.

In these later tests, primary emphasis was placed on solid-fuel rocket booster (SRB) motor case fragments. Much additional information was obtained on the basic response of the heat source components to various accident conditions. These data provided calibration points for computer codes used to model the response of the GPHS-RTG to accident environment conditions outside the limited test ranges.

Following is an overview of the GPHS-RTG test program (GE 1990b). More details are available in Bennett et al. (1988 and 1989).

Pre-1986 Safety Test Program

The test series in the GPHS-RTG safety test program for accident environments related to the Shuttle/Centaur launch vehicle before 1986 included the following tests:

- Shock tube tests,
- Fragment/projectile tests,
- Solid propellant fire test,

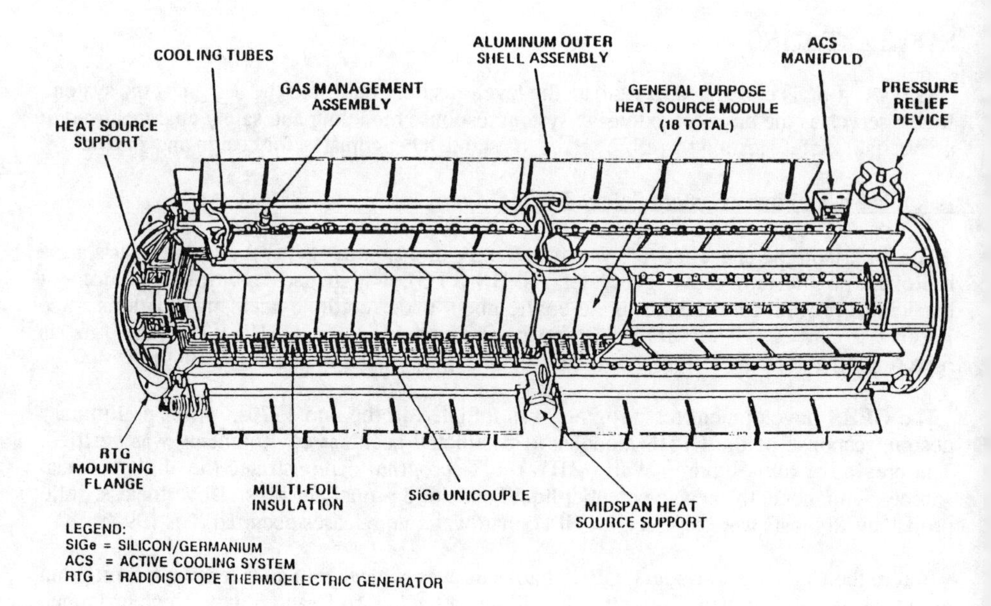

FIGURE 2. Diagram of GPHS-RTG Assembly (Cutaway).

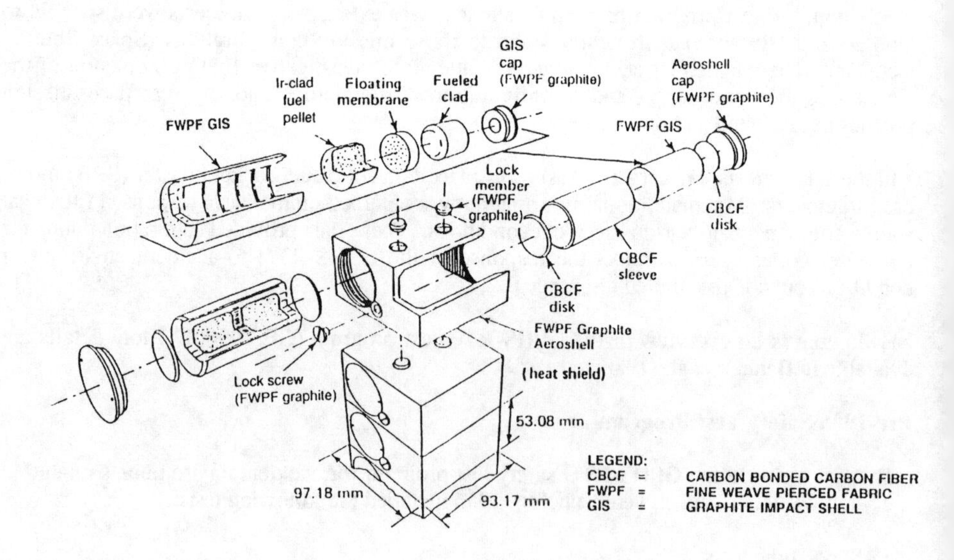

FIGURE 3. Diagram of GPHS Modules.

- Bare Clad Impact (BCI) tests, and
- Safety verification tests (SVTs).

In addition to these tests, the results of the previous design iteration test series were included. These were conducted during the development program using flight-qualified hardware.

Shock Tube Tests

Hydrogen-oxygen propellant explosions could result from accidents involving the Centaur and the Shuttle External Tank (ET). Shock tube tests simulated the shock wave conditions expected from such explosions. Bare module, converter segment, and GPHS-RTG cylinder tests were included in this series. The shock tube tests included the following:

- Four bare module tests at 1.4 to 7.4 MPa (200-1070 psi) static overpressure;
- Three converter segment tests (CSTs) at 6.9 to 12.1 MPa (1000-1750 psi) static overpressure; and
- Five GPHS-RTG cylinder tests at 3.0 to 15.3 MPa (430-2210 psi) static overpressure.

The tests showed that modules would survive overpressures in excess of 3.5 MPa (500 psi) and fueled clads would remain essentially undamaged at static overpressures up to 15.3 MPa (2210 psi).

Fragment/Projectile Tests

Aluminum bullets were fired into GPHS half-module targets with depleted-urania-fueled clads. Titanium bullets were fired into bare fueled clads with fuel simulant to establish margins to failure. Other tests were run on fine-weave pierced fabric (FWPF) specimens representative of the aeroshell and graphite impact shell (GIS). Another test was run in which an aluminum plate (representative of a Shuttle or Centaur tank section) was propelled edge-on into a GPHS module with fuel simulant. Aluminum bullet tests caused no clad failures, but some clad penetration was noted from the higher velocity titanium bullet tests. Potential small aluminum fragments were demonstrated not to lead to fuel release for the GPHS-RTG.

Solid Propellant Fire Tests

A fueled clad and an impact assembly with urania fuel simulant were exposed to a 10.5-minute fire from a cube of solid propellant. This represents the maximum possible solid-propellant fire duration for the Space Shuttle. No significant clad damage and no fuel release occurred during these tests.

Bare Clad Impact (BCI) Tests

Fueled clads were fired against concrete, steel, and sand targets. This test series included the following:

- Nine plutonia-fueled clad impacts on hard targets; and
- Eighteen urania-fueled clad impacts on hard and sand targets.

Bare fueled clads were not designed to survive an impact on hard targets, and a number of failures were recorded. These results helped establish the failure criteria that subsequently were used in evaluating SRB fragment impacts later in the program. However, bare fueled clads did not fail on soft targets, such as sand, at impact velocities to 250 m/s.

Safety Verification Tests (SVTs)

In this series, GPHS modules were fired against concrete and steel targets. These tests were designed to look primarily at module survivability for terminal velocity impacts on hard surfaces. Fuel releases were observed to be retained within the graphite components. These tests also determined the internal fuel particle size distribution generated.

Post-Challenger Safety Test Program

After the Challenger accident, accident environments were scrupulously reevaluated. As a consequence, a substantial supplemental test program was established. Tests included in this test program were as follows:

- Bare Clad Impact (BCI) tests,
- Solid Rocket Booster (SRB) fragment tests in gas gun (FGTs),
- Large SRB fragment impact tests (LFTs), and
- SRB fragment/orbiter fuselage tests (FFTs).

Bare Clad Impact (BCI) Tests

This BCI test series provided multiple impact data in areas where none existed. Indications were that releases could potentially occur as a result of multiple impacts; that is, from an initial SRB fragment impact followed by a subsequent impact on a hard surface. However, this testing did not result in any releases, even with clad deformations as large as 48%.

SRB Fragment Tests in Gas Gun

A 178-millimeter-diameter gas gun using a pneumatically propelled, upward firing sabot with provision for containing the impacted test article in a catch tube was used for this test series. In these tests, a simulated GPHS-RTG section was propelled into an SRB case fragment.

Tests were run with simulant (urania) and plutonia fuel to determine response to SRB fragment impacts and to compare clads containing fuel and those containing simulant. A comparative evaluation between the fuel and simulant response was provided by direct replication of impact tests. The plutonia basic response data and urania response relationship were to calibrate the analytical models and to evaluate the large SRB fragment field tests in which only urania-fueled clads could be tested. These tests provided a good basis for comparison of urania to plutonia fuel. No clads failed.

Large SRB Fragment Impact Tests (LFTs)

An SRB plate fragment was propelled by rocket sled into a flight configuration GPHS-RTG housing containing two simulant fueled test modules and six mass simulant modules. The purpose of these tests was to determine the response to impact by an SRB fragment. Three tests have been completed: two SRB face-on tests at 112 m/s and one SRB edge-on test at 95 m/s.

No fueled clad failures were observed in the face-on SRB fragment tests. Clad failures were observed in the leading fueled clads of the edge-on SRB fragment test but not in the trailing fueled clads.

SRB Fragment/Orbiter Fuselage Tests (FFTs)

Between the SRBs and the GPHS-RTGs are substantial Space Shuttle Orbiter structural components that are expected to attenuate a fragment's linear and rotational velocity. This test series, aimed at investigating SRB fragment velocity attenuation, was performed using actual Orbiter components. SRB fragments were propelled through sections of an Orbiter wing and/or payload bay wall structure.

The general conclusion from all the safety testing is that the GPHS-RTG is very rugged in severe accident situations. Specifically, fuel release is not expected except for terminal or near terminal velocity impacts on hard surfaces or high-momentum impacts from SRB fragments, particularly if they involve shearing action.

Reactor Space Power and Propulsion System Safety Testing

Safety testing of space reactors intended for power or propulsion is focused on assuring that safety design requirements are satisfied and that risks are kept as low as reasonably achievable. Test data are obtained to confirm that design features important to safety meet their requirements, or to support risk analysis. The scope of issues for space reactor safety testing tends to be broader than that for isotopic systems, especially RTGs, because reactor systems include more complex active control and safety features. The type and extent of reactor safety testing required depends strongly on the specifics of the mission. Testing important to safety includes selected performance testing that contributes to probabilistic evaluation. Because differences between space reactors for thermal propulsion and those for electric power generation or electric propulsion are significant, the safety testing for these two categories is discussed separately.

The two major programs that have contributed most of the safety testing results are the ROVER/NERVA nuclear rocket program (Koenig 1986) and the SNAP-10A space reactor program (Voss 1984). Current reactor nuclear power and propulsion programs have not yet involved substantial safety testing; however, program plans and approaches to safety testing are discussed.

Reactor Space Power Systems Safety Testing

Historical Programs

The Systems for Nuclear Auxiliary Power (SNAP) program was initiated in 1955 and terminated in 1973. The program produced more than six test reactors; one space reactor was launched and operated in space during 1965. Five were zirconium hydride moderated, NaK cooled reactors, and a sixth was a uranium carbide fueled, lithium cooled fast reactor. The SNAP 2, SNAP 8, and SNAP 10 programs operated ground test reactors and contributed operational information useful to safety assessments. However, the greatest source of safety data is the Aerospace Safety Program (Otter et al. 1973), which was conducted from 1961 to 1972 and included the following elements:

- Reactor disruption,
- Fuel rod reentry burnup,
- Criticality configurations,
- Reactor transient behavior,
- Mechanical and thermochemical incidents,
- End-of-life shutdown, and
- Disposal mode studies.

A significant portion of the Aerospace Safety Program focused on reentry characteristics. During that time, high-altitude breakup and dispersal was the preferred reactor response. Testing included half-scale model SNAP 8 and SNAP 10A wind tunnel tests, and arc jet heating and ablation tests of components and fuel. The program also conducted a suborbital reentry flight test of a full-scale simulated reactor assembly. Test data provided evidence regarding the difficulty of dispersing a ZrH-type space reactor during an inadvertent reentry accident.

Another major portion of the program involved transient testing of the reactor to destruction. The experiments were conducted at the Idaho National Engineering Laboratory. These tests demonstrated the effective destruction of the reactor with limited energy and fission product generation and releases. The rapid heating produced hydrogen release from the ZrH, which disrupted the reactor and terminated the transient. These tests were important to safety evaluation because SNAP reactors were not designed to be subcritical in all accident scenarios.

Other portions of the program included extensive critical tests of various configurations of ZrH reactors and tests of the reactor under mechanical and thermochemical environments in prelaunch accidents. Techniques were developed and used for safety issues associated with fabrication, testing, transportation, handling, and launch of a SNAP reactor system. The program developed techniques for making probabilistic risk assessments and conducted testing that provided reliability data needed to complete analyses.

Current Programs

The SP-100 program is a major U.S. space nuclear power system technology development program and is implementing the safety philosophy that has evolved since the SNAP program. Early in the SP-100 effort, the SNAP program safety data and evaluations were reviewed along with recent RTG experience. These reviews were used along with evaluations of proposed design approaches to establish the SP-100 program safety philosophy and approach. Key safety and design selections that influenced development of the SP-100 safety and safety test program were intact reentry and, for selected missions, retention of reactor geometry for permanent disposal. In addition, it was established that inadvertent reactor criticality should be precluded for all credible accidents. This decision caused the SP-100 safety design approach and planned testing to differ from that of the SNAP 10A program. These selections and supporting safety analyses led to the inclusion of a reentry cone and internal safety rods that must be retained during accidents. In systems above about 500 kW_t, an auxiliary cooling loop is provided if the mission requires high confidence for the retention of core geometry for disposal.

The special safety testing remaining for SP-100 is focused on two areas: (1) retention of safety rods during launch pad accidents and during inadvertent reentry and impact, and (2) demonstration that credible solid booster fires do not cause reactor criticality. The important area of criticality testing for various core configurations is complete, and validated methods are available for assuring adequate shutdown margins in design.

Within the performance testing important to safety, it is necessary to complete controller subsystem and mechanism testing that supports the probabilistic risk analysis. The probability of control subsystem failures that produce uncontrolled reactivity insertion and the probability of successful shutdown when required are key issues. For missions requiring an auxiliary cooling loop, it is necessary to confirm the hydraulic performance of the loop.

Confirmation of the reentry shield performance during inadvertent reentry may require testing, which, if required, is expected to be accomplished with scale models or key component tests in wind tunnels or arc jet facilities.

The planned SP-100 safety testing program is anticipated to demonstrate key safety features in a broad range of designs including reactors used for electric propulsion.

Reactor Space Propulsion Systems Safety Testing

Historical Programs

In 1955, the U.S. initiated Project ROVER to develop a nuclear rocket engine. The program was terminated in 1973 prior to flight engine development. Testing had not identified any technological barriers to development of a safe and reliable flight system based on solid core fuel and hydrogen propellant. The development testing included operational safety testing as an integral part of the development program.

Project ROVER's safety experience relating to the fabrication, assembly, shipping, and ground testing of a nuclear thermal rocket engine was extensive and well documented. Most of the testing demonstrated operational limits of the fuel. During the NERVA engine development phase of the ROVER program, safety practice and analysis techniques useful to both ground and space safety were developed by Aerojet General and Westinghouse Electric Corporations. These techniques included probabilistic risk assessment as well as reliability testing and data development to support risk assessment.

The program built and operated 20 reactors. One reactor, KIWI-TNT, was deliberately destroyed by conducting a nuclear excursion. This test validated analytical models of a reactor power excursion. The test supported flight safety analysis and has been used to calibrate safety analysis tools for both space and terrestrial reactor accident evaluations.

An important safety accomplishment of the program was the demonstration of the ability to reliably control the rapid start and restart of the nuclear rocket engine. Introducing hydrogen into the core as a coolant introduces a positive reactivity effect that, along with control drum rotation positive reactivity, must compensate for the negative reactivity effect of the rapid increase in temperature during startup. A large portion of the testing focused on developing fuel and core structures that resisted erosive and corrosive effects of the high temperature hydrogen coolant.

Current Programs

Nuclear rocket propulsion in support of military applications and the Space Exploration Initiative (SEI) is the subject of renewed U.S. interest; the program is known as the Space Nuclear Thermal Propulsion Program (SNTP). Current focus is on evaluating testing requirements and test facility needs for nuclear thermal rocket system hardware. More stringent environmental policies for the candidate test sites make this a priority consideration. The facilities required for safety testing associated with transportation and launch accidents are not expected to be significantly more extensive than those currently in place at the national laboratories. Critical configurations and impact testing of transportation and core features may be required unless existing equipment can be shown to be adequate.

Preliminary safety assessments indicate that there may be no need to conduct another full reactor destruction test like KIWI-TNT. Improvement of the analytical tools available for this analysis has been made possible by the growth in computer capability, and these improved models have been validated against reactor transient tests and the KIWI-TNT test.

The initial focus of testing that is important to safety and environmental evaluations involves fuel performance with emphasis in the area of fuel particle and fission product retention. The demand for retaining or capturing fission products appears to be more critical to ground testing than to space operations. Test facilities must accommodate normal operating release of radioactivity and credible upsets. This type of nuclear rocket fuel testing has been limited and is expected to be one of the first areas of testing to proceed as the program moves forward. These tests are obviously important to establishing environmentally safe ground test facilities and ultimately safe designs for use in space.

QUANTITATIVE RISK ASSESSMENT

This section, which serves as an overview, describes the approach and important concepts for quantitative risk assessment of nuclear-powered space missions.

Overview

To perform a risk assessment for a given mission, it is advantageous to segment the mission into phases and identify accident scenarios and environments that could occur in each phase. The system response to these accidents and their associated environments is then evaluated using empirical test data supplemented by modeling and analysis. This evaluation identifies those accident sequences and accident environment levels that can lead to projected radiological releases (source terms). Evaluations of source-term release probability distributions are completed at this point for each accident sequence determined to lead to a release. Source terms are next characterized by, for example, amount, radionuclide, chemical form, and physical form (vapor or particulate, with a particle-size distribution). Each identified potential radiological release is then analyzed to evaluate its dispersion and potential population exposures through various pathways. Potential radiation doses and associated health effects consequences are subsequently calculated. Finally, these results are combined with their associated probabilities to characterize mission radiological risk and the uncertainties associated with the risk estimates (Bartram 1992). Some examples of recent risk assessments for space nuclear systems include the Galileo FSAR (GE 1989a and 1989b), the Galileo SER (INSRP 1989), the Ulysses FSAR (GE 1990a and 1990b), the Ulysses SER (INSRP 1990), and the SP-100 Mission Risk Assessments (Bartram and Weitzberg 1988, Damon et al. 1989, and Brown et al. 1990).

Potential Accident Scenarios and Environments

Each space nuclear system is analyzed with regard to its application to a particular mission. Typically, each mission phase is considered to ensure that all important procedures as well as normal mission and accident events are included so that their effects can be systematically analyzed.

The systematic analysis typically begins with a failure modes and effects analysis (FMEA) of the launch vehicle, upper stage booster, and spacecraft configuration. The FMEA includes all failures and the resultant vehicle condition (such as thrust termination, propellant tank rupture, loss of altitude control) as well as the occurrence probability for that vehicle failure or accident condition.

For each accident scenario and failure identified in the FMEA, a sequence of accident environments is then defined. In general, accident environments include explosion overpressure, fireball, projectile/fragment impacts, hard surface impacts, propellant fires, and reentry. The severity of each of these accident environments is usually characterized in terms of a probability distribution.

System Accident Response and Consequence Analyses

Evaluation of nuclear system responses to accident scenarios and environment sequences enables the prediction of which accident sequences would result in adverse consequences (such as release of radioactive material or nuclear criticality). The occurrence probability of conditions that lead to threatening environments and the conditional probability of undesirable nuclear system responses to these conditions can be combined to predict the overall probability of experiencing adverse consequences. The full spectrum of potential accident sequences and environments is sometimes displayed in logic trees (referred to as failure and abort sequence trees or FASTs) depicting the possible failure sequences for each mission phase. These logic trees include the probabilities of all sequence branches and end points. Resulting consequences can be characterized in terms of probability distributions.

Several approaches can be used to perform this systematic determination of the consequences and their probability distributions, depending on the mission phase, the availability of test data, and the analysis methods used to model the physical phenomena involved. These methods, primarily developed for RTGs, include the following:

- Launch phase accident analysis;
- Orbital debris/meteor impact analysis;
- Reentry/breakup analysis;
- Post-reentry impact analysis; and
- Radiological dispersion and health effects consequence analysis.

For reactor systems, it is also necessary to evaluate the potential for, and consequences of, an inadvertent nuclear criticality, on-orbit operational accidents, and loss of effective nuclear material safeguards control.

Launch Phase Accident Analysis

Launch Accident Scenario Evaluation Modeling

An analysis method recently developed by the General Electric Company involves a Monte Carlo computer code called the Launch Accident Scenario Evaluation Program (LASEP). LASEP was used to assess the response of the GPHS-RTG to potential Space Shuttle launch accidents and was used for the safety analyses of the Galileo and Ulysses space missions (GE 1989a, 1989b, 1990a, 1990b). The overall approach used in LASEP is shown in Figure 4. LASEP determines the sequential response and cumulative damage of the GPHS-RTG to launch accidents and defines source terms and particle size distributions for all projected fuel releases. LASEP uses a Monte Carlo approach by running many RTG response trials for each accident scenario or subscenario identified. The source terms that result from the LASEP runs are in the form of a probability-mass distribution for each source term. LASEP is specific to the nuclear system involved and the mission accidents and accident environments that can occur. Consequently, a new code must be developed, or the existing code must be modified, for any other nuclear-powered space mission. Some of the more important aspects for propellant explosion, fire, and criticality analyses that would be incorporated into such a launch accident computer program is briefly addressed in the following three subsections.

Propellant Explosion Analysis

An accident scenario involving a propellant explosion might (1) provide thermal radiation heating that could weaken important components, (2) create a blast wave that could distort the system geometry and rip the system from its mount, (3) distort the system, (4) rupture the system so that nuclear fuel is released, (5) create conditions so that released material rises in a plume and is dispersed into the atmosphere, (6) threaten shutdown features, and (7) pose the potential (generally unlikely) for compaction-induced criticality. Damage mechanisms include nuclear system impact by fragments resulting from the explosion, overpressure, impulse, and the possibility of Earth surface impact caused by ejection of the nuclear system. Multiple, rugged containment barriers are used in isotopic and reactor systems to prevent fuel release. In addition, a large shutdown margin and protected withdrawal devices are built into reactor systems to prevent an inadvertent nuclear criticality.

The magnitude of an explosion of propellant is often expressed in terms of an equivalent weight of trinitrotoluene (TNT). TNT weight equivalence is used because of the large amount of experimental data on blast waves and damage produced by TNT explosions. However, it is only useful for far-field effects. Two important parameters for characterizing the blast wave are the peak overpressure and the positive phase impulse.

Blast wave boundary conditions are obtained from blast calculations on rigid bodies. These are then applied to an object and distortion is computed. In actual propellant explosions, distortion of the body during the diffraction and drag phases influences the blast wave and subsequent flow and, in turn, the damage mechanisms. Acceleration, overturning, tumbling, and other factors can all influence breakup of the nuclear system. It is conceivable that coupled blast and structural response calculations could improve predictive capability.

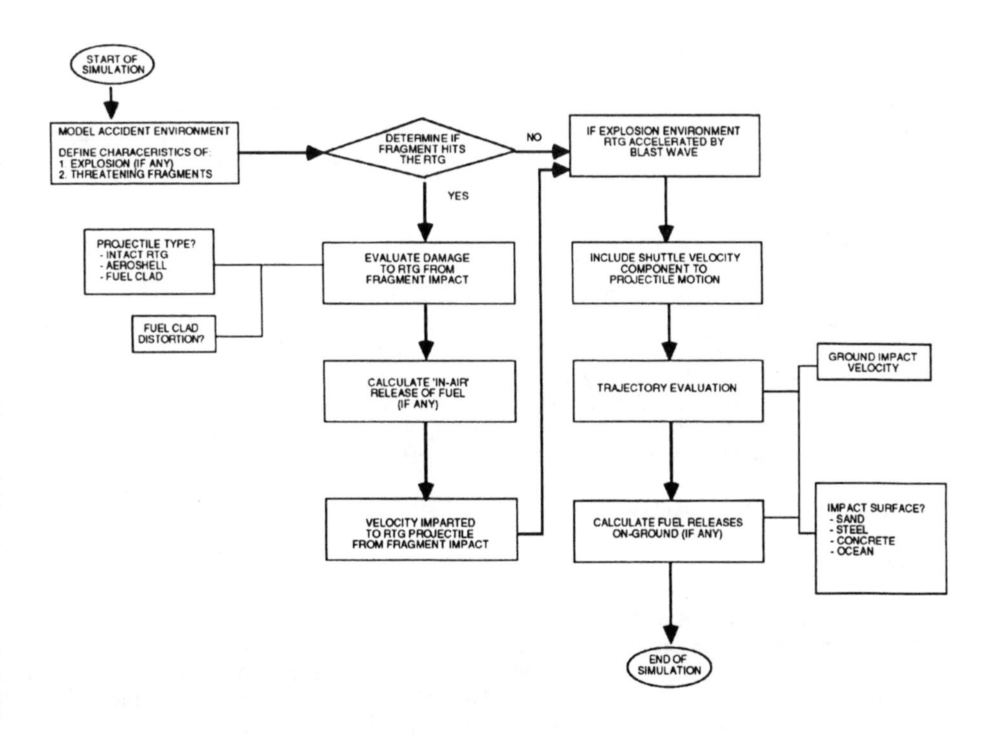

FIGURE 4. Launch Accident Scenario Evaluation Program Logic (from GE [1990b]).

Fire Analysis

Fire is a natural outcome of an explosion that can ultimately have two consequences. The first is structural damage, and the second is accidental release. Ideally, fire analysis requires a single code that can perform these analyses concurrently.

Isotopic systems like the GPHS-RTG will not release fuel in solid or liquid propellant fires. During the launch phase, solid propellant fires in the vicinity of a space reactor could conceivably damage structures sufficiently to cause a nuclear critical configuration or release of unirradiated nuclear fuel. The severity and effects of the fire depend upon factors such as the reactants involved, location with respect to the reactor, and fire duration.

The first step in fire analysis is to model the nuclear system. Properties for most materials can normally be obtained from a number of sources (for example, Touloukian and Ho [1976]). Flame temperatures can be obtained from calculations or experimental data. Empirical formulas are available for calculating various parameters such as fireball diameter, time to liftoff and fireball duration. Finally, the heat flux input to the analysis code can be calculated from fireball temperature measurements and the surface area of the fireball (Erdahl et al. 1988).

Criticality Analysis

Fires, explosions, and impact could induce an inadvertent criticality for reactor systems. Accident scenarios can also be postulated in which reactor system barriers are breached from impact and subsequently flooded with water (or some other fluid), increasing neutron moderation and inducing criticality. Water immersion and wet soil or sand burial could increase neutron reflection and also induce an inadvertent criticality. These reactor accident scenarios are evaluated using either neutron transport or Monte Carlo calculational tools. Benchmarking neutronics computer models against test data is usually required.

Operational Phase Accident Analysis

Internally Initiated Accidents

Internally initiated accidents are a safety concern for all reactors; this is particularly true for terrestrial reactors. These accident types can be grouped into reactivity initiating accidents and cooling failure accidents. Although these types of accidents must be addressed for space reactors, a reactor accident in space generally exhibits a small safety risk to the Earth's population and environment. A variety of system dynamics calculational models may be used to predict the effect of these types of accidents on the reactor system.

Orbital Debris/Meteor Impact Analysis

Meteoroids are part of the interplanetary environment, and sweep through earth orbital space at an average speed of about 20 km/s. Most of the mass is in meteoroids about 0.01 cm in diameter. The particle flux is well known as a result of a number of measurements conducted since the early phases of space flight. The hazard from meteoroid impact is understood well enough to provide adequate design criteria for spacecraft shielding (NASA 1991). All U.S. manned spacecraft have incorporated some degree of protection against meteoroid impacts.

The man-made debris environment was initiated by the first explosion in orbit, which occurred on 29 June 1961, when the Thor-Ablestar stage used for launch of a Transit 4A satellite (1961-Omicron) exploded. Not only was this the first explosion in orbit, it was also the first launch of a nuclear power service (SNAP-3B7 RTG) into space. The upper stage explosion yielded 292 fragments cataloged by U.S. Space Command. Many of these fragments are still being tracked. Following this event, explosions continued to occur at a rate of 3 or 4 per year. To date, more than 130 breakups have occurred in space, for the most part in low earth orbit (LEO) at altitudes below 5500 km. This fact has important implications, because explosions generate a wide range of fragments. The U.S. Space Command currently tracks about 7,000 objects in Earth orbit with sizes ranging down to about 10 cm, and with a total mass of between two and three million kilograms. About half the objects being tracked are fragments from explosions. Reynolds et al. (1991) developed a model (EVOLVE) of the low earth orbit (LEO) debris environment to predict the population of debris smaller than 10 cm. The model was based on estimates of the number and sizes of debris fragments resulting from all known on-orbit breakups, and includes the effects of atmospheric drag and solar cycle effects on the removal of debris from orbit. The EVOLVE model has been used to develop an engineering model of the debris environment; this model is being used by the Space Station Freedom program to design shielding for the Station (NASA 1991). The engineering model has been verified down to the 1-cm-debris-size level by radar measurements with the Haystack radar. The result is that there are approximately 140,000 objects 1 cm or larger in low earth orbit (Stansbery et al. 1992).

The probability of collision with debris and the probable consequences of the collision can be calculated for any specific mission using the models described above. An example is given by Anz-Meador and Potter (1991), in which the lifetimes of Soviet reactor cores against destruction by orbital debris impacts is estimated.

If launch rates and breakups continue at the current rates, spontaneous, runaway growth of the small debris environment in LEO will begin in the latter half of the next century (Eichler and Rex 1990; Kessler 1991). This is the result of collisions, which generate debris, which in turn leads to even more collisions.

Geosynchronous Earth orbits (GEO) are the most commercially valuable regions of space. There are currently about 300 satellites and about 100 upper stages in GEO. One projection predicts that new objects will be added to GEO at a rate of about 60 per year (ESA 1988). This value includes active satellites, spent upper stages, and other residual hardware. Collisions between intact payloads and upper stages in GEO are statistically very unlikely at the current population density. However, since atmospheric drag is ineffective for removal of satellites from GEO, every object put into GEO will remain there for millennia. Consequently, substantial accumulation and crowding of objects in GEO could eventually occur. In time, collision probabilities could become significant.

Very little is known about orbital debris in GEO. Two breakups have been known to occur in GEO (McKnight 1992). It is likely that others have occurred unobserved. Orbital debris in GEO could shorten the useful life of satellites there, even though collision velocities in GEO are much less than in LEO.

Because the average relative collision velocity of objects in orbit is 10 km/s, collisions with even small objects can have a dramatic effect. For space nuclear systems, severe collisions can mean premature inadvertent reentry, although this is extremely unlikely, or the release of radioactive materials in space, which ultimately could potentially impact the Earth's population and environment. Although radioactive material release into space is possible, its likelihood is remote, particularly for isotopic systems, and its consequences are not severe. Multiple rugged barriers provided for isotopic systems and highly reliable shutdown and cooling features for reactors help ensure that the likelihood of such a radioactive material release is extremely remote. In fact, for reactors, so long as the system can be shut down effectively and adequately cooled, no release would be expected.

Reentry/Breakup Analysis

History

Analytical and experimental means to determine the sequential disassembly of a space nuclear system undergoing atmospheric reentry were developed in the late 1950s and early 1960s. They resulted from the efforts to develop ballistic missiles and from the growth of the civilian space program. The manned space program and the ICBM program had to solve the problem of ensuring the survival of an object returning to Earth. Reentry was one of the most difficult problems that confronted aerospace engineers of that time (von Karman 1956). The aerothermal environment is so severe that structural materials normally cannot survive without a thermal protection system. One such means is a reentry heat shield. A reentry heat shield must be lightweight, of sturdy construction, and capable of being securely bonded to the substructure. RTGs are designed to survive atmospheric reentry intact; this is the preferred reentry response for many U.S. space reactor concepts as well.

Reentry Analysis Methodology

The methodology of determining the breakup of a space reactor is described below. The reentry analysis for an RTG can use the same basic approach, but the details of the breakup would vary because of the differences in design.

The top-level procedure to predict the aerothermal reentry behavior of a space reactor is as follows:

1. Compute the trajectory (time, altitude, speed, flight path angle);
2. At selected trajectory points, compute the heat transfer rate a various points on the body;
3. With the heat transfer known, determine the thermal response (the time-dependent temperature profile of the body) and predict the disassembly of a key component (for example, reactor separation from the radiation shield) based on a failure criterion established by structural analysis or testing; and
4. Go back to step 1 and repeat for the new configuration. Continue looping until complete disassembly has been predicted or the aerothermal environment ceases to drive the disassembly.

The output from these calculations is a picture of the breakup process that will include, at a minimum, the altitude points along the trajectory at which important nuclear system components break free. For a reactor system, the breakup of components could potentially include separation of items such as the thermal radiator, radiation shield, and reflector system, as well as failure of the pressure vessel, core structural components, fuel elements, and fuel pellets. The analysis would also normally include a thermal history of key components including a calculation of the ablation mass loss. The ablation data could then be used as input to atmospheric dispersal codes that predict depositions and doses. Current ablation codes do not predict the form of the ablation products such as the fraction of mass stripped off as a liquid, the fraction vaporized, and the distribution of particle sizes.

Trajectory Analysis

Although six-degree-of-freedom codes are available, a point-mass (three-degree-of-freedom) trajectory code may provide an adequate level of detail for safety assessments. A point mass computation is performed for each segment of the flight. The desired output from this computation is the altitude and speed of the object versus time. The required inputs are the initial conditions (speed, altitude, and flight path angle) and the ballistic parameters (drag coefficient versus Mach and Knudsen numbers, mass, and reference area). If drag coefficient data are not already available, a flow field code will determine the pressure distribution about the body and from this predict drag. Trajectory analysis tools are very well developed and this segment of the analysis is fairly routine. Many trajectory codes are available, but one of the best written and documented is the Trajectory Simulation and Analysis Program (TSAP) (Outka 1990).

Thermal Response

Prior to the arrival of high-speed computers, extensive testing would be performed to obtain data on in-depth temperature and surface mass loss. Scale models of the reentry object could be tested in rocket nozzle exhausts, shock tubes, and plasma arc-jet test facilities. With the development of the Charring Material Ablation (CMA) code (Moyer and Rindal 1967), accurate simulation of the ablation and in-depth thermal response of reentry bodies could be performed. CMA computes the one-dimensional time-dependent temperature profile in the reentry body and the surface recession rate. A two-dimensional code was also developed, Axi-Symmetric Transient Heat Conduction and Material Ablation (ASTHMA) (Moyer et al. 1970), but is not as popular because of its complexity. A companion code, Aerotherm Chemical Equilibrium (ACE), is also often needed to characterize the chemical composition of the gas at the surface of the body caused by the chemical reactions occurring between the atmosphere and the body.

The availability of these codes has not obviated the usefulness and importance of ablation testing. Many cases arise where the surface chemistry is not well understood and ablation tests must be performed. These tests provide empirical factors such as the heat of ablation and ablation temperature, which are used in CMA in lieu of the ACE data.

Post-Reentry Impact Analysis

The consequences of post-reentry impacts can be evaluated based on test data and analyses. For GPHS-RTG components, this determination is based on test data coupled with an estimate of hard rock impact probability as a function of orbital inclination. For reactor systems, further analyses are necessary to evaluate the potential for inadvertent criticality due to impact damage deformation and burial or submersion. An analysis of the footprint of any fuel particles or components is also performed.

Radiological Dispersion and Health Effects Consequence Analysis

To determine the potential environmental and health effects consequences of each accident scenario identified for the risk analysis, direct radiation exposure, environmental dispersion, and the subsequent human uptake must be estimated.

Many scenarios are possible for the release of radioactivity, dispersion through the environment, and subsequent population exposure by various pathways. Released radioactivity could be dispersed into various environmental media (air, soil, and water).

Interactions with people through various exposure pathways could lead to inhalation, ingestion, and external radiation doses and associated health effects. Intact components could also lead to direct radiation exposure doses.

The evaluation of the radiological impact of postulated scenarios involving releases of radioactive material into the environment includes the following steps, as shown in Figure 5:

1. Identification of postulated release modes, including the probability of release and the release location;
2. Definition of the source term, including radioactivity of each radionuclide and the corresponding chemical form and particle size distribution;
3. Analysis of time behavior and dispersion of released radioactivity to determine concentrations in environmental media (air, soil, and water) as a function of time;
4. Analysis of interaction with man, including ingestion, inhalation, and direct external doses through each environmental exposure pathway; and
5. Evaluation of radiological impact on man in terms of individual and population doses received and their associated health effects.

Radiation doses to the general population following a postulated release of radioactivity can be calculated in various levels of detail depending on the amount and quality of input data available, the geographic area affected by the release, and the time scale of interest. For time scales on the order of one year, in which the released radioactivity is restricted to a well-defined localized geographical area, detailed dose calculations can be performed to determine the maximum individual dose, the population-versus-dose distribution, and the number of persons receiving a dose above a specified level for each environmental pathway leading to inhalation, ingestion, and external doses.

Uncertainty Analysis

An important consideration in performing a risk assessment is consideration and treatment of uncertainties. Principal factors affecting mission risk are:

- Accident scenario
  - initiating accident probability,
  - accident progression, and
  - accident environment;
- Release characteristics
  - system accident response,
  - conditional source term release probability,
  - source term quantity and form,
  - particle size distribution,
  - initial plume configuration, and
  - release location;
- Meteorological conditions
  - wind speed and direction,
  - atmospheric stability,
  - mixing height,
  - sea- or land-breeze recirculation, and
  - space and time variation;
- Exposure pathway parameters
  - population distribution,
  - resuspension factor,

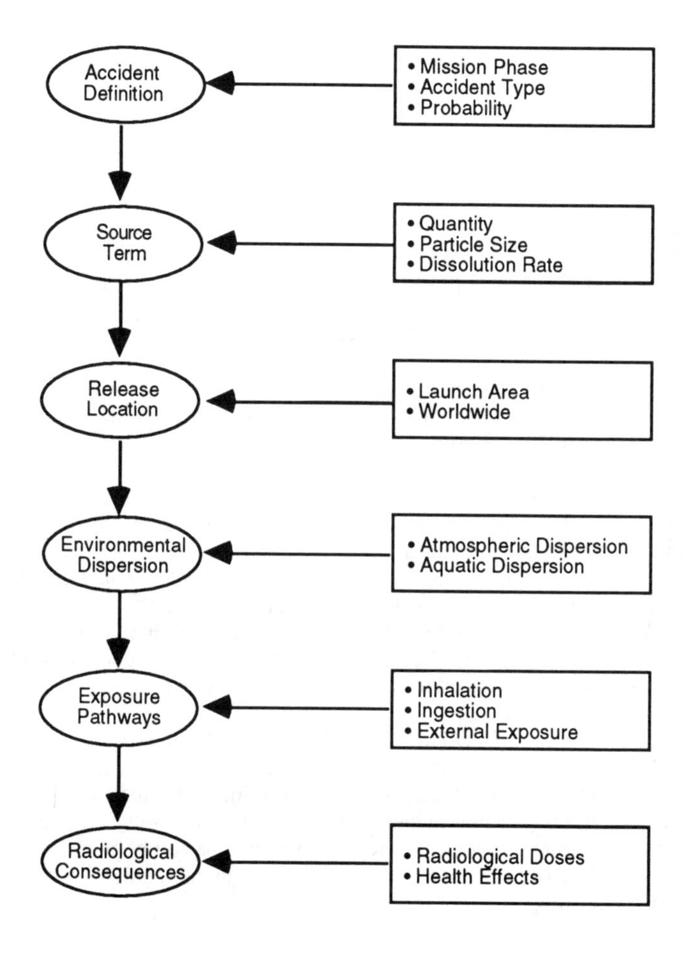

FIGURE 5. Elements of Radiological Risk Analysis for Radioactive Material Releases.

  - deposition velocity,
  - food ingestion, and
  - protective action; and
- Radiation dose and health effects
  - internal dose factors and
  - health effects estimator.

The effect of each factor on mission risk is subject to both variability and uncertainty. For purposes of this discussion, variability encompasses the range of possibilities exhibited by nature or reality. Uncertainty reflects how well variability can be characterized for the purpose of analysis. From this viewpoint, the use of any model to reflect reality has inherent uncertainties. Uncertainty also reflects a lack of complete knowledge. Thus, whenever variability exists or analytical models are used, uncertainty is present. To fully characterize the state of knowledge about risk for informed decision making, variability and uncertainty must be reflected in the risk assessment. This is done by systematically estimating variability and uncertainty at each step and properly propagating them through the analysis.

SAFETY ASSURANCE IN THE FUTURE

The prospects for application of nuclear space power and propulsion include closed cycle reactor power systems for power production and electric propulsion, open cycle nuclear thermal rockets, and RTGs for primary and auxiliary electrical power.

The safety policies, practices, and information that have been developed over the past 30 years provide a very substantial base for supporting these anticipated projects. The specific changes and activities that can be anticipated for safety work will be dependent on the number and scope of the space nuclear projects. The safety review and launch approval process conducted under Presidential Directive/National Security Council Memorandum 25 (PD/NSC-25) has served well and should continue if the number of space nuclear missions are relatively few in number and frequency. If and when the application of space nuclear systems becomes extensive, a need may develop for a different formalism.

Under the continued use of PD/NSC-25 and established safety practice, one can anticipate that safety design and assessment will be determined largely on the basis of the specific project needs. These will vary depending on mission specifics, such as launch vehicle, specific nuclear system (thermal nuclear rocket or nuclear power), destination, trajectories, and manned or unmanned system. However, observations can be made on the trends in the following four safety topical areas: accident consequence analysis, risk analysis, safety design, and safety testing.

Accident Consequence Analysis

Based on advances in numerical methods and high-speed computer capabilities, one can anticipate refinements in the analytical predictions of accident consequences. This will include nuclear, thermal, aerodynamic, structural dynamic, and radiological computations important to accident consequence prediction. The computer codes will receive further validation through benchmark calculations and expanded application. Consequence analyses that are probabilistic in nature, such as atmospheric transport and explosion fragment impacts, have received considerable attention in recent years and are less likely to have significant improvement. This can also be said for neutronics calculations. The specific improvements and code selection will depend on the project needs and design features that contribute to controlling the safety risk.

Risk Analysis

Risk analysis methods and computational code refinements that permit accident sequences to be expanded to a larger number with acceptable computational cost can be anticipated. Methods for pre- and post-processing of the computations will make these expanded analyses more easily used to support design decisions. Methods for economically incorporating the assessment of uncertainties and propagating these uncertainties in both probability and consequence analyses can be expected to improve and make it possible to obtain consistent and comprehensive assessments of the importance of uncertainties. These analytical results will allow improved judgment concerning the need for specific safety testing.

Safety Design

The improvements anticipated in both accident consequence and risk analysis can be expected to support improvements in specification of safety requirements and in demonstrating the importance of safety features to controlling risk. The more specific coupling of safety requirements with their role in risk reduction should allow more cost-effective designs. The confidence in safety design features and requirements should also improve as a result of testing and more timely, iterative coupling of design and risk analysis.

Safety Testing

The need for safety testing is likely to continue even with improvements in analysis. This is due, in part, to the desire to provide a high level of confidence in safety-feature performance. Validation of accident consequence and analysis methods also requires testing. Selected key safety features may also require reliability evaluations supported with testing. The specific testing can be expected to be highly dependent on project needs, as well as those that develop during the independent safety review of a specific project. Because there is a need to satisfy a broad spectrum of safety perspectives, it is reasonable to expect a continuing requirement for selected safety testing.

SUMMARY

A focus on safety from the outset has permitted the U.S. to launch a variety of nuclear systems. Although failures have occurred, they were anticipated and specifically accounted for in the design of the on-board nuclear systems to prevent harmful effects. The historical record, including those failures, indicates that the U.S. space nuclear safety program has worked extremely well.

Safety policies and philosophies have evolved for space nuclear programs. The most recent policy effort was provided by the Nuclear Safety Policy Working Group (NSPWG) for nuclear propulsion in support of the Space Exploration Initiative. The fundamental safety philosophy of this safety policy is to reduce risks to levels as low as reasonably achievable. One of the most important aspects of this policy statement is that it involves a hierarchical safety approach. The safety policy also requires provisions for independent safety oversight. Independent safety/risk evaluation by an Interagency Nuclear Safety Review Panel (INSRP) is also required by Presidential directive to support the launch approval process. The results of INSRP's work are used by the Office of the President to assure that the benefits of potential missions outweigh the risks before launch approval can be granted.

Space nuclear systems are usually designed so that no single credible event will result in significant radiological exposure to the public or workers. Safety strategies have been developed for radioisotopic and reactor systems. For radioisotopic systems, the strategy of providing several rugged containment barriers to protect against fuel dispersal is the most fundamental design safety principle. The prime safety concern for space reactors as a result of accidents during the prelaunch, launch, ascent, and the orbital injection phases of the mission is inadvertent criticality. To prevent inadvertent criticality, space reactors typically incorporate substantial amounts of neutron absorbers in the core with features to assure retention under credible accidents.

Safety tests for radioisotopic systems include shock tube tests, fragment/projectile tests, solid propellant fire tests, bare clad impact tests, solid rocket booster fragment tests, large solid rocket booster fragment impact tests, orbiter fuselage tests, and safety verification tests. Early tests in support of space reactor programs focused a significant effort on reentry response. Another major portion of the program involved transient testing to destruction. Testing to destruction is not expected to be required for current or future U.S. space reactor programs. Other portions of the program included extensive critical tests of various reactor configurations.

Quantitative risk analysis is an important aspect of safety assessments for space nuclear programs. Mission phases are first defined and potential accident scenarios and environments are delineated. System response to potential accidents is explored. These accident scenarios can include fires, explosions, and reentry of space nuclear systems. Occurrence probabilities and dispersion of radioactive materials are predicted to determine potential consequences of postulated accident scenarios.

Although a great deal has been done to assure the safe design and use of space nuclear systems, refinements can still be made. They can be expected to add new perspective, enhance fidelity, and further reduce risks; thus, they should be pursued. In particular, improvements in design by iterative safety analysis and risk assessment show great promise.

Through the continued application of sound engineering and safety principles, guided by analysis and supported by testing, nuclear systems can continue to forge new, ever-expanding horizons for humankind in space.

Acknowledgments

The authors acknowledge the pioneering efforts in space nuclear safety carried out by the Department of Energy, the Department of Defense, the National Aeronautics and Space Administration, and their contractors. Their support of, and contributions to, this survey article are greatly appreciated. Special thanks go to Mr. Bart W. Bartram of Haliburton NUS Environmental Corporation for his contributions in the area of safety technical analyses and to Mr. Dan Scott of Tech Reps, Incorporated, for initial drafting, setup, updating, editing, and final production of this manuscript.

References

Anz-Meador, P. D. and A. E. Potter (1991) "Radioactive Satellites: Intact Reentry and Breakup by Debris Impact," *Adv. Space Res.*, 11(12): 37-42.

Bartram, B. (1992) *A Reference Guide to the Safety Aspects of Space Nuclear Systems*, (draft), prepared for U.S. Department of Energy.

Bartram, B. W. and A. Weitzberg (1988) "Radiological Risk Analysis of Potential SP-100 Space Mission Scenarios," Report No. 5125, NUS Corporation, Gaithersburg, MD, August 1988.

Bennett, G. L. (1987) "Flight Safety Review Process for Space Nuclear Power Sources," in *Trans. 22nd Intersociety Energy Conversion Engineering Conference*, held in Philadelphia, PA, 10-14 August 1987.

Bennett, G. L., et al. (1988) "Development and Implementation of a Space Nuclear Safety Program," in *Space Nuclear Power Systems 1987*, M.S. El-Genk and M.D. Hoover, eds., Orbit Book Co., Malabar, FL.

Bennett, G. L. et al. (1989) "Update to the Safety Program for the General-Purpose Heat Source Radioisotope Thermoelectric Generator for the Galileo and Ulysses Missions," in *Trans. of Sixth Symposium on Space Nuclear Power Systems*, held in Albuquerque, NM, 8-12 January 1989.

Bennett, G. L. (1991) "The Safety Review and Approval Process for Space Nuclear Power Sources," *Nuclear Safety*, 32(1): 1-18.

Brown, N. W., D. R. Damon, M. A. Smith, and M. I. Temme (1990) "Space Reactor Power System Nuclear Safety," *Nuclear Safety*, 31(4): 460-483.

Damon, D. R., M. I. Temme, C. Pupek, N. W. Brown, and J. Garate (1989) "SP-100 Mission Risk Analysis," GESR-00849, GE Aerospace, San Jose, CA, August 1989.

Eichler, P. and D. Rex (1990) "Debris Chain Reactions," Paper AIAA-90-1365, presented at *AIAA/NASA/DOD Orbital Debris Conference: Technical Issues and Future Directions*, 16-19 April 1990. Also published in *Orbital Debris: Technical Issues and Future Questions*, NASA Conference Publication 10077, September 1992, 187-195.

Erdahl, D. et al. (1988) *Space Propulsion Hazards Analysis Manual, Volumes 1 and 2*, AFAL-TR-88-096, Air Force Astronautics Laboratory, June 1988.

ESA (1988) *Space Debris*, ESA Space Debris Working Group, ESA SP-1109, November 1988.

GE (1989a) *Final Safety Analysis Report for the Galileo Mission, Volume I, Reference Design Document*, GE-87SDS4213, prepared for U.S. Department of Energy by the General Electric Company, Astro-Space Division, May 1988.

GE (1989b) *Final Safety Analysis Report for the Galileo Mission, Volume II, Books 1 and 2, Accident Model Document*, GE-87SDS4213, prepared for the U.S. Department of Energy by the General Electric Company, Astro-Space Division.

GE (1990a) *Final Safety Analysis Report for the Ulysses Mission, Volume I, Reference Design Document*, ULS-FSAR-002, prepared for the U.S. Department of Energy by the General Electric Company, Astro-Space Division, March 1990.

GE (1990b) *Final Safety Analysis Report for the Ulysses Mission, Volume II, Books 1 and 2, Accident Model Document*, ULS-FSAR-003 and -004, prepared for the U.S. Department of Energy by the General Electric Company, Astro-Space Division, March 1990.

Kessler, D. J. (1991) "Collisional Cascading: Limits to Growth in Low Earth Orbit," *Adv. Space Res.*, 11(12): 63-66.

Koenig, D. (1986) *Experience Gained from the Space Nuclear Reactor Rocket Program (ROVER)*, LA-10062-H, May 1986.

McKnight, D. (1992) "Breakup in Review - Two Geosynchronous Breakups," *Orbital Debris Monitor*, 5(2): 35-36.

Marshall, A. C. et al. (1992) *Nuclear Safety Policy Working Group Recommendations on Nuclear Propulsion Safety for the Space Exploration Initiative*, NASA TM-105705.

Marshall, A. C. (1992) "A Philosophy for Space Nuclear Systems Safety," in *Proceedings of the American Nuclear Society Topical Meeting on Nuclear Technologies for Space Exploration*, 16-19 August 1992, Jackson Hole, WY, NTSE-92, 3: 870-875.

Martin Marietta (1992) "Titan IV CRAF/Cassini EIS Databook," MCR-91-2580, December 1991; Section 10, "System Failure Probability Analysis," SLS-92-20435, Martin Marietta Space Launch Systems, September 1992.

Moyer, C. B., B. Blackwell, and P. Kaestner (1970) *A User's Manual for the Two-Dimensional Axi-Symmetric Transient Heat Conduction Material Ablation Computer Program (ASTHMA)*, SC-DR-70-510, Sandia National Laboratories, Albuquerque, NM, December 1970.

Moyer, C. B. and R. A. Rindal (1967) *Finite Difference Solution for the In-Depth Response of Charring Materials Considering Surface Chemical and Energy Balances*, Report No. 66-7, Aerotherm Corporation, Mountain View, CA, March 1967.

NASA (1988) *Space Shuttle Data for Planetary Mission Radioisotope Thermoelectric Generator (RTG) Safety Analysis*, NSTS 08116, Revision B, National Aeronautics and Space Administration, September 1988.

NASA (1991) *Space Station Program Natural Environment Definition for Design*, National Aeronautics and Space Administration Document SSP 30425, Revision A, July 1991.

Otter, J., K. Buttrey, and R. Johnson (1973) *Aerospace Safety Program: Summary Report*, Atomics International Division, Rockwell International, report of work performed under contract AT(04-3), July 1973.

Outka, D. E. (1990) *User's Manual for the Trajectory Simulation and Analysis Program*, SAND88-3158, Sandia National Laboratories, Albuquerque, NM, July 1990.

Reynolds, R. C., P. D. Anz-Meador, and G. W. Ojakangas (1991) "Impact of Alternative Mission Models on the Future Orbital Debris Environment," *Adv. Space Res.*, 11(12): 29-32.

Sholtis, J. A., Jr. (1989) "Nuclear Space Power Systems: Ensuring Safety from Beginning to End," Space Power, 8(3): 406-407, *Proc. International Astronautics Federation International Conference on Space Power,* held in Cleveland, OH, 5-7 June 1989.

Sholtis, J. A., Jr., D. A. Huff, L. B. Gray, N. P. Klug, and R. O. Winchester (1991) "Technical Note: The Interagency Nuclear Safety Review Panel's Evaluation of the Ulysses Space Mission," *Nuclear Safety*, 32(4): 494-501.

Sholtis, J. A., Jr. and R. O. Winchester (1992) "Obtaining Launch Approval for Nuclear-Powered Space Exploration Missions," *Trans. Nuclear Technologies for Space Exploration*, held in Jackson Hole, WY, 16-19 August 1992.

Stansbery, E. G., G. Bohannon, C. C. Pitts, T. Tracy, and J. F. Stanley (1992) *Characterization of the Orbital Debris Environment Using the Haystack Radar,* National Aeronautics and Space Administration Report JSC-32213, April 1992.

Touloukian, T. S. and C. Y. Ho (1976) *Thermophysical Properties of Selected Aerospace Materials*, Thermophysical and Electronic Properties Information Center, CINDAS, Purdue University, West Lafayette, IN.

von Karman, T. (1956) "Aerodynamic Heating: The Temperature Barrier in Aeronautics," in *Proceedings of the High Temperature Symposium*, Stanford Research Institute, Berkeley, CA, June 1956.

Voss, S. (1984) *SNAP Reactor Overview*, AFWL-TN-84-14, August 1984.

White House (1977) "Scientific or Technological Experiments with Possible Large Scale Adverse Environmental Effects and Launch of Nuclear Systems into Space," *Presidential Directive/National Security Council Memorandum 25,* PD/NSC-25, White House, Washington, DC, December 1977.

DYNAMIC POWER CONVERSION SYSTEMS FOR SPACE NUCLEAR POWER

SECTION I: STIRLING CYCLE

James E. Dudenhoefer,
James E. Cairelli, Jeffrey
G. Schreiber, Wayne A.
Wong, Lanny G. Thieme, Roy
C. Tew and Steven M. Geng
NASA Lewis Research Center
21000 Brookpark Road
Cleveland, OH 44135
(216) 433-6140

Donald L. Alger and
Jeffrey S. Rauch
Sverdrup Technology, Inc.
Lewis Research Ctr. Group
2001 Aerospace Parkway
Brook Park, OH 44142

SECTION II: RANKINE CYCLE

Harvey S. Bloomfield
NASA Lewis Research Center
21000 Brookpark Road
Cleveland, OH 44135
(216) 433-6131

SECTION III: BRAYTON CYCLE

David M. Overholt
Allied-Signal Aerospace Company
1300 West Warner Road
Tempe, AZ 85285-2200
(602) 893-7356

Abstract

A review of Stirling, Rankine, and Brayton cycle dynamic power conversion systems for space nuclear power applications is presented in three separate stand-alone sections. The purpose of this review is to provide a summary of the current technology database of the candidate dynamic power conversion systems for future space nuclear power applications. The important features of each dynamic conversion system type are presented along with discussion of their general characteristics, history, current state of technology, and research and technology issues.

INTRODUCTION

Direct thermal-to-electric conversion solar and nuclear-heated devices and chemical energy storage systems have provided the power for all U.S. space flights to date. However, future long life power and propulsion missions requiring multi-kilowatt and higher power levels will likely rely on solar and nuclear-heated dynamic conversion power systems. These conversion systems are based on Brayton, Stirling, and Rankine thermodynamic cycles. As mission power requirements rise, dynamic conversion systems can offer significant cost, efficiency, mass, and performance advantages over direct conversion systems.

Of the dynamic conversion options, alkali liquid metal Rankine cycle conversion has historically been of primary interest for power generation and electric propulsion applications where potential for low mass, small heat rejection area and scalability to very high power levels are paramount. However, closed Brayton cycle technology has a significantly more extensive database primarily because of operational simplicity, high reliability, and potential mass advantage at lower power levels. Brayton cycle conversion has also been of particular interest for solar-heated terrestrial and space applications. The Free-Piston Linear Alternator engine variant of the Stirling cycle is a relatively recent innovation that has seen significant technology development for space applications over the last decade. Interest in the Stirling cycle is based on the potential for achieving low mass and high efficiency at the lowest required peak cycle temperature of any thermodynamic cycle.

SECTION I: STIRLING CYCLE

INTRODUCTION

Technology for The Free-Piston Stirling Space Power Converter is being developed under NASA's Civil Space Technology Initiative (CSTI). The goal of the CSTI High Capacity Power element is to develop the technology base needed to meet the long duration, high capacity power requirements for future NASA space initiatives. Efforts are focused upon increasing system thermal and electric energy conversion efficiency at least fivefold over current SP-100 technology (Thermoelectrics), and on achieving systems that are compatible with space nuclear reactors.

This article reviews the salient features of the Stirling space power system including its general characteristics, history, current state of technology, and research and technology issues.

GENERAL CHARACTERISTICS

Cycle Description

The alpha configuration (Walker 1973) ideal Stirling cycle is useful for getting an introductory appreciation of how a Stirling cycle works. Alpha configuration ideal Stirling cycle components are shown in Figure 1 and include: a hot expansion space, an expansion space piston, a porous matrix regenerator with negligible volume, a cold compression space, and compression space piston.

The alpha ideal Stirling engine cycle can be described as follows with reference to Figure 1: (1→2) with the compression piston fixed at its innermost position relative to the regenerator, the expansion piston moves from its intermediate position to its outermost position; during this piston motion, the gas expands from minimum to maximum volume, isothermally, at the hottest cycle temperature; (2→3) the next phase of the cycle involves movement of expansion and compression piston together to the right; thus the gas is cooled at constant volume as it passes through the regenerator from the hot to the innermost position while the compression piston moves from innermost to outermost position; heat from the gas is stored in the regenerator matrix to be recovered later; (3→4) the compression phase of the cycle is next; with expansion piston fixed, the compression piston moves from its outermost position to its intermediate position; during this period gas is compressed isothermally at the coldest cycle temperature, from maximum to minimum volume; and (4→1) the final phase, constant volume heating, returns the pistons to their original positions; during this phase, the two pistons move together to the left--the compression piston from intermediate to innermost position and the expansion piston from innermost to intermediate position; the gas recovers the heat that was stored earlier in the regenerator matrix.

P-V and T-S diagrams corresponding to the process described above are shown in Figure 2. Each portion of these diagrams can be correlated with the process described above. The useful work produced by the cycle is represented by the areas inside the P-V and T-S diagrams. In order to achieve positive work production, the work produced by the gas during expansion (equal to the heat in through the expansion space) must exceed the work absorbed by the gas during compression (equal to the heat rejected from the compression space).

The constant volume heating and cooling processes (4->1 & 2->3), during which the regenerator matrix exchanges heat with the gas, have no impact on the work production. However, if there were no regenerator matrix on the interior of the cycle for storage and recovery of heat, this heat would have to be supplied from and rejected to the environment during each cycle. For an ideal engine with regenerator heat storage per cycle equal to 4 times that entering the heater (a reasonable assumption for practical engines), replacing the ideal regenerator with an ideal heat exchanger which interchanges heat between the working gas and the environment (with all other features of the cycle remaining ideal) will reduce the cycle efficiency by a factor of 5. That is the work production would be the same, but heat in would have to increase from Q_{in} to (Q_{in} + $4Q_{in}$).

Figure 3 shows a beta configuration ideal Stirling cycle, which also corresponds to the P-V and T-S diagrams shown in Figure 2. This configuration, with a displacer and a power piston, is closer to the configuration of the current space power designs (although the regenerator is normally in a separate heat exchanger circuit rather than inside the displacer).

Only five parameters are needed to define the performance (for example, the work per cycle and the efficiency) of an ideal Stirling cycle. These are: the expansion-end temperature, T_e; the compression-end temperature, T_c; the ratio of maximum-to-minimum volume, $r = V_{max}/V_{min}$; the gas constant, R, of the working gas; and the mass, m, of the working space gas. The expansion end--for engines, refrigerators and heat pumps--is where net heat is taken into the cycle and converted into expansion work; the compression end is where net compression work is converted into heat to be rejected from the cycle.

For an engine, the useful output product of the device is the net work per cycle (expansion minus compression work). This work is produced with efficiency identical to the Carnot cycle efficiency:

$$\eta = \frac{\textit{heat supplied-heat rejected}}{\textit{heat supplied}} = \frac{\textit{work done}}{\textit{heat supplied}} = \frac{mRT_e \ln r - mRT_c \ln r}{mRT_e \ln r} = 1 - \frac{T_c}{T_e}$$

Walker (1973) shows that the ideal Stirling cycle produces greatly enhanced work per unit swept volume, compared to the Carnot cycle. To get useful estimates of power and efficiency from the ideal cycle, however, it is necessary to multiply the idealized work and efficiency by experience factors, F_{work} and F_{eff}.

Therefore--

$$Useful\ Work = F_{work} mR \ln r (T_e - T_c)$$

$$Useful\ Efficiency = F_{eff}\left(1 - \frac{T_c}{T_e}\right)$$

The latest space power designs are getting very close to achieving an F_{eff} of 0.5 based on electrical power out and net heat into the engine. This is for an engine with a temperature ratio of T_e/T_c = 2.0. For an engine, $T_e > T_c$ and T_c is typically close to ambient temperature.

For a refrigerator the useful output product of the device is the "cooling" or absorption of heat from outside the cycle which occurs at the expansion end. To achieve useful cooling, it is necessary to put net work into the cycle. The coefficient of performance of a refrigerator is:

$$COP_{cooling} = \frac{heat\ supplied}{work\ done} = \frac{mRT_e \ln r}{mrT_c \ln r - mRT_e \ln r} = \frac{T_e}{T_c - T_e}$$

where, for the case of the cooler, $T_e < T_c$ and T_c is typically ambient temperature.

For a heat pump, the useful product of the device is the heat rejected from the cycle at the compression-end. The heat pump has the useful effect of magnifying the heating effect of the work that is put into the cycle. Therefore, the coefficient of performance of a heat pump is:

$$COP_{HP} = \frac{heat\ rejected}{work\ done} = \frac{mRT_c \ln r}{mRT_c \ln r - mRT_e \ln r} = \frac{T_c}{T_c - T_e}$$

where, again, $T_e < T_c$, as for a refrigerator, but, unlike the refrigerator application, T_e is typically the ambient temperature.

Work (or heat) is converted into power (or rate of cooling/heating) by multiplying by the frequency of the machine, for example,

$$\frac{Joules}{cycle} x \frac{cycles}{sec} = \frac{Joules}{sec} = Watts$$

In addition to providing useful electrical power in space, Stirling cycle machines are capable of providing on board refrigeration (for food, biological samples). Several studies have also been conducted to study the potential of using Stirling machines as heat pumps to boost waste heat to a higher temperature than would otherwise be possible--and thus reduce the size and weight of the radiators in space power systems.

More complete discussions of the ideal Stirling cycle can be found in West (1986), Walker (1973) and Urieli and Berchowitz (1984). Many general thermodynamic texts contain brief discussions of the ideal Stirling cycle. It should be remembered that the ideal cycle neglects all thermodynamic irreversibilities and to get useful estimates of real engine performance using ideal cycle calculations one must use "experience factors". The ideal Stirling cycle implies two additional assumptions. These are: (1) piston velocities are discontinuous, whereas piston velocities in the current space power designs, and in many other Stirling designs, are very nearly sinusoidal; and (2) the dead volume of the regenerator, which significantly impacts the volume and pressure ratios for most practical machines, is neglected.

SYSTEM ARCHITECTURE

The SP-100 Program

SP-100 is a joint DOE/DOD/NASA program to develop nuclear power for space power applications. The SP-100 program is directed toward the development and validation of technology for a versatile space nuclear reactor power system having the capability to generate from tens to hundreds of kilowatts of electrical power for at least seven years at full power. Stirling power converters are currently a growth option for SP-100; Figure 4 compares Stirling power systems at two different hot-end temperatures and various efficiencies to the baseline thermoelectric system. Stirling technology development at NASA Lewis is being done as part of NASA's Civil Space Technology Initiative (CSTI) High Capacity Power Program, a program to complement and enhance SP-100.

Figure 5 shows a basic configuration for a 1050 K (converter hot-end temperature) Stirling/nuclear reactor power system. A sodium heat pipe is used to transport the heat input into the Stirling power converter. An intermediate heat exchanger is used to transfer the heat from the reactor lithium pumped loop to the heat pipe. Both directly coupled and radiatively coupled heat exchangers have been studied to make this transition between the refractory lithium loop and the superalloy Stirling heat pipe (Schmitz et al 1993). Figure 6 shows a preliminary layout of a directly coupled intermediate heat exchanger. The heat pipe provides uniform temperature distribution around the Stirling heater head and isolates the reactor pumped loop from the Stirling high pressure working fluid.

The Stirling converter is a free-piston engine with a linear alternator to convert reciprocating motion to electrical output. Dual-opposed converters would be used for balancing purposes (only a single cylinder is shown in Figure 5). A NaK pumped loop connects the cold end of the converter to the radiator; the radiator consists of water heat pipes. Higher temperature systems may require an additional radiator to cool the Stirling linear alternator to maintain the permanent magnets at an acceptable temperature. A higher-temperature system will also require either changing heat pipe fluids or switching to a liquid metal pumped loop radiator.

The electrical output of the linear alternator is single-phase AC with an output voltage of around 240 volts. Tuning capacitors are used to achieve a stable system with minimum overall system mass. A parasitic electrical load system ensures that the converters run at a constant operating condition.

Possible applications for Stirling/nuclear reactor power systems include lunar and Mars bases, electrical propulsion power for science and unmanned cargo missions to the outer planets, power for air and ocean traffic radar control systems, higher power communication platforms and earth observing platforms, and in-space materials processing facilities. A conceptual design for a nuclear Stirling lunar base is shown in Figure 7. (Schmitz and Mason 1991, and Nainiger and Mason 1992) discuss the lunar base application. (Schmitz et al. 1992) describes near term nuclear power system technology options available for space flight demonstrations. (Angelo and Buden 1985) gives an overview of space nuclear power technology and reviews its history.

Radioisotope Power Systems

The Stirling power converter is also a leading candidate for isotope power systems producing multihundred watts to tens of kilowatts. (Bents et al. 1992, Bents et al. 1992, Bents et al. 1991 and Schmitz et al. 1991) discuss the application of free-piston Stirling converters to dynamic isotope power systems (DIPS). DIPS have applications in a variety of unmanned deep space and planetary surface exploration missions and, eventually for manned lunar and Mars missions. The RTG (radioisotope thermoelectric generator) is the power source currently available for these mission requirements. The Stirling converter is able to achieve higher efficiencies at these lower power levels, thus, allowing a significant reduction in the amount of isotope required. Preliminary system analysis has shown that, for a multihundred watt output, the Stirling DIPS is equivalent in size and mass to the next generation Mod RTG but requires only about one-third of the radioisotope (Figure 8). The high efficiency of the Stirling converter leads to the following advantages: significant reduction in cost due to the high cost of the radioisotope, reduced dependence on the limited amount of available radioisotope, reduced launch hazard with decreased amount of radioisotope, and reduced waste heat loading on the spacecraft thermal management system.

Heat input for these Stirling DIPS systems would be provided by the U.S. Department of Energy (DOE) General Purpose Heat Source (GPHS). For the multihundred watt applications, the GPHS blocks can be arranged to radiate directly to the heater head of the Stirling converter. Higher-power converters would be heated by a heat pipe connecting the GPHS blocks and the Stirling heater head. Figures 9 and 10 show a layout for a Stirling DIPS system directly heated by the GPHS blocks and using dual opposed converters for balancing.

Radioisotope Stirling generator have been studied for the Pluto Fast Flyby Mission (Schock 1993) as directed by JPL and the Department of Energy. The study showed that the Stirling systems do offer great multi-mission benefits in mass and cost reduction and basically concurs with the NASA LeRC studies.

Stirling Technology Company (STC) has been developing a ten watt Stirling power convertor for radioisotope applications (Ross et al 1993) and reports over two thousand hours of endurance testing through August 1993 on the prototype endurance test engine. STC uses flexure technology to achieve non-contact bearings in their small power converters.

DESIGN AND OPERATIONAL FLEXIBILITY

Free-piston Stirling power converters have been built and tested to produce power outputs varying from a few watts to about 12.5 kWe per cylinder. Scaling study designs have been completed for free-piston Stirling converters producing up to 500 kWe per cylinder.

The Stirling power converter can operate with various heat sources, including nuclear, solar, radioisotope, and combustion. Utilizing a heat pipe for heat input into the Stirling heater head, a single power converter design should be able to operate with any of the various heat sources. Alternately, the power converter design could be optimized for the heat source such as for direct radiative heating from a radioisotope source. If desired, the Stirling heater head could also be designed to be heated directly by a pumped loop.

The Stirling power converter not only has a high efficiency at its design point but maintains that efficiency over a wide range of part-load powers. This can be seen in Figure 11 which shows the piston PV efficiency for the Space Power Research Engine (SPRE) over a wide range of powers for a temperature ratio of 2.0. The current Stirling space power converter, the Component Test Power Converter (CTPC), can be operated over a range of power levels from about half-power to full-power. The CTPC uses internally-pumped hydrostatic gas bearings that are pressurized by the working-fluid cycle; the operation of these bearings sets the possible operating range for the stroke. The SPRE tests were run with externally-pressurized gas bearings allowing operation over a wider range of output powers.

Currently, the nuclear space power systems are expected to operate at a fixed operating point using a parasitic load. However, Stirling power converters can be controlled with various methods to match part-load power requirements. Control methods include varying displacer stroke, displacer-piston phase angle, mean working fluid pressure, temperature level, and engine dead volume. Controlling displacer stroke or displacer-piston phase angle would be most appropriate for the free-piston Stirling space power converters.

The alternator output voltage can be varied in the converter design depending on the number of coil turns and the coil wire diameter selected. Output voltages from about 100 volts to 1000 volts or somewhat greater should be possible. Also, it may be possible to achieve 3-phase power output with a system of multiple power converters. Optimum electrical output and load system design will depend on the overall system requirements. NASA Lewis is currently researching, both experimentally and analytically, the coupling of the Stirling converter to various types of load systems with both single and multiple converters.

SYSTEM OPTIMIZATION

Stirling power systems can be optimized for a variety of mission drivers, including minimum mass, minimum radiator area, maximum efficiency, or a combination of these. For the system optimization, a wide variety of parameters are modeled for each component of the system allowing flexibility in modeling different types of systems and components. Nuclear, solar, and radioisotope Stirling systems can all be optimized. The Stirling converters are modeled in terms of converter efficiency as a function of converter power level, converter temperature ratio (Stirling hot-end temperature/Stirling cold-end temperature), and converter specific mass. Other inputs to the optimization for a nuclear system include, but are not limited to, power level per converter, Stirling hot-end temperature, shielding requirements for the reactor, thermal input from the reactor, radiator specific mass, radiator emissivity, sink temperature, and radiator and Stirling redundancy.

The parameters which usually dominate Stirling nuclear power system optimization are the reactor and its associated shield, the Stirling power converter, and the radiator which rejects the waste heat from the cycle. Typically, temperature ratio and converter specific mass are varied to perform mass optimization. Lower temperature ratios provide lower converter efficiencies, larger reactor and shield masses, and a higher-temperature, lower-mass radiator. As the temperature ratio increases, converter efficiency increases, reactor and shield mass decreases, and radiator mass increases (lower-temperature radiator). For a constant temperature ratio, converter efficiency varies directly with the specific mass of the converter with a low-specific mass converter providing a lower converter efficiency.

For typical missions, SP-100 power systems with Stirling converters optimize for minimum mass at low (below 6 kg/kWe) specific mass converters and at temperature ratios around 2.0. Solar Stirling power systems tend to optimize at relatively higher converter efficiencies and, correspondingly, higher temperature ratios (around 2.5-2.7); concentrator and receiver mass vary strongly with converter efficiency due to the relatively low power density provided by the sun at one earth radii. Radioisotope power systems for unmanned systems optimize at temperature ratios of about 2.2-2.5 while manned systems can optimize at much higher temperature ratios depending on shielding requirements.

A typical optimization surface for a Stirling nuclear power system is shown in Figure 12. Table 1 lists a mass breakdown for a 100-kWe SP-100 Stirling power system using five 25-kWe dual opposed Stirling converters (four operating and one spare) with a Stirling hot-end temperature of 1050 K. Tuning capacitor mass is included in the mass of the Stirling power converters/alternators.

As previously stated, the efficiency of the Stirling converter varies with the converter specific mass. High-efficiency converters tend to operate at lower frequencies and have larger heat exchangers leading to greater specific mass. The efficiency of the converter also increases with increasing temperature ratio, due to a higher Carnot efficiency and to achieving a greater percentage of that Carnot efficiency. Converter specific mass tends to decrease with increasing temperature ratio. (Dochat and Dhar 1991) gives projections for Stirling specific mass and efficiency as functions of temperature ratio and power level for systems optimized for minimum mass and for best efficiency at reasonable specific mass.

A large number of parameters can be optimized for the Stirling converter to achieve the design goals of converter power output, efficiency, and specific mass. These include, but are not limited to, cooler and heater dimensions (passage length and size, no. of passages), regenerator dimensions (frontal area, length, porosity, wire diameter), stroke, frequency, working fluid pressure, phasing between piston and displacer, cylinder bore, and linear alternator magnet, coil, and stator dimensions. The working fluid for a Stirling space power converter is normally helium.

RELIABILITY

Free-piston Stirling converters are a relatively new technology for space power conversion. Proof of long-term reliability remains as possibly the main issue in development of Stirling converters. Fundamentally, the free-piston Stirling offers very long lifetimes based on a hermetically sealed system and only two moving parts per cylinder that are supported on non-contacting bearings.

Overall, numerous Stirling machines have been built and operated. (Ross and Dudenhoefer 1991) summarizes the operating experience with a number of Stirling machines, both free-piston and kinematic. This operating experience largely substantiates the claim that Stirling machines are capable of reliable operation over long lifetimes. In addition, many of these machines were meant for proof-of-concept and, as such, were not expected to operate for the long lifetimes that they achieved.

A Stirling thermomechanical generator using an edge-clamped flat diaphragm for the piston ran for over 100,000 hours in the laboratory before being taken out of service. This engine was built and tested by the Atomic Energy Research Establishment in Harwell, England and was designed for producing remote power for marine applications. Several field trials providing power for buoys and a lighthouse were also completed and operated in excess of

10,000-23,000 hours. Table 2 (Ross and Dudenhoefer 1991) summarizes the test experience on these machines. Stirling Technology Company has operated a Stirling artificial heart assist engine with a flexural bearing at the hot end of the displacer for 60,000 hours, including a continuous run of 36,000 hours. Under a NASA contract, Mechanical Technology, Inc. completed an endurance test of 5385 hours on a 2-kWe free-piston Stirling engine using hydrostatic gas bearings. This included low-power and full-power conditions over 262 planned starts/stops. At the end of the testing, only minor scratches were found and no debris was generated.

Stirling cryocoolers have already flown in space. Table 3 Ross and Dudenhoefer 1991) includes a list of past experience of Stirling space flight cryocoolers. These, generally, were engineering development models that were not specifically designed for space, yet were provided space mission opportunities. The Table also shows further planned long-life cryocooler missions. The long-life cryocoolers will include either hermetic sealing or improved sealing arrangements; non-contacting parts through the use of clearance seals and gas bearings, magnetic bearings, or flexural bearings; and improved material selection and processes to minimize outgassing. All the long-life units shown in table 3 use a flexure bearing that was demonstrated by Oxford University in the early 1980's. For terrestrial applications, Philips has sold over 3000 kinematic Stirling cryocoolers; a number of these machines have run over 150,000 hours including regular replacement of the contact bearings and seals used in these designs.

HISTORY

The Stirling engine dates back to the early 1800s. It is named after its inventer Robert Stirling, a Scottish minister. During the 19th century, the hot air engine, as it was called at that time, received considerable attention as an alternative to the steam engine. These early engines were much more fuel efficient than steam engines and did not present the hazard of high pressure boilers. However, since they operated at near atmospheric pressure, they ran at low speed and tended to be large and heavy. With the introduction of the electric motor and the gasoline engine near the end of the 19th century, the hot air engine became obsolete.

There was little interest in the Stirling engine until N.V. Philips Co. of the Netherlands revived hot-air engine R&D about 1938; they found that the power-to-weight ratio could be increased by more than a hundredfold by raising the operating pressure and by using helium or hydrogen as a working fluid instead of air. Recognizing the reversibility of the Stirling Cycle, Philips developed and was commercially successful in producing large cryogenic cooling machines as well as small machines for military and other applications. In the 60's General Motors teamed up with Philips to explore Stirling engine use for a variety of applications such as prime movers for ground transportation, power generators, and even for space power generation. Philips and several of its European Stirling licensees built many engines and demonstrated Stirling engines in buses, cars, and boats. Concerns about air pollution and the oil embargo of the 70's triggered U.S. Stirling R&D as an alternative to the automobile I.C. engine. A program to develop Stirling engines for the automotive application was initiated under U.S. DoE sponsorship and NASA Lewis Research Center management. Philips teamed with Ford Motor Company, and the Swedish firm, United Stirling, teamed with Mechanical Technology Inc. and American Motors in parallel efforts to develop automotive Stirling engines. The resulting engines, tested through 1989, demonstrated clean emissions, good fuel economy, and were potentially cost competitive. In an effort coordinated with the Automotive Stirling Engine (ASE) Project, NASA funded a vehicle demonstration project. This project put Stirling powered vehicles in the hands of non-technical end users, and successfully demonstrated the viability of the Stirling engine as a vehicle prime mover. As documented in (Richey et al. 1989), the cross country journeys of the vehicle demonstration project clearly illustrate the technical success of the ASE project. Simultaneous improvements in I.C. engines kept pace with Federal emissions and fuel economy standards, essentially negating need for an alternative auto engine. However, development of Stirling engines continues, throughout the world, for applications such as power generation, heat pump and total energy systems, hybrid buses, and submarine power.

The Free-Piston Engine: The Early Years

By the late 1950's and early 1960's, several independent efforts to develop the resonant, free-piston Stirling Engine (FPSE) were underway. The lack of mechanical linkages in the FPSE offer several advantages over the kinematic counterpart: no lubricated linkages or mechanical seals were needed, thus eliminating the problem of regenerator contamination and blockage by oil which was often encountered in kinematic engines; hermetic sealing, to eliminate gas leakage became feasible; and the elimination of wear caused by contact between moving parts offered the potential for very long life and high reliability.

The early FPSE's, many of which were designed and built by Sunpower Inc., tended to be small demonstration units of low power, eventually evolving into small water pumps and electric generators. Other applications for which free-piston Stirling has been developed include heat pumps, artificial hearts, and cryocoolers. Thus far, only the Stirling cryocooler has become a commercial success.

The Evolution of the Free-Piston Stirling Engine

By the late 1970's, power output from the FPSE was reaching the 1 kW level. Since the power output from the FPSE results from linear motion, attention was now directed toward integration of the FPSE with a linear alternator to form a Stirling power conversion system. By the early 1980's, a 3 kW power output level was not uncommon, and by the mid-1980's activities were directed toward a 2 piston, 25 kWe free-piston Stirling power convertor via the NASA/DOD/DOE SP-100 Program.

The first generation of hardware in this program was the Space Power Demonstrator Engine (SPDE), which is a free-piston Stirling engine coupled to a linear alternator. The SPDE is a double cylinder, opposed-piston converter which was designed to produce 25 kW of electrical power at 25% overall efficiency. After successful demonstration, the SPDE was modified to form two separate, single cylinder power convertors called the Space Power Research Engines (SPRE) which serve as test beds for Stirling technology development and design code validation.

The SPRE has a design operating point of 15.0 MPa using Helium as the working fluid, 650K hot-end temperature and 325K cold-end temperature, providing a temperature ratio of 2.0, a piston stroke of 20 mm, and an operating frequency of approximately 100 Hz. The SPRE incorporates gas springs, hydrostatic gas bearings, centering ports and close clearance noncontacting seals. Research conducted using the SPRE includes: hydrodynamic gas bearings tests; high efficiency linear alternator tests; centering port tests; displacer clearance seal tests; and tests with a varying number of cooler tubes. These tests are discussed in greater detail in (Spelter et al. 1989, Rauch et al. 1990, Cairelli et al. 1991, and Wong et al. 1992)

Another project, sponsored by DoE and managed by NASA Lewis Research Center, is the Advanced Stirling Conversion System (ASCS). The purpose of this project is to develop a dish-mounted, solar Stirling power conversion system for terrestrial commercial application. NASA LeRC has interest in this project because of the synergism in the technologies used between the nuclear space power program and the solar terrestrial project. The ASCS project is discussed in (Shaltens et al. 1992).

Stirling Space Power Converter (SSPC) Development

Because of the success of the SPDE and SPRE projects, in 1988 NASA funded the Stirling Space Power Converter Development (SSPC) under the Civil Space Technology Initiative (CSTI) High Capacity Power Program. The objective of the Stirling Space Power Converter (SSPC) project is to develop technology for the SSPC and to demonstrate that technology in a full-scale power converter test. A NASA contract was awarded to Mechanical Technology Inc. (MTI) to carry out the power converter development. Under this contract, two power converters are to be fabricated and tested. The first converter, identified as the Component Test Power Converter (CTPC), has been fabricated from easily machineable and weldable Inconel 718 alloy. The purpose of this converter is to develop converter components that will enable the CTPC to achieve the performance goals of full converter design power of 12.5 kWe and overall efficiency of greater than 20%. The 60,000 hour converter life goal will be achieved in the SSPC by material substitution of Udimet 720 for the Inconel 718 of the converter heater head.

The Component Test Power Converter (CTPC)

The CTPC is a 25 kWe modular design consisting of two 12.5 kWe/piston opposed power converters. Only one-half of the CTPC has been fabricated and tested. Details of the design, fabrication and early testing of the CTPC have previously been reported. (Dochat and Dhar 1991, and Dochat 1992). A schematic of the power converter, configured as an opposed power converter, is shown in Figures 13 and 14.

The 12.5 kWe/piston CTPC was fabricated to develop critical technologies for the SSPC. These critical technologies are identified as: bearings, materials, coatings, linear alternators, mechanical structural issues, and heat pipes.

Both CTPC and SSPC converters are similar except for the material of the heater head. The mean helium working gas pressure of both converters is 150 bar, heater temperature is 1050 K, and cooler temperature is 525 K. The dynamic components (power piston and displacer piston) of the converters oscillate at 70 Hz. The heater of the SSPC, now being designed, will be fabricated from Udimet 720 and will have a design life of 60,000 hours. The heater of the CTPC, which has been fabricated and is currently under test, is made from Inconel 718 alloy and has a relatively short design life (100-1000 hours at 1050 K).

CURRENT STATE OF THE ART

In terms of space power, the CTPC embodies the current

state-of-the-art. Some of the technologies being investigated in support of FPSE power conversion systems for space include hydrostatic gas bearing systems, advanced linear alternator concepts, heat pipe technology, super-alloy materials and advanced control systems.

Hydrostatic Gas Bearing System

The CTPC hydrostatic bearing design has been demonstrated with self-pumped operation over a power range of 50% to 100%. The hydrostatic gas bearing is a simple, passive system that relies upon the piston gas springs to supply gas to bearing plenums that are located within the pistons. The plenums are charged through ports in the piston and cylinder walls that open at appropriate times as the piston traverses its stroke. Each piston is supported by gas that flows from the supply plenum, through orifices in the piston wall, along a gas gap between the piston and its mating cylinder, and then into the drain plenum.

The CTPC bearing is a new design that is efficient because it uses the gas spring full pressure wave to charge the supply plenum and to discharge the drain plenum. In contrast, the SPRE hydrostatic bearing system used only half of the pressure wave to charge the supply plenum and thus required a greater gas spring pressure amplitude with resultant greater hysteresis power loss. The CTPC bearing losses are only 570 watts compared to bearing losses of 1700 watts for the SPRE. The radial stiffness of the new bearing is 1.48×10^6 lbf/in. of radial deflection. A schematic of the CTPC power piston gas bearing system is shown in Figure 15.

The new hydrostatic bearing incorporates other mechanical improvements compared to the SPRE. For example, bearing plenums were previously placed in the cylinder and involved tortuous flow paths which could neither be inspected nor cleaned. The new bearing system incorporates straight passages that can easily be inspected and cleaned.

Permanent Magnet Linear Alternator

Background - The SPRE Alternator: The SPRE linear alternator was built and tested prior to the CTPC. Though there are significant differences between the two alternators, they are functionally very similar. Low temperature materials used extensively for the SPRE were replaced in the CTPC. The SPRE alternator has served as a test bed for verifying analysis and evaluating design improvements applied to the CTPC. Tests with the SPRE showed that significant alternator losses can occur if magnetic structure is used for the alternator support and surrounding pressure vessel. Other smaller, but no less important, losses were also identified including: magnet eddy current and stator core losses.

Description of the Alternator Design: The CTPC linear alternator, Figure 15, consists of three major parts: the moving plunger, the inner stator and the outer stator. The plunger magnets create a full reversing flux in the stator coil as it moves from one stroke extreme to the other. The axisymetric magnetic path encircles the alternator coil located in the outer stator.

The plunger is directly connected to the engine power piston and carries four cylindrical, radially magnetized, samarium cobalt permanent magnet rings. This material has high magnetic strength but is structurally weak. The plunger structure is provided by spacer rings and axial tie bolts which compressively preload the magnets. Both rings and tie bolts are made of Inconel 718 which is structurally strong and non-magnetic. The spacer rings have a non-conductive gap which virtually eliminates eddy current losses. Each magnet ring is built up of 18 segments of 3 pieces, each with a non-conductive joint to reduce eddy current losses. The magnet segments are both mechanically fixtured and bonded in place by a high temperature bismaleimide adhesive. A kapton film surrounds each magnet segment, electrically insulating them from both the structure and each other. This eliminates currents which can circulate in the cylindrical magnet assemblies and double the losses in the plunger assembly. A partly assembled plunger showing the magnets is shown in Figure 16.

The outer stator is radially laminated electrical steel and contains the alternator coil. Radial, aluminum orthophosphate coated laminations reduce induced eddy currents orthogonal to the direction of magnetic flux, which for the outer stator is both radial and axial. The electrical steel is a high permeability, high resistivity material (Permendur V). Stator laminations are welded to rings at each end, which provide for attachment to the engine and attachment of other engine components. Though these circumferential rings and attachment welds result in paths for circulating currents, tests have shown that the resulting losses are small. The alternator coil is assembled from two small and two large coils. The coils can be connected in either series or parallel to provide for nominal 240/120 volt output. The coils are wound from flat copper to improve packing and enhance thermal conduction. Fiberglass tape impregnated with bismaleimide is wound on the flat "wire" and around each coil assembly. High temperature life of this assembly has not yet been demonstrated.

The inner stator is a cylinder of radially laminated

electrical steel which completes the magnetic flux path. As with the outer stator, there are both radial and axial sections of the flux path in the inner stator. The inner stator laminations are welded to a ring at one end, with which it is attached to the engine. The laminations are welded to each other at the other end. Though these circumferential welds result in possible paths for circulating currents, tests have shown that the resulting losses are small.

High Temperature (525K) Operation: The cold end temperature (525K) of the CPTC and SSPC creates difficult materials problems principally for magnets and insulation used in linear alternators.

Magnets lose residual magnetism, coercivity and energy product as their operating temperature increases. This loss is reversible up to about 573 K (Niedra 1992).

Development of a long life, high temperature alternator is one goal of the SSPC program. Conventional insulation materials have short operating life or break down entirely at the alternator design temperature (525-550 K). High temperature ceramic insulated wires are being developed which are capable of operating at up to {1030K (1400F) firing temp.}, well above the design temperature. Its use will impact the geometry and performance of the linear alternator and is the subject of ongoing work.

The first CTPC alternator has completed initial tests and has performed well both mechanically and electrically. The alternator efficiency of 88% is only slightly below the design goal of 90%. Part of the efficiency deficit has been traced to circulating currents in the plunger magnet assembly. The magnets of the first plunger assembly were simply bonded together, this bond joint was found to have less than satisfactory electrical resistance. A second plunger with Kapton film insulation between each magnet is expected to reduce this loss.

The Starfish Heater

Both the CTPC and SSPC use a novel heater design that was proposed by MTI. The heater is known as the Starfish heater because of its configuration in which 50 fins extend radially outward from the inner annular radius. Each fin contains 38 one-millimeter diameter gas passageways. These holes are drilled by a STEM (shaped tube electrochemically machining) process. The external fin surface will become the condenser surface of a heat pipe connected to a heat source.

The reliability of the Starfish heater is expected to improve, in comparison to the SPRE heater, because the Starfish is machined from a single piece of superalloy material. The SPRE heater required 3200 braze joints. The walls of the Starfish are of nearly constant thickness, which will minimize thermal shock effect. Gas passages through the fins add very little dead volume to the working space.

The SSPC heater head will be fabricated from Udimet 720. Udimet 720 is a relatively new alloy that is very difficult to join and has not been fully characterized. Therefore, a development program was initiated to better characterize the material. Such characterization includes determination of the most suitable product form: cast wrought or powdered; evaluation of creep and fatigue properties; evaluation of sodium compatibility; and evaluation of joining methods (Mittendorf and Baggenstoss 1992).

Heat-Pipe Heat Transport System

This system consists of an annular enclosed ring that attaches to the outer radius of the Starfish heater as shown in Figures 14 and 17. The inner surface of this annular heat pipe is covered with screen wick with arteries that extend radially outward from the root of each condenser slot to the outer radius of the evaporator section. For test cell operation, the evaporator is heated by electrical radiant heaters that heat the external surface on one side of the annular heat pipe. The CTPC heat pipe structural members are fabricated of Inconel 718 and the wick and artery material are 316 stainless steel. Inconel 617 alloy has been chosen for the SSPC heat pipe structural members. Wick and artery material have not yet been selected.

An assessment of the compatibility of both Udimet 720 and Inconel 718 with sodium has been performed (Alger 1992). It was learned that to minimize oxygen-induced corrosion, the oxygen level in the sodium of the heat pipe must be kept below 10 ppm. Therefore, early in the program, small heat pipes were fabricated to develop a process for filling heat pipes with sodium from a high-purity sodium loop. The procedures for this process were prepared by NASA LeRC, MTI, Thermacore, and ETEC and carried out by ETEC personnel. A 1/10 segment of the CTPC heat pipe is shown in Figure 17 attached to a sodium loop in preparation for filling with sodium. Even if corrosion by oxygen-related mechanisms can be minimized, there remains the probability that some metal alloy components of Udimet 720 and Inconel 718 or Inconel 617 can dissolve directly into the sodium. Analyses to address the dissolution of nickel in sodium have been performed (Tower 1992 and Alger 1992).

The heat pipes that were fabricated by Thermacore and loaded with sodium from ETEC's high-purity sodium loop, will be placed on long-term test at MTI to evaluate the corrosion rates of these heat pipes.

Power Converter Control

Engine/Alternator System Dynamics: FPSE/LAs fall into a matrix of classifications, one dimension of the matrix characterizing the dynamics and the second dimension characterizing the thermodynamics, Table 4. The dynamic subclasses [categories] include: free oscillators, which are self-excited and forced oscillators, in which the displacer is driven to control system frequency and amplitude. The forced oscillator can also be thought of as a thermally powered amplifier in which a small electrical power to a displacer drive motor is thermally amplified by the Stirling cycle to obtain a large electrical power output from the alternator. The thermodynamic types include: alpha, beta and gama types as defined by (Walker 1973). Alpha type Stirling engines have not emerged as a practical FPSE. This is probably due to the necessity of transferring large fractions of the gross power flow to the compression piston at essentially 100% efficiency. This may be practical in a multi-cylinder {cycle} engine, however, it is unlikely in single-cylinder engines. Beta and gama types have, on the other hand, been reduced to practical engines. In terms of their operation and control, there is very little difference between them, other than the gama type having slightly larger mean volume and therefore lower pressure amplitude and specific power. The space power FPSE/LAs being developed by NASA Lewis (SPDE, SPRE & CTPC) have been predominantly free, self-excited, gamma type Stirling engines. The following discussion of FPSE/LA control is limited to self-excited free-piston Stirling engines of the beta or gamma type.

The behavior of a free-piston Stirling engine/linear alternator (FPSE/LA) connected to a constant load can be described by a system of "slightly" non-linear 2nd-order differential equations (Kankam and Rauch 1991). Most practical engines are sufficiently linear that useful solutions may be obtained by the method of harmonic linearization (MHL) (Elgerd 1967). The linearized equations can be solved by well established frequency domain techniques to obtain the characteristic frequencies (eigenvalues) and mode shapes (eigenvectors). The eigenvalues are the most important result of this analysis and are typically plotted on the complex frequency plane, Figure 18. Eigenvalues in the left hand plane (LHP) indicate stable modes that decay with time. Eigenvalues in the right hand plane (RHP) indicate unstable modes that grow with time. Solutions on the imaginary axis are metastable and oscillate with a constant amplitude. Solutions on the ordinate or alpha axis correspond to state variables with no oscillating components, and are not usually observed for practical FPSE/LAs. A typical engine/alternator (no case motion) has 3 degrees of freedom (DoF) corresponding to displacer and piston motion and alternator current. Therefore, the system will typically have 3 complex conjugate pairs of eigenvalues. One pair of eigenvalues, corresponding to the system's steady-state operating frequency, must fall on the imaginary axis, and the other two pairs of eigenvalues must fall in the stable LHP for the system to be stable. The eigenvector corresponding to the least stable eigenvalue determines the relative amplitudes and phases of the displacer, piston and current. However, the frequency domain analysis does not, by itself, determine the absolute component amplitudes. In general, because of internal or external nonlinearities, the frequency domain analysis must be iterated with a system power balance to find the steady-state solution and the absolute component amplitudes.

There are two requirements for stable FPSE/LA system operation. First, all eigenvalues must fall either in the LHP or on the imaginary axis. In the case of a forced system all eigenvalues must be in the LHP, otherwise unstable or meta-stable modes will dominate or beat with the forced response. For a free, unforced, system one eigenvalue (pair) must lie on the imaginary axis. Second, the FPSE/LA must exhibit stable static operating characteristics relative to the load characteristics. Static stability is indicated when the system net load power vs voltage characteristic is "stiffer" (for example, steeper positive slope) than the engine/alternator characteristic at the desired operating point.

Most, if not all, FPSE have internal non-linearities, such as heat exchanger flow losses, which tend to stabilize the system. These may be sufficient to stabilize the system when it is operating in the general vicinity of a desired operating point. However, external controls are generally required to insure global stability and maintain static output over a wide range of user load.

The purpose of a FPSE/LA control system is to maintain the output frequency and voltage (amplitude) of the linear alternator at constant levels. The FPSE/LA's output frequency is inherently fixed (within a few percent) by its design for given operating pressure and temperature. It therefore needs little or no active control. Voltage amplitude is more difficult to control and generally requires active feedback control. From the perspective of a frequency domain analysis, voltage control is achieved by maintaining the average location of the dominant (least stable) eigenvalue on the imaginary axis. The control system must force the dominant eigenvalue to make small excursions from the imaginary axis to correct for voltage errors due to random variation in the system parameters. The dominant eigenvalues (on the imaginary axis) must be moved slightly into the RH plane when the output is below the set-point, held on the jw axis when the output is at the set-point and moved slightly into the LH plane when above the set-point. This control can be accomplished by either changing the overall system damping or the generated thermodynamic power.

One approach to effecting this control is to put an active (controllable) damper on the displacer. An advantage of this approach is that it is relatively efficient compared to others. This reduces the displacer amplitude relative to that of the piston, thus, reducing both cycle power output and input in addition to increasing total system damping. Further, since this is independent of the piston amplitude, the output voltage can be maintained at off design power. The major disadvantage of this control mechanism is that it adds to the internal complexity of the engine.

The control which has, largely by default, been implemented on the existing space power engines (SPDE, SPRE & CTPC) is to provide an actively controlled parasitic electrical load on the alternator output. The parasitic load is controlled via voltage feedback to maintain constant voltage at the user load terminals. Thus, the combined user and parasitic loads absorb the total power produced by the engine/alternator. For a system with large variation in power demand and a fuel cost proportional to thermal energy usage, this is a relatively inefficient control. However, for unregulated nuclear or solar space power systems where the thermal power is a capital (versus an operating) cost, low efficiency at less than design power is not a disadvantage. The major advantage of this control mechanism is that it is entirely external to the engine proper. Since this method is typically used, its operating characteristics are described in detail below.

Figure 19 shows the power versus voltage characteristics of a typical FPSE/LA, where power output (and input) increase with voltage, other operating parameters (such as temperature ratio and pressure) being constant. Also shown is the steady-state behavior of a voltage controlled parasitic load. The parasitic load power is zero for voltages below the set-point voltage. Above the set-point the load increases sharply as the voltage error increases up to the point where this load saturates, this is the proportional band within which the system is designed to operate. Beyond saturation, the resistance is constant and power increases in proportion to voltage squared. Once the load saturates, the system is potentially out of control and may be unstable.

Heater head temperature control via engine load: Stirling Heater temperature control using nuclear (or solar) fueled systems will be accomplished by control of the engine/alternator power. In this case, where thermal power input is assumed to be quasi-constant, temperature control is achieved by increasing engine power to reduce heater temperature or conversely reducing power to raise temperature.

Temperature control of the heater head is not thought to be a major problem since the response of an engine/alternator/load is several orders of magnitude faster than the reactor time constant. This may not be the case for a solar heat source where the thermal mass may be significantly smaller. Present parasitic load controls are designed to control voltage by regulating total power output which directly affects power input. In order to control both load voltage and power input an additional degree of freedom needs to be added to the system. Mechanisms internal to the engine/alternator are expected to be more difficult to incorporate into the system than external mechanisms. A conceptually simple external control could use an unregulated voltage parasitic load to control total engine power (and thus heater head temperature), followed by a voltage regulator to maintain the user bus voltage.

Performance Testing of the CTPC

The CTPC is the first power converter developed in which the cold-end hardware must operate from ambient to 525 K and still maintain clearances between the pistons and their cylinders. Therefore, the first phase of development was planned to verify the mechanical design at 525 K, and mechanical, structural and electrical design of the linear alternator. The second phase of converter development consists of fully-functional testing of the CTPC at design conditions.

CTPC Cold-End Testing

The purpose of the cold-end test is to verify mechanical operation, internally pressurized bearing operation, and alternator performance at the operating temperature of 525 K before adding heat exchangers to complete the assembly of the CTPC. Heating of the cold-end was accomplished by a dummy heater head which contained an SPRE cooler connected to a hot oil pumped loop to heat the Cold-End to 525 K. Motion of the piston and displacer was provided by the linear alternator which functioned as a linear motor.

Full-stroke operation with power and displacer pistons at 28 mm was achieved at design operating conditions: at a temperature of 525 K, a helium working gas pressure of 150 bar, and at a frequency of 70 Hz. Successful operation of the Cold-End at 525 K demonstrated major accomplishments in bearing and alternator design. The power piston and displacer were supported on internally supplied hydrostatic gas bearings and were capable of dry starts and smooth mechanical operation at temperature levels from ambient temperature to 525 K. Alternator stator coil and insulation materials operated at a temperature of 573 K with minimum performance penalty. The integrity of the Sm_2-Co_{17} permanent magnets was maintained at their operating temperature of 548 K.

Testing of CTPC at Design Power

During the first fully-functional test of the CTPC, the design goals of 12.5 kWe electrical output and 20% overall efficiency were easily surpassed. The converter operated flawlessly for over three hours. Heater wall temperature was held at 950 K, cooler wall temperature at 475 K, helium working gas pressure was at 150 bar, and converter frequency was 70 Hz. First testing of the CTPC was accomplished with direct heat input by electrical radiant heaters. This was necessary because of the longer time needed to develop the heat pipe.

FUTURE DEVELOPMENTS

Stirling technology has made significant advances during this past decade which have demonstrated the advantages and applicability of Stirling power conversion devices for use in long duration space missions. Additional development needs to continue to extract the highest performance from these power converters at the lowest specific mass.

Materials Development

Stirling power converters, using superalloy materials of construction, can operate in space at hot end temperatures to 1050K. The best candidate material to date is Udimet 720 for strength and life. Udimet 720 is a relatively new alloy that is very difficult to join and has not been fully characterized. Therefore, a development program should be initiated to better characterize the material. Such characterization should include further evaluation of creep and fatigue properties; sodium compatibility; and joining methods.

Additional reductions in system mass may be obtained by increasing hot end temperatures to 1300-1400K. These operating temperatures would require the use of refractory materials. Negligible work has been done in this area with Stirling systems.

Alternators

As hot end temperatures continue to rise, the cold end temperatures likewise rise; this is primarily due to the requirement to minimize radiator size and mass. Projects have begun to investigate various types of glass encapsulation and insulation for the linear alternator components. These new materials solve two problems: (1) they can provide long life at high temperature (525K), and (2) they do not outgas and contaminate the working fluid (Helium).

Coatings

The probability exists that some metal alloy constituents of heat pipe and heater head materials can dissolve directly into liquid metals used as working fluids in heat pipes. Although dissolution is expected to be very slow, coatings should be developed, if possible, to retard dissolubility. Characterization of dissolubility at operating conditions needs to continue.

Life Testing

Although free piston Stirling systems have the potential and the pedigree for long life, components and full-up engine systems should be put on life test, under real operating conditions. Testing should include "shake and bake" pre-flight testing.

CONCLUSIONS

Free-piston Stirling power converters have the potential to significantly enhance the state-of-the-art of nuclear space power systems, providing low mass, compact systems for a wide range of NASA future missions from tens of watts to hundreds of kilowatts. Technology development and demonstration under the CSTI High Capacity Power Program has shown the flexibility and ultimate performance capability of free piston linear alternator machinery in the 25kW category. Stirling machines in the form of kinematic cryocoolers have already flown successfully on long term space missions thus demonstrating the viability of the Stirling cycle in space. What remains is the accumulation of statistical data to verify the long life potential and reliability which the Stirling system promises.

References

Alger, D. (1992) "Heat-Pipe Heat Transport System for Stirling Space Power Converter", Proceedings of 27th Intersociety Energy Conversion Engineering Conference, Society of Automotive Engineers, Inc., San Diego, CA.

Angelo, J. A., Jr. and D. Buden (1985) "Space Nuclear Power," Orbit Book Company, Inc., Malabar, FL.

Bents, D. J., B. I. McKissock, C. D. Rodriguez, J. C. Hanlon, and P. C. Schmitz (1992) "Dynamic Isotope Power System Design Considerations for Human Exploration of the Moon and Mars," Proceedings of the 27th IECEC, 2: 439-444.

Bents, D. J., J. G. Schreiber, C. A. Withrow, B. I. McKissock, and P. C. Schmitz (1992) "Design of Small Stirling Dynamic Isotope Power System for Robotic Space Missions," *NASA TM-105919*, January 1992.

Bents, D. J., S. M. Geng, J. G. Schreiber, C. A. Withrow, P. C. Schmitz, and T. J. McComas (1991) "Design of Multihundred-Watt Dynamic Isotope Power System for Robotic Space Missions," Proceedings of the 26th IECEC, 204-209.

Cairelli, J. E. et al. (1991) Update on Results of SPRE Testing at NASA Lewis, *NASA TM-104425*.

Dochat, G. R. and M. Dhar (1991) "Free-Piston Stirling Engine System Considerations for Various Space Power Applications," Eighth Symposium on Space Nuclear Power Systems, II: 598a-604a, January 1991.

Dochat G. R. and M. Dhar (1991) "Free-Piston Stirling Component Test Power Converter." 26th Intersociety Energy Conversion Engineering Conference, Boston, Massachusetts, August 1991.

Dochat, G. R. (1992) "Free-Piston Stirling Component Test Power Converter Test Results and Potential Stirling Applications." 27th Intersociety Energy Conversion Engineering Conference, San Diego, CA, August 1992.

Elgerd, O. I. (1967) Control Systems Theory, McGraw-Hill Book Co., New York, NY.

Kankam, M. D. and J. S. Rauch (1991) "Comparative Survey of Dynamic Analyses of Free-Piston Stirling Engines, 26th IECEC, Boston, MA.

Mittendorf, D. L. and W. G. Baggenstoss (1992) "Transient Liquid Phase Diffusion Bonding of Udimet 720 for Stirling Power Converter Applications", 27th IECEC, Vol. 5, 5.393-5.397, 1992.

Nainiger, J. J. and L. S. Mason (1992) "Nuclear Reactor Power Systems for Lunar Base Applications," NTSE-92, Nuclear Technologies for Space Exploration; Jackson, Wyoming; 247-256, August 1992.

Niedra, J. M. (1992) "M-H Characteristics and Demagnetization Resistance of Samarium-Cobalt Permanent Magnets to 300 C", 27th IECEC, San Diego, CA.

Rauch, J. S. et al. (1990) <u>SPRE Alternator Dynamometer Test Report</u>, *NASA CR-182251*.

Richey, A. et al. (1989) <u>Upgraded Mod I Stirling Engine Field Evaluation Program</u>, ASME Energy Sources Technology Conference and Exhibition, paper 89-ICE-7, 22-25 January 1989.

Ross, A. (1977) <u>Stirling Cycle Engines</u>, Solar Engines/Phoenix.

Ross, B. A., D. Ritter, I. Williford, D. Lewis, and K. Colenbrander (1993) "Performance of a Laboratory 10 Watt Stirling Generator Set", Stirling Technology Company, Richland, WA, 1993.

Ross, B. A. and J. E. Dudenhoefer (1991) "Stirling Machine Operating Experience," Proceedings of the 26th Intersociety Energy Conversion Engineering Conference, 278-283.

Schmitz, P. C. and L. Mason (1991) "Space Reactor/Stirling Cycle Systems for High Power Lunar Application," *NASA TM-103698*, January 1991.

Schmitz, P. C., B. H. Kenny, and C. Fulmer (1991) "Preliminary Design of a Mobile Lunar Power Supply," Proceedings of the 26th IECEC, 216-222.

Schmitz, P. C., L. Tower, R. Dawson, B. Blue, and P. Dunn (1992) "Preliminary SP-100/Stirling Heat Exchanger Designs", Tenth Symposium on Space Nuclear Power and Propulsion, Albuquerque, N.M., January 1993.

Schmitz, P., H. Bloomfield and J. Winter (1992) "Near Term Options for Space Reactor Power," NTSE, Nuclear Technologies for Space Exploration: Jackson, Wyoming, 313-321, August 1992.

Schock, A., (1993) "Radioisotope Stirling Generator Options for Pluto Fast Flyby Mission", Fairchild Space and Defense Corporation, Germantown, MD, 1993.

Shaltens, R. K. et al. (1992) Update on the Advanced Stirling Conversion System Project for 25 kW Dish Stirling Applications, Proceedings from the "27th Intersociety Energy Conversion Conference," August 1992.

Spelter, S. et al. (1989) Space Power Research Engine Power Piston Hydrodynamic Bearing Technology Development, *NASA CR-182136*.

Tower, L. K. (1992) "Simple Analysis of Nickel Transport in Heat Pipes of Stirling Engine Power Conversion Systems," Proceedings of 27th Intersociety Energy Conversion Engineering Conference, Society of Automotive Engineers Inc., San Diego, CA.

Walker, G. (1973) Stirling-Cycle Machines, Oxford University Press.

West, C. D. (1986) Principles and Applications of Stirling Engines, Van Nostrand Reinhold Company, Inc.

Wong, W. A. et al. (1992) NASA Lewis Stirling SPRE Testing and Analysis With Reduced Number of Cooler Tubes, *NASA TM-105767*.

Urieli, I. and D. M. Berchowitz (1984) Stirling Cycle Engine Analysis, Adam Hilger Ltd.

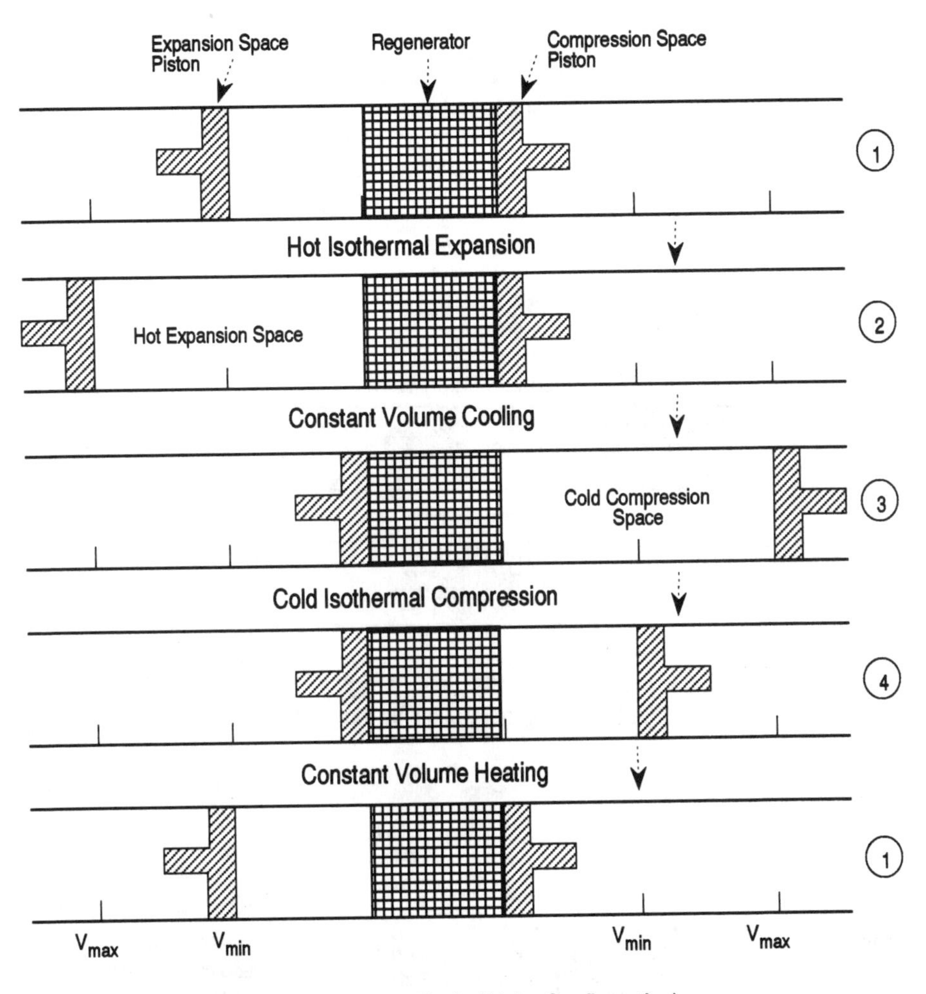

FIGURE 1. Ideal Stirling Cycle (Alpha Configuration).

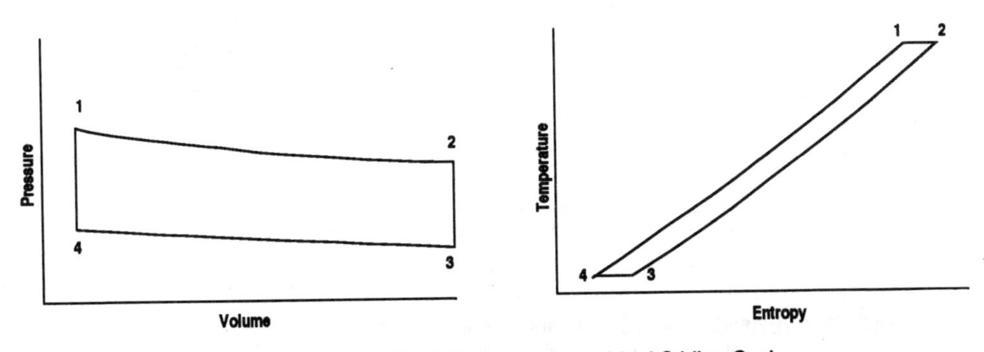

FIGURE 2. P-V and T-S Diagrams for an Ideal Stirling Cycle.

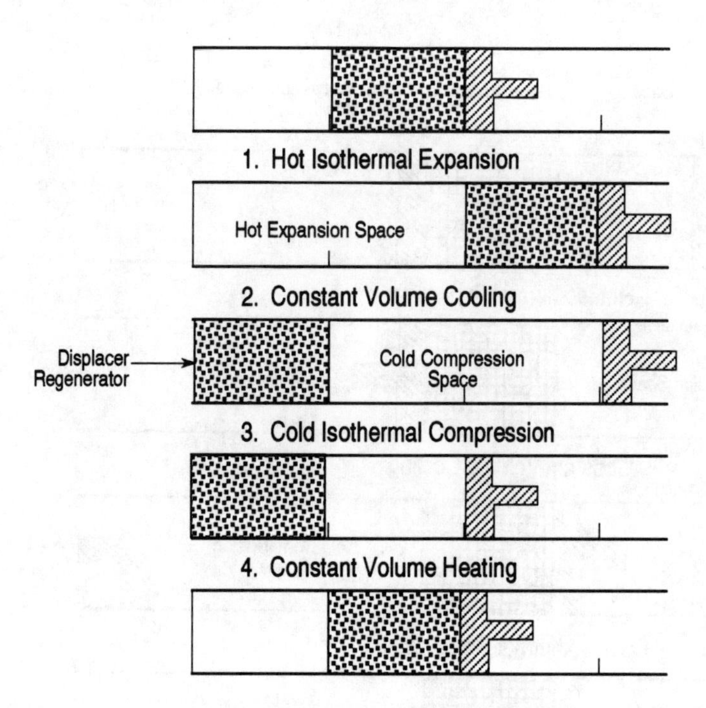

FIGURE 3. Ideal Stirling Cycle Beta Configuration (With the Regenerator Inside the Displacer).

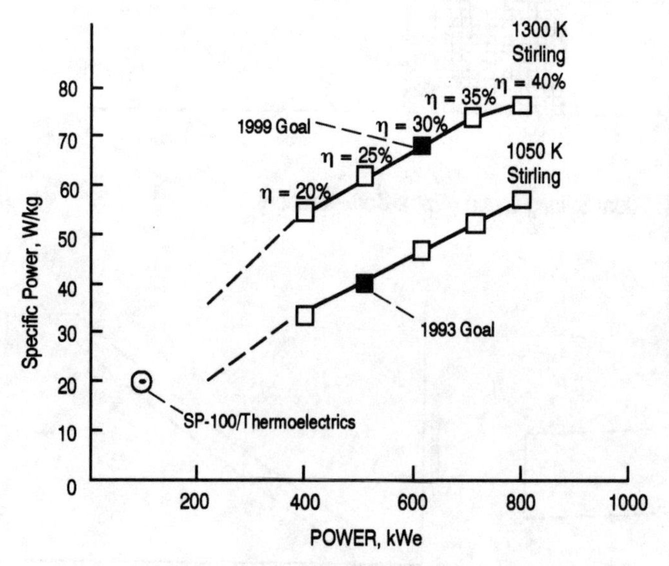

FIGURE 4. Extending SP-100 Reactor Power Systems Capability; Thermoelectrics (TE) and Stirling.

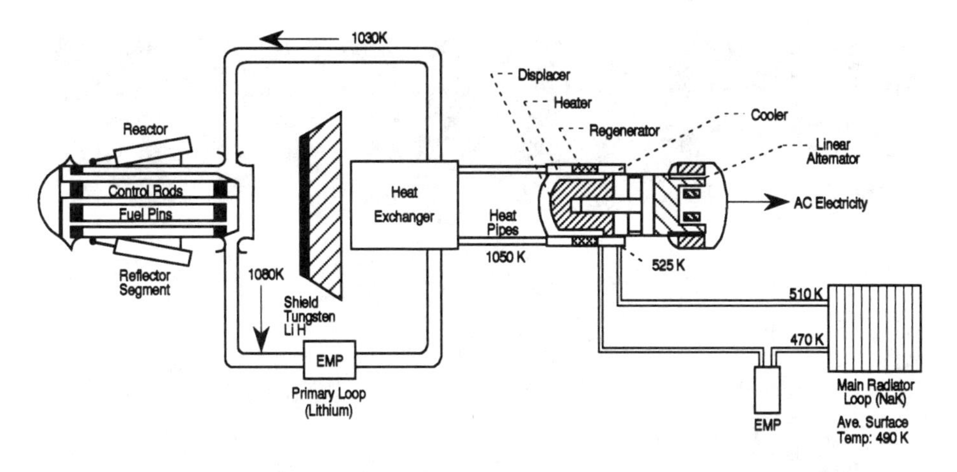

FIGURE 5. SP-100 System Schematic, Stirling Cycle, Superalloy.

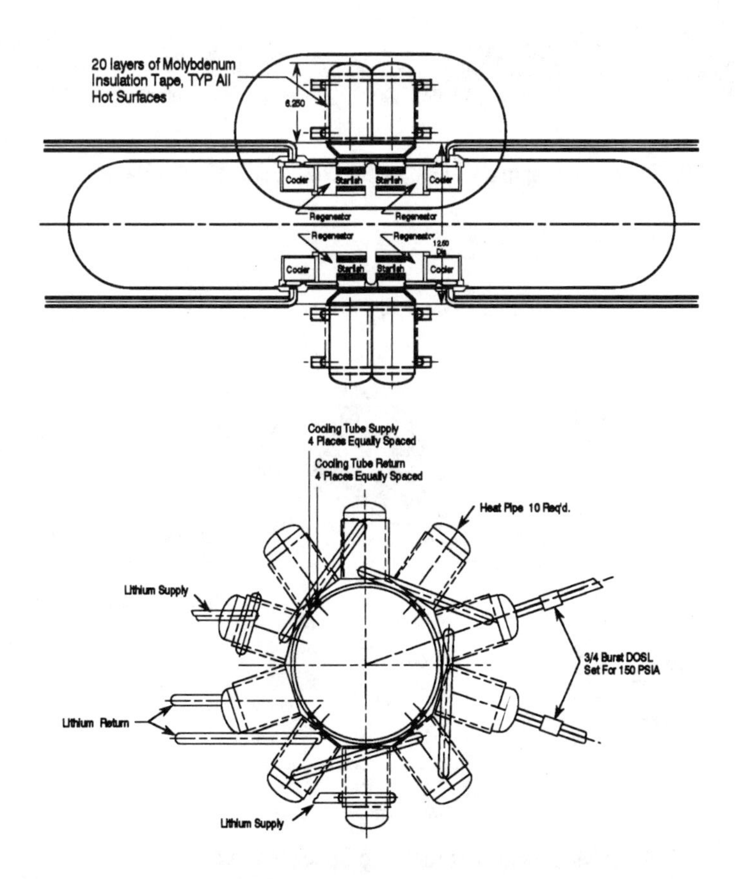

FIGURE 6 : Intermediate Heat Exchanger Between Reactor Pumped Loop and Stirling Converter

FIGURE 7. Conceptual Design for Nuclear Stirling Lunar Base.

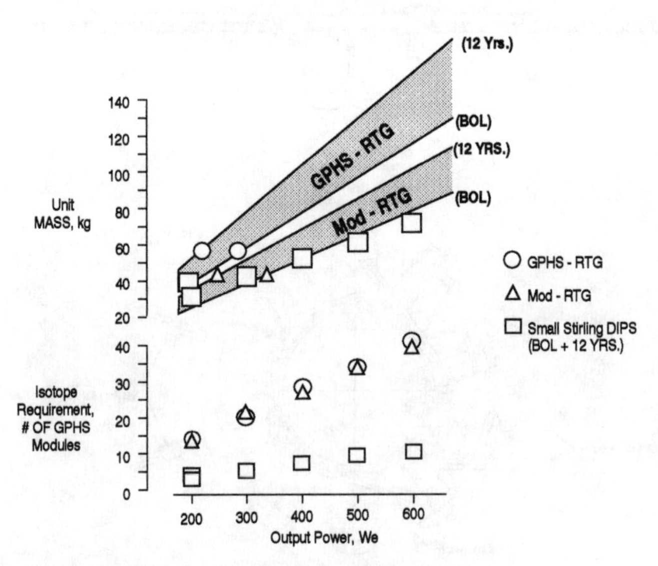

FIGURE 8. Small Stirling Dips for Deep Space Platform
Comparison of Unit Mass and Isotope Requirement with RTG's.

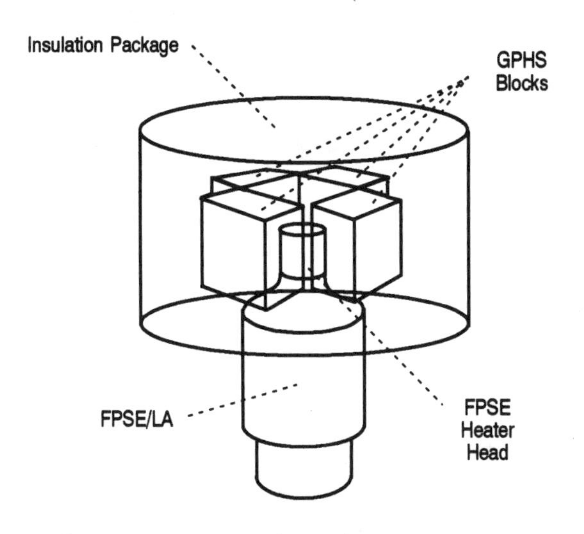

FIGURE 9. Direct Heat Source/Heater Head Integration

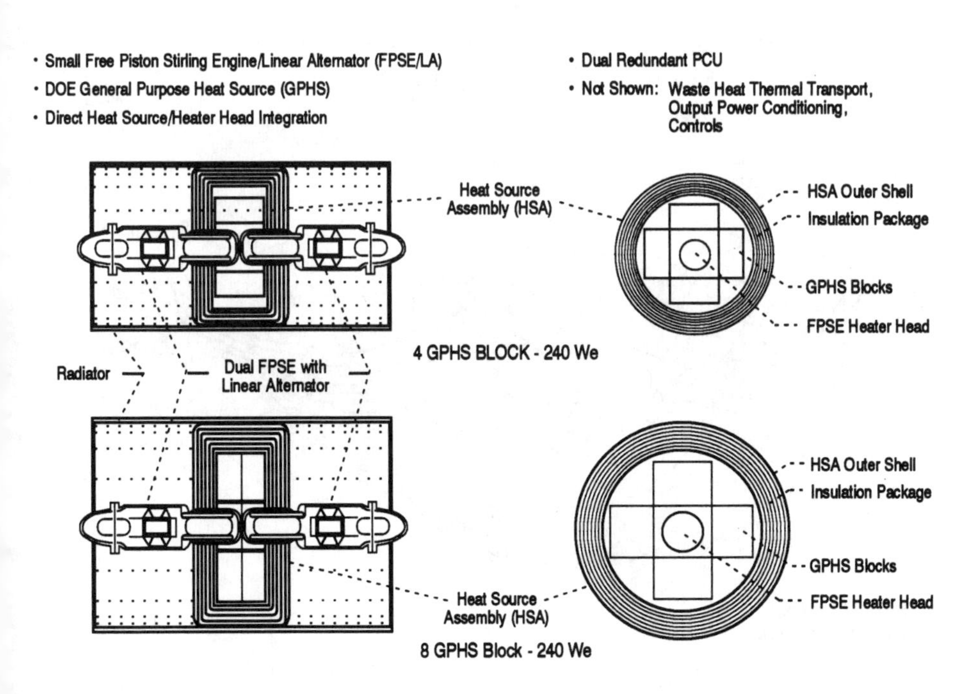

FIGURE 10. Multihundred Watt Stirling DIPS Configuration.

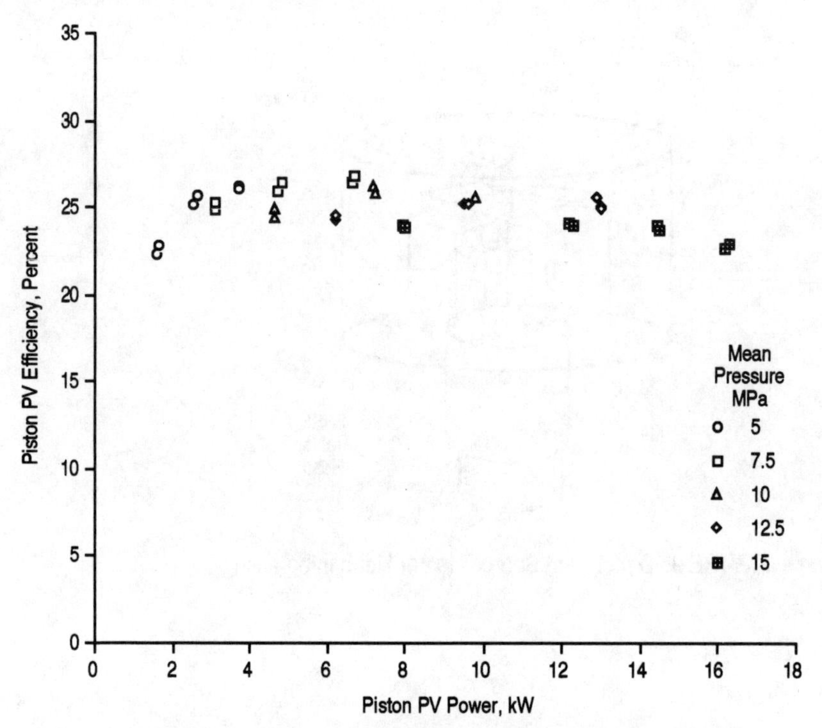

FIGURE 11. Piston PV Efficiency vs Piston PV Power for SPRE Data (1056 Cooler Tubes. TR = 2).

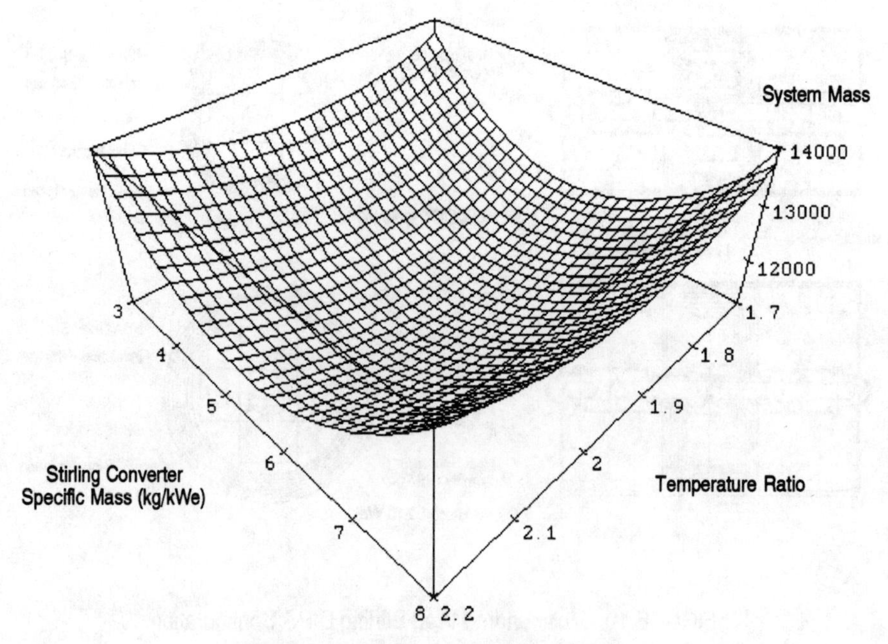

FIGURE 12. System Optimization Surface.

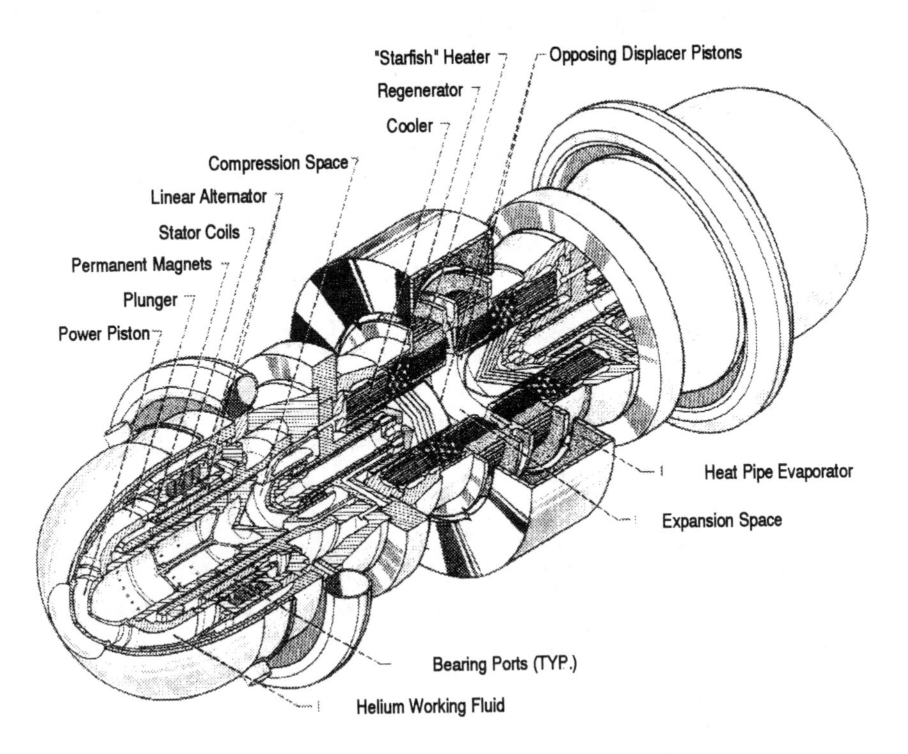

FIGURE 13. Stirling Power Converter 1050 K Superalloy - 25 kWe.

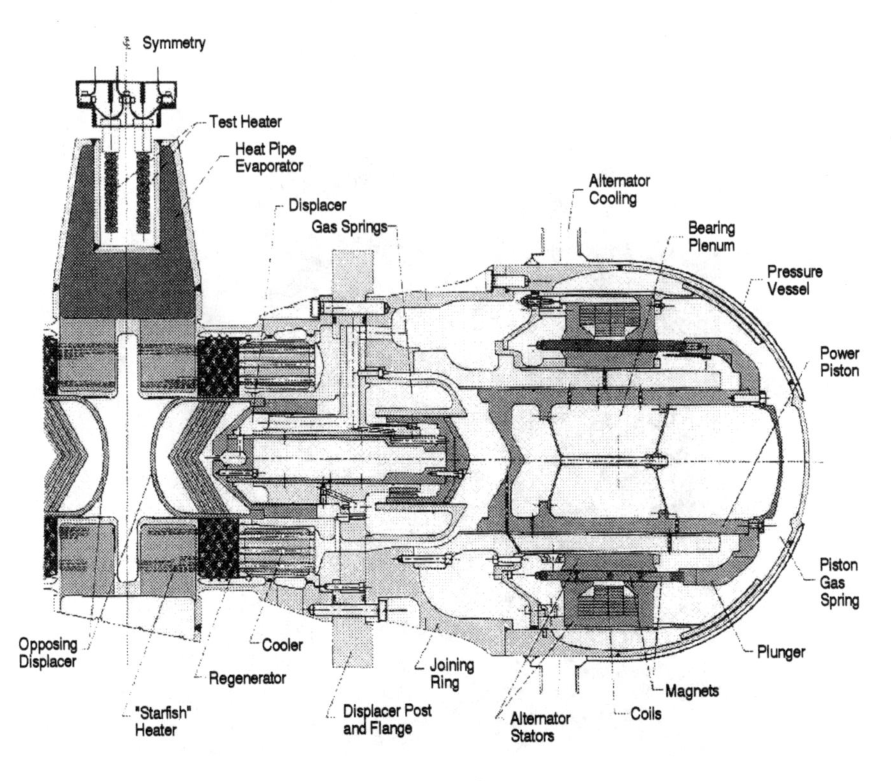

FIGURE 14. Stirling Power Converter.

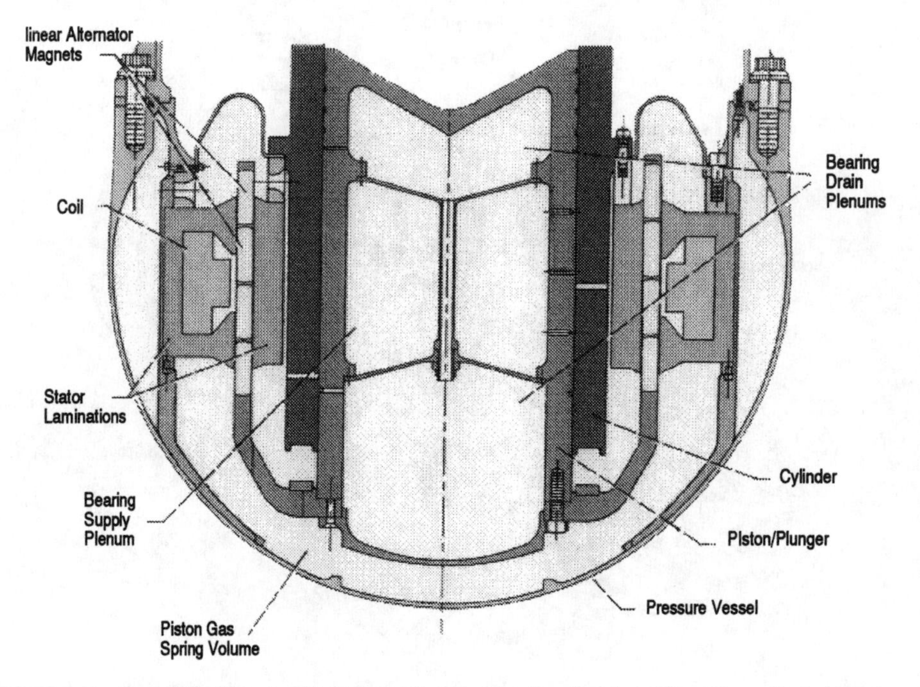

FIGURE 15. Schematic Showing Hydrostatic Gas Bearing Concept for a Free-Piston Stirling Engine.

FIGURE 16. A Partly Assembled CTPC Alternator Plunger Showing the Magnets.

FIGURE 17. 1/10 Segment Heat Pipe Mounted in ETEC's Sodium Loop in Preparation for Loading with Sodium.

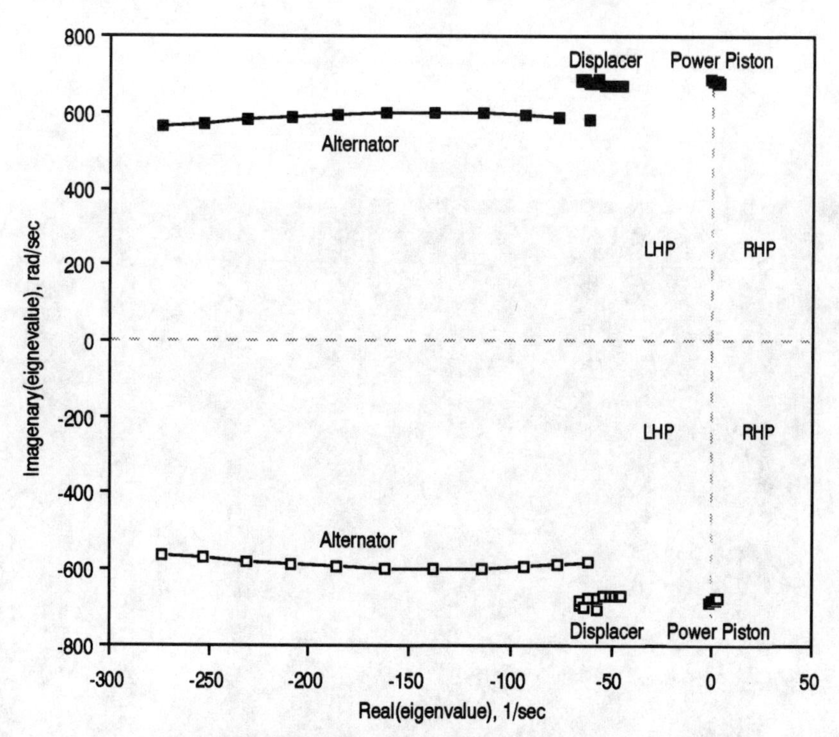

FIGURE 18. Root Locus Plot for Variable Load Resistance of a FPSE/LA.

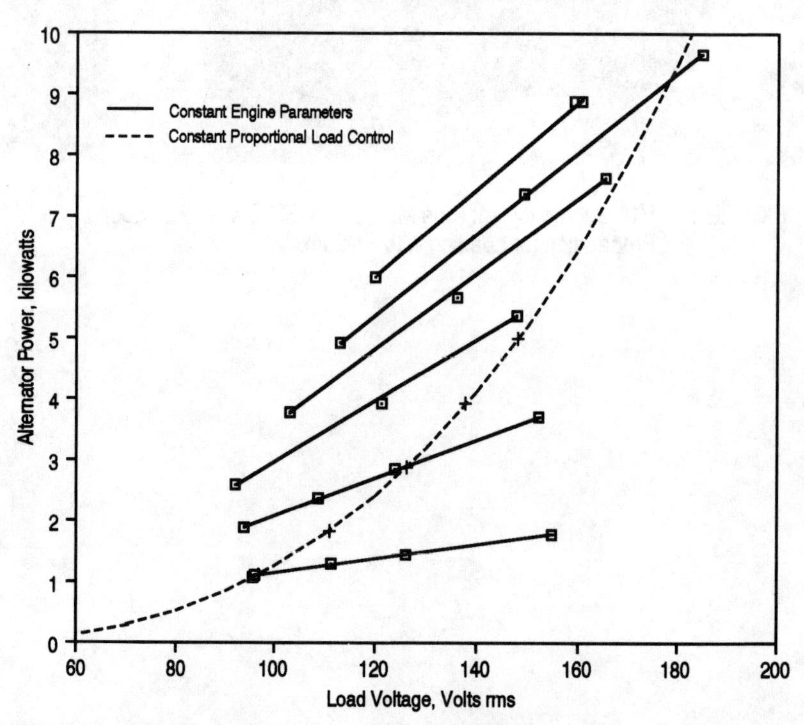

FIGURE 19. FPSE/LA Power Output vs. Load Voltage Characteristics for Various Constant Engine Parameters. Load Control Characteristics for Constant Proportional Gain and Set Point Voltage

TABLE 1. 100 kWe Stirling Lunar Base

Heater Head Temperature	1050 K
Temperature Ratio	2
Cycle Efficiency	27%
Radiator Area	127 m^2
	Mass Breakdown of System (kg)
Reactor	568
Shield	159
Intermediate Liquid Metal Loop	289
Liquid Metal Pumps	170
Primary Piping, Accum, etc.	441
Stirling PCU/Alternator	632
Main Radiator	636
Structure	318
Reactor Bulkhead	783
Upper Scatter Shield	23
Reactor Support Stand	20
Excavation Sleeve	52
Excavation Cooling System	56
Transmission Line (250 meters)	303
Power Conditioning	
Rectifier	61
Filter	147
Enclosure	66
Radiator	53
Misc.	25
Total Base Mass (kg)	**4802**

TABLE2. Performance of Thermomechanical Generators (significant hours).

Application	Power (Watts)		Propane Used	Efficiency		Operating Hours to April.
	AC	DC	kg/year	to AC	to DC	
	Development (D) Machines					
D1 Lab	31.7		196	10.2		9,000
D2 Lab	10.7		Isotope	7.7		72,000
	Field-Trial (F) Machines					
Date Buoy F1	24.5	18.9	190	8.1	6.25	10,000
Date Buoy F2		27	166		10.7	21,000
AGA Lighthouse Up-rated	65	58	450	9.1	8.1	23,000
					Total Hours:	135,000

TABLE 3. Cumulative Past Experience of Space Flight Cryocoolers Through 1990

Mission	Cooler Type	Operating Point	Comments
RM-19 (1970) IR Telescope	(2) Malaker Integral Stirling	1.7 W at 100 K (in parallel)	Accomplished 6-month mission. Unit operated over two years before degradation and outgassing effects were observed.
SESP 71-2 Celestial Mapping Program (1971)	(1) Hughes Rotary VM (2-stage)	0.15 W at 15 K, 3.5 W at 55 K	Failure of ambient cooling loop after 3 weeks; there were 690 ground hours of pre-flight testing
SKYLAB Series (1971)	(3) Malaker Integral Stirling	77 K - 90 K 1 W (each)	Fully successful, 10 hours operation over 90 days, many starts and stops
P78-1 X-Ray Spectrometer (1979)	(4) Philips Rhombic Drive Stirling (2-stage)	0.3 W at 90 K, 1.5 W at 140 K (each) 90 K on station	Accomplished 1-year mission with 1 unit failure. Degradation and outgassing effects observed. Continued functioning through 1985
ATMOS Atmospheric Measurement on Space Lab 3 (1985)	(2) CTI Split Stirling	1.6 W at 75 K	Reflight planned on future missions
SALYUT	2-Stage Stirling with J-T	Not Available	
STS 61-C (1986)	(1) Cryo-dynamics Integral Stirling	1 W at 80 K	RCA IR Camera, 6-day mission
3-Color Experiment (1989)	(2) Magnavox Split Stirling	1 W at 105 K (each)	2 years to date, but contamination caused degradation after 200 hours

Number in ( ) refers the number of units

Planned Long-Life Cryocooler Missions

Mission	Cooler Baseline	Operating Point	Comments
X-Ray Spectrometer (XRS)	(TBD) Flexure Stirling	0.4 W at 65 K (each)	Planned for STS launch in 1996 (NASA AXAF)
Earth Observing System (EOS) Instruments	(TBD) Flexure Stirling	0.8 W at 80 K, 0.5 W at 55 K, 0.3 W at 30 K	Projected launches 1996 (NASA), 1997 (ESA), 1998 (NASA)

Number in ( ) refers the number of units

TABLE 4. Classification of Free-Piston Stirling Engines.

		Dynamics	
		Free-	Forced -
Thermodynamics	Alpha	None	None
	Beta	SPIKE	?
	Gamma	SPRE, CTPC	MTI - EM

SECTION II: RANKINE CYCLE

INTRODUCTION

In general, nuclear heat sources, both radioisotope and reactor, are of major interest for space applications because of their ability to operate in earth orbit shadow, lunar night, or deep space where the solar flux is too low for solar powered systems. While isotope heat sources have seen significant application in NASA solar system exploration missions, only one reactor heat source has achieved Earth orbit. However, none of these missions required power levels above 500 watts, and they all used static thermoelectric conversion for electricity generation. There is, therefore, no existing operational space power database for high power reactor heat sources or dynamic conversion. This is to be contrasted with the existing extensive operational terrestrial technology data base and familiarity with fossil-fired and nuclear-heated steam and gas stationary powerplants. Both Rankine and Brayton cycle technology have been used for commercial and military electricity generation, cogeneration, and shaft power and propulsion applications. In particular, the widespread use of Rankine cycle dynamic conversion for terrestrial commercial electrical power generation has created a public acceptance of this technology. However, the application of familiar Rankine cycle technology to electric power generation in space represents a significant departure from terrestrial technology.

The application of Rankine cycle technology to space power applications represents a significant departure from familiar terrestrial technology because of differing requirements. For example, unique space power generation requirements include: (1) closed loop operation; (2) compactness and light weight; (3) high specific power; (4) capability for operation in the vacuum and microgravity space environment; (5) highly reliable autonomous operation with minimal maintenance or repair; (6) radiant heat rejection; and (7) relatively low power levels compared to terrestrial powerplants. These special requirements have led to major technology and design variations for space nuclear Rankine cycle powerplants compared to terrestrial applications.

Consideration of operational temperature regimes for various space reactor heat source options and practical limits of both upper and lower cycle pressure levels have historically led to the selection of a wide variety of different working fluids. For example, toluene has been the working fluid of choice for peak cycle temperatures ranging up to about 650 K. Liquid metals such as mercury have been selected for temperatures up to about 950 K, and any of the alkali metals - sodium, potassium, rubidium, and cesium - are potential fluids of choice for peak cycle temperatures above 950 K. Working fluid selections and applications for typical space and terrestrial Rankine cycle turbine inlet conditions are shown in Table 1. In addition to the unique space requirements previously discussed, there are specific design variations peculiar to space nuclear reactor Rankine cycle powerplants. These include: (1) the use of an indirect cycle to provide isolation of the Rankine cycle working fluid from the reactor coolant; (2) compact, high heat transfer rate; once-through boilers; (3) electromagnetic and canned motor pumps; and (4) condensing heat rejection radiators.

HISTORY

Application studies of Rankine cycle technology to electricity generation in space began in the early 1950's with government sponsored feasibility and conceptual design studies of space nuclear power systems for electric propulsion and spacecraft auxiliary electric power for Lunar and interplanetary exploration (Dix and Voss 1984, Voss 1984, and Bloomfield and Sovie 1991). The launch of Sputniks I and II in 1957, and the subsequent formation of NASA in 1958, greatly increased the nation's interest in space and led to the initiation of many government-sponsored technology development programs that covered a wide variety of nuclear heated dynamic and static space power systems.

Technology development of nuclear-heated liquid metal vapor Rankine cycle space power systems began in 1957 with the Space Nuclear Auxiliary Power (SNAP) programs sponsored by the Atomic Energy Commission (AEC). The mercury Rankine cycle was selected as the prime dynamic conversion candidate for space power generation because of it's potential for high efficiency and low pressure operation in a compact configuration with high temperature waste heat rejection to space (Lubarsky 1969). Additional potassium Rankine cycle development programs aimed at higher peak cycle temperatures were carried out into the early 1970's, at which time all government sponsored efforts were terminated.

A summary discussion of the technology database generated by the major Rankine cycle technology programs carried out in the United States since the mid-1950's is presented. The paper concludes with a brief discussion of key technology issues requiring further resolution.

SNAP Mercury Rankine Cycle Technology

SNAP-1 was the first dynamic space power system development program. The 500 We single mercury loop system design used a Cerium-144 radioisotope combined heat source/boiler coupled to a 1300 F (977 K) superheated mercury Rankine cycle energy conversion subsystem (Dick 1959 and Tapco Group 1960). This program was sponsored by the Atomic Energy Commission (AEC) and the Rankine cycle energy conversion development was carried out by the Tapco Group Thompson Ramo Wooldridge, Inc., Cleveland, Ohio. The major components of the system, shown schematically in Figure 1, included an isotope heat source-boiler, a combined rotating unit (turbine, alternator, and pump), and a condensing radiator. Unique design features included: (1) mercury lubricated hydrosphere bearings were used to support the single rotating shaft; and (2) a single circulating mercury liquid metal loop that removed the heat from the radioisotope through a stagnant intermediate fluid, lead. The heat transferred is utilized to preheat, boil, and superheat the mercury to 1300 F (977 K) in a once-through boiler design concept. The estimated overall system efficiency was about 7-8 percent. Technology development was successfully carried forward through a 2500 hour test of the single shaft combined rotating unit, however, further development of this system was terminated because parallel development of simpler isotope thermoelectric conversion systems such as SNAP-1A and SNAP-3 had advanced at a more rapid pace. Although these systems were considerably less efficient and produced much less power, they met U.S. Navy navigation satellite mission requirements. In 1961 two plutonium 238 SNAP-3 units generating 2.7 kWe each, were successfully launched into orbit.

In parallel with the odd-numbered isotope heated SNAP programs, the AEC sponsored development of the even-numbered reactor-heated SNAP series. The first space nuclear reactor-heated Rankine cycle development efforts began in the mid-1950's under AEC sponsorship of the SNAP-2 power system which used a two-loop mercury Rankine cycle power generation subsystem with an integrated mercury condensing radiator (Southam 1960, Wallerstadt and Ono 1966, and Jarrett 1973). This thermal-to-electric conversion subsystem was coupled to a small (50-80 kWt) thermal fission reactor to provide a 3-5 kWe power output. Figure 2 schematically indicates the system concept which incorporates two liquid metal loops, and Figure 3 provides key state points for the cycle. A liquid metal eutectic mixture, NaK-78, is heated by the reactor to 1200 F (930 K) and circulated through the boiler where it boils and superheats the mercury to 1150 F (895 K). Mercury vapor drives a turbine which is part of combined rotating unit comprised of the mercury vapor turbine, the alternator, and the mercury pump on a single shaft. Bearings for the combined rotating unit consisted of a double-acting thrust bearing and two journal bearings lubricated with liquid mercury. The alternator of the SNAP-2 system was a six-pole permanent magnet machine with a hermetically sealed stator to prevent entrance of mercury. The alternator operated at temperatures on the order of 700 F (650 K) to prevent mercury condensation in the rotor area. The overall efficiency of the system was about 6 percent with an unshielded system specific weight of 300 lb/kWe (136 kg/kWe).

The SNAP 2 Rankine cycle development effort which was also conducted by the Thompson Ramo Wooldridge Company (TRW) in Cleveland, OH, resulted in the successful performance and duration demonstration of a space prototype design that employed a once-through boiler, mercury lubricated bearings, a permanent magnet and vapor cooled alternator, and an isothermal hermetically sealed combined rotating unit (CRU) that contained the pump, bearing, turbine, and alternator on a single shaft. A total accumulated testing time of 21,196 h was completed on flight-type CRU's with one unit operating 4759 h. A capability for 10,000 h of successful operation at turbine inlet design conditions of 1150 F (900 K) and 115 psia (0.8 MPa), with a shaft rotational speed of 36,000 rpm and 3-5 kWe electrical output at 1800 hz was indicated. No loss of performance or hardware damage was seen after 37 injection-type startups and launch environment shock vibration tests. In addition, capability for power output levels in excess of 9 kWe were possible with the substitution of a Lundell type alternator for the permanent magnet type.

The SNAP-1 and SNAP-2 mercury Rankine cycle system concepts did not mature into useful space power systems primarily because the power levels were too low to be competitive with static energy conversion concepts. However, the technology generated and lessons learned in these programs were utilized in the SNAP-8 program, the third and last investigation of nuclear-heated dynamic space power systems based on mercury Rankine cycle conversion.

Initiated in 1959, under joint AEC and NASA sponsorship, the SNAP-8 program was aimed at power levels exceeding 30 kWe and a capability for unattended operation for a minimum of 10,000 hours. By 1971, after 11 years of development, the SNAP-8 capability had evolved to a restartable human-rated system operating at 90 kWe with a 5 year life, and a substantial Rankine cycle technology base was developed by the NASA Lewis Research Center, Cleveland, OH and the Aerojet General Corp., Azusa, CA. During it's development, the SNAP-8 mercury Rankine cycle system underwent considerable evolution and modification. By late 1962, the power conversion system design had evolved from a 35 kWe instrument-rated two-loop concept similar to the SNAP-2 design to a 35 kWe human-rated four-loop concept with new components based on state-of-the-art technology in order to overcome design and operational difficulties encountered with the original concept. A comparison of the two design concepts are shown in Figures 3 and 4. The introduction of a SNAP-8 phaseout program in 1964 resulted in a sharp curtailment of development and test activities. However, consideration of a human-rated system for space station/space base applications led to the reactivation of SNAP-8 activities in late 1966. The final phase of the program, conducted until termination in 1971, was focussed on significantly increasing system performance and efficiency. The final design, shown in Figure 5, incorporated a six loop configuration, redundant components and power conversion systems, and generated 90 kWe at 15 percent efficiency from a 600 kWt reactor operating at a reduced outlet temperature of 940 K (1200 F), with capability for 120 kWe at 20 percent efficiency.

In summary, the technology status of mercury Rankine cycle dynamic conversion established by the SNAP programs includes a substantial materials, component, and subsystem ground test database. The SNAP-8 program, which grew out of the SNAP-1 and SNAP-2 programs, is the primary source of the relevant database. All critical mercury Rankine cycle components are well understood and each was endurance tested for at least 7,300 hours. The relevant technology database is valid for reactor outlet, or maximum cycle, temperatures up to 1300 F (977 K), system power levels up to about 100 kWe, and operational system lifetimes of at least 10,000 hours.

Potassium Rankine Cycle Technology

During the 1960's a large and varied technology database for potassium, cesium, and rubidium Rankine cycles was generated. However, only the potassium Rankine cycle technology database will be discussed herein since it was the primary focus of these efforts. The major attraction of the alkali metal fluids is their compatibility with higher reactor outlet temperatures than mercury. This capability provides the potential for higher system efficiency and/or higher heat rejection temperature with a resulting lower system mass and/or radiator area.

In contrast to the focused development of the SNAP mercury Rankine programs, the potassium Rankine cycle effort was more diversified in terms of system design concepts and participants. The potassium Rankine cycle technology effort grew out of the AEC sponsored SNAP-50 reactor development program which began in 1957 at Pratt and Whitney, Middletown, CT, in support of the Aircraft Nuclear Propulsion (ANP) program. Upon termination of the ANP supersonic flight program in 1961, the project was consolidated with the Air Force's SPUR (Space Power Unit Reactor) program and managed by the AEC under a triagency Air Force/AEC/NASA memorandum of agreement. The SNAP-50/SPUR concept was based on advanced high temperature reactor and power conversion technology aimed at providing 300 to 1200 kWe for space power or electric propulsion applications. During the early 1960's, at least twelve different organizational participants were involved in Rankine cycle development alone (Anderson et al. 1983). However, in 1965 the AEC terminated the SNAP-50/SPUR program, and by 1970 the only remaining original participants were the NASA Lewis Research Center, Cleveland, OH, and it's principal contractor, General Electric Co., Evendale, OH. The NASA advanced Rankine cycle program maintained a strong materials and component technology development until program termination in 1973.

The principal technologies evolved in the NASA program were based on the design concept shown in Figure 6. Lithium, at a temperature of 1422 K (2100 F), is pumped through an advanced reactor and heated to 1477 K (2200 F). The lithium then flows through a counterflow Li/K heat exchanger/boiler. Boiling potassium at 1350-1395 K (2000-2050 F) is then superheated to 1420-1450 K (2100-2150 F) and expanded through the turbine. Separate cooling loops and radiators cool the condenser and major electrical components. Although a complete power system of this class was never built, an extensive materials and component technology database was generated (Peterson et al. 1971 and English 1982).

The technology base includes: (1) long term material compatibility and corrosion loop testing of niobium alloys (5,000 h) and tantalum alloys (10,000 h) at temperatures up to 1500 K with virtually no corrosion; (2) successful fabrication and performance and endurance tests of heat transfer components (once-through boilers and condensers); (3) fabrication and 10,000 hour endurance test of the potassium boiler feed EM induction pump; (4) 5,000 hour performance and endurance testing of a two stage potassium vapor turbine and an additional 5,000 hour performance and erosion test of a three stage turbine; and (5) successful performance and 10,000 hour duration tests of high temperature (980 K, 1300 F) electrical components (solenoid, transformer, and stator).

Additional potassium Rankine technology was also generated at the Oak Ridge National Laboratory (ORNL) under the Medium Power Reactor Experiment (MPRE) program (Anderson et al. 1983). The primary goal of this effort was to develop a simplified single loop space power system based on a boiling potassium reactor heat source. The majority of the technology development was for a near term stainless steel system with a peak cycle temperature of 1100 K (1540 F). Significantly less effort was expended on an advanced design operating at 1365 K (2000 F). The low temperature development efforts showed that at temperatures up to 1140 K (1600 F), potassium is less corrosive to stainless steel than either mercury at 750 K (900 F) or water at 800 K (1000 F). Several boiling potassium systems constructed of stainless steel were operated for 16,000 hours at up to 1150 K (1600 F) and showed virtually no corrosion.

TECHNOLOGY ISSUES

There are still a number of technology issues requiring additional development and demonstration in order to achieve a space flight readiness level. These include: (1) characterization of processes and fabrication methods for high temperature, high strength refractory metal alloys such as T-111 and ASTAR-811C; (2) fabrication and endurance testing of large high temperature potassium turbines to determine the limits of turbine blade tip speed and potassium erosion; (3) extension of test temperatures and pressures for turbine bearings and seals; and (4) development of non-magnetic generator winding seals.

References

Anderson, R.V., et al. (1983) "Space Reactor Electric Systems Subsystem Technology Assessment," *ESG-DOE-13398*, Rockwell International, Energy Systems Group, Canoga Park, CA.

Bloomfield, H.S., and R.J. Sovie (1991) "Historical Perspectives: The Role of the NASA Lewis Research Center in the National Space Nuclear Power Programs," *AIAA-91-3462*, AIAA/NASA/OAI Conference on Advanced SEI Technologies, Cleveland, OH.

Dick, P.J. (1959) "SNAP-1 Radioisotope-Fueled Turboelectric Power Conversion System Summary," Martin Co.

Dix, G.P., and S. Voss (1984) "The Pied Piper - A Historical Overview of the U.S. Space Power Reactor Program," in *Trans, 1st Symposium on Space Nuclear Power Systems*, CONF-840113-Summs., M.S. El-Genk and M.D. Hoover, edts., University of New Mexico's ISNPS, Albuquerque, NM.

English, Robert, E. (1982) "Power Generation from Nuclear Reactors in Aerospace Applications," NASA Technical Memorandum 83342.

Jarrett, A.A. (1973) "SNAP 2 Summary Report," AI-AEC-13068, Atomics International, Canoga Park, CA.

Lubarsky, B. (1969) "Nuclear Power Systems for Space Applications," in *Advances in Nuclear Sciences Technology*, V.5.

Peterson, J.R., J.A. Heller, and M.U. Gutstein (1971) Status of "Advanced Rankine Power Conversion Technology," *GESP-623*, and ANS Seventeenth Annual Meeting, Boston, MA.

Southam, D.L. (1960) "SNAP-2 Power Conversion Status," ARS Space Power Systems Conference.

Tapco Group (1960) "SNAP-1 Power Conversion System Development," New Devices Lab., Thompson Ramo Wooldrodge, Inc., *Engineering Report 4050*.

Voss, Susan S. (1984) "SNAP Reactor Overview," *AFWL-TN-84-14*, Air Force Weapons Laboratory, Kirtland Air Force Base, NM.

Wallerstedt, R.L. and Ono, G.Y. (1966) "A Summary of SNAP Mercury Rankine System Status," in *Progress in Astronautics and Aeronautics*, Vol. 16, Space Power Systems Engineering.

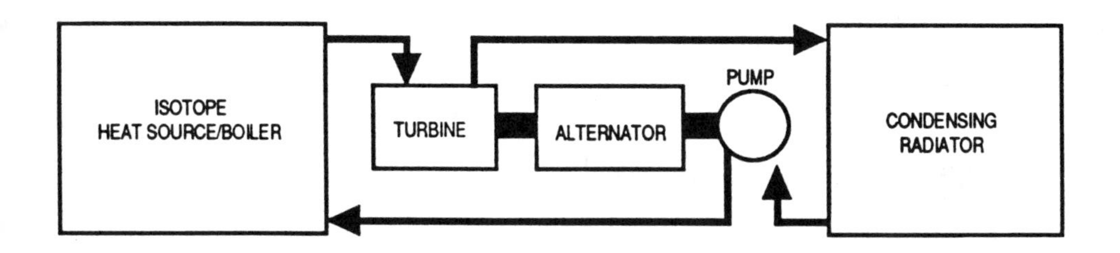

FIGURE 1. Schematic Diagram of SNAP-1 Mercury Rankine System

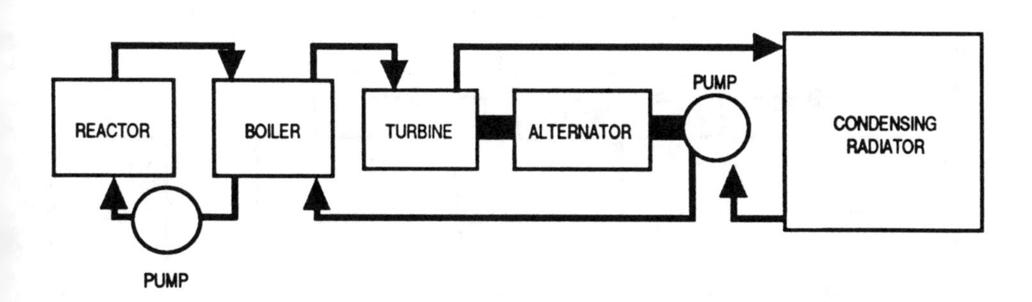

FIGURE 2. Schematic Diagram of SNAP-2 Mercury Rankine System

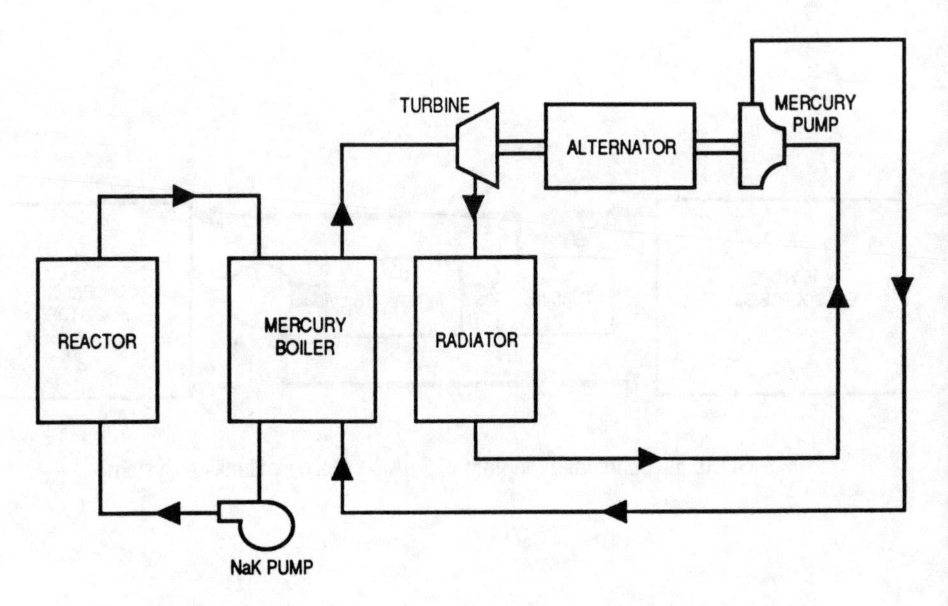

FIGURE 3. Two-Loop SNAP-8 Power Conversion System Schematic.

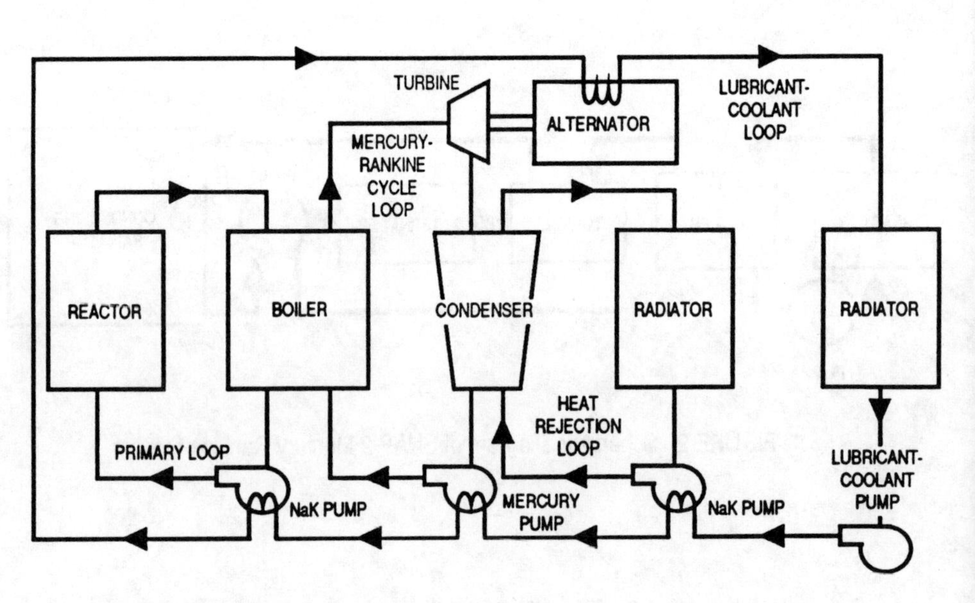

FIGURE 4. Four-Loop SNAP-8 Power Conversion System Schematic.

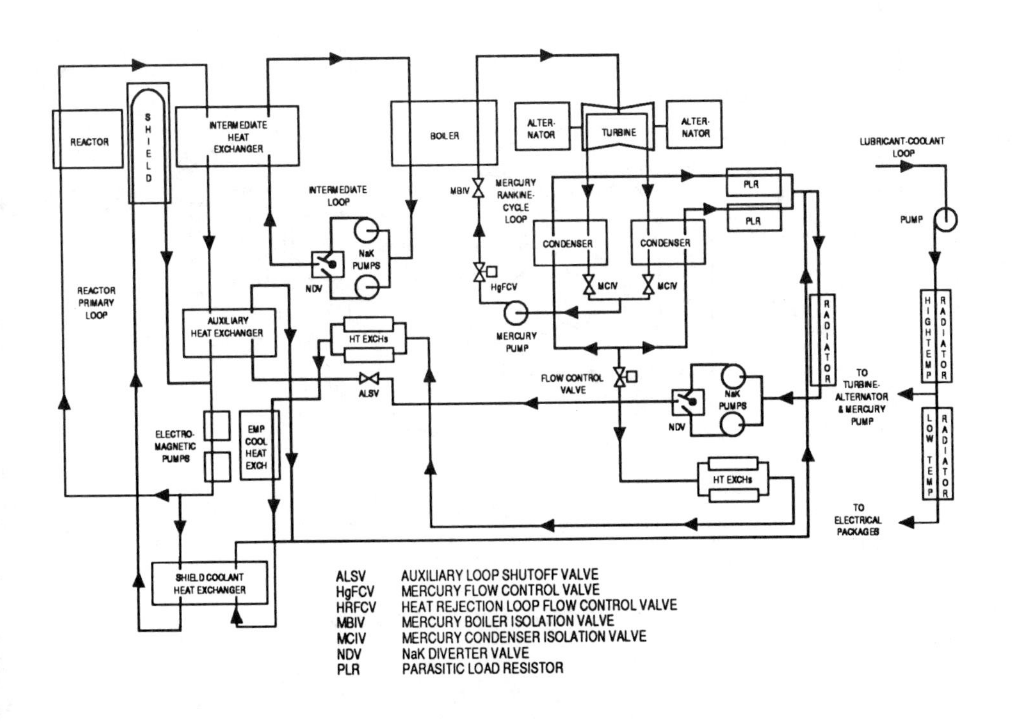

FIGURE 5. 90-kWe Electrical Generating System Schematic.

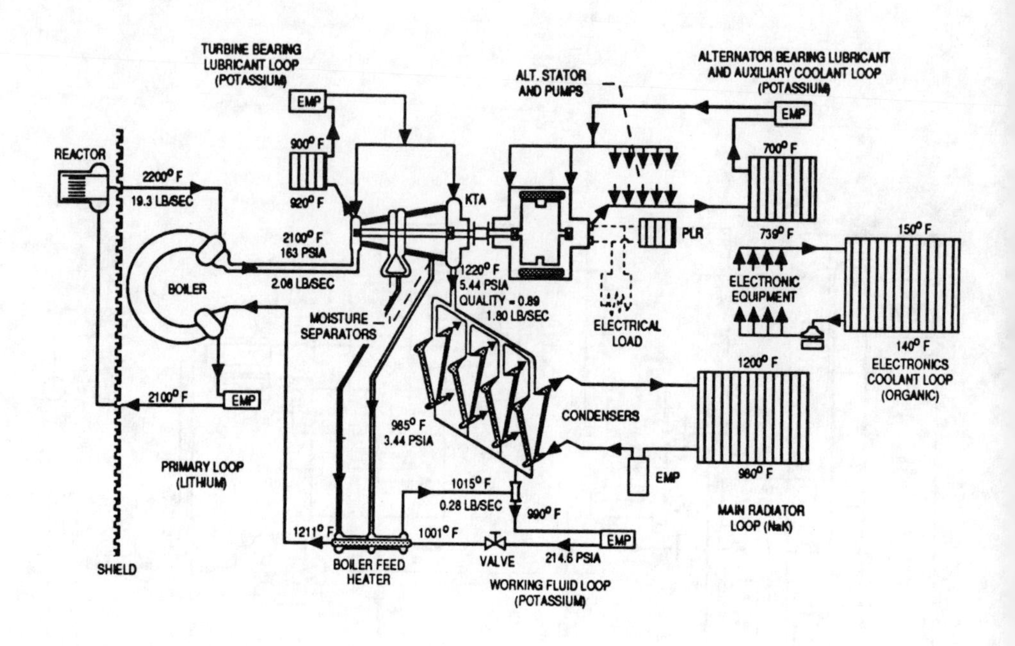

FIGURE 6. Advanced Rankine System Schematic.

TABLE 1. Typical Rankine Cycle Turbine Inlet Conditions

FLUID	APPLICATION	TEMPERATURE		PRESSURE	
		K	(F)	kPa	(psia)
ORGANIC	WASTE HEAT RECOVERY	500	(450)	2200	(320)
	SPACE POWER	650	(700)	2200	(320)
WATER	TERRESTRIAL POWER	800	(1000)	16600	(2400)
MERCURY	SPACE POWER	950	(1250)	1830	(265)
POTASSIUM	SPACE POWER	1420	(2100)	1120	(160)

SECTION III: BRAYTON CYCLE

INTRODUCTION

The closed Brayton cycle (CBC) is well suited to the task of converting thermal energy from isotopes or nuclear reactors to electrical power. CBC has an extensive development history and has had significant operational ground test success. It is the most mature of the technologies available for application to dynamic space power systems.

This article reviews the salient features of the CBC space power system including its general characteristics, history, current state of technology, and research and technology issues.

GENERAL CHARACTERISTICS

Cycle Description

The CBC temperature entropy diagram is shown in Figure 1. Energy from the heat source heats the gaseous working fluid to the maximum cycle temperature. The high-temperature gas is expanded through a turbine to produce useful output power and drive the compressor. The turbine discharge gas is cooled, first in the recuperator heat exchanger and then the heat rejection heat exchanger (HRHX), before being introduced into the compressor. Following compression, the gas is heated at constant pressure in the recuperator before being directed to the heat source, thus completing the closed loop.

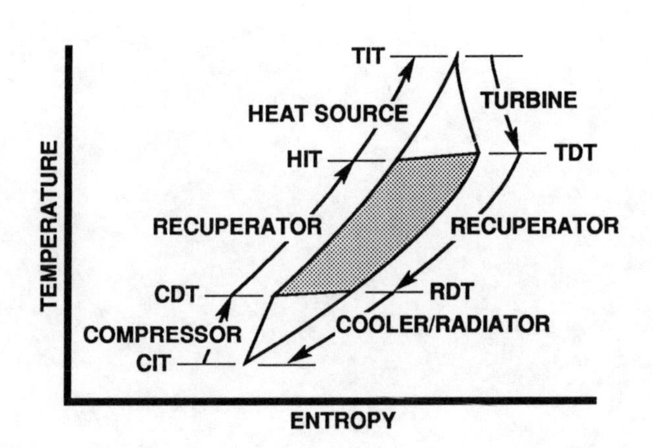

FIGURE 1. CBC Temperature Entropy Diagram

Recuperation is a key feature of the closed Brayton cycle. The recuperator increases efficiency by recovering a significant portion of the turbine waste heat for preheating the compressor discharge gas prior. This reduces the heat that must be added by the heat source. Because of this, recuperated cycles optimize at relatively high turbine exit temperatures, reducing the turbine pressure ratio and therefore the overall cycle pressure ratio. A low-pressure ratio has the benefit of reducing the complexity of the turbomachinery.

Any gas or gas mixture can be used for the cycle working fluid. However, the long life and cleanliness requirements of space power systems benefit from the use of inert gas to eliminate corrosion and corrosion by-products. For space power CBC, a mixture of helium and xenon gasses has most commonly been used.

The effective molecular weight of the inert gas working fluid has a strong effect on the optimized cycle parameters and system mass. A system operating with helium (MW of 4) would have optimum heat transfer properties and thus the lightest heat exchangers. However, because of its low molecular weight, helium gives the maximum number of turbomachinery stages and the largest rotating component mass for a given power level. Thus, system mass is not optimum with pure helium. When xenon (MW of 131) is used to reduce the mass of the turbomachinery, its poor heat transfer characteristics increase the size of the heat exchangers. Again, system mass is not optimized.

When the heat exchangers and turbomachinery are considered as a system, the optimum working fluid molecular weight is normally somewhere in between pure helium and pure xenon. A mixture of these two gasses to a specific MW optimizes to a lower mass system than the use of a single pure gas at the same MW, because of the superior heat transfer properties of this mixture over pure monatomic gases.

Analysis has shown that in a direct-cooled gas reactor, xenon coolant will become activated. No macroscopic changes in the coolant's physical properties occur from this process. The main result will be the presence of radioactive sources in the CBC loop. This may or may not be a concern depending on mission profile. Other mixtures can be used (for example, He/Kr) with little impact on performance if the activation of cycle gas is not allowed.

System Architecture

Figure 2 represents the basic configuration of a space nuclear reactor using CBC power conversion. The diagram given is for a liquid metal cooled core.

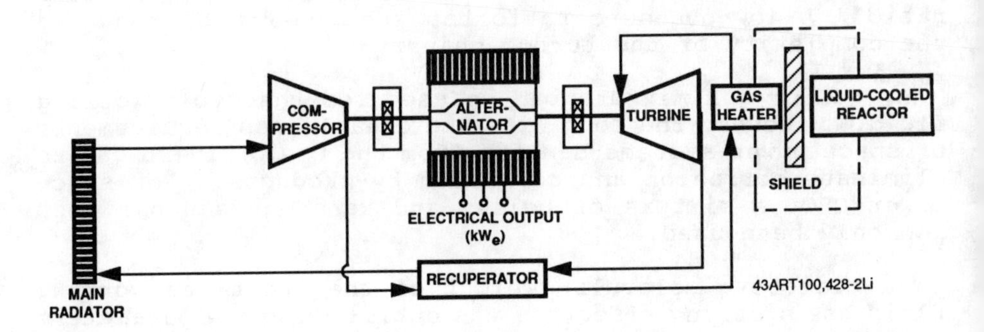

FIGURE 2. Schematic Diagram of Nuclear CBC Power Conversion System.

The inert gas working fluid means that a direct gas-cooled reactor is feasible. Such a system would utilize the CBC working fluid for core cooling and eliminate the necessity for the heat source liquid-to-gas heat exchanger and reactor primary loop coolant circulators. The direct gas-cooled reactor architecture gives the lightest weight for a nuclear reactor CBC system. For adaptation to a radioisotope heat source, the cycle gas flows directly through a helical tube heat source heat exchanger surrounding the isotope.

The basic schematic given is applicable to designs from the sub-kilowatt to several hundred kilowatt range; however, the individual component technologies will change for different operational power ranges. At lower power levels, (hundreds of kilowatts and below) single-stage radial turbomachinery is used. Radial turbomachinery has limited scalability due to the weight of large rotors and multiple stages. At higher power levels, axial compressors and turbines give the lowest weight and highest efficiency systems. Gas-bearing systems shift from compliant to solid geometry at power levels greater than approximately 100 kilowatts. The single-shaft turboalternator-compressor (TAC) arrangement shown for lower power levels is not practical with higher power axial machinery. In these systems, a turbocompressor will drive a separate alternator shaft.

Design and Operational Flexibility

The CBC system is adaptable to a variety of system requirements and affords a great deal of flexibility in design and operation. The single-phase gas working fluid provides a simple interface that is adaptable to any type of heat source (radioisotope, gas-cooled, and liquid-cooled reactor).

Because the cycle uses circulating gas as the working fluid, there is a wide range of heat source pressure, flow and temperature characteristics with which the system can be integrated. This capability is aided by the absence of a working fluid phase change. Because there is no phase change, the cycle is insensitive to gravitational conditions, allowing equal performance in a surface application as in a microgravity or zero-g environment.

CBC power system operational flexibility is provided by good off-design capabilities. The system could be required to operate off-design due to environmental influence factors such as heat sink temperature or due to changing user load requirements. Obviously each off-design condition has a different effect on the cycle; however, the design features of the system enable it to respond well to changes. In addition, the effects of off-design conditions can be reliably predicted and reactive measures such as changes in speed, flow and inventory (amount of gas in the loop) can be taken if necessary.

Power turndown ratio is often a key design driver for a power conversion system. For space reactor operation, significant peak power requirements are often specified above steady state operation. The typical CBC system can achieve effective turndown ratios greater than 6:1 with minimal effect on cycle efficiency.

System Optimization

System and resultant component configurations can be optimized using several independent cycle parameters to satisfy a wide variety of mission drivers. Cycle thermodynamic statepoints are tailored within a wide range of possibilities to meet the system requirements. The determination of the necessary performance characteristics and physical configuration of each component is an integral part of this process. For example, compressor inlet temperature (CIT), the lowest temperature in the cycle, is lowered to improve cycle efficiency. However, to reduce

CIT, radiator size must increase. The satisfaction of relevant mission design drivers will dictate the optimum CIT.

CBC power systems for nuclear reactors are generally optimized to provide minimum system mass. Conversion efficiencies range from 12 to 50 percent depending on the application and system architecture. Manned systems with low allowed radiation levels demand high efficiencies, which are achieved by recuperated systems with relatively large radiators. The higher mass of these components necessary to gain extremely high efficiency is offset by a reduction in overall reactor power level, reactor mass, and shielding mass.

Alternatively, unmanned nuclear systems with limited shielding requirements do not demand high efficiencies and will have the lowest mass when unrecuperated. An unrecuperated cycle will reject heat at elevated temperatures (turbine exit), requiring less radiator area for a given heat flux.

A CBC system design can be optimized relative to any parameter that is key to a given mission. Thus, systems can be designed for minimum mass, minimum radiator area, maximum efficiency, or any weighted combination of these. The major independent cycle parameters selected during the optimization process are

- Net output power,
- Compressor inlet temperature (CIT),
- Turbine inlet temperature (TIT),
- Recuperator effectiveness (E_r),
- Compressor pressure ratio (CPR),
- Pressure loss parameter (β),
- Engine shaft speed (N),
- Compressor-specific speed (N_s),
- Working fluid, and
- Molecular weight of working fluid.

As discussed, mission requirements drive the optimized system. This is illustrated by comparing 100 kWe nuclear power system for fixed base and a rover application (Baggenstoss and Ashe 1991). In this study, an SP-100 type liquid metal reactor is mated with a fully redundant CBC power conversion system with an assumed turbine inlet temperature of 1250 K.

The two systems are governed by different mission design drivers. The rover mission requires a man-rated shield and a one-sided radiator which is limited to 100 m^2. For a fixed base, only an instrument-rated shield is necessary and a double sided radiator can be utilized. Although each system was optimized for minimum mass, the different design drivers for the fixed base and rover missions give dramatic differences in system parameters.

The results of the mass optimization are given in Table 1. The large shield of the rover system drives the system to optimize at a very high conversion efficiency of 37 percent. Contrast that to the fixed base, where the actual CBC hardware constitutes a greater percentage of system mass. In this case, the components are sized to yield a moderate efficiency of 23 percent. Figure 3 graphically compares the differences in system mass fractions between the two missions.

<u>Reliability</u>

Space power systems have to meet the high reliability requirements of space missions. Although there are failure rates associated with direct energy conversion systems, very few, if any, involve moving parts. Because CBC is dynamic, the reliability of a system with moving parts must be carefully evaluated.

Because CBC components are derivatives of aircraft open-cycle gas turbine machinery, actual operational failure data is readily available. An important factor to note is that in a comparison between the open-cycle aircraft and the closed-cycle space system operating environment, the closed-cycle environment is extremely benign. An example of this is the foil bearing and turbine.

Foil gas bearings are currently employed in 45 different aircraft environmental control unit applications. Over 14,000 production units are in service and over 250 million operating hours have been accumulated. The number of operational failures that have occurred on these production units is the result of foreign object damage (ice crystals or dirt as a consequence of the open cycle operation) causing damage to the foil coating, which is used for lubrication during startup. With particulate matter in the bleed air of these open cycle machines, bearing life maybe reduced to 5000 to 10,000 test/stop cycles. Considering the above operational data with respect to the high gas purity and the low number of start/stop cycles (on the order of 100) of space system operation, failure modes revealed in the operational database are eliminated. This is also the case for turbine failure modes in which aircraft turbine failures are primarily due to corrosion and low cycle fatigue (start-stop cycling). In the inert operating environment of CBC systems these modes are eliminated.

TABLE 1. 100 kWe Nuclear CBC System Optimization Results.

	Case	
	1	2
Sink temperature, K	194	194
Radiation dose	Man	Inst
Total mass, kg	18,116	2,256
Shield mass, kg	13,173	391
Radiator area, m^2	100.0	174.9
Compressor inlet temperature, K	383	401
Efficiency	0.374	0.227
Recuperator effectiveness	0.969	0.750
Pressure loss ratio	0.970	0.940
Compressor pressure ratio	1.71	2.03
Rotor speed, rpm	26,662	28,207

Open cycle failure data was used on the Dynamic Isotope Power System (DIPS) program to conduct a detailed reliability study of the CBC power system. DIPS had a 10-year life requirement with a specified reliability (nonfailure probability) of 0.9950. Actual failure rate data was obtained, then derated based on the differences in operating conditions and environment between the open cycle and closed cycle machinery. This derating was done in a conservative manner based on reduced operating temperatures, inert environment, reduced stress levels, and reduced operational and thermal cycling requirements. This failure data was then used in a CBC system reliability model that is capable of evaluating a variety of CBC configurations.

Table 2 gives reliability results for several levels of system redundancy. In DIPS program terminology, the PCU consists of the CBC power conversion mechanical hardware. The PCCU consists of the electronics that conditions the electrical output and controls PCU operation. It is interesting to note that electronics, necessary for all space power systems, are the limiting factor in the reliability model. Thus, more redundancy is necessary in the PCCU than in the mechanical systems to achieve the required overall system reliability.

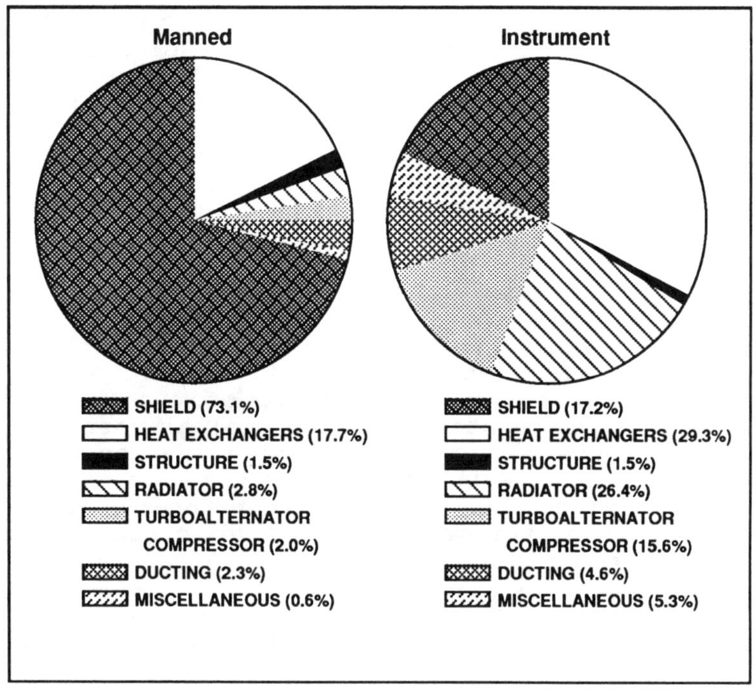

FIGURE 3. 100 kWe Nuclear CBC Mass Comparison for Instrument and Man-Rated Shielding.

TABLE 2. DIPS Reliability.

DESCRIPTION	RELIABILITY
Mechanical Components, PCU	0.946
Electronic Components, PCCU	0.789
System - 1 PCU, 2 PCCU	0.920
2 PCU, 2 PCCU	0.965
2 PCU, 3 PCCU	0.993
2 PCU, 4 PCCU	0.9985

HISTORY

The first operational closed Brayton gas turbine cycle power plant was a 2 MWe fossil fuel demonstration plant completed in 1939. This plant was constructed by Escher Wyss of Switzerland and operated at 32 percent conversion efficiency with a TIT of 973 K. Following the success of this plant, approximately 20 others were built in Europe and Japan throughout the 1950s and 1960s. The most recent CBC plant was a 5 MWe cogeneration plant operated in Torrance, California, until 1988.

Operationally these plants were very successful. However, there are no CBC systems in power plant service today. The economics of fossil-fueled CBC power plants are not competitive with open cycle gas turbines. There are two main reasons for the predominance of open cycle plants. CBC plants have relatively large capital costs due to the number and size of components, including the heat source, ducting, and heat exchangers. Also, CBC cannot as easily achieve the higher turbine inlet temperatures made possible in open cycle turbines by active cooling techniques developed by the aircraft industry. As a result, although the open cycle power plants do not achieve the efficiency levels possible with CBC systems, their overall cost is lower due to the availability of relatively inexpensive fuel.

The CBC power plants discussed so far were all fossil fueled. The U.S. Army built a 330 kWe gas-cooled nuclear reactor with CBC power conversion in the early 1960s (McDonald 1992). This plant, designated ML-1, was water moderated and nitrogen cooled. It is the only nuclear CBC system to have operated to date. Although the project was technically a success, it was terminated in 1965 for programmatic reasons.

Because the need for a high efficiency scalable power conversion system was recognized early in planning for the nation's future role in space, CBC systems began receiving industry and government development funding in the early 1960s. CBC space power study and demonstration programs have focused on the utilization of nuclear reactor, radioisotope, and solar heat sources.

The 3 kWe Brayton Cycle Demonstrator (BCD) program was initiated by Garrett, now a part of Allied-Signal Aerospace, in 1962. This unit was tested subsequently by NASA with an electrical resistance heater to simulate a radioisotope heat source. The working fluid was argon gas. The compressor and turbine were scaled from an aircraft auxiliary power unit. The heat exchangers were manufactured using the plate-fin concept, which subsequently found extensive use in the aircraft industry. The alternator was adapted from a three-phase induction motor, and the rotor was supported on wick-lubricated ball bearings.

Following successful demonstration of the system, the machine was upgraded to a Rice alternator and foil gas bearings. These improvements eliminated oil lubrication requirements and all mechanical contact with the rotating shaft during operation. Although testing on the BCD was completed at NASA in the 1960s, the unit has recently been restored and is again operational.

In 1963, NASA initiated funding of development work on a 10 kWe solar powered system designated Engine A. This was a recuperated dual shaft engine with a turbocompressor gas generator and a separate, turbine-driven alternator. The first use of tilting pad solid geometry gas bearings in a CBC engine was demonstrated here. Aerodynamic component evaluation determined that demonstrated radial aerodynamic components were optimum but that the dual-shaft arrangement should be changed to an integrated single-shaft configuration.

On the basis of the Engine A experience, NASA embarked on Engine B, later designated the Brayton Rotating Unit (BRU). The BRU was also a 10 kWe unit but was configured similar to the earlier BCD. It had a single shaft consisting of the rotor of a brushless Rice alternator mounted between a centrifugal compressor and a radial inflow turbine supported on tilting pad gas bearings (see Figure 4). Shaft speed was 36,000 rpm.

This was a highly successful program culminating in 52,000 hours of closed loop system operation. A significant portion of this testing was accomplished in a vacuum chamber (see Figure 5). These hours were accumulated on two units, one of which ran for 41,000 hours. There was no determinable degradation in performance or mechanical condition of the turboalternator-compressor. The long life capability of CBC dynamic power conversion for space systems was demonstrated. A BRU unit was later modified to operate on foil bearings and the alternator upgraded to an output power rating of 15 kWe. This unit was designated the BRU-F. An air-driven simulator version of the BRU-F was used as a test bed in the late 1980s for Space Station Freedom (SSF) Solar Dynamic Power Module development testing.

In 1972, development was begun of a 100 kWe CBC system to be powered by a SNAP 8 (Systems for Nuclear Auxiliary

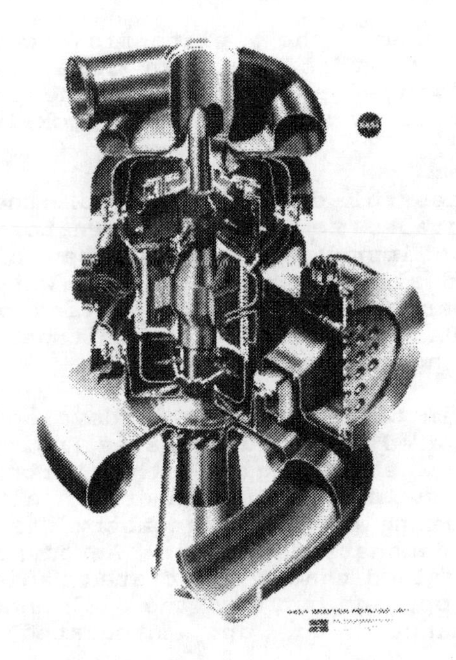

FIGURE 4. Brayton Rotating Unit.

FIGURE 5. BRU Vacuum Chamber Test Configuration.

Power) nuclear reactor. This program was halted during the component design phase due to a presidential decision to halt work on nuclear systems for space. However, an important demonstration was made of foil gas bearing operation at 24,000 rpm with a full-scale 100 kWe rotor. The Brayton Isotope Power System (BIPS) was designed as a radioisotope power source for military surveillance satellites. This project was jointly funded by NASA and DOE from 1972 to 1978. The turboalternator-compressor, designated the Mini-BRU, was designed for variable power output up from 500 to 2100 We. The Mini-BRU was a single-shaft design with a Rice alternator and a high-efficiency, backward-curved radial compressor and a radial turbine, both with splitter vanes. The shaft speed was 52,000 rpm, and the shaft was supported on foil gas bearings.

The BIPS was a complete 1.3 kWe power conversion loop utilizing the Mini-BRU as the heart. To reduce the amount of isotope needed in the heat source, the BIPS was designed for high conversion efficiency. Significant development effort was expended to optimize the efficiency of the compressor, turbine and recuperator. As a result, the compressor and turbine achieved efficiencies of 0.77 and 0.83, respectively, which are exceptionally high considering the diminutive size of the machinery. The recuperator effectiveness was demonstrated at 0.98.

The successful BIPS test program operated a laboratory configured power conversion loop in a vacuum chamber (see Figure 6). This workhorse loop simulated the radioisotope heat source with electric heaters and the radiator/cooler with a laboratory heat exchanger. This unit accumulated 1000 hours of endurance testing, achieving an overall power conversion efficiency of 28 percent. To reach even higher efficiencies, a refractory metal hot section of the Mini-BRU was designed to allow higher TIT of 1144 K and still meet the turbine static structure creep life requirements. The high temperature version was designed and fabricated but never tested due to program cancellation.

During the late 80s, work began on a 25 kW_e CBC Solar Dynamic Power Module to provide growth power capability for the NASA Space Station Freedom. The system consisted of a solar receiver, which accepted sunlight from a multifaceted collector (the concentrator), a CBC heat engine that converted the solar heat to electricity, and a radiator to radiate the system waste heat to space. A key design aspect of the SSF system was the thermal energy storage (TES) capability. Canisters containing an eutectic mixture of calcium and lithium fluoride salts were integrated into the receiver/heat source heat exchanger. Thermal energy was stored for the eclipse part of LEO by melting the salts during the sunlight periods and using the heat of fusion to run the cycle during the eclipse. Work on the CBC Dynamic Power Subsystem had progressed to the preliminary design phase when it was dropped from the Space Station Program due to budget constraints in 1991.

FIGURE 6. Brayton Isotope Power System Workhorse Loop.

The DIPS program, being conducted under DoE contract, is intended to develop the technology for dynamic isotope power systems in the 1-to-10 kW_e class for emerging civilian and military space missions. The DIPS technology has been identified by NASA as a key technology for the space exploration initiative (SEI) surface power applications. The current focus of the DIPS program is on a 2.5 kW_e standardized power module for use in a man-tended environment.

Ground testing of an integrated solar dynamic power system in a simulated space environment is the logical next step toward demonstrating the availability of solar dynamic technology for application to Space Station Freedom (SSF) and other space platforms for future missions. The Solar Dynamic Ground Test Demonstration (SDGTD) Program being conducted for the NASA Lewis Research Center is taking that next step. In order to minimize cost, schedule and technical risk, the SDGTD Program will utilize a combination of designs from the SSF program and existing hardware from the BIPS Program.

CURRENT STATE OF THE ART

CBC is a mature technology which utilizes proven design concepts and technology from an extremely broad base open-cycle gas turbine engines. From this open Brayton cycle technology base come proven materials, and mechanical and electrical components with millions of hours of accumulated operation.

Turboalternator-Compressor

The turboalternator-compressor (TAC) arrangement has been used on all space-configured CBC systems built to date. The CBC radial turbine and compressor are mounted on a single rotating shaft which also includes the alternator rotor. Turbine shaft power is split between the compressor and the alternator. This high-speed shaft is the only rotating component in the power convertor. Because the shaft is supported on gas bearings and the alternator is brushless, metal-to-metal contact during operation is eliminated.

Within the TAC, the alternator and gas bearings require cooling flows. Compressor discharge bleed is used to cool the bearings, whereas the alternator can either be cooled with compressor discharge gas directly or with a separate liquid loop, depending on operating conditions.

Alternator

All space-configured CBC systems to date have used the Rice (modified Lundell) alternator. The Rice is a brushless, nonrotating coil synchronous alternator. The stator has a conventional three-phase winding. To improve windage losses, a solid rotor is used. The magnetically conductive portions of the rotor are separated by a nonmagnetic material which gives structural strength, pole separation, windage loss reduction and electromagnetic damping. The field excitation coil is stationary and located in the frame of the machine. The field flux is carried to the rotor through two auxiliary air gaps at each end of the rotor. Rotating salient poles are created by preferential magnetic conduction of the field flux from the auxiliary to the main air gaps by the steel portions of the rotor. By designing fully redundant field coils, the brushless field circuit meets high reliability requirements of space systems.

Advances made in the 1970s and 1980s in rare earth permanent magnet technology have made the use of permanent magnet generators (PMG) in CBC systems competitive. Use of a permanent magnet generated field allows the elimination of field windings and much of the stator iron. At power levels above 100 kWe, this weight advantage is considerable and the Rice alternator is not competitive with the PMG. PMG technology will be available for future CBC space power systems.

Compressor/Turbine

The CBC aerodynamic components are derived from well characterized open cycle gas turbine engine designs. Closed cycle turbomachinery operates well within open cycle proven aerodynamic and structural performance parameters and entails low technical risk in development.

All space-configured CBC systems to date have used compact, single-stage radial compressors and turbines. Figure 7 shows several examples of radial compressors and turbines and achieved performance.

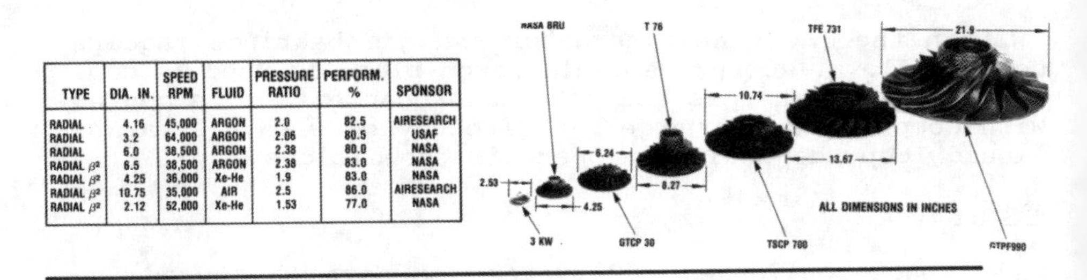

TYPE	DIA. IN.	SPEED RPM	FLUID	PRESSURE RATIO	PERFORM. %	SPONSOR
RADIAL	4.16	45,000	ARGON	2.0	82.5	AIRESEARCH
RADIAL	3.2	64,000	ARGON	2.06	80.5	USAF
RADIAL	6.0	38,500	ARGON	2.38	80.0	NASA
RADIAL β^2	6.5	38,500	ARGON	2.38	83.0	NASA
RADIAL β^2	4.25	36,000	Xe-He	1.9	83.0	NASA
RADIAL β^2	10.75	35,000	AIR	2.5	86.0	AIRESEARCH
RADIAL β^2	2.12	52,000	Xe-He	1.53	77.0	NASA

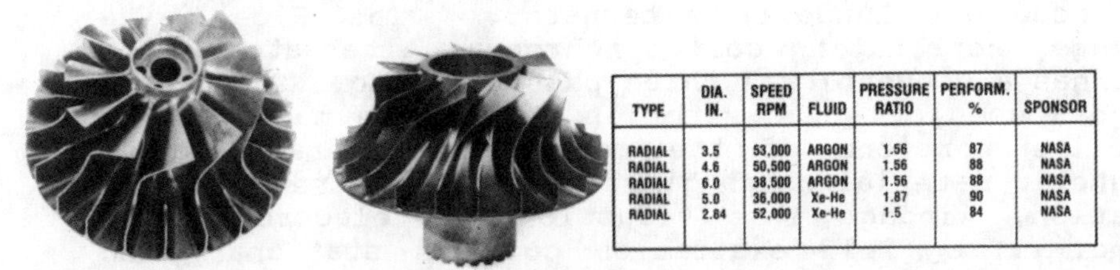

TYPE	DIA. IN.	SPEED RPM	FLUID	PRESSURE RATIO	PERFORM. %	SPONSOR
RADIAL	3.5	53,000	ARGON	1.56	87	NASA
RADIAL	4.6	50,500	ARGON	1.56	88	NASA
RADIAL	6.0	38,500	ARGON	1.56	88	NASA
RADIAL	5.0	36,000	Xe-He	1.87	90	NASA
RADIAL	2.84	52,000	Xe-He	1.5	84	NASA

FIGURE 7. Compressors and Turbine.

With the larger mass flow requirements of higher power level systems, the size and weight of the radial flow turbomachinery increases. Somewhere between 0.5 and

1.0 MW, depending on system parameters, a crossover point exists in which axial flow devices are lighter than radial aerodynamic components. At power levels greater than about 1 MW, nuclear Brayton systems will employ a multistage axial flow compressor and turbine.

Gas Bearings

Up to approximately 100 kWe, self-actuated, compliant foil gas bearings (Figure 8) are used to support the TAC rotor. While at rest, the foil segments are in contact with the thrust runner or shaft. The foils are formed with a specific radius of curvature and thickness, and the moving element is centralized in the housing. When the machine is started, the foils remain in contact with the moving surface until the boundary layer forms and the journal becomes airborne.

The most significant advantage offered by gas bearings over conventional journal or rolling element bearings is the elimination of a separate lubricant system. Working fluid supplied from the compressor discharge is used as the bearing lubricant and coolant. Thus, gas bearings require no seals to separate lubricant from working fluid and no lubricant pump.

The bearing loads imposed by the larger rotors of 100-plus kilowatt machines result in unacceptable length-to-diameter ratios for foil bearings. As a result, higher power machines use solid geometry gas bearings. These sophisticated, close-tolerance bearings have more load carrying capability than foil bearings at the cost of greater system complexity. This type of hydrodynamic gas bearing was developed and extensively tested for CBC application during the 1960s and 1970s. This effort resulted in a gas bearing system design utilizing tilting pad journal bearings and a Raleigh stepped thrust bearing.

Heat Exchangers

The CBC recuperator/cooler technology is also a mature technology available to the system. This technology was demonstrated on past CBC development programs and is founded on heat-transfer equipment used in commercial gas turbine pipelines, refinery applications, and thousands of aircraft and high-performance automotive applications.

CBC heat exchangers are of the plate-fin design. The plate-fin design is a brazed assembly of stacked, alternating low and high pressure counterflow panels. The BIPS recuperator, Figure 9, is an example of a high effectiveness CBC recuperator. In test, the BIPS recuperator operated at an effectiveness of 0.975. This unit was subjected to an extensive accelerated life test. Two hundred thermal shock cycles, equivalent to thousands of actual start cycles, were run with no appreciable increase in measurable internal or external leakage.

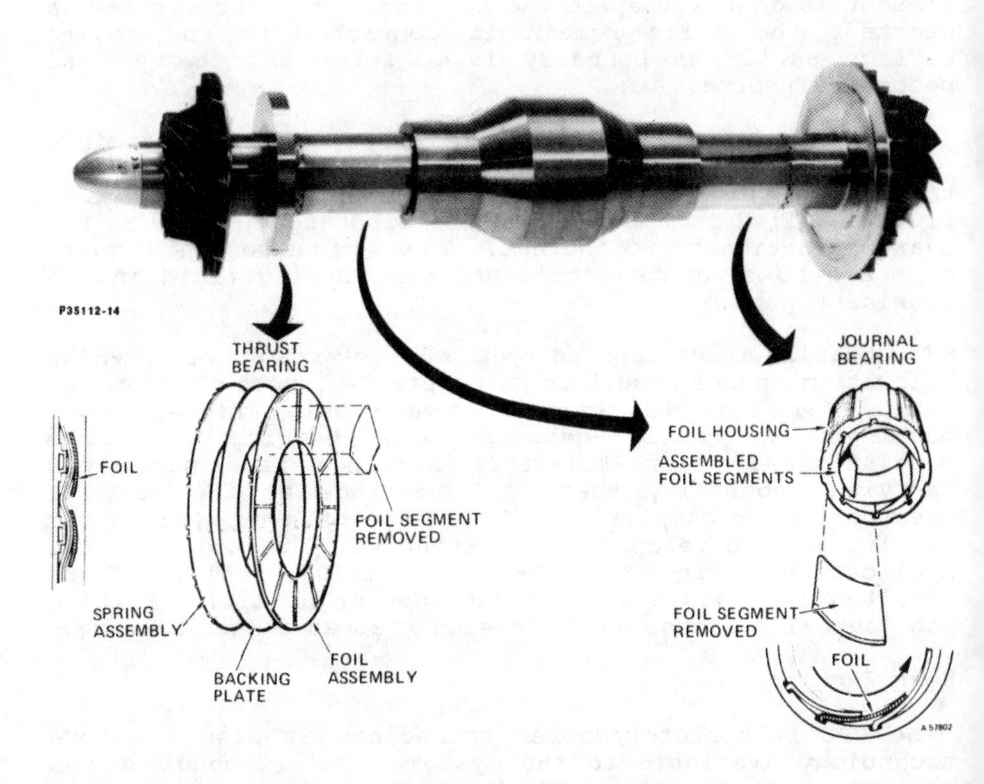

FIGURE 8. Foil Gas Bearing.

FIGURE 9. CBC Recuperator/Cooler.

FUTURE DEVELOPMENTS

Although closed Brayton cycle dynamic power conversion is a well-developed and proven technology, there are several avenues of research which can lead to improvements in system weight and efficiency and a more detailed understanding of system behavior under diverse operating conditions.

TIT Improvements

For a given heat sink temperature, as the heat source temperature is increased, ideal Carnot conversion efficiency goes up: $\eta_T=1-(T_L/T_H)$. An increase in conversion efficiency allows a proportional reduction in reactor thermal power capability and shielding thickness. Less heat rejection for a more efficient system also requires a smaller radiator.

The CBC space power system has undergone significant test and development throughout the past 25 years. These systems are based upon well-proven superalloy metal technology that limits turbine inlet temperatures to approximately 1150 K. Dramatic improvements can be attained in power system specific mass and specific radiator area by increasing TIT above the limits of current technology. Recent developments in gas turbine and space technology in high-temperature refractory metals, ceramics, and composites provide the capability of increasing the CBC TIT.

Figure 10 shows the relative trends of mass and required radiator area for increasing reactor exit temperature. TIT is equal to reactor exit temperature minus duct temperature losses. Temperature breakpoints for component material selection are also shown in the figure. As component operating temperatures are increased, materials technology becomes increasingly advanced. Acceptable materials technology temperature ranges are given assuming typical space reactor system reliability and life requirements.

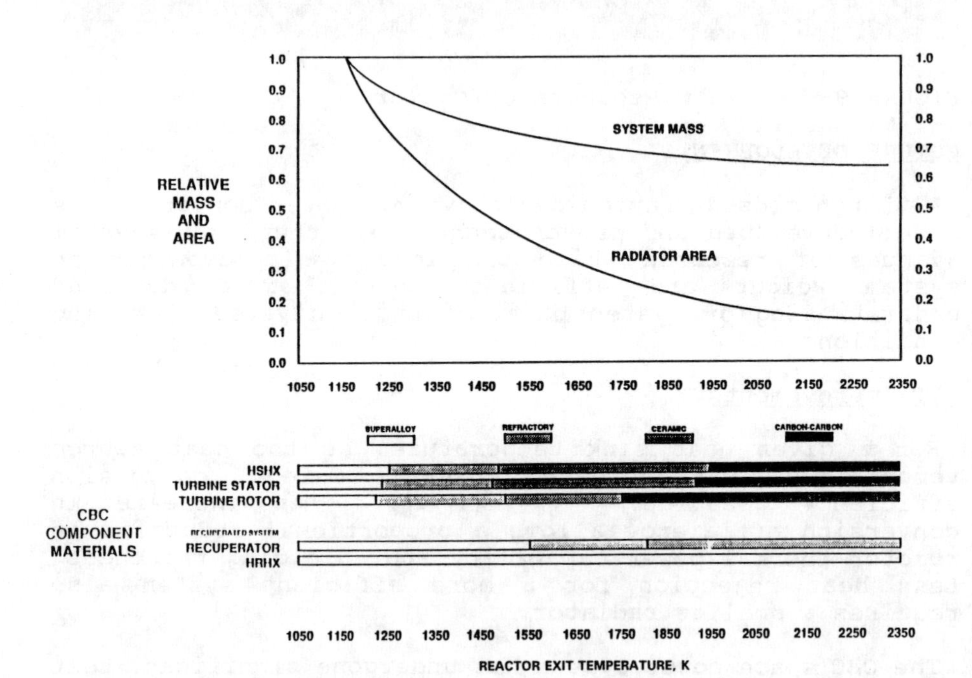

FIGURE 10. CBC Component Materials.

The use of refractory metals in terrestrial applications is difficult due to oxidation problems. The inert gas working fluid of the CBC means that the high temperature capabilities can be taken advantage of in space.

In recent years, significant progress has been made in applying ceramic technology to the gas turbine application. High-strength and high-toughness ceramics such as silicon nitride and silicon carbide, have been used with considerable success in demonstration engines. Future challenges involve the practical application of ceramics and ceramic composites to long-life, high-reliability space applications.

High-temperature carbon composites offer the highest strength-to-weight ratio of any material at temperatures above 1000 K. Although applications in an oxidizing atmosphere have hinged on the development of a workable coating, carbon-carbon use on the CBC inert environment allows the potential elimination of coating development issues.

Active Magnetic Bearings

The practical application of active magnetic bearing technology is in its infancy. However, the basic principles are well understood and have been demonstrated in operation on rotating machinery. An electromagnetic stator produces a rotating attractive magnetic field which levitates a ferritic rotor. Active electronics sense rotor position and control the magnetic field to compensate for varying loads and speed.

Advantages shared with gas bearings include no lubrication requirements and no mechanical contact during operation. Magnetic bearings have lower drag losses, but these may be offset by electrical power requirements. Load capacity of these bearings is variable and independent of speed, temperature, and fluid characteristics.

Magnetic-bearing research has proven their viability in high-speed rotating equipment. They have also found flight applications in control moment gyros in which their absence of drag is an important advantage. Future research towards applying magnetic bearings to CBC systems should focus on weight and envelope reductions and controls reliability.

Megawatt CBC Systems

The experience base gained over the last 30 years on CBC space power technology has been created on units of 100 kWe and under. Although CBC fossil-fueled power plants have operated at several megawatts, the scalability of a space-configured CBC system to the higher power ranges has some technology issues remaining to be fully investigated.

Large-scale CBC terrestrial power plants have utilized conventional bearing technology. Large scale CBC space power systems will utilize gas or magnetic bearings. Gas bearing technology in the megawatt class is conceptually sound, but the determination of detailed design variants required to meet system requirements will require development effort.

An additional development issue involved in the design of multimegawatt CBC power systems is the optimum rotor arrangement. The optimum rotor arrangement for the small CBC systems to date has been a single rotor with the brushless alternator rotor straddled between the bearings. With the use of axial flow turbomachinery in the megawatt range, the optimum rotor configuration may consist of separate alternator and turbocompressor rotors. In this arrangement, the development issues of a gas-bearing-mounted alternator or a rotating shaft seal must be addressed.

CONCLUSIONS

The closed Brayton cycle has the characteristics and technology readiness necessary for use as the power conversion system for space nuclear power applications. Desirable characteristics include low power-to-weight ratio, high conversion efficiency, operational flexibility, and reliability. The maturity of the component technologies has been demonstrated conclusively in open cycle aircraft and commercial applications as well as in a number of successful closed cycle systems. Significant government and industrial funding has been dedicated in the last 30 years to prove the viability of CBC technology using solar, isotope and nuclear heat sources.

Review of this body of knowledge pertaining to the suitability of use of CBC in space reactor systems leads to the conclusion that future nuclear and radioisotope space power systems can gain advantageous operating capabilities by the incorporation of closed Brayton cycle power conversion.

Acknowledgments

The writing of this article was sponsored by Garrett Fluid Systems Division, Allied-Signal Aerospace Company.

REFERENCES

Allied-Signal (1989) "Closed Brayton Cycle Technology and Experience," *41-8550*, Allied-Signal Aerospace Company, Garrett Fluid Systems Division, Tempe, AZ.

Allied-Signal (1991) "Space Power 1991 and Beyond," *41-9270A*, Allied-Signal Aerospace Company, Garrett Fluid Systems Division, Tempe, AZ.

Amundsen, P. C. and W. B. Harper (1992) "BIPS Turboalternator-Compressor Characteristics and Application to the NASA Solar Dynamic Ground Test Demonstrator Program," in *27th IECEC Proceedings*, San Diego, CA, 5: 239-244

Ashe, T. L., W. G. Baggenstoss, R. Bons (1990) "Nuclear Reactor Closed Brayton Cycle Power Conversion System Optimization Trends for Extra-Terrestrial Applications," in *25th IECEC Proceedings*, Reno, NV, 1:125-134

Baggenstoss, W. G. and Ashe, T. L. (1991) "Mission Design Drivers for Closed Brayton Cycle Space Power Conversion Configuration," *91-GT-139*, International Gas Turbine and Aeroengine Congress and Exposition, Orlando, FL.

Brandes, D. J. (1991) "High Temperature Nuclear CBC Power Conversion System for the Space Exploration Initiative," in *Proc. 8th Symposium on Space Nuclear Power Systems*, CONF-910116, M.S. El-Genk and M.D. Hoover, eds., American Institute of Physics, New York, 2:561-566

Davis, K. A. (1987) "Flexibility of the Closed Brayton Cycle for Space Power," *87-GT-101*, 2: 561, Gas Turbine Conference and Exhibition, Anaheim, CA.

Harper, W. B., A. Pietsch, and W. G. Baggenstoss (1989) "The Future of Closed Brayton Cycle Space Power Systems," *IAF-ICOSP89-5-4*, International Astronautical Federation Space Power Conference, Cleveland, OH.

Harper, W. B., R. V. Boyle and C. T. Kudija (1990) "Solar Dynamic CBC Power for Space Station Freedom," 35th ASME International Gas Turbine Conference and Exhibit, Brussels, Belgium.

McDonald, C. F., (1992), "The Future of the Closed-Cycle Gas Turbine - A Realistic Assessment," *929013*, 3:51, 27th IECEC, August 3-7, 1992, San Diego, CA.

McDonald, C. F., (1988), "Active Magnetic Bearings for Gas Turbomachinery in Closed-Cycle Power Plant Systems," ASME Gas Turbine and Aeroengine Congress, Amsterdam, The Netherlands.

NUCLEAR ELECTRIC PROPULSION: STATUS AND FUTURE

James H. Gilland and Roger M. Myers
Sverdrup Technology, Inc.
NASA Lewis Research Center Group
Brook Park, OH 44142
(216) 977-7426

James S. Sovey
NASA Lewis Research Center
21000 Brookpark Road
Cleveland, OH 44135
(216) 977-7454

John R. Brophy
Jet Propulsion Laboratory
4800 Oak Grove Drive
Pasadena, CA 91103
(818) 354-7765

Abstract

Electric propulsion, and in particular, Nuclear Electric Propulsion (NEP), can enable or enhance advanced space missions of the present and into the future. Mission applications range from near-Earth orbital transfer to robotic and human missions to the Moon and beyond. System designs have ranged from 30 kWe to 200 MWe nuclear power systems in conjunction with arcjet, ion, or magnetoplasmadynamic (MPD) thruster systems. NEP technology is interdisciplinary, spanning nuclear reactors, power conversion, heat rejection, power electronics, and electric propulsion. The technology review portion of this paper focusses upon the propulsion technologies under development in order to avoid repetition with space nuclear power reviews. Challenges and benefits specific to NEP systems and technology development are identified and discussed.

INTRODUCTION

Electric propulsion (EP) systems and technology have followed a steady, continuous path of development and application since the inception of research in the 1950s. Pacing elements have been mission requirements and power system development. Until the 1980s space power and electric propulsion needs were adequately met by solar photovoltaic and/or battery systems and thruster concepts in the sub-kilowatt power range. These solar electric propulsion systems have flown and are continuing to fly on both domestic and international space missions. Chief domestic users of these technologies are the National Aeronautics and Space Administration (NASA), the Department of Defense, and commercial satellite firms. Internationally, the Commonwealth of Independent States (C.I.S.), Japan, Germany, Italy, and England have active programs in electric propulsion technology. Both the U. S. and the C.I.S. also have active programs in space nuclear power systems. The systems, missions, and technologies specific to Nuclear Electric Propulsion (NEP) will be reviewed herein.

Electric propulsion systems utilize electromagnetic fields to provide propulsive capabilities unattainable by existing propulsion technology. The electrical acceleration of gases generates propellant exhaust velocities that are one or even two orders of magnitude higher than existing chemical rockets. Chemical propulsion is capable of exhaust velocities of up to 5 km/s, whereas electric thrusters have achieved values up to 100 km/s. This increased exhaust velocity, or specific impulse, allows for significant reductions in propellant mass requirements, as expressed by the rocket equation:

$$M_f/M_0 = e^{-(\Delta V/c)}$$

where M_f is the final spacecraft mass, M_0 is the initial mass, ΔV is the mission characteristic velocity, and c is the exhaust velocity of the propulsion system.

Electric propulsion systems are commonly referred to as "low thrust" propulsion systems. The more accurate term would be low acceleration, as the vehicle's acceleration is limited by the mass of the on-board power supply that must be carried. For reasonable assumptions of power and propulsion system masses, NEP systems generate accelerations of 10^{-3} to 10^{-4} m/s^2. As a result, electric propulsion systems operate continuously over a significant portion of the mission time. At one extreme, operation in relatively high planetary gravity fields results in long mission times, continuous operation, and spiral trajectories. At the other extreme, in heliocentric space, electric propulsion trajectories are similar to impulsive trajectories, with the exception that the electric propulsion systems must operate during most of the mission instead of generating impulsive bursts at the beginning and end of a coast period. Traditionally, electric propulsion systems have been stereotyped as requiring trip times longer than high thrust systems for any mission; however, the increase in trip time occurs only for near-planet orbital transfer missions. As will be seen in this review, electric propulsion systems can reduce trip time as well as vehicle mass for interplanetary missions, depending upon the power and propulsion technology used.

NEP possesses additional benefits over solar electric propulsion in that nuclear power systems have a higher power density, with reduced surface area for heat rejection compared to that required by photovoltaic arrays. In addition, the nuclear source allows operation at full power regardless of distance from the sun, providing the capability for effective exploration of the outer planets. For near Earth operations, NEP systems are not subject to shadowing and are much less susceptible to Van Allen radiation belt degradation effects. In terms of power system mass, nuclear power systems provide mass benefits over solar options at power levels of 75 kWe or higher (Jones and Scott-Monck 1984). Space nuclear power is a more immature technology relative to solar power systems, particularly in the United States. A single nuclear power system has been flown in space, the Space Nuclear Auxiliary Power (SNAP) 10A flight in 1968. The SNAP 10A flight also tested electric propulsion in the form of cesium ion thrusters, although results of both the power system and thruster tests were inconclusive (Sovey et al. 1992). The C.I.S. has space experience with several types of thermionic and thermoelectric nuclear power systems at low powers.

MISSION APPLICATIONS

NEP has potential applications over a wide range of missions. These can be grouped as station-keeping and maneuvering, orbital transfer, interplanetary robotic exploration, and interplanetary human exploration.

Station-keeping and Maneuvering

Communications and observation satellites have stringent requirements on attitude control and pointing. In particular, geosynchronous communication satellite lifetime is determined by the satellite's capability to maintain a given orbital position. Current commercial satellite lifetimes are on the order of 7 to 10 years, using chemical monopropellant or bipropellant rockets (Noyret et al. 1990 and Sovey and Pidgeon 1990). The use of high specific impulse electric propulsion promises to dramatically increase the possible lifetime of such satellites without an increase in satellite launch mass. Similarly, the capability to perform orbital maneuvers such as change of inclination is enhanced (Janson et al. 1992 and Pollard et al. 1993). If sufficient onboard power is available, electric propulsion can be utilized with no added power system mass penalty.

Orbital Transfer

The transportation of payloads from a low Earth orbit (LEO) to higher orbits such as geosynchronous orbit (GEO) is currently done using chemical upper stages. The use of high specific impulse NEP systems for such missions could allow either reduced vehicle LEO mass or increased payload delivered to the target orbits, through the reduction in propellant mass. If the payload requires significant amounts of on-board power, then the same power supply can be used for both orbit transfer and satellite operation, further improving the mass savings of NEP. The low thrust orbital transfer mission has a spiral trajectory, with the vehicle incrementally increasing its altitude by increasing orbital velocity. Typically, the optimal thrust vector is tangential to the orbital velocity.

Many studies of NEP orbital transfer vehicles have been performed since the initiation of the SP-100 program. Some missions considered include space based radar deployment and station-keeping, and geostationary platform deployment (Miller et al. 1984). The typical mission involves launch of the payload/NEP upper stage to a low Earth or Nuclear Safe Orbit, where the NEP system is activated to raise the orbit and change inclination as desired. Two options can be considered: that of the reuseable NEP transfer stage which delivers the payload and returns for another, or the integrated NEP upper stage, which remains part of the spacecraft in the target orbit and performs station-keeping and maneuvering functions as well as provides power to the payload.

The prime benefit of NEP for orbital transfer is the reduction in vehicle mass due to decreased propellant requirements. This benefit may be translated to the user in two ways. First, for the same launch vehicle, the NEP stage is capable of delivering a greater payload than the standard chemical system. Secondly, for the same payload, the upper stage payload mass to orbit requriement is decreased, allowing the use of a smaller, less expensive, more easily launched booster. The second option is promising for the deployment of large constellations of satellites. The reduced spacecraft mass can allow for multiple spacecraft on a single launch vehicle, or more rapid launch of smaller launch vehicles. In either case, the total time required for deploying the full constellation is reduced (Deininger and Vondra 1988).

Most NEP orbital transfer studies of the past decade have assumed the development of the SP-100 power system at a range of power levels. Power levels of 100 kWe or less, at specific masses of 30 kg/kWe or higher have traditionally been assumed. Mission requirements have varied depending upon the user, with defense missions emphasizing minimizing trip time and civil or commercial missions emphasizing increased payload or reduced spacecraft mass. Several thruster options have been considered: the arcjet (Deininger and Vondra 1986, Deininger and Vondra 1989, Zafran 1989, and Zafran and Bell 1990), the ion engine (Hardy et al. 1987), and to some extent, the MPD thruster (Rudolph and King 1982 and Auweter-Kurtz and Schrade 1985). The arcjet's relatively low specific impulse (8-15 km/s) makes it most applicable to the shorter trip time mission requirements (Caveny and Vondra 1989 and Caveny and Vondra 1990), while the ion or MPD thruster can operate at specific impulses of 20 to 100 km/s, providing reduced propellant mass at the expense of longer trip times. Recent interest has arisen in the use of Soviet technology (Pollard et al. 1993), such as the TOPAZ II thermionic power system and the Stationary Plasma Thrusters (SPT); however, mission and system planning have not fully resolved the role of these systems in orbital transfer vehicle (OTV) analysis.

Typical results of a LEO-GEO NEP orbital transfer vehicle are shown in Figures 1 and 2 (Deininger and Vondra 1988). Figure 1 shows launch vehicle requirements versus deployed platform mass for a 50 spacecraft constellation using chemical and NEP OTV systems. The use of either arcjet or ion propulsion, in concert with a 100 kWe SP-100 power system, allows reductions in launch vehicle numbers on the order of up to 50% compared to the chemical upper stage option. Figure 2 shows the comcomitant reduction in deployment time for the entire constellation through the use of NEP for the OTV mission.

Interplanetary Robotic Exploration

The NASA Office of Space Science and Applications (OSSA) has identified exploration of the outer planets of our solar system - Jupiter, Saturn, Uranus, Neptune, and Pluto - as a future objective of its program (Solar System Exploration Committee Workshop 1991). Flyby observations of all but Pluto have been achieved using chemical propulsion in conjunction with complex gravity assist maneuvers. The most recent example of this type of mission is the Galileo probe. The goal of this mission is a Jupiter orbiter/probe mission, which utilizes a Venus-EarthEarth-Gravity Assist (VEEGA) in order to reach Jupiter. Most other missions envisioned by OSSA for the outer planets also require complex gravity assist maneuvers. Because these trajectories depend on precise planetary alignments, there are only narrow windows of opportunity for mission launches. Swing-by maneuvers also tend to increase trip time over direct routes. NEP systems are capable of performing a direct trajectory mission for comparable vehicle masses, as well as reduced trip times. Without any need for gravity assists from other planets, the launch window for NEP systems is much broader than that for the chemical systems. An additional benefit is the availability of larger amounts of on-board power for instruments or data transmission. A second class of missions also enabled or enhanced by NEP are the multiple rendezvous and sample return missions, such as the Multiple Main Belt Asteroid Rendezvous and Comet Nucleus Sample Return missions envisioned by NASA. These missions require a great deal of propulsive capability in that they involve interplanetary travel, multiple target transits, and rendezvous maneuvers.

The use of NEP for such demanding missions has been considered and assessed for as long as the mission concepts themselves have been conceived (Nock and Garrison 1982, Garrison and Nock 1982, and Palaszewski and Frisbee 1988). In addition, missions that can only be achieved through the use of NEP have been identified, such as the Thousand Astronomical Unit Probe (Lyman and Reid 1989). Recently, the application of NEP to many of the NASA objectives has been revisited for updated mission timeframes and launch dates (Hack 1990, Yen & Sauer 1991, and Kelley, Boain, and Yen 1992). The use of SP-100 class nuclear power sources, in the 50 to 100 kWe range, in conjunction with near term ion propulsion, was assessed for the following missions:

Neptune Orbiter/Probe
Jupiter Grand Tour
Uranus Orbiter/Probe
Pluto Orbiter/Probe
Multiple Main Belt Asteroid Rendezvous (MMBAR)
Comet Nucleus Sample Return

Of these missions, the Jupiter Grand Tour had previously been identified by OSSA to be enabled by NEP. The Pluto Orbiter mission had not been previously conceived by the science community due to the trip time (40 years) and initial mass penalties imposed by projected launch vehicle and chemical upper stage limitations. An example NEP Jupiter Grand Tour Trajectory is shown in Figure 3. All missions used common launch assumptions: Titan IV/Centaur or Shuttle C/Centaur launch vehicle/upper stage combination, with NEP system activation at Earth escape. Key results of the studies revealed that power system specific masses of <70 kg/kWe, at power levels of 50 - 100 kWe, enhanced trip times savings over chemical for the Uranus, Neptune, and Pluto missions, as shown in Table 1.

Other results of the study identified the need for power systems capable of full power operation over periods of 4 to 12 years, depending upon mission scenario. The capability for shutdown and restart is also needed, especially for the multiple rendezvous missions such as MMBAR or the Jupiter Grand Tour. The improvements in space science that might be derived from the availability of 100 kWe of power for instruments or communications at the target were not identified.

Interplanetary Human Exploration

Interest in human travel beyond Earth orbit burgeoned into the Space Exploration Initiative (SEI). As stated by President George Bush in 1990, the goal is "for the new century - back to the Moon. . . . And then - a journey into tomorrow - a journey to another planet - a manned mission to Mars." (Synthesis Group, 1991). Human planetary exploration imposes unique requirements upon propulsion systems: Mars poses a significantly more distant destination than ever before attempted by humans: Payloads are massive, on the order of tens to hundreds of metric tons; and for piloted missions, short trip times are important to reduce radiation and zero-gravity exposure. The large payloads and reduced trip times lead to high propulsion powers and increased propellant, putting a premium on increased specific impulse propulsion systems. NEP has been shown to allow mass efficient transfer of cargo to the moon and Mars (Palaszewski 1989a,b, Hack et al.1990, Mason et al. 1989, Frisbee 1991, Galecki 1987, Holdridge et al. 1990, Coomes et al. 1986, Gilland 1991, Gilland 1991, and English 1991).

Cargo missions to the moon represent the nearest exploration class application of NEP. Lunar cargo transfer is actually a subset of the orbital transfer mission class, set apart from more typical LEO - GEO missions by payload mass requirements. As with the other OTV missions the trajectory is a spiral in earth's gravity field to lunar vicinity, followed by a spiral in the lunar orbit to the desired parking orbit. Because the destination is near, and trip time is not of primary interest, mission requirements on system power levels, specific mass, and specific impulse are lower than for planetary missions (Gilland 1992). For payload masses of 25 to 100 Mg and round trip transit times of one year or less, as assumed in most lunar mission studies, power levels on the order of 0.5 to 5 MWe are adequate (Gilland 1991, Hack 1990, Palaszewski 1989a). Some representative trends in vehicle power and initial mass requirements are shown in Figures 4 and 5. Specific masses as high as 20 kg/kWe can provide a 40% to 50% reduction in intial mass compared to chemical propulsion on either a single vehicle basis or over multiple reuse. In fact, due to the nature of NEP performance, mass savings increase for the reuse scenario, due to the low propellant resupply involved. NEP vehicles that are more massive than comparable chemical vehicles may still provide net mass savings when reused. This behavior is shown in Figure 6, where a 5 year reuse option is considered for the lunar NEP cargo mission.

Mars cargo missions are also characterized by a reduced emphasis on trip time. Typical mission times are from 1 to 2 years. The standard split/sprint mission scenario assumes a cargo vehicle which leaves for Mars one opportunity (2.14 years) prior to the piloted vehicle. This insures safe arrival of the cargo before the crew leaves Earth. NEP performance requirements for the Mars cargo mission are similar to those for the lunar cargo; 1 - 5 MWe, specific impulse of 40000 m/s or greater, specific masses of 20 kg/kWe or less. Unlike the lunar missions, a coast period is generally used in the Mars cargo trajectory. NEP vehicle masses are 50% lower than the chemical/aerobrake option (Mason 1989, Hack 1990, Gilland 1991, Palaszewski 1989b, and Frisbee 1991), and 30 - 40% lower than the Nuclear Thermal Propulsion (NTP) cargo vehicles (Dudzinski et al. 1992). In addition, reuse becomes an economically desirable feature, as the empty cargo vehicle can be returned to Earth for an additional 20 to 30 Mg of propellant (Hack 1990, Hack 1991, and Palazewski 1989a), which is on the order of 10% of the one way propellant mass.

The study of NEP for piloted space exploration has evolved steadily in the past decade. Although piloted Mars missions with NEP have been proposed since the 1950's, only in the last five years has mission designs considered alternative scenarios for the application of NEP. The key issue in piloted planetary missions has been the achievement of short transit times with acceptable vehicle and propellant masses. Typical NEP Mars mission scenarios have dealt with opposition class, short stay time missions (NASA 1989). Emphasis was placed upon single vehicle ("All-up") configurations, with fully reuseable vehicles returning to Earth (Hack 1990, Hack 1991, George 1991, Frisbee 1991, and Sercel 1986). The opposition class mission is the most energetic round trip trajectory, as one of the legs of the trip is by necessity non-optimal. In general, the most difficult portion is the inbound return leg, which requires a trajectory passing within the Earth's orbit. Propulsive capture at Earth in a reasonable time imposes energetic braking requirements, the effects of which are multiplied throughout the entire trajectory. The combination of propulsive braking at Earth and an All-up vehicle configuration drove NEP systems to power levels of 25 to 200 MWe, at specific masses of 1 to 10 kg/kWe in order to compete with high thrust missions which were utilizing expendable propulsion systems and tanks, as well as Earth fly-by return trajectories in conjunction with an Apollo-capsule type Earth Crew Capture Vehicle (ECCV) to minimize braking at Earth (Borowski, et. al. 1990). Recent mission studies applying both split/sprint and Earth fly-by/ECCV mission scenarios have shown that near term and advanced NEP systems have the capability of comparable mission round trip times and reduced vehicle mass compared to high thrust missions (Hack 1991 and George 1992). A recent comparison of NEP mission performance relative to NTP for a typical opposition class mission is shown in Figure 7.

SYSTEM STUDIES

Historically, electric propulsion implementation has been paced by the availability of space power. Therefore, systems studies of the past ten years have tended to focus primarily on application of the SP-100 power system to NEP. The SP-100 reactor is a fast spectrum, uranium nitride fueled, lithium cooled power system capable of generating 2.5 MWt. The initial power conversion system chosen is a conductively coupled thermoelectric system, capable of 100 kWe power output over a nominally lifetime of 7 years. The SP-100 has been designed with a primary lithium coolant loop that may be connected to a variety of secondary power conversion loops, allowing the system to generate power levels from 10 to 1000 kWe. Specific mass of the reference thermoelectric system is currently projected to be 46 kg/kWe (Mondt 1991); previous projections had extended to values as low as 30 kg/kWe .

Two forms of SP-100 NEP flight system have been conceived: the arcjet propelled, orbital transfer vehicle for Department of Defense applications; and the ion engine propelled, interplanetary science probe for NASA space science. Power levels ranging from 30 to 100 kWe have been considered, including the SP-100 Reference Flight System for demonstration of the technologies. A representative vehicles consisted of the 100 kWe power system, nine to twelve 30 kWe ammonia arcjet engines, power processors, and instrumentation. The power system was the nominal 7 year life system; however, the baseline mission was a 6 month orbital transfer mission from nuclear safe orbit (NSO) to GEO or some other higer orbit (Deininger and Vondra 1989). This six month mission was considered an acceptable demonstration of NEP technologies in space. Because arcjet lifetimes have been demonstrated to about 1500 hours (Polk et al. 1992), 3 sets of 3 thrusters were required for the mission. Later studies have considered both arcjet and ion propelled systems for operational orbit raising and deployment of military payloads. System specific masses ranged from 50 to 60 kg/kWe.

Ion propulsion has been identified as the thruster of choice for more the demanding interplanetary missions, for reasons of technology readiness as well as performance. As with the Reference mission, SP-100 power system parameters were considered. Over the years, system parameters have varied in assumptions of power level (100 - 300 kWe), specific mass (30 - 50 kg/kWe), and thruster propellant (mercury, xenon, krypton, and argon)(Jones and Scott-Monck 1984, Nock and Garrison 1982, and Palaszewski and Frisbee 1988). The most recent studies have assumed conservative SP-100 power system specific mass projections of 35 kg/kWe, in conjunction with 50 cm xenon, krypton, or argon ion thrusters. Total vehicle specific masses ranged from 57 to 70 kg/kWe, including thrusters and power processing (Yen and Sauer 1991).

More advanced space power system concepts have been developed in the past 5 years under the aegis of the Multimegawatt (MMWe) Program (Buden 1990). The system requirements for these systems were set by for orbital, burst power applications, which were spanned by three categories:

Category I: Open Cycle, 10's of MWe for 100's of seconds.
Category II: Closed Cycle, 10's of MWe for 100's of seconds, 1 MWe-year steady state lifetime.
Category III: Open Cycle, 100's of MWe for 10's of seconds.

Emphasis was on the reactor concepts that met the above requirements. A single concept was selected in each category for further study; this program has since been discontinued. The results of these studies, although not directly applicable to the multiyear lifetime, 1 - 20 MWe requirements of NEP, have provided impetus for later development of such concepts. In particular, conclusions and ideas on the importance of reactor fuels and materials, power conversion materials, power management and distribution components, and radiator mass have been found to be common in designing high performance NEP power systems. Perhaps the most significant bridge not crossed by these studies is the integration of the power system with the propulsion systems.

An overall overview of possible MWe NEP systems concepts was gathered at a Nuclear Electric Propulsion Workshop in June 1990. In this workshop, the power and electric propulsion community was solicited to provide data on promising system concepts for use in a reference piloted Mars mission. Presentations were given to a variety of expert panels, focussing on safety, technology, mission performance, and technical maturity. The full range of reactor, power conversion, and thruster options is given in Table 2. It is important to note that very few concepts were presented as complete systems, from reactor to thruster. Instead, different presenters addressed different aspects of the system - either power or propulsion, but not both. System integration issues, such as matching power supply output to thruster power input requirements, transmission line masses and efficiencies, and shielding effects for piloted missions were not addressed (Barnett 1991).

Recent studies of MWe NEP systems suitable for SEI missions have yielded system specific mass estimates ranging from 3 to 15 kg/kWe, depending upon power level and mission requirements such as crew shielding. Low values of specific mass were obtained for system power levels of 5 to 10 MWe using assumptions of high temperature reactors (1500 K liquid metal cooled or 2000 K gas cooled), high temperature potassium Rankine or helium/xenon Brayton power conversion (George 1991), lightweight carbon-carbon composite heat pipe radiators, high efficiency and baseplate temperature power electronics, and MWe electric thrusters operating at efficiencies greater than 50%, and specific impulses of 40000 m/s or more (Gilland 1990 and Gilland 1992). An example human rated NEP system developed by the NASA Nuclear Propulsion Office was based upon SP-100 fuel and lithium cooled reactor technologies, in conjunction with Rankine power conversion, advanced heat pipes, high temperature power distribution and control, and MWe argon ion engines, yielded specific mass values of 7.3 kg/kWe for 2 year life and 10 kg/kWe for 10 year life at 5 to 10 MWe. System reliability and redundancy were enhanced through the use of multiple 5 MWe power modules to configure a 10 to 15 MWe vehicle (George 1990 and 1991). Projected system performance is shown in Figure 8.

NEP TECHNOLOGY DEVELOPMENT

Space nuclear power systems are applicable in a myriad ways to space exploration. NEP is a subset of these applications, set apart by its use of electricity to generate thrust. The unique technologies of NEP to be discussed here are therefore the thrusters and power processing that must be linked with the nuclear power supplies. Power processing consists of the electronics and controls necessary to configure the power output from the nuclear power system into acceptable voltage, current, and frequencies for use by the electric thrusters. The thrusters then create a propellant plasma and accelerate it through interaction with electrical and magnetic fields.

Electric propulsion thrusters can be grouped into 3 types: Electrothermal, Electrostatic, and Electromagnetic. Electrothermal thrusters are the resistojets and arcjets, which use ohmic heating to heat a propellant and exhaust it out a nozzle, much like a chemical rocket. Electrostatic thrusters include the ion engines and stationary plasma thrusters (SPT's). The ion engine is a gridded system which uses an electric field between the grids to accelerate ions to high velocities. A hollow cathode at the thruster exit provides electrons to neutralize the ion beam. The SPT uses a radial magnetic field to provide impedance to electrons produced by an external hollow cathode. An axial electric field, normal to the magnetic field, is produced and is used to accelerate ions from the discharge chamber up to 16000 m/s. More than 50 SPT's have been flown by the Former Soviet Union (Brophy 1992). Electromagnetic thrusters such as the magnetoplasmadynamic (MPD) thruster use the Lorentz body force - the cross product of current and magnetic field - to accelerate a quasineutral plasma (Jahn, 1968). All of these concepts have been or are currently being developed for a wide range of missions (Sovey et al. 1992, Barnett 1992, and Pollard et al. 1993).

Electric Propulsion Heritage

During the last three decades more than 60 spacecraft using electric propulsion have been deployed in Earth-orbit (Sovey et al. 1992 and Pollard et al. 1993). All steady-state electric thrusters developed to date for flight applications have had power levels less than 2 kW because of the availability of spacecraft power and the state of space power system development. With the exception of NASA's Skylab spacecraft flown in 1973, nearly all US spacecraft have a power capability of less that 5 kW (Sovey et al. 1992). Near term applications of electric propulsion involve the use of hydrazine resistojets and arcjets for North-South stationkeeping of geostationalry satellites (Sovey et al. 1992), stationary plasma thrusters for near-Earth orbit adjustments (Bober et al. 1991), and pulsed plasma thrusters to compensate for spacecraft drag caused by solar radiation pressure (Ebert et al. 1989). As larger power capabilities become available, electric thrusters will provide primary propulsion for orbit transfer, spacecraft maneuvering, and planetary missions.

Ion Thrusters

Tables 3 and 4 review some of the flight programs and technology demonstrations of high specific impulse electric propulsion (Sovey et al. 1992, Rawlin et al. 1990, Sovey and Mantenieks 1991, Brophy et al. 1992). Table 3 itemizes the seven experimental ion propulsion flights using mercury or cesium thrusters. The longest test was the SERT II, with two thrusters accumulating more than 6700 h of operation in a 1000 km polar orbit (Kerslake and Ignaczak 1993). In 1994 Japan's National Space Development Agency (NASDA) will demonstrate the first operational capability of ion thrusters: the thrust level is 23 mN at a power of about 780 W. As shown in Table 4 flight qualification life tests of six xenon thrusters have accumulated over 59,000 h of operation (Yoshikawa 1991 and Shimada et al. 1993). The European Space Agency (ESA) has flown a radiofrequency ion thruster experiment on a retrievable platform using the Space Shuttle (Bassner et al. 1988). The experiment was launched in 1992 and retrieved in 1993. The thruster operated for more than 200 h (Bassner et al. 1993), and detailed results will be reported in late 1993.

Table 4 shows that there have been major ground demonstrations of mercury and xenon ion thrusters in the United States. Tests of 0.12 kW mercury, 2.6 kW mercury, and 1.3 kW xenon ion thruster systems have accumulated test times of over 34,000 h, 30,000 h, and 4300 h, respectively (Francisco, Low and Power 1988, Bechtel et al. 1982, Beattie et al. 1987). Life testing of many thrusters for long periods of time has helped develop a strong data base for cathode, neutralizer, and ion optics wear rates and potential life-limiting mechanisms.

The Solar Electric Propulsion System (SEPS) technology program, circa 1980, included the development of "technology ready" 2.7 kW thrusters, power processors, gimbals, thermal control systems, and propellant management systems (Anon. 1979). In addition to the demonstration of long thruster life, eight SEPS power processors accumulated more that 64,000 h of operation under various loads. Many of the thruster component technology and structural design concepts have been transferred to present inert-gas ion thruster programs.

NASA's ion thruster technology program is focussing on 0.5 to 5 kW xenon thrusters for stationkeeping and other auxiliary propulsion functions and 5 to 10 kW xenon and krypton thrusters for missions in Earth and planetary space. The major efforts are on developing long-life, light-weight thrusters and efficient, light-weight power processor components. A "derated" xenon ion thruster is being developed for stationkeeping applications with a view to eliminate known life limiting issues, increase thrust-to-power, and reduce qualification life-test times (Patterson and Foster 1991). Since the "derated" thruster operates at low voltages and current densities, the erosion rates of the positive and negative grids are estimated to be 16 and 41 times lower than those experienced by smaller stationkeeping thrusters. An engineering model pathfinder, designed to operate in the 0.5 to 5 kW range, is now being assembled with a mass estimate of 7 kg. Detailed performance evaluations and vibration test diagnostics are underway.

In the 5 to 50 kW power range, ion thruster specific impulse and efficiency requirements for NEP systems are expected to be 4000 to 10,000 s and 0.65 to 0.75 (Patterson and Verhey 1990, Patterson and Williams 1992 and Patterson and Rawlin 1988), respectively depending on power and propellant selection. Technology efforts have primarily dealt with the development of long-life ion optics, cathodes, and neutralizers. In the last four years at least five cathode life tests have been conducted by NASA for periods of 500 to 5000 h (Brophy and Garner 1988, Sarver-Verhey 1992, Sarver-Verhey 1993, and Brophy and Garner 1991). Results of these tests have led to improved cathode thermal designs, handling/starting procedures as well as the definition of low-risk operating envelopes.

Over the last four years there have been three wear tests of ion thrusters operating from 5.5 to 10 kW (Brophy, Garner and Pless 1992,Patterson and Verhey 1990, and Rawlin 1988). Test periods were 567, 890, and 900 h. Tests of xenon thrusters indicated that internal erosion of the positive grid was not a life-limiting factor for discharge voltages of ~28V and average ion current densities < 8 mA/cm^2 (Patterson and Verhey 1990 and Rawlin 1988). Many of the tests indicated there was transport of metals and metal oxides within the hollow cathode with some deposition on the orifice plate and sometimes in the region of the orifice (Sarver-Verhey 1992). Detailed efforts are underway to understand the sensitivity of cathode life to oxygen bearing contaminants from the feed system and propellant purifiers are on going.

Close-spaced ion optics have been developed for 30 cm diameter ion thrusters that have provided xenon ion current densities of ~ 13 mA/cm^2 at specific impulse and power levels of 4600 s and 16.7 kW, respectively (Patterson and Rawlin 1988). The ion extraction with krypton and argon would be expected to exceed the xenon results by factors of 1.2 and 1.8 respectively (Rawlin 1992). Because of the sensitivity of the negative grid to charge exchange ion erosion, two-grid systems will have maximum ion current densities in the 5 to 10 mA/cm^2 range in order to insure lifetimes > 10,000 h. This estimate is based on a negative grid voltage that is 300 V or more. To overcome the thrust density limitation with current molybdenum grids, work is underway at JPL and the Boeing Company to fabricate carbon-carbon grids which for example may provide a factor of about six reduction in the xenon ion sputter yield (Garner and Brophy 1992 and Hedges and Meserole 1992). The factor of six reduction in negative grid erosion by using carbon over molybdenum is consistent with the sputter yield ratios and was demonstrated experimentally (Anderson and Bay 1981). JPL is also investigating a 3-grid extraction system where the two downstream grids are negatively biased at values less than a conventional 2-grid system, thereby reducing the inert gas ion sputter yields and potentially extending grid system life (Brophy, Garner, and Pless 1992).

Technology efforts for 5 to 50 kW thrusters have focussed on the development of long-life carbon-carbon ion optics designed to operate in the 5000 to 10,000 s specific impulse range with ion extraction capabilities that meet or exceed those of molybdenum grids. Hollow cathode design criteria and procedures are also being developed, and lifetimes > 10,000 h will be validated for main cathodes and neutralizers. Lightweight thrusters (< 1kg/kW) will be developed. Two thruster concepts have been evaluated. One concept will simply scale the size of a single thruster to obtain the required performance consistent with life requirements. The other concept, called segmented ion thruster, clusters smaller thrusters as one unit to give the performance required with a single power supply providing the positive high voltage (Brophy 1992b).

MPD Thrusters

MPD thrusters have shown potential for high power, high specific impulse capability. In their simplest form, MPD thrusters consist of a central cylindrical cathode, a coaxial anode, and and insulating backplate at the rear of the chamber through which propellant is injected. An additional solenoidal magnetic field may also be applied to enhance performance. While simple from a component standpoint, these devices have yet to demonstrate performance levels and lifetimes required for demanding missions; although, as shown in Table 3 there have been two flight tests of pulsed MPD systems, and a flight of a hydrazine MPD system is planned in 1994 by Japan's NASDA (Yoshikawa 1991). The only ground-based extended test of a steady-state MPD thruster was performed for 500 h in 1969 using ammonia propellant at 33 kW (Esker et al. 1969).

Sovey and Mantenieks (1991) and Myers, Mantenieks, and LaPointe (1991) have reviewed most of the results of technology programs for steady-state and pulsed MPD thrusters. The maximum performance measured to date with a non-condensable propellant is less than 4000 seconds and 0.3 efficiency; lithium MPD thrusters demonstrated efficiencies greater than 0.6 at over 5000 seconds specific impulse in the 1960's. Because of potential NEP missions, there has been renewed interest in MPD thruster technology with a near-term focus on understanding concepts and design that provide high specific impulse and efficiency compatible with long electrode life (Myers, Mantenieks, and LaPointe 1991). Hydrogen and lithium MPD thrusters have demonstrated relatively high performance (See Figure 9), and activities are now underway to assess the feasibility and practicality of such devices.

Recently, research emphasis has been placed on thruster performance at steady-state power levels less than 100 kWe. The effects of electrode geometry, propellant choice, and applied field strength upon thruster efficiency and specific impulse are being examined at NASA Lewis Research Center and the Jet Propulsion Laboratory. Tests have been performed at NASA Lewis at power levels up to 220 kWe, yielding 3700 seconds specific impulse and a thrust efficiency of 0.2. Performance improvements were observed for increasing applied magnetic field strengths. Electrode geometry effects have been examined in relation to the interplay between thermal and flow efficiency effects. The scaling of MPD thrusters with power also continues in pulsed operation research funded by NASA at national laboratories (Los Alamos) and universities (Princeton). Thruster lifetime research has concentrated on cathode erosion, which has been identified as the life limiting component of the MPD thruster. The effects of propellant choice, dispenser cathodes, cathode geometry, and the use of high current hollow cathodes are all under investigation. Modelling of the physics of cathode erosion and energy flow is being carried on concurrently. An extensive, multi-organization effort in MPD thruster modelling is underway in order to better understand the processes observed in experiment.

Power Processing

As electric thrusters reach maturity for a specific application, power processor breadboards (PPB) have been developed and integrated with the thrusters (Kerslake and Ignaczak 1992, Yoshikawa 1991, Beattie et al. 1987, Anon. 1979, Smith et al. 1990, Wong et al. 1991, and Hamley, Pinero, and Hill 1992). Major PPB efforts have dealt with ion thrusters and thermal arcjets. As shown in Table 5 early ion PPB's were rather heavy and complex. More recently a 1.4 kW Xenon ion PPB was developed with an efficiency of 92% and a specific mass of ~ 7.9 kg/kW (Beattie et al. 1987). More importantly, the power electronics parts count was only - 400 which was a vast improvement over previous systems.

Although thermal arcjets are not high specific impulse devices, the PPB concept and technology are very applicable to both ion and MPD systems. Thermal arcjet PPB's and flight systems have recently been developed with an efficiency > 91% and a specific mass of 2.3 kg/kW for a 1.8 kW input power level (Smith et al. 1990). Since the arcjets usually require only one major power supply which does not have stringent load regulation requirements, the power processor can be simple, efficient, and light weight. The recently developed ion and thermal arcjet power electronics makes use of new switching topologies and a high level of circuit integration to reduce parts count and mass as well as increasing reliability (Hamley et al. 1992).

Simplifications to power processors for xenon ion thrusters at a nominal power of 2.5 kW have been identified (Rawlin et al. 1993 and Hamley et al. 1993). A dual-use power supply concept has been validated via integration tests with xenon ion thrusters. In this concept, three power supplies provide al the functions for thruster operation. Using inverter frequencies of ~50 kHz, a packaged specific mass of 5.4 kg/kW and an efficiency of ~93% are estimated.

CONCLUSIONS

Nuclear electric propulsion has a long history of mission assessment and technology development. Electric propulsion development and implementation has been paced by power system development, with the result that system power levels less than 5 kWe have been considered both domestically and abroad. The development of the SP-100 space nuclear power system, capable of producing 10 - 1000 kWe, and the growing need for improved space vehicle maneuvering and orbital transfer allow consideration of a range of NEP options including LEO - GEO orbital transfer. State of the art or next generation arcjet and ion propulsion systems have been shown through analysis to allow significant savings in mass, lifetime, or deployment time.

The promise of NEP for more demanding missions such as outer planet exploration or exploration of the Moon and Mars indicates the direction of both system and technology development. High specific impulse (> 40000 m/s), high efficiency (>0.5), increased power (1 - 10 MWe), and low specific mass (<20 kg/kWe) NEP systems could open the solar system to science and industry. Technologies currently under development, such as liquid metal cooled fuel pin reactors, ion and MPD thrusters, and high temperature heat rejection, provide the capability for scaling to higher powers in order to meet or exceed the needs of future missions.

Thruster technology efforts have focussed to date upon inert gas ion engines and argon or hydrogen propellant MPD thrusters. In the case of ion thrusters, specific impulse and efficiency have been demonstrated that are more than adequate for mission needs; scaling to powers greater than 10 kWe at lifetimes of 10,000 hours or more is the challenge. MPD thrusters have shown the ability to process MWe of power in a small volume; however, thruster specific impulse, efficiency, and lifetime are below requirements for the more demanding missions. An additional challenge in developing any NEP system is the system integration, optimization, and demonstration of an "end-to-end" reactor-to-thruster system for use in space. Recent studies in which researchers have focussed on either thrusters or power system, without acknowledging the interactions between them, have shown the sensitivity to system design for NEP systems. A vital, active, progressive development program addressing all aspects of NEP performance and development will result in effective NEP systems capable of revolutionizing space exploration in the coming decades.

ACKNOWLEDGEMENT

This work was performed by Sverdrup Technology for the NASA Lewis Research Center Nuclear Propulsion Office under contract number NAS3-25266.

REFERENCES

Anderson, H. H. and H. L. Bay, (1981) "Sputtering Yield Measurements," *Topics in Applied Physics*, Sputtering by Particle Bombardment, Springer Verlag.

Anon., (1979) "30-Centimeter Ion Thrust Subsystem Design Manual," *NASA TM 79191*.

Auweter-Kurtz, M., H. L. Kurtz, and H. O. Schrade, (1985) "Optimization of Propulsion Systems for Orbital Transfer with Separate Power Supplies Considering Variable Thruster Efficiencies," *AIAA Paper 85-1152*.

Barnett, J. W., (1992) "Nuclear Electric Propulsion Technologies: Overview of the NASA/DOE/DOD Nuclear Electric Propulsion Workshop," *Transactions of the Ninth Symposium on Space Nuclear Power Systems*, 2:511-523.

Bassner, H., et al., (1993) "Flight Test Results of the RITA Experiment on EURECA," *IEPC Paper 93-102*.

Bassner, H., et al., (1988) "Status of the Rita - Experiment on EURECA," *IEPC Paper 88-029*.

Beattie, J. R., et al., (1987) " Status of Xenon Ion Propulsion Technology," *AIAA Paper 87-1003*.

Bechtel, R. T., G. E. Trump, and E. L. James, (1982) "Results of the Mission Profile Life Test," *AIAA Paper 82-1905*.

Bober, A. S., et al., (1991) "State of Work on Electrical Thrusters in USSR," *IEPC Paper 91-003*.

Borowski, S. K., T. J. Wickenheiser, and J. J. Andrews, (1990) "Nuclear Thermal Rocket Technology and Stage Options for Lunar/Mars Transportation Systems," *AIAA Paper 90-3787*.

Brophy, J. R., (1992) "Stationary Plasma Thruster Evaluation in Russia," *JPL Publication 92-4*.

Brophy, J. R., C. E. Garner, and L. C. Pless, (1992) "Ion Engine Endurance Testing at High Background Pressures," *AIAA Paper 92-3205*.

Brophy, J. R. and C. E. Garner, (1991) "A 5000 Hour Xenon Hollow Cathode Life Test," *AIAA Paper 91-2122*.

Brophy, J. R. and C. E. Garner, (1988) "Tests of High Current Hollow Cathodes for Ion Engines," *AIAA Paper 88-2913*.

Buden, D., (1990) "Possible Nuclear Power Applications for Strategic Defense," *Transactions of the Seventh Symposium on Space Nuclear Power Systems*, Albuquerque, N. M., 1:10-15.

Caveny, L. H. and R. J. Vondra, (1990) "Ion Propulsion Goals for Earth Orbit Transfer, or the Pursuit for Low Specific Impulse," *AIAA Paper 90-2621*.

Caveny, L. H. and R. J. Vondra, (1989) "SDIO Electric Propulsion Objectives and Programs." *AIAA Paper 89-2491.*

Coomes, E. P., D. Q. King, J. M. Cuta, and B. J. Webb, (1986) "PEGASUS: A Multi-Megawatt Nuclear Electric Propulsion System," *AIAA Paper 86-1583.*

Deininger, W. D. and R. J. Vondra, (1986) "Development of an Arcjet Nuclear Electric Propulsion System for a 1993 Flight Demonstration," *AIAA Paper 86-1510.*

Deininger, W. D. and R. J. Vondra, (1989) "A Baseline Spacecraft and Mission Design for the SP- 100 Flight Experiment," *AIAA Paper 89-2594.*

Deininger, W. D. and R. J. Vondra, (1988) "Electric Propulsion for Constellation Deployment and Spacecraft Maneuvering," *AIAA Paper 88-2833.*

Dudzinski, L. A., L. P. Gefert, and K. J. Hack, (1992) "Nuclear Electric Propulsion Benefits to Piloted Synthesis Missions," *NASA TM 105738.*

Ebert, W. L., S. J. Koval, and Sloan, R. F., (1989) "Operational NOVA Spacecraft Teflon Pulsed Plasma Thruster System," *AIAA Paper 89-2497.*

English, R. E., (1991) "Evolving the SP-100 Reactor in Order to Boost Large Payloads to GEO and to Low Lunar Orbit via Nuclear Electric Propulsion," *AIAA Paper 91-3562.*

Esker, D. W., R. J. Checkley, and J. C. Kroutil, (1969) "Radiation Cooled MPD Arc Thruster," *NASA CR-72557.*

Francisco, D. R., Low, C. A., and Power, J. L., (1988) "Successful Completion of a Cyclic Ground Test of a Mercury Ion Auxiliary Propulsion System," *IEPC Paper 88-35.*

Frisbee, R., (1990) "Advanced Propulsion Options for the Mars Cargo Mission," *AIAA Paper 90-1990.*

Frisbee, R. H., J. J. Blandino, and S. D. Leifer, (1991) "A Comparison of Chemical Propulsion, Nuclear Thermal Propulsion, and Multimegawatt Electric Propulsion for Mars Missions," *AIAA Paper 91-2332.*

Galecki, D. L., and M. J. Patterson, (1987) "Nuclear Powered Mars Cargo Transport Mission Utilizing Advanced Ion Propulsion," *AIAA Paper 87-1903.*

Garner, C and Brophy, J. R., (1992) "Fabrication and Testing of Carbon/Carbon Grids for Ion Engines," *AIAA Paper 92-3149.*

Garrison, P. W., and K. T. Nock, (1982) "Nuclear Electric Propulsion Spacecraft for the Outer Planet Orbiter Mission," *AIAA Paper 82-1276.*

George, J. A., (1990) "Multi-Reactor Configurations for Multi-Megawatt Spacecraft Power Supplies," *AIAA Paper 90-2111.*

George, J. A., (1991) "Multimegawatt Nuclear Power Systems for Nuclear Electric Propulsion," *AIAA Paper 91-3607.*

Gilland, J. H., (1991) "Mission and System Optimization of Nuclear Electric Vehicles for Lunar and Mars Missions," *IEPC Paper 91-038.*

Gilland, J. H., R. M. Myers, and M. J. Patterson, (1990) "Multimegawatt Electric Propulsion System Design Considerations," *AIAA Paper 90-2552.*

Gilland, J. H., (1992) "NEP Mission Sensitivities to System Performance," *Transactions of the Ninth Symposium on Space Nuclear Power Systems,* Albuquerque, N. M., 3:1192-1197.

Hack, K. J., J. A. George, and L. P. Gefert, (1991) "Nuclear Electric Propulsion for Fast Piloted Mars Missions," *AIAA Paper 91-3488.*

Hack, K. J. et. al., (1990) "Evolutionary Use of Nuclear Electric Propulsion," *AIAA Paper 90-3821.*

Hamley, J. A. et al., (1993) "Development of a Power Electronics Unit for the Space Station Plasma Contactor," *IEPC Paper 93-052.*

Hamley, J. A., Pinero, L. R., and Hill, G. M., (1992) "10-kW Power Electronics for Hydrogen Arcjets," *NASA TM 105614.*

Hardy, T. L., V. K. Rawlin, and M. J. Patterson, (1987) "Electric Propulsion Options for the SP-100 Reference Mission," *NASA TM 88918.*

Hedges, D. and Meserole, J. S., (1992) "Demonstration and Evaluation of Carbon/Carbon Ion Optics," *AIAA Paper 92-3150.*

Holdridge, J., K. Shepard, U. Hueter, and P. Sumrall, (1990) "An Infrastructure Assessment of Alternative Mars Transfer Vehicles," *AIAA Paper 90-1999.*

Jahn, R. M., (1968) *The Physics of Electric Propulsion*, McGraw-Hill.

Janson, S. W., M. W. Crofton, and R. B. Cohen, (1992) "Electric Propulsion for Post-NATO IV Station-Keeping," *Symposium on Concepts for the Post-NATO IV SATCOM System,* The Hague, Netherlands.

Jones, R. M. and J. Scott-Monck, (1984) "Status of Power Supplies for Electric Propulsion," *Proceedings of the 17th International Electric Propulsion Conference, IEPC Paper 84-83,* pp. 614-629.

Kelley, J. H., R. J. Boain, and C. L. Yen, (1992) "Robotic Planetary Science Missions with NEP," *Transactions of the Ninth Synposium on Space Nuclear Power Systems,* Albuquerque, N. M., 1:78-80.

Kerslake, W. R. and L. R. Ignaczak, (1993) "Development and Complete Flight History of the SERT II Spacecraft and Experiments," *J. of Spacecraft and Rockets,* 30:3:258-290.

King, D. Q. and L. K. Rudolph, (1982) "100 kWe MPD Thruster System Design," *AIAA Paper 82-1987.*

Lyman, P.T. and M. S. Reid, (1989) "TAU as Tao," *IAF Paper 89-662, 40th Congress of the International Astronautical Federation,* Malaga, Spain.

Mason, L. M., K J. Hack, and J. H. Gilland, (1989) "Nuclear Electric Propulsion for Mars Cargo Missions," *Transactions of the Seventh Symposium on Space Nuclear Power Systems,* Albuquerque, N. M., pp. 32-37.

Miller, B. P., S. W. McCandless, and L. S. Dreifus, (1987) "Parametric Analysis of Power Requirements for Space-Based Radar," *Transactions of the Fourth Symposium on Space Nuclear Power Systems,* Albuquerque, N. M.

Mondt, J. F., (1991) "Overview of the SP-100 Program," *AIAA Paper 91-3585.*

Myers, R. M., M. A. Mantenieks, and M. R. LaPointe, (1991) "MPD Thruster Technology," *AIAA Paper 91-3568.*

NASA, Report of the 90-Day Study on Human Exploration of the Moon and Mars, National Aeronautics and Space Administration, Washington, D. C., November, 1989.

Nock, K. T. and P. Garrison, (1992) "Nuclear Electric Propulsion Mission to Neptune," *AIAA Paper 82-1870.*

Palaszewski, B., (1987) "Geosynchronous Earth Orbit Base Propulsion: Electric Propulsion Options," *AIAA Paper 87-0990.*

Palaszewski, B., (1989a) "Lunar Transfer Vehicle Design Issues with Electric Propulsion Systems," *AIAA Paper 89-2375.*

Palaszewski, B., (1989) "Electric Propulsion for Manned Mars Exploration," *Transactions of the 1989 JANNAF Propulsion Conference,* Cleveland, OH, pp. 421-436.

Palaszewski, B. and R. Frisbee, (1988) "Advanced Propulsion for the Mars Rover Sample Return Mission," *AIAA Paper 88-2900.*

Patterson, M. J. and Foster, J. E., (1991) "Performance and Characterization of a "Derated" Ion Thruster for Auxiliary Propulsion," *AIAA Paper 91-2350.*

Patterson, M. J. and G. Williams, (1992) "Krypton Ion Thruster Performance," *AIAA Paper 92-3144.*

Patterson, M. J. and V. K. Rawlin, (1988) "Performance of 10-kW Class Xenon Ion Thrusters," *AIAA Paper 88-2914.*

Patterson, M. J. and T. R. Verhey, (1990) "5-kW Xenon Ion Thruster Lifetest," *AIAA Paper 90-2543.*

Patterson, M. J., (1992) "Low-Isp Derated Ion Thruster Operation," *AIAA Paper 92-3203.*

Polk, J. E., K. D. Goodfellow, and L. C. Pless, (1992) "Ammonia Arcjet Engine Behavior in a Cyclic Endurance Test at 10 kW," *AIAA Paper 92-0612.*

Pollard, J. E., et al., (1993) "Electric Propulaion Flight Experience and Technology Readiness," *AIAA Paper 93-2221.*

Rawlin, V. K., L. R. Pinero, and J. A. Hamley, (1993) "Simplified Power Processor for Inert Gas Ion Thrusters," *AIAA Paper 93-2397.*

Rawlin, V. K., (1992) "Characterization of Ion Accelerating Systems on NASA's Ion Thrusters," *AIAA Paper 92-3827.*

Rawlin, V. K., (1988) "Internal Erosion Rates of a 10-kW Xenon Ion Thruster," *AIAA Paper 88-2912.*

Rawlin, V. K., Patterson, M. J., and Gruber, R. P., (1990) "Xenon Ion Propulsion for Orbit Transfer," *AIAA Paper 90-2527.*

Sarver-Verhey, T. R., (1993) "Extended Test of a Xenon Hollow Cathode for a Space Plasma Contactor," *IEPC Paper 93-020.*

Sarver-Verhey, T. R., (1992) "Extended Testing of Xenon Ion Thruster Hollow Cathodes," *AIAA Paper 92-3204.*

Sercel, J., and S. Krauthamer, (1986) "Multimegawatt Nuclear Electric Propulsion; First Order System Design and Performance Evaluation," *AIAA Paper 86-1202.*

Shimada, S., et al., (1993) "Ion Thruster Endurance Test Using Development Model Thruster for ETS VI," *IEPC Paper 93-169.*

Shimada, S. et al., (1991) "Ion Engine System Development of ETS VI," *IEPC Paper 91-145.*

Smirnov, B. M. and Chibisov, M. I., (1965) "Resonance Charge Transfer in Inert Gases," *Soviet Physics-Technical Physics,* 10:88-92.

Smith, R. D., et al., (1990) "Development and Demonstration of a 1.8 kW Hydrazine Arcjet Thruster," *AIAA Paper 90-2547.*

Solar System Exploration Subcommittee, (1991) "Mission Fact Sheets for NASA/Solar System Exploration Division Program Elements." *Presented at the Solar System Exploration Workshop,* San Diego, CA.

Sovey, J. S. and David J. Pidgeon, (1990) "Advanced Propulsion for LEO and GEO Platforms," *AIAA Paper 90-2551.*

Sovey, J. S. and Manenieks, M. A., (1991) "Performance and Lifetime Assessment of MPD Arc Thruster Technology," *Journal of Propulsion and Power,* 7:1:71-83.

Sovey, J. S. , et al., (1992) "The Evolutionary Development of High Specific Impulse Electric Thruster Technology" *AIAA Paper 92-1556.*

Switick, D., et. al. (1992) "SP-100 First Demonstration Flight Mission Concept Design," *Transactions of the Ninth Symposium on Space Nuclear Power Systems,* 2:532-543.

Synthesis Group, "America at the Threshold," U.S. Government Printing Office, Washington, D.C., May 1991.

Wong, S., et al., (1991) "Operational Testing of a Power Conditioning Unit for a 30-kWe Arcjet." *Transactions of the Eighth Symposium on Space Nuclear Power,* pp. 982-987.

Yen, C. L. and C. G. Sauer, (1991) "Nuclear Electric Propulsion for Future NASA Space Science Missions," *IEPC Paper 91-035.*

Yoshikawa, T., (1991) "Electric Propulsion Research and Development in Japan," *IEPC Paper 91-004.*

Zafran, S. and M. Bell, (1990) "Conceptual Spacecraft and Arcjet Propulsion System Design for the SP-100 Interim Reference Mission," *AIAA Paper 90-2549.*

Zafran, S., (1989) "Conceptual Arcjet System Design Considerations for the SP-100 Mission," *AIAA Paper 89-2596.*

TABLE 1. Relative Trip Time Capabilities for Outer Planet Missions.

Mission	Heliocentric Trip Time (years) Chemical	NEP/Tital IV	NEP/Shuttle C
Uranis	15	-	10.5-14
Neptune	>18	-	12-15
Pluto	40	14.5	11.5-14
MMBAR	8(2 asteroids)	13.5(6 asteroids)	11 (6 asteroids)
JGT	5 Titan IV	5-7	5-6.5

TABLE 2. Megawatt NEP System Concepts from the 1990 NASA/DOE/DoD Nuclear Electric Propulsion Workshop.

Reactor	Power Conversion	Thrusters
Liquid Metal Cooled	K-Rankine	Electrostatic
Fuel Pin	Brayton	Ion
Prismatic Cermet	Thermionic	Steady Electromagnetic
In-Core Boiling K	Alkali Metal Thermo-electric Conversion (AMTEC)	MPD
Gas Cooled	Magnetohydrodynamic	Electron Cyclotron Resonance
NERVA Derived		Ion Cyclotron Resonance
Particle Bed		Pulsed Electromagnetic
Pellet Bed		Pulsed MPD
Vapor Core		Pulsed Inductive
Thermionic		Pulsed Plasmoid
In-Core		Deflagration
Out-of-Core		Pulsed Electro-thermal

TABLE 3. Ion and Plasma Propulsion Flights.

Spacecraft	Propellant	Power per Thruster, kW	Specific Impulse, s	Thrust, N
Ion Systems[a]				
SERT I, 1964	Hg	1.6	-	0.020
SNAP 10A, 1965	Cs	-	-	0.009
ATS 4, 1968	Cs	-	-	~0.02
ATS 5, 1969	Cs	-	-	-
SERT II, 1970	Hg	1.0	420	0.028
ATS 6, 1974	Cs	0.15	2500	0.004
ETS III, 1982	Hg	0.11	2600	0.002
EURECA, 1992	Xe	0.44	4800	0.10
ETS VI, 1994	Xe	0.78	2900	0.023
ARTEMIS, 1996	Xe	0.58	~4800	0.015
Pulsed MPD Systems[a]				
MS-T4. 1980	NH3	0.03	-	-
Spacelab, 1983	Ar	0.20	-	-
Spaceflyer Unit, 1994	N2H4	0.43	-	23,pk

[a]See Pollard et al. 1993 and Sovey et al. 1992.

TABLE 4. Ion and Plasma Technology Demonstration Tests

Ion Thruster[a]	Test Period(s), h	Propellant	Power, kW
Ion Auxiliary Propulsion System, (NASA LeRC)	4 tests for a total of 34,000 h	Mercury	0.12
Solar Electric Propulsion Tech., (NASA LeRC)	7 tests for a total of 30,000 h	Mercury	2.7
Xenon Ion Propulsion System (Hughes Aircraft and INTELSAT)	4350	Xenon	1.3
Ion Engine System, (NASDA, MELCO)	9 tests for a total of > 59,000 h	Xenon	0.78
Xenon Ion Thruster, (NASA LeRC)	900	Xenon	5.5
Argon Ion Thruster, (JPL)	900	Argon	6.2
MPD Thruster[a]			
MPD Lifetest (McDonnell Douglas)	500	Ammonia	33

[a]See Pollard et al. 1993 and Sovey et al. 1992.

TABLE 5. Power Electronics for Electric Propulsion Systems

System	Power, kW	Efficiency, %	Specific Mass, kg/kW
Hydrazine Arcjet, (Smith et al. 1990)	1.8	>91	2.3
Ammonia Arcjet, Wong et al. 1991)	30	95	1.8
Hydrogen Arcjet, Hamley et al. 1992)	10	94	
Mercury Ion Engine, (Kerslake and Ignaczak 1993)	0.98	87	16.9
Mercury Ion Engine, (Anon. 1979)	3.1	87	12.3
Xenon Ion Engine, (Beattie et al. 1987)	1.4	92	7.9
Xenon Ion Engine, (Shimada et al. 1991)	0.79	86	-

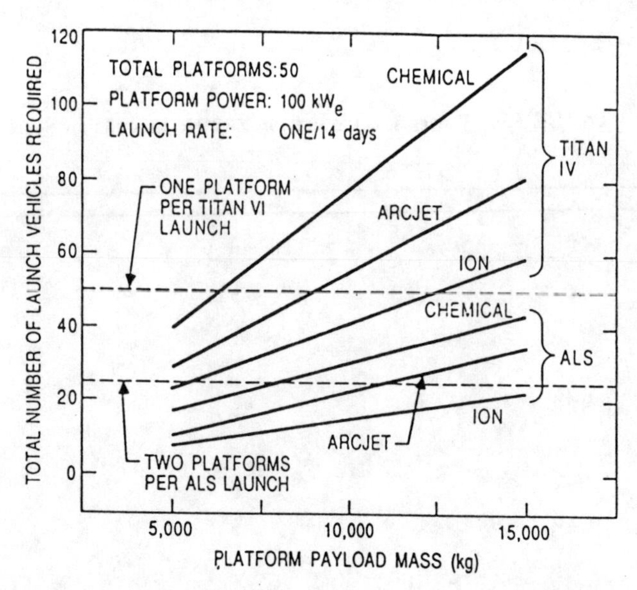

FIGURE 1. Launch Vehicle Savings from the use of Nuclear Electric Propulsion for Satellite Constellation Deployment. (Deininger and Vondra 1988)

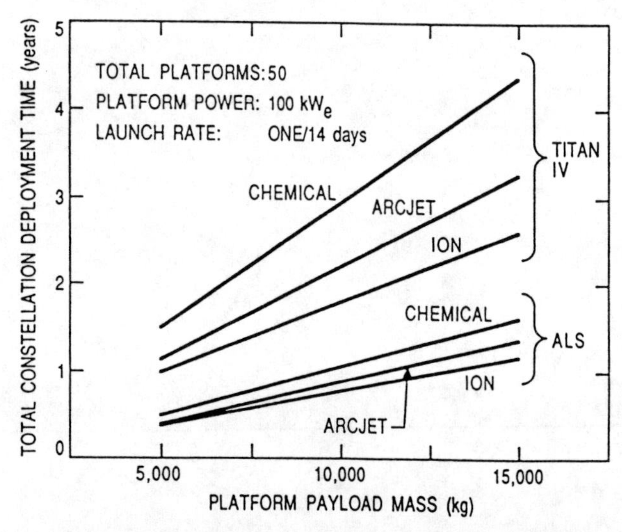

FIGURE 2. Deployment Time Savings from the use of Nuclear Electric Propulsion for Satellite Constellation Deployment. (Deininger and Vondra 1988)

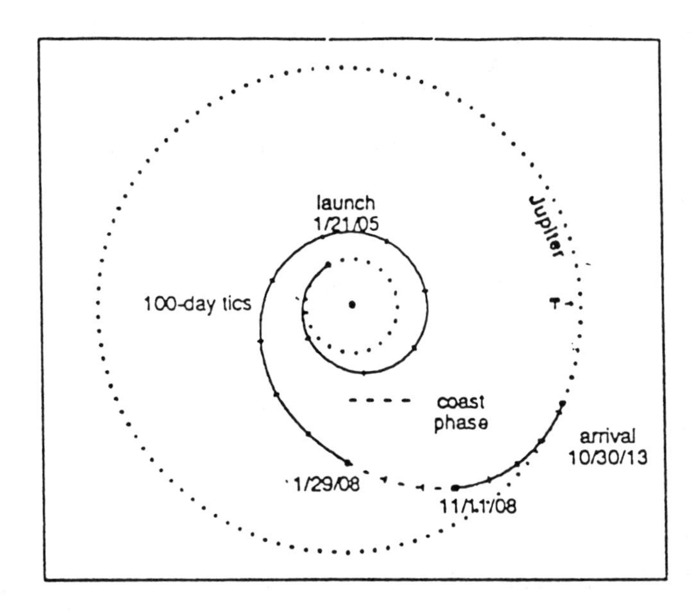

FIGURE 3. NEP Jupiter Grand Tour Interplanetary Trajectory.

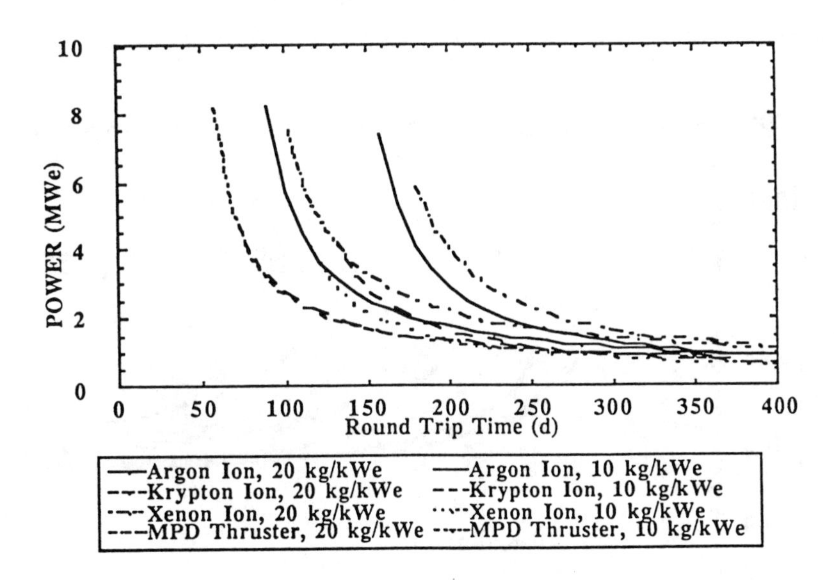

FIGURE 4. NEP Power Requirements for Lunar Cargo Delivery. Round-trip Lunar Mission. 58 MT Payload to Low Lunar Orbit. (Gilland 1991)

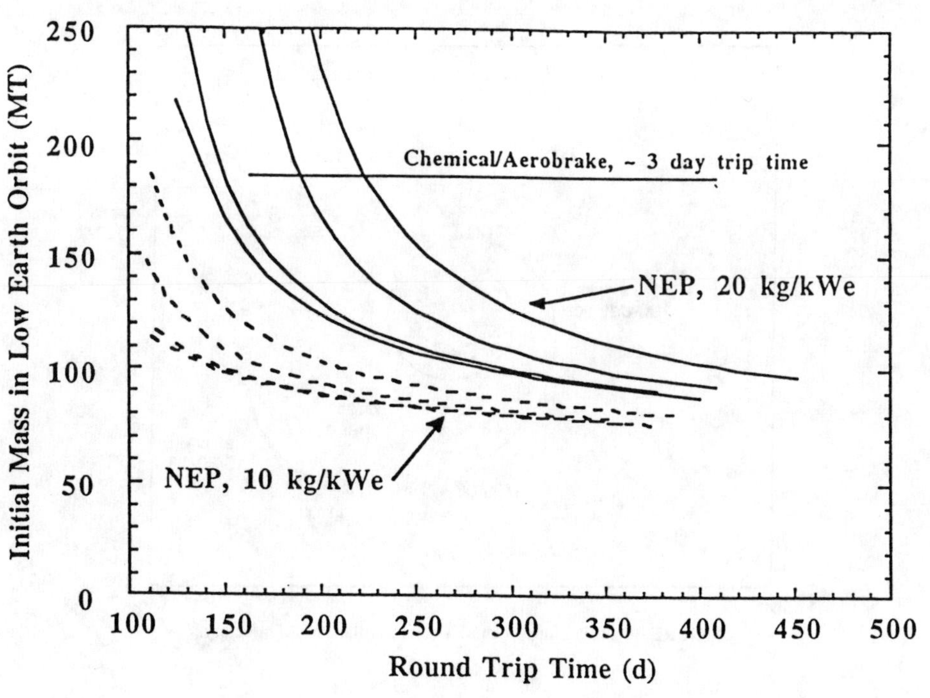

FIGURE 5. NEP Vehicle Initial Mass Requirements for Lunar Cargo Delivery. Round-trip, Reuse Option Compared to Chemical Aerobrake Option.

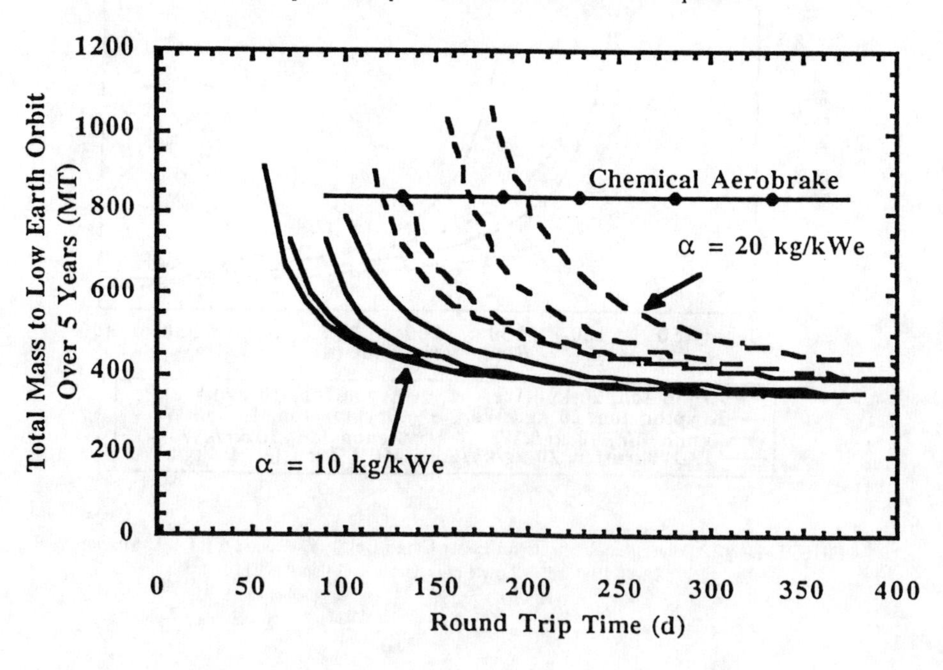

FIGURE 6. Effects of NEP Reuse for the Lunar Cargo Mission Using Ion or MPD Thrusters. Round-trip, 5 Year Reuse Option Compared to Chemical Aerobrake Option.

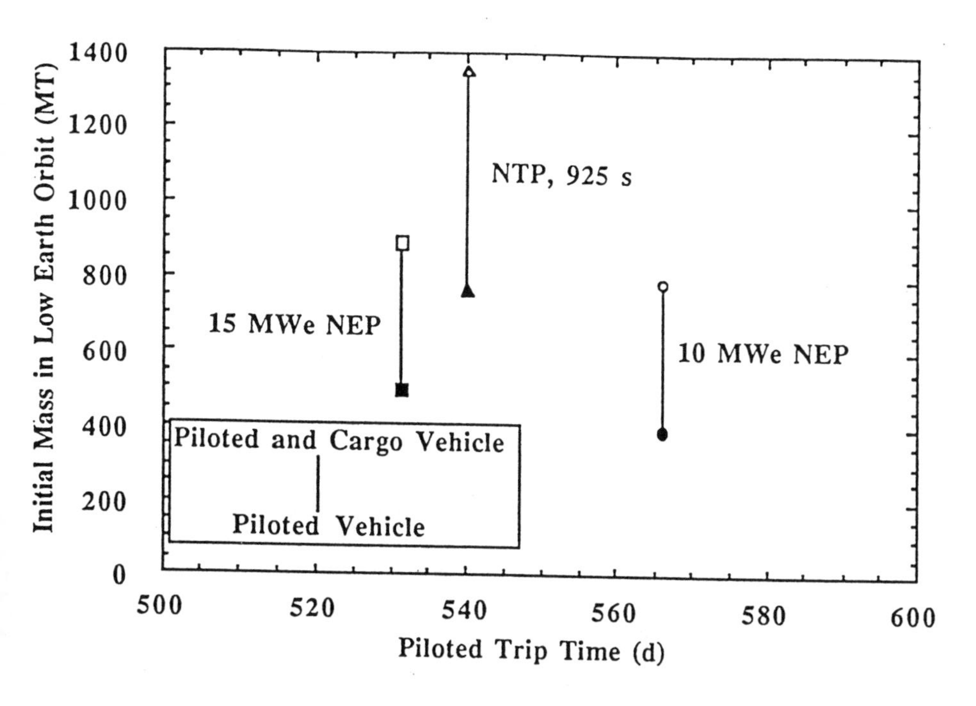

FIGURE 7. Comparison of NEP and NTP Options for the Piloted Mars Opposition Mission. 2014. (Dudzinski et al. 1992)

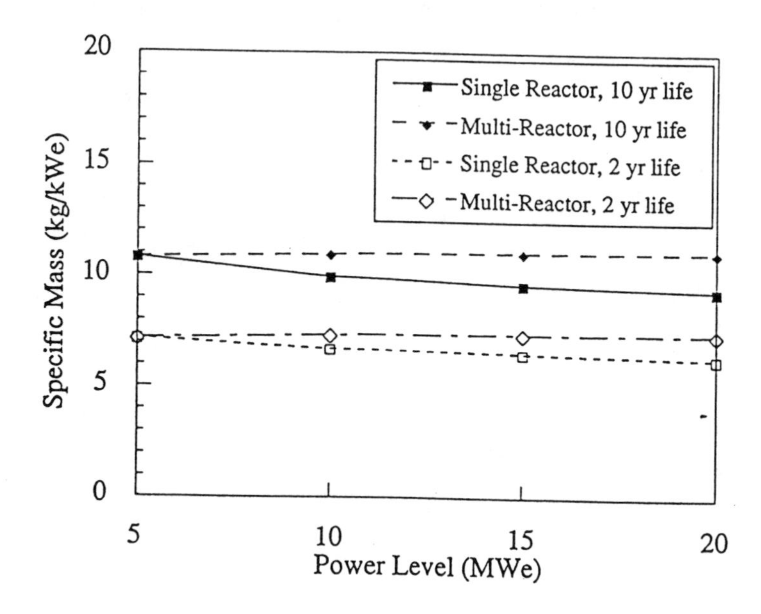

FIGURE 8. "Growth" SP-100 Multimegawatt System Characteristics Using Potassium Rankine Power Conversion. (George 1991)

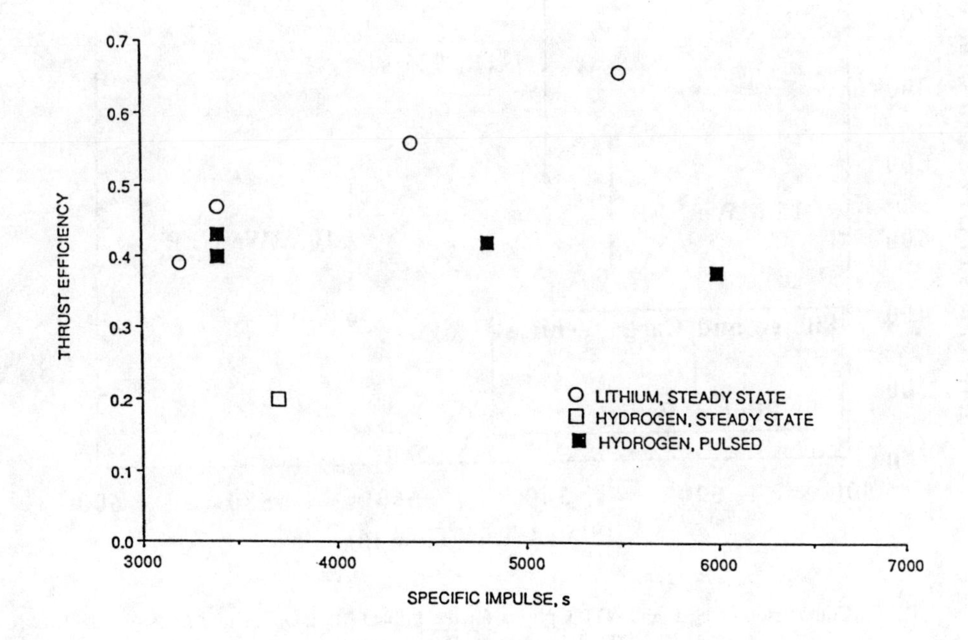

FIGURE 9. Experimental Hydrogen MPD Thruster Performance.

A REVIEW OF ADVANCED RADIATOR TECHNOLOGIES FOR SPACECRAFT POWER SYSTEMS AND SPACE THERMAL CONTROL

A. J. Juhasz
NASA-Lewis Research Center
21000 Brookpark Rd.; MS 301-5
Cleveland 44135, Ohio
(216) 433-6134

G. P. Peterson
Department of Mechanical Engineering
Texas A&M University
College Station, TX 77843
(409) 845-5337

Abstract

Thermal management of manned spacecraft traditionally has relied primarily on pumped single-phase liquid systems to collect, transport, and reject heat via single-phase radiators. Although these systems have performed with excellent reliability, evolving space platforms and space-based power systems will require lighter, more flexible thermal management systems because of the long mission duration, large quantities of power system cycle reject heat, and variety of payloads involved. One of the critical elements in these thermal management systems are the radiators. Presented here is a two part overview of progress in space radiator technologies during the eighties and early nineties. The first part is a review and comparison of the innovative heat rejection system concepts proposed during the past decade, some of which have undergone preliminary development to the breadboard demonstration stage. Included are space constructable radiators with heat pipes, variable surface area radiators, rotating solid radiators, moving belt radiators, rotating film radiators, liquid droplet radiators, Curie point radiators, and rotating bubble membrane radiators.

The second part contains a summary of a multi element project effort including focussed hardware development under the CSTI (Civil Space Technology Initiative) High Capacity Power program carried out by the NASA Lewis Research Center and its contractors for the purpose of light weight space radiator development in support of SEI (Space Exploration Initiative) power systems technology. Principal project elements include both contracted and in-house efforts conducted in a synergistic environment designed to facilitate accomplishment of project objectives. The contracts with Space Power Inc. (SPI) and Rockwell International (RI) are aimed at advanced radiator concepts (ARC) while the in-house work guides and supports the overall program by system integration studies, heat pipe testing, analytical code development, radiating surface emissivity enhancement and composite materials research aimed at development and analysis of light weight high conductivity fins.

PART I. INNOVATIVE RADIATOR TECHNOLOGIES

INTRODUCTION

Traditional means for rejecting heat from manned spacecraft is through the use of heat rejection systems comprised of single-phase fluid loops (Peterson 1987). These single-phase fluid loops use a mechanically pumped coolant to transfer heat from the habitation portion of the spacecraft to the radiators where it is rejected to the space environment. Although these systems have performed with excellent reliability in the past, evolving space platforms and space based power systems will require more flexible thermal management systems due to the multi-year mission duration, large quantities of heat to be rejected, long physical distances, and large variety of payloads and missions which must be accommodated (Mertesdorf et al 1987).

In general, space thermal management systems consist of three separate subsystems:

(1) a heat acquisition subsystem to collect heat from the various payload or power system heat rejection interfaces,

(2) a heat transport subsystem to transport heat from the acquisition sites to the radiating surfaces, and

(3) a heat rejection subsystem composed of the radiating surfaces which form the space radiator.

As an example of a typical space thermal management system proposed for large space platforms is a two-phase heat rejection system consisting of the subsystems identified above, was presented by Edelstein (1984). These three subsystems comprise a thermal "utility" which would utilize the high latent heat of a working fluid to transport heat from the heat sources to the radiators, where it would be radiated to the space environment.

The last of the three subsystems, the radiators, are a critical component of virtually all proposed space-borne installations. In most current designs, the radiator is composed of an array of tubes or tube-fin structures through which a liquid coolant is circulated. The tube wall must be sufficiently thick, and hence massive, to minimize micrometeoroid penetration. As a result, the radiator mass could comprise as much as half of the total system mass (Juhasz et al. 1986).

The technical challenges associated with the development of heat rejection systems capable of meeting future requirements have been described previously (Ellis 1989).

Presented here is a review and comparison of the heat rejection systems currently under development for space platforms and space based power systems including: space constructable radiators, variable surface area radiators, rotating solid radiators, moving belt radiators, rotating film radiators, liquid droplet radiators, Curie point radiators, and rotating bubble membrane radiators.

Space Constructable Heat Pipe Radiator (SCR)

The heat rejection system, presently in use on the space shuttle orbiter, consists of over 250 small parallel tubes, embedded within a honeycomb structure. Warm single-phase Freon from the heat collection and transport circuit, is circulated through these tubes as shown in Figure 1. Heat is transferred from the coolant by convection to the tube walls, by conduction through the honeycomb structure, and finally through radiation to space. Application of this technology to Space Station would require over 1,000 interconnected tubes. If only a single redundant loop were used, a puncture in any single tube could disable the entire system, making this type of system infeasible for long term missions.

A space-constructable radiator, comprised of a series of individually sealed heat pipe elements similar to that shown in Figure 2, has been proposed (Ellis 1989) and several advantages of this type of system over pumped single-phase fluid loops have been identified. These advantages include a significant reduction in weight due to the reduction in fluid inventory, increased heat rejection capacity due to the uniform temperature of the radiating surface, and increased reliability since penetration by a single micrometeoroid or piece of space debris would result in the failure of a single heat pipe element and therefore only a slight degradation in performance.

High capacity space constructable radiator elements have been investigated in several recent Shuttle flight experiments, including the STS-3 flight of the Thermal Canister (Harwell 1983), the STS-8 Heat Pipe Radiator Experiment (Alario 1984), and the recent Space Station Heat Pipe Advanced Radiator Element (SHARE) flight test (Rankin et al. 1989 and Kossan et al. 1990). These three flight tests along with numerous ground tests, have demonstrated that heat pipe radiators present a feasible alternative to pumped single-phase systems.

In addition to Space Station, such space constructable heat pipe radiator systems could be utilized in solar dynamic (SD) power systems (Brandhorst et al. 1986 and Gustafson et al. 1987). In this application, the radiators must reject the non-convertible thermal energy supplied to the power system. They thus represent a critical component of the overall development of space based power systems. Figure 3 shows a typical solar dynamic power module design which incorporates a space constructable heat pipe

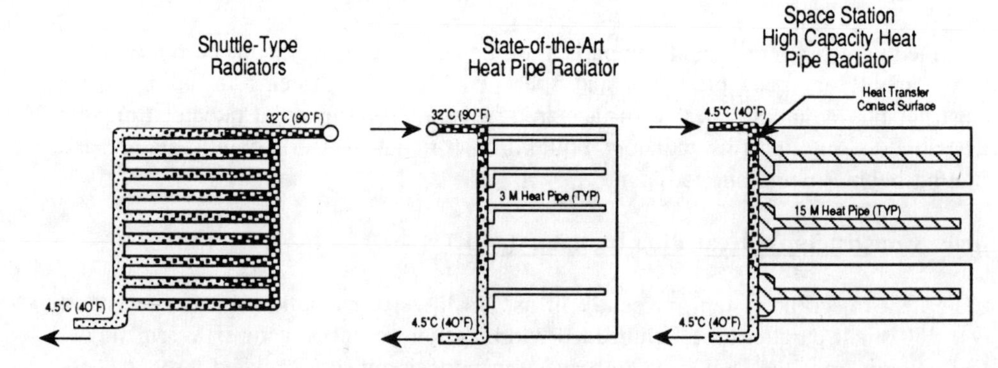

Shuttle-Type Radiators	State-of-the-Art Heat Pipe Radiator	Space Station High Capacity Heat Pipe Radiator
• Fluid Must be Pumped Over Entire Radiator Area	• Fluid Needs to be pumped only thru radiator manifold	• Two-phase fluid interfaces radiator only at manifold
• Over 250 Parallel Tube Required for 25 kW	• Less than 75 heat pipes required for shuttle type system	• Only 50 heat pipes required for 75 kW station
• Puncture of any Tube Destroys Radiator	• Each heat pipe independent of all others, thus a puncture of one tube only reduces efficiency	• Each heat pipe independent of all others
		• Contact heat exchanger allows on orbit assembly and repair of radiator

FIGURE 1. Evolution of Heat Rejection in Spacecraft .

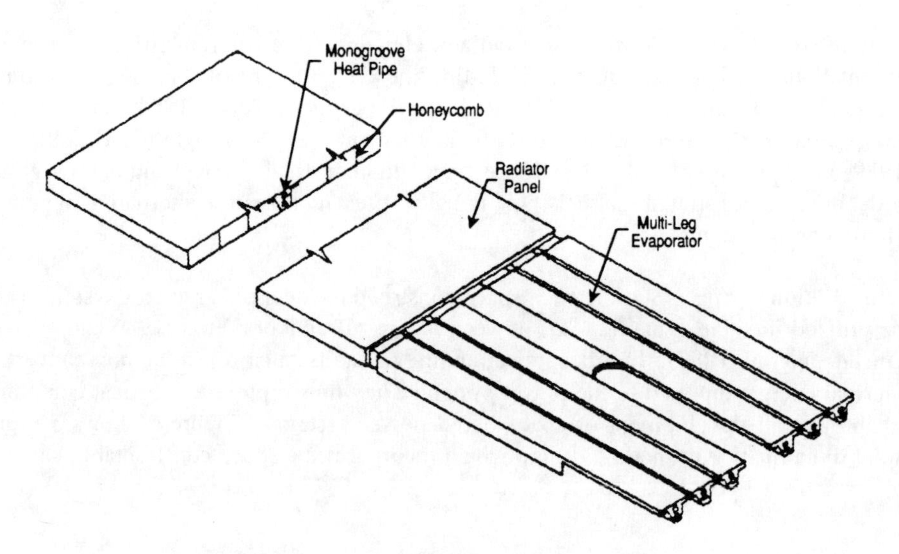

FIGURE 2. Space Constructable Radiator Panel Configuration.

radiator system. As illustrated, the radiator could be segmented into several panels for redundancy with a small number of excess panels incorporated to reduce maintenance.

Although these space constructable heat pipe radiators have performed adequately under realistic thermal/vacuum test conditions and several shuttle flight tests, the application of this technology to an SD power system for Space Station, for example, would require about 50 heat pipes, each 10 to 15 meters long, to reject the 75 kW required by the 1989 space station design (Ellis 1989).

Variable Surface Area Radiator

The concept of a flexible, variable surface area radiator, capable of absorbing high peak heat loads for brief time intervals, was first introduced in 1978 (Leach et al. 1978). These types of radiators can be classified into two major categories: (1) those in which no phase change occurs, and (2) those which utilize a phase change of the working fluid. Oren (1982) presented the results of an investigation involving two types of flexible roll out fin (ROF) radiators in which no phase change was required. The first of these utilized a rolled up fin with a plastic or elastomeric tube attached to both sides. When gas pressurization was allowed to inflate the two tubes, the fin "unrolled" and provided a substantial increase in the surface area. The second utilized aluminum radiator tubes wound in a helical spring configuration to form a cylinder covered by the fin material. This variable surface area radiator used the inherent spring force (similar to a jack-in-the-box) for deployment and was intended to meet heat rejection needs of up to 12 kW (Leach et al. 1978 and Oren 1982). Because no phase change is required, the working fluid proposed was an ethylene glycol/water solution.

Several types of variable surface area radiators that utilize liquid/vapor phase change and the associated increase in volume have been proposed. The concept of operation for these types of radiators is illustrated in Figure 4. As shown, in their simplest form, these types of radiators are comprised of two thin walled sheets sealed along the edges and formed into a concentric roll. This roll extends or rolls out due to the increased vapor pressure generated by heating the wick structure in the evaporator. This wicking structure can in some cases line the inside of the entire fin to assist in liquid return. Once the vapor condenses, the fin curls back or retracts to the stowed position. Figure 5 illustrates the principle of operation for this type of radiator. Initially the working fluid within the fin exists as a subcooled or saturated liquid, Figure 5a. Because of the flexibility of the fin, heat addition and rejection occur at constant pressure. Heat added to the evaporator vaporizes the working fluid, which expands, thereby causing the fin to extend, Figure 5b. This permits the entire external surface of the fin to radiate heat to space, Figure 5c. As the vapor condenses, the longitudinal stiffness causes the fin to curl into its original spiral shape thereby squeezing the liquid droplets toward the evaporator, Figure 5d, where they can be stored in the capillary wick structure. In this system, maximum heat rejection occurs when the radiator is fully expanded. In a steady-state mode, the fin spring constant could be designed so that the length of the fin would automatically adjust to balance the heat input and rejection. From a construction

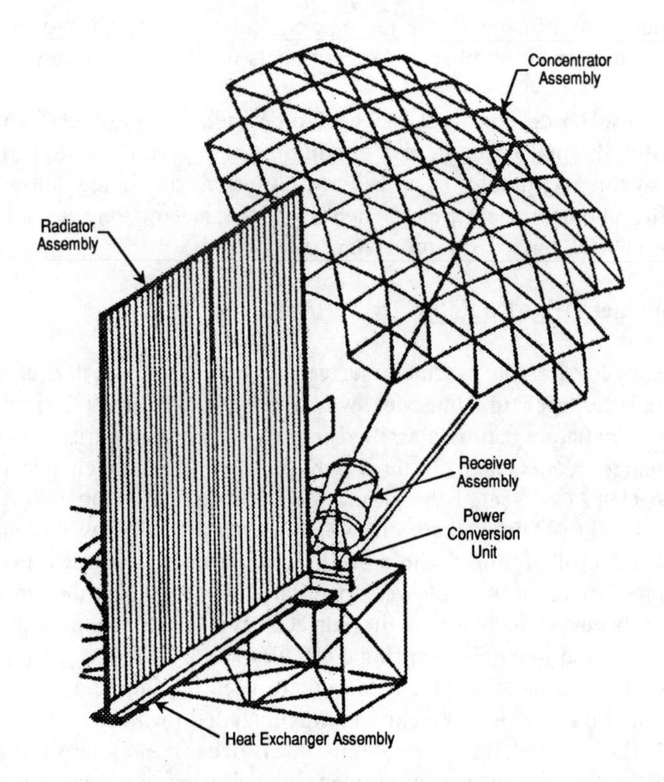

FIGURE 3. Solar Dynamic Power Module.

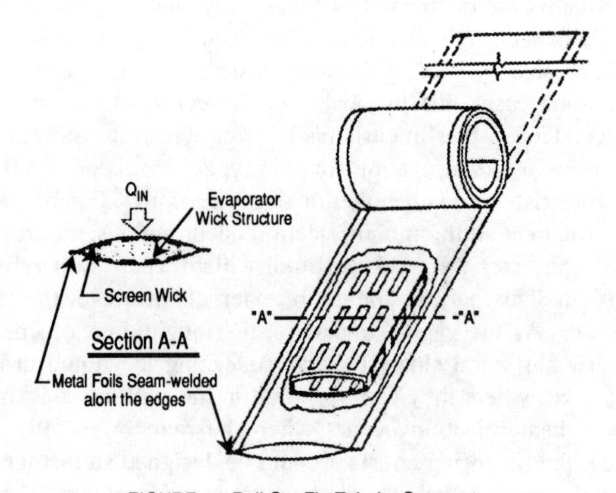

FIGURE 4. Roll Out Fin Tubular Segment .

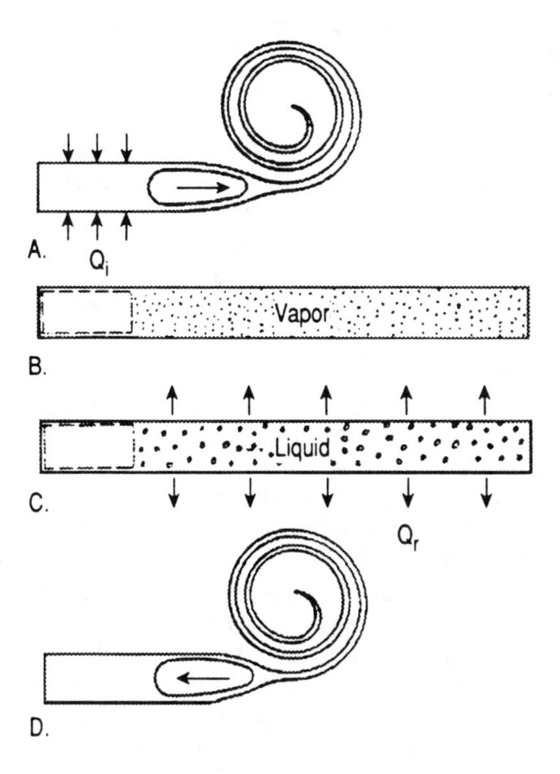

FIGURE 5. Operation of a Roll Out Fin Radiator.

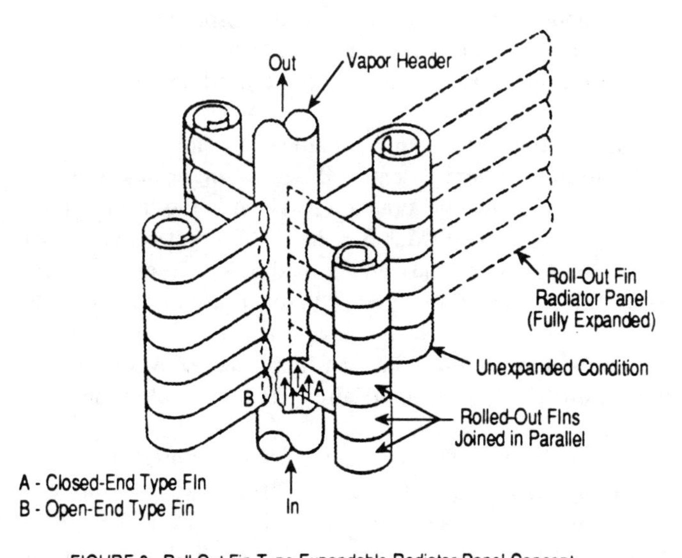

FIGURE 6. Roll Out Fin Type Expandable Radiator Panel Concept.

perspective, roll out fins could be made from either a thin metallic foil or plastic film with an internal spring. In the case of metallic foil, the metal itself could be heat treated to act as a spring and provide the retraction force.

Several different variations of this device have been investigated, including radiators that employ the previously described principle (Ponappan et al. 1984), larger multi-component expandable radiators (Chow et al. 1985), and inflatable-expandable pulse power radiators. Conceptual design of a 1 m long roll out fin capable of accommodating modest peak to average (10/1) heat loads by varying the projected surface area has been presented and discussed (Ponappan 1984). This concept has been expanded to include radiator panels which utilize several of these roll out fins in parallel to form panels as shown in Figure 6. In this application, each of the four segments could function independently for a given pulsed or steady heat input condition, or the four panels could be arranged around a common vapor header and act jointly to reject heat. Figure 7 illustrates a concept similar to the roll out fin, however in this situation, an inflatable bag is utilized in place of the roll out fin (Chittenden et al. 1988). As illustrated, during the operation phase, the radiator is extended out of the spacecraft and filled with vapor produced by a power pulse. The inflatable bag proposed for this concept would be made of a thin, strong, lightweight, internally lined or coated fabric with water as the working fluid. As was the case with the roll out fin, this concept is characterized by a high condensation heat transfer rate inside the radiator, low operating fluid mass due to the large latent heat of vaporization, and high radiator effectiveness due to near isothermal operation. This type of radiator is capable of absorbing substantial quantities of heat during the peak power phase of the duty cycle, and rejecting the stored heat during the cooling and retraction phase. During the high heat absorption phase, the radiator would be extended out of the spacecraft and filled with steam generated by vaporization of the working fluid. Then, as the spacecraft continues orbiting, the steam condenses as heat is radiated to space. The radiator would be retracted during condensation, so as to maintain a constant saturation pressure, and it would be folded into the spacecraft, ready to be extended again during the next pulse.

Studies of space SDI (Strategic Defense Initiative) missions showed requirements for peak electric power in the megawatt range. Based on these missions and their orbital cycles, systems were required that generated reject heat in the form of pulses with peak-to-average ratios of 10,000 or more (Mahefky 1982). Present conventional radiators are sized to reject peak heat loads and they are "turned down" to reject off peak loads. As a result, these conventional radiators are capable of near constant load thermal control over a range of nominally 10/1 peak-to-average heat loads. However, for high heat load, weight constrained applications with very high peak-to-average ratio, conventional radiator designs would be of limited applicability (Chow et al. 1986).

Elliott (1984) and Koenig (1985) suggested the use of expandable balloon radiators to provide ultra-lightweight surfaces. Utilization of expandable surfaces for cooling imposes a fundamental limit on operation time. In addition, there is a severe mass penalty associated with periodic heat release. Since late 1983, a collectable expandable,

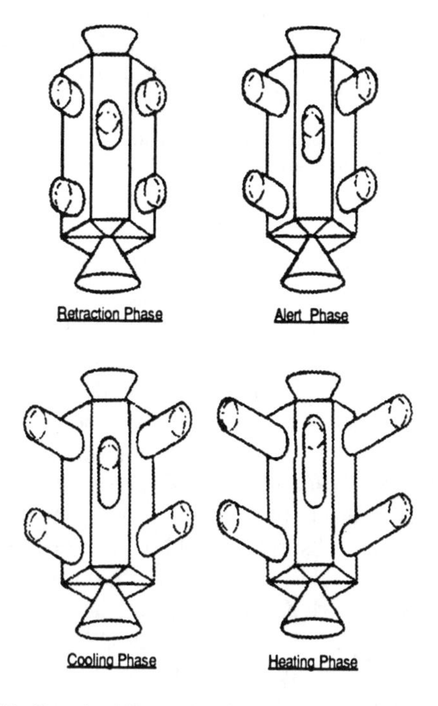

FIGURE 7. Operational Phase of High Power Inflatable Radiator System.

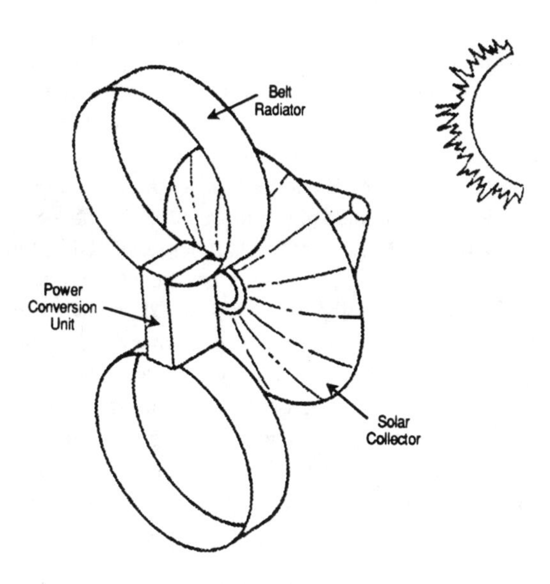

FIGURE 8. Moving Belt Radiator Concept .

also known as the expandable pulse power radiator, has been investigated at the Air Force Aero Propulsion Laboratory (Chow et al. 1986). Basically, in the collectable expandable radiator concept, utilization of a phase change material is used to cool high power density devices through flash evaporation. The vapor is collected on an expandable variable surface area, thin metallic or plastic inner liner, on which it is allowed to condense during the time between pulses. The condensate is then pumped back or brought back by other means to the coolant reservoir to be recycled.

Several possible expandable containers have been proposed. For high peak heat load pulsed radiators, low surface to volume inflatable bags or bellows radiators appear promising, due to the large energy storage capacity (Chow et al. 1985). The radiator would be constructed from a thin, low-density, flexible material that could be collapsed and stored in a compact form, ready for expansion during times of high peak heat loads. Because of the large volume to mass ratio, large amounts of vapor could be contained during the pulse period and rejected through condensation and radiation during the inter-pulse period. This design results in a lightweight radiator which is very compact in the stowed position and can be easily protected from micrometeoroid damages, except when in use. Due to their high heat absorption and low heat rejection rates, for pulsed systems of this type, the duration of the pulse period must be shorter than the time interval between the pulsed high heat load cycles. This calls for stringent restrictions on the time response characteristics. For cases requiring higher energy pulses, an expandable bellows concept has also been proposed (Chow et al. 1985). The bellows concept differs from the roll out fin and inflatable bag concept in that a significant amount of heat energy is stored in the expanding structure.

Rotating Solid Radiators (RSR)

Sensible heat capacity heat rejection systems were first proposed in 1960 (Weatherston et al. 1960). These systems proposed to rotate a solid material past an internal heat source and then to space where the heat could be rejected through radiation. In a majority of these systems heat was transferred to the solid material through either conduction or convection. An extension of this concept has been proposed for high temperature ranges such as those found in reactor cores and is referred to as the radiatively-cooled, inertially-driven nuclear generator (RING) heat rejection system (Apley et al. 1984). In this system, reactor waste heat is radiatively transferred from a cavity heat exchanger to the rotating ring. Although at low temperatures conduction and/or convection can reduce the size of the primary/secondary interface, at higher temperatures radiative heat transfer becomes attractive. The RING power system takes advantage of the need to offset the reactor from the mission platform (for radiation field reduction) by using the reactor to mission platform separation space and the boom structural assembly to support four (4) counter-rotating, 90 degree offset, coolant-carrying rings. The proposed rings are segmented, finned, thin-walled pipes, filled with liquid lithium.

The enclosed cavity heat exchanger allows a higher emissivity material to be used and since the configuration is protected, the primary coolant tubes can be placed closer to the heat transfer surface. The cavity configuration also results in an increased hemispherical emissivity for the wall material (Siegel et al. 1980).

Moving Belt Radiators (MBR)

Another advanced radiator concept is that of a moving belt radiator (Teagan et al. 1984). This concept was being developed under contract to the NASA Lewis Research Center during the latter part of the eighties. The basic operation of a moving belt radiator, is illustrated in Figure 8, where a cylindrical belt is rotated about a fixed center through some type of driving mechanism attached to the spacecraft. Heat collected inside the spacecraft is transferred from a primary heat transport loop to the belt through solid-to-solid conduction or directly through convection. As the belt rotates into the spacecraft heat exchanger it absorbs heat, and while rotating through space it rejects heat by radiation. Several materials have been proposed for the belt material, including homogeneous solids, or two solid belts with a phase change material between them. In a follow on report, analytical and experimental investigations of the rotational dynamics of this type of system along with methods for transferring heat to the moving belt, deployment and stowage, and fabrication are presented and discussed. Also, life-limiting factors such as seal wear and micrometeoroid resistance are identified (White 1988).

The moving belt radiator was projected to be only 10 to 30% as massive as advanced heat pipe radiators, and it could operate without exposing the working fluid to space, thereby reducing vaporization losses. The major technological challenge appears to be maintaining the stability of a rotating belt in the presence of spacecraft attitude maneuvers. Although other issues, such as long term reliability of the roller drive mechanism, must be solved, this concept compares favorably with the 5 to 8 kg/m^2 space constructable heat pipe radiator at both 300 K and 1000 K.

A concept similar to the moving belt radiator is that of a liquid belt radiator (LBR) also proposed by Teagan (1984). In the liquid belt radiator, illustrated in Figure 9, a thin screen or porous mesh structure supports a low vapor pressure liquid by capillary forces. This screen is drawn through a liquid bath where warm liquid is picked up and retained in the screen material. The screen and liquid form a ribbon which is then rotated through space where heat is rejected through radiation. The advantages and disadvantages with this type of radiator system are similar to those of the moving belt radiator, except that in this case, the liquid must have a very low vapor pressure (less than 10^{8} torr) over the entire operating temperature range, to prevent evaporative losses. Several materials have been proposed, including diffusion pump oils, gallium, lithium, and tin. The material selection depends primarily on the temperature range of interest with the oils limited to about 350 K and the liquid metals being applicable over a wide temperature range, as high as 2000 K. For space radiator applications the maximum operating temperature is expected to be in the 1000 K range for thermionic power systems. Although the proposed mode of operation is in the sensible heat mode, in some situations, it may be desirable for the liquid belt radiator to operate in the latent heat

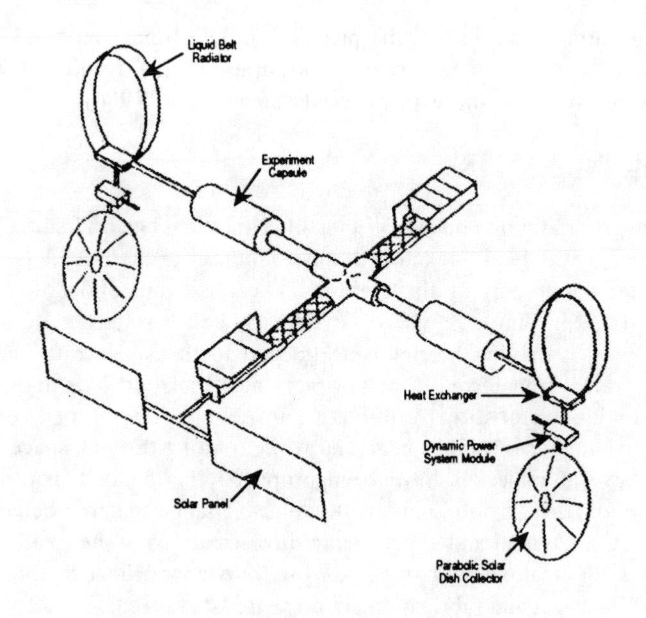

FIGURE 9. Artist's Schematic of the Liquid Belt Radiator.

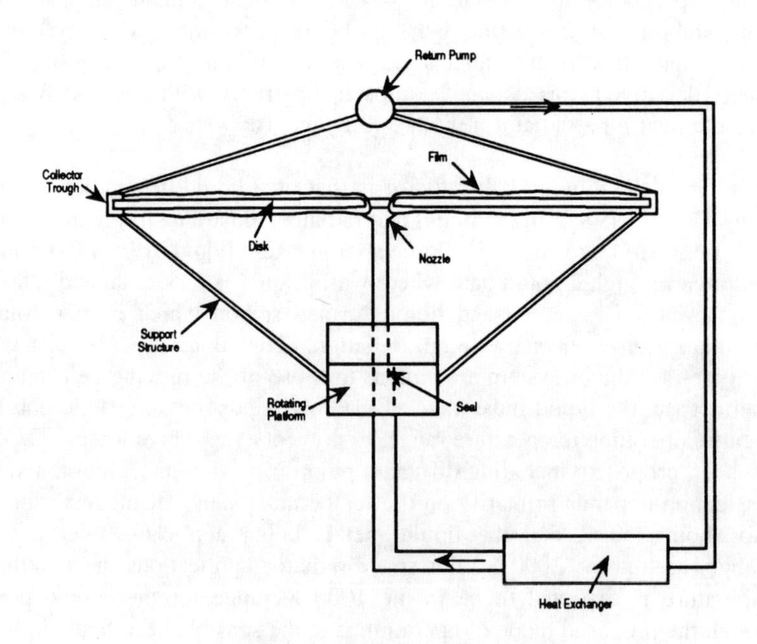

FIGURE 10. Rotating Film Radiator Schematic.

mode. When this is done, liquid changes phase during its transverse through space. Clearly, the mode of operation would depend upon material selection, operating temperatures, and heat rejection requirements. Parametric analyses (Teagan 1984) indicated that the liquid belt radiator could reduce the radiator mass by as much as 70% of space constructable state-of-the-art (SOA) heat pipe radiators. However, the Advanced Radiator Concepts (ARC) program, to be discussed in the second part of this report, has demonstrated reductions in heat pipe specific mass by a factor of 3 to 4 over the SOA heat pipe technology used by Teagan for comparison.

Rotating Film Radiator (RFR)

Figure 10 depicts another advanced radiator concept which uses a thin liquid film, the rotating film radiator (Song et al. 1988). As shown, this concept utilizes a rotating disk with a thin film of liquid flowing radially. Initially, the proposed working fluid, toluene, is injected at the center and the flow is split equally between the two surfaces of the disk. The fluid then spreads into thin films where it can radiate heat to space. The rotational speed of the disk, controls the thickness, velocity, and flow regime of the film. Upon reaching the outer circumference of the disk, the fluid is collected and returned to the center as illustrated. Preliminary analysis for this concept (Prenger et al. 1982), indicates that the rotating film radiator can achieve a specific mass of 5.5 kg/kW or 3.5 kg/m^2, based upon total emissivities in excess of 0.3. Hence it cannot compete with advanced heat pipe radiators (to be discussed in part 2), which achieve equal specific mass at surface emissivities of 0.85 to 0.9, since such radiators would require only a third of the surface area to reject the same amount of heat.

Liquid Droplet Radiator (LDR)

The liquid droplet radiator concept retains the low-mass advantages of a disk radiator. As illustrated in Figure 11, a warm low vapor pressure working fluid is projected from a droplet generator, where the liquid absorbs heat, to a droplet collector which collects the radiatively cooled droplets in a rotating drum (Mattick et al. 1981). Because of the large surface area of the droplets, this type of system has the additional advantage of greatly reduced mass especially in the case of paired modules, which eliminates the need for a long return loop for the liquid. The generator is a pressurized plenum with an array of holes or nozzles to form liquid jets, which break up into droplets via surface tension instability. A provision for rapidly varying the pressure of the fluid (a piezo-electric vibrator) may be employed to control the drop size and spacing. The generation and collection of the droplets, as well as heat transfer to the liquid, can be achieved with modest extensions of conventional technology. While it is true that LDRs have a large surface area unit per unit mass, or low mass per unit radiating area, this advantage is offset to some extent by the lower effective emissivity of the droplet sheet than that of advanced heat pipe radiators. However, proponents during the last decades argued that with low vapor pressure liquids that are available over a wide radiating temperature range 250 K-1000 K, with negligible evaporation loss (silicone oils: 250 K-350 K; liquid metal eutectics: 370 K-650 K; liquid tin: 550 K-1000 K), the LDR could be adapted for

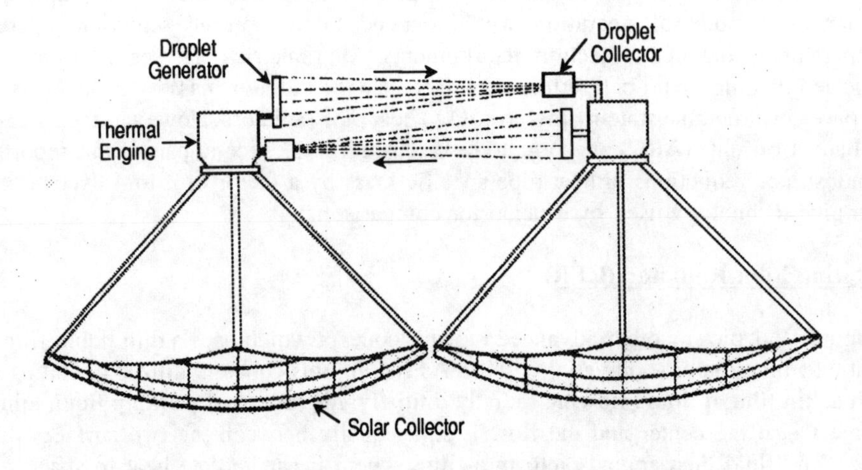

FIGURE 11. Dual Module Solar Power Satellite with Liquid Droplet Radiator.

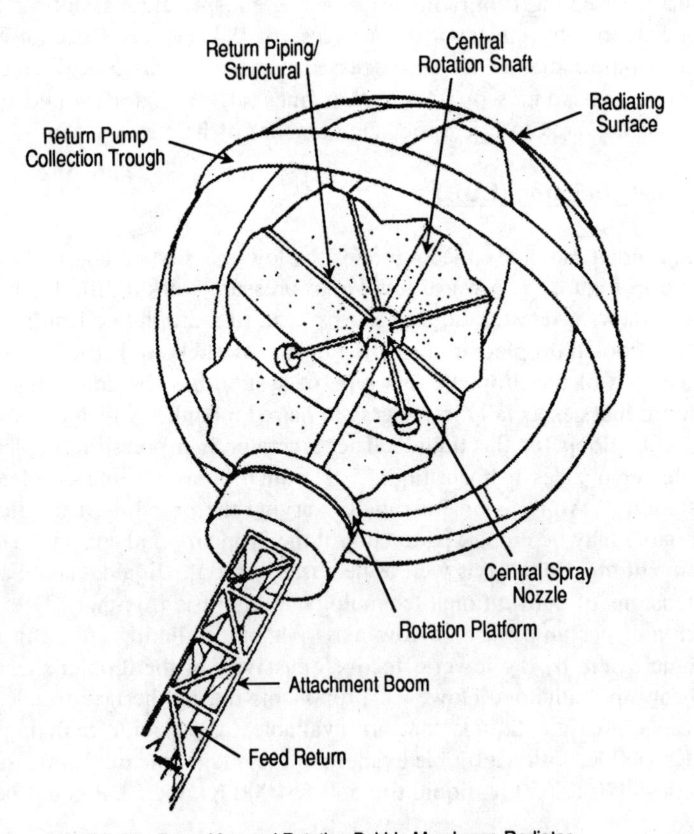

FIGURE 12. Boom Mounted Rotating Bubble Membrane Radiator.

a wide range of heat rejection applications (Elliot 1984).

A governing factor in design of liquid droplet radiators is the mass loss via evaporation. The mass required to replenish the evaporation must be included in the overall radiator mass for comparison with other systems. It has been found, however, for rejection temperatures between 300 K and 1000 K liquids are available with low enough vapor pressures that evaporation losses can be considerably smaller than the radiator mass, even with operational lifetimes of 30 years. Thus droplet radiators were considered suitable for a wide range of applications, from heat rejection in high-temperature thermal engines, where rejection temperatures might be in the range 500 K-1000 K, to cooling of photovoltaic cells and heat rejection from refrigerators where rejection temperatures would be in the 250 K-350 K range (Mattick et al. 1981).

An extension of the liquid droplet radiator, referred to as the liquid sheet radiator (LSR) has also been proposed. The operation of this type of system is similar to that of the liquid droplet radiator with the exception, that a continuous liquid sheet, rather than a multitude of individual droplets is used to accomplish the required heat rejection. Because fabrication of the narrow slits used in producing sheet flow do not require the precision machining techniques required for small orifices, this system reduces the level of technology development required. In addition, the liquid sheet radiator requires lower pumping power due to the reduced viscous losses and offers a simplified collection system due to a self focussing feature (Chubb et al. 1987). Both, the LDR and the LSR are compatible with power systems that have near constant heat rejection temperatures (Juhasz et al. 1991). However, they are not compatible with closed cycle gas turbine (CCGT) power systems, which must reject heat over a broad temperature range (Juhasz et al. 1991 and Juhasz et al. 1993). A recent status report on LSR development (Chubb et al. 1993) summarizes the work done on sheet stability and points out the need to conduct sheet emissivity measurements and to solve the problem of developing a sheet fluid collector before a viable LSR can be demonstrated.

Curie Point Radiator (CPR)

The Curie Point Radiator maintains the low mass advantage of the liquid droplet radiator and is similar in operation with one major exception. The Curie point radiator utilizes a large number of small solid ferromagnetic particles (Carelli et al. 1986). These particles are heated to a temperature above the Curie point, that point at which a ferromagnetic material loses its magnetic properties, and are ejected from the heat source towards a magnetic field. As the particles radiate heat to space and cool, they regain their magnetic properties and are collected by a magnetic field collector. The Curie point radiator has all of the advantages of the liquid droplet radiator, such as low mass to radiating area ratio, reduced mass for meteoroid protection, and mass of radiating particles which is only a minor fraction of the total. In addition, the unique characteristics of the ferromagnetic particles result in several other significant advantages, including (1) a particle inventory which can be actively controlled thereby reducing the loss of particles, (2) particles which can be coated to increase surface emissivity (a value as high as 0.9 can be achieved with SiC coating), and (3) elimination of the need for strict temperature control. The key disadvantage is the possibility of magnetic perturbations to other components of the spacecraft. Also, the problem of transferring spacecraft reject heat to the particle stream and the actual mass transport of the particles through the power system heat exchanger has not been resolved.

Rotating Bubble Membrane Radiator (RBMR)

Perhaps the most promising alternative to space constructable heat pipe radiators after liquid droplet radiators technologies, is the rotating bubble membrane radiator (Webb et al. 1988). Rotating bubble membrane radiators function as a two-phase direct contact heat exchanger. This hybrid radiator design incorporates the high surface heat fluxes and isothermal operating characteristics of conventional heat pipes along with the low system masses normally associated with liquid droplet radiators. As pictured in Figure 12, a two-phase working fluid enters the bubble through a central rotation shaft where it is sprayed radially from a central spray nozzle. This combination of liquid droplets and vapor moves from the central portion of the bubble towards the outside, transferring heat by both convection and radiation. As the droplets move outward, they increase in size due to condensation of vapor upon the droplet surface and by collision with other droplets. Upon striking the inner surface of the radiator the droplets form a thin surface film. This film then flows towards the equator due to the rotationally induced artificial gravity. Heat transfer between the fluid and bubble radiator then becomes a combination of conduction and convection. As the fluid reaches the equator of the sphere, it is collected in a gravity well and pumped back to repeat the process.

To operate effectively in space, the rotating bubble membrane radiator will include design features to minimize damage and mitigate coolant losses that may result from meteoroid and space debris impact. New high-strength, low-weight fiber and metallic alloy cloths show excellent promise for inhibiting micrometeoroid penetration of the rotating bubble membrane radiator (Webb et al. 1988). In addition, design options are being considered that would seal membrane penetrations and reduce coolant losses from micrometeoroid penetrations. Selection of materials for the thin film membrane is dictated by the desired operating temperature. Candidate materials include carbon-epoxy compounds, silica, alumino-borosilicate, or silicon carbide cloth with metallic liners (Swako 1983), and niobium-tungsten composites, with final selection of the envelope material depending upon the radiator fluid and its intended operating temperature. Pump selection will also be determined by the working fluid. Electromagnetic pumps are possible candidates for liquid metal coolants, and mechanical or electric pumps are favored for other applications.

SUMMARY

Among the various heat pipe technologies considered for advanced heat rejection systems (see Table 1), external artery and conventional axially grooved heat pipes have the greatest heat rejection capabilities. For high peak load pulsed heat absorption expandable roll out fin radiators offer a considerable weight savings over conventional tube and fin radiators. However, further study is necessary to improve the micrometeoroid vulnerability and header/fin heat exchanger design and operating characteristics (Ponappan et al. 1984). Among the inflatable radiator concepts previously developed, a retractable cylinder configuration with a stationary sponge appears to be the best candidate and although previous analyses has indicated that this type of system has good dynamic stability and excellent thermal behavior, additional investigations are required.

Several other advanced radiator systems have been proposed to reduce the mass requirement. Among these the most actively pursued through the late eighties was the liquid droplet radiator, where sub-millimeter liquid particles constitute the radiating surface. While the liquid droplet radiator has the potential for substantial reduction in mass versus the conventional radiator systems, problems associated with inventory losses due to vaporization, aiming inaccuracies and splashing on the collector, which tend to increase the total system mass must be addressed in future work. Another disadvantage of the liquid droplet radiator which must be resolved is that it is not suitable for missions requiring high maneuverability during full power operation. Similar observations apply to the liquid sheet radiator, albeit it would be much easier and cheaper to fabricate the coolant fluid injectors for this concept. The moving belt radiator concept compares favorably with heat pipe radiators on the basis of specific mass, however, the biggest advantage is achieved with the hybrid belt systems that exploits the phase change potential of a liquid belt radiator, yet offers high surface emissivities over a broad range of temperatures. But control of the belt shape under micro-gravity operating conditions and long term reliability of roller drive system is a serious problem. According to its proponents, the Curie point radiator, utilizing the ferromagnetic properties of the radiating particles, represents a unique concept which offers significant advantages in mass, high reliability, and practically unlimited temperature range (Carelli 1986). However, some key drawbacks, including heat transfer to and actual mass transport of the particles through the power system heat exchanger, and the possibility of magnetic perturbations to other spacecraft components, had not been resolved at the time of program termination.

Although the space constructable radiator is the most developed of the concepts and is a proven reliable technology, additional investigations into the behavior during startup from a frozen state and the effect of on orbit accelerations needs to be addressed. While rotating bubble membrane radiators and rotating film radiators presently lack the technical maturity of heat pipe radiators, rotating machinery and shaft vapor seals to space have proven effective and do not require additional development. Moving belt radiators require seals to wipe off the working fluid without leakage or damage to the belt as the belt exits the heat exchanger. In contrast, the liquid droplet radiator, Curie point radiator and rotating bubble membrane radiator require non-existent technologies which must be completely developed, tested, and qualified. Liquid droplet radiators require a droplet generator/collector combination with a high degree of aiming accuracy. Curie point radiators require a collector with a magnetic field generator and an effective heat exchanger to transfer heat to the solid particles from the working fluid.

TABLE 1. COMPARISON OF ADVANCED RADIATOR CONCEPTS

CRITERION	(SCR)	(ROF)	(RSR)	(MBR)	(LBR)	(RFR)	(LDR/ LSR)	(CPR)	(RBMR)
Weight	Mod.	Mod.	High	Mod.	Mod.	Mod.	Low	Low	Low
Reliability	Mod.	Avg.	Mod.	Mod.	Mod.	Good	Mod.	Exc.	Good
Maintenance Required	Low	Mod.	Low	Mod.	Mod.	Low	Mod.	Mod.	High
Technology Readiness	Exc.	Mod.	Good	Poor	Poor	Mod.	Mod.	Poor	Poor
Life Expectancy	Good	Good	Mod.	Mod.	Poor	Mod.	Exc.	Unk.	Mod.
System Complexity	Low	High	Mod.	High	Mod.	Mod.	High	High	High
Area Required	High	High	Mod.	Mod.	Mod.	Mod.	Mod.	Mod.	Low
Performance	Exc.	Unk.	Good	Unk.	Unk.	Mod.	Good	Unk.	Mod.
Life Cycle Cost	Low	Unk.	Unk.	Unk.	Unk.	Unk.	Unk.	Unk.	Low
Micrometeoroid Vulnerability	Mod.	High	Low	Mod.	Mod.	Mod.	Low	Low	High

Legend

SCR = Space Constructable Radiator
RSR = Rotating Solid Radiator
LBR = Liquid Belt Radiator
LDR = Liquid Droplet Radiator
LSR = Liquid Sheet Radiator

ROF = Roll Out Fin Radiator
MBR = Moving Belt Radiator
RFR = Rotating Film Radiator
CPR = Curie Point Radiator
RBMR = Rotating Bubble Membrane Radiator

PART II. Highlights of the LeRC-CSTI Thermal Management Program

Ongoing "CSTI Thermal Management" related work at LeRC, to be concluded during 1994, has been an integral part of the NASA CSTI High Capacity Power Program, and specifically the Tri-Agency (DOD/DOE/NASA) SP-100 nuclear reactor space power program (Winter 1991).

The goal of the LeRC thermal management effort is to develop near term space radiator and heat rejection system concepts, optimized for a spectrum of space power conversion systems for planetary surface (lunar base) and nuclear propulsion applications for deep space long duration missions needed for the Space Exploration Initiative (Bennett et al. 1991). The power, or energy, conversion system concepts range from static systems such as thermoelectric or thermionic, to dynamic conversion systems based on heat engines such as the Stirling engine or the closed cycle gas turbine also known as the Closed Brayton Cycle (CBC). Although the principal heat sources for these systems are nuclear (Juhasz et al. 1986), the technology being developed for the heat rejection subsystem is also applicable to low earth orbit (LEO) dynamic power systems with solar energy input, using a concentrator and heat receiver, studied as alternatives to photovoltaic power systems (Brandhorst et al. 1986). The performance goals for the advanced radiator concepts being developed are lower radiator mass (specific mass of 5 kg/m^2, or lower), at a surface emissivity of at least 0.85 over the operating temperature range, greater survivability in a micro-meteoroid or space debris environment (up to 10 years), and a sub-system reliability of 0.99 or higher. These performance goals may be realized by radiator segmentation and parallel redundancy, using a large number of heat pipes. Achievement of these goals may lead to a reduction of the SP-100 radiator specific mass by a factor of two or more over the original baseline design, and even greater mass reductions for radiators used in contemporary spacecraft.

The project elements, shown in Figure 13, include development of advanced radiator concepts under LeRC managed contracts and a NASA/DOE interagency agreement, as well as in house work directed at radiator design for optimum power system matching and integration. Also, in house and university supported heat pipe research and development is being carried out, comprising analytical computer code development for predicting heat pipe performance, both under steady state and transient operating conditions, along with experimental testing for the purpose of validating analytical predictions.

Contributing to the in-house advanced development program is continued research on radiator surface treatment techniques (surface morphology alteration), aimed at

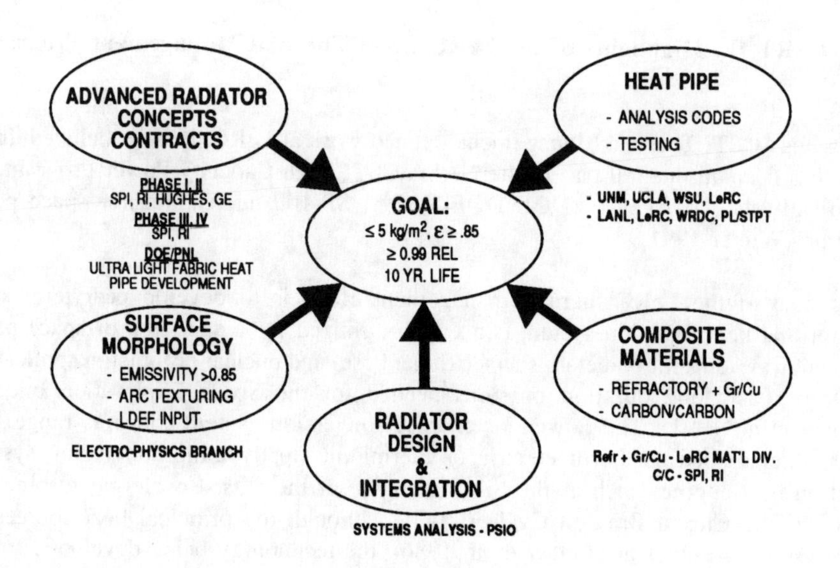

FIGURE 13. LeRC Thermal Management Project Elements

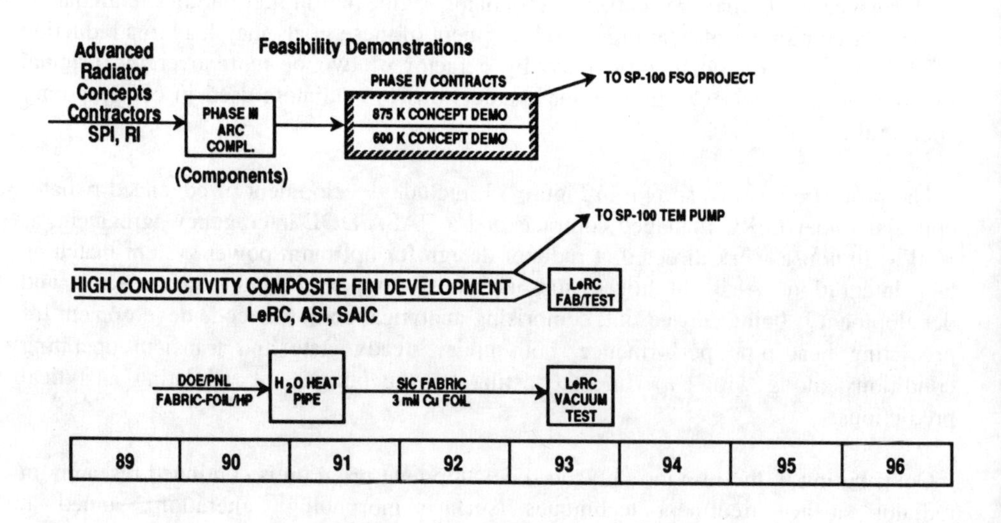

FIGURE 14. Thermal Management Project Plan

enhancing surface emissivity and resistance to atomic oxygen attack.

The development of new radiator materials showing high strength-to-weight ratio, and high thermal conductivity, such as carbon-carbon composites for light weight radiator fin applications is another major objective. The project plan to accomplish the above program objectives is shown in Figure 14. Note that due to funding constraints, the development of far term innovative radiator concepts covered in Part 1, such as liquid droplet radiator (LDR) or moving belt radiator (MBR), is not being actively pursued at the present time. Instead, technologies capable of development before the end of the decade are being concentrated on for both surface power and nuclear electric propulsion (NEP) applications. Near term applications to small spacecraft and technology transfer to terrestrial uses are also being considered.

This remainder of this paper will be devoted to a review of the major project elements, concentrating on the contracted efforts which account for the major portion of the baseline budget.

Advanced Radiator Concepts Development Contracts (ARC)

The Advanced Radiator Concepts (ARC) contractual development effort is aimed at the development of improved space heat rejection systems, with special emphasis on space radiator hardware, for several power system options including thermoelectric (T/E) and Free Piston Stirling (FPS). The targeted improvements will lead to lower specific mass (mass per unit area), at high surface emissivity, higher reliability and higher survivability in a natural space environment, thereby leading to longer life for the power system as a whole. As stated above, specific objectives are to achieve specific mass values < 5 kg/m^2, achieved with radiator surface emissivities of 0.85 or higher, at typical radiator operating temperatures, and reliability values of at least 0.99 for the heat rejection subsystem over a ten year life. Although the above specific mass figure does not include the mass of pumps and the heat transport duct, it represents about a factor of two mass reduction over the baseline SP-100 radiator, and even greater mass savings for state of the art heat rejection systems used in current spacecraft applications.

Phases I, II, and III of the ARC contracts have been completed by both contractors, Space Power Inc. (SPI) of San Jose, Ca., and Rockwell International (RI) of Canoga Park, Ca. Based on phase III results both contractors were selected to proceed into component level development, fabrication, and demonstration, to be accomplished under phase IV over a two year period, to be concluded by early 1994.

Both, a high temperature heat rejection option (800 K to 830 K) applicable to T/E power conversion systems, and a low temperature option (500 K to 600 K) applicable to Stirling power conversion systems, are being developed by SPI, while the focus at RI has been on heat rejection technology for the higher temperature T/E power systems.

Contract NAS 3-25208 with SPI

Among the advanced concepts proposed by SPI are the "Telescoping Radiator" (Begg et al. 1989 and Koester et al. 1991) for multi-megawatt thermoelectric (TE) or liquid metal Rankine (LMR) power systems, shown in Figure 15, and the "Folding Panel Radiator" (Koester et al. 1991), shown in Figure 16, for the 500 K to 600 K heat rejection temperature range. The latter concept was based on a pumped binary lithium/sodium potassium (Li/NaK) loop, motivated by a desire to avoid the need for mercury heat pipe radiators originally planned for a "free piston Stirling (FPS)" power system which rejects heat in the above temperature range.

Lithium-NaK Binary Pumped Loop

A detail of a typical Li/NaK heat rejection loop, which also utilizes high conductivity fins for the heat transport is illustrated in Figure 17. As indicated in Figure 18 (a), the advantage of using a lithium-NaK mixture, rather than NaK alone (melting point 261 K), lies in its combining the high heat capacity and low pumping power of Li (melting point 452 K) with the liquid pumping capability of NaK, down to its freezing temperature of 261 K.

To illustrate operation of this binary loop during system start-up and shutdown, a brief explanation is in order. During startup, (frozen Li) liquid NaK would be pumped through the inner cores of radiator tube passages in hydraulic contact with the frozen layers of Li coating the inner passage surfaces. As the NaK is heated during power system startup, it will eventually melt the solid Li annuli by direct contact forced convection heat transfer, progressively mixing with the NaK to form the all liquid Li/NaK coolant. Conversely, on shut-down of the power system, the molten lithium with its higher freezing point will selectively "cold trap" or freeze on the inner passage surfaces as their temperatures drop below the 452 K, while the NaK continues to be pumped in its liquid state through the inner cores of the radiator passages.

Tests conducted thus far, using a trunnion mounted test loop which can be rotated about pitch and roll axes to isolate gravity effects (Figure 18 b), have demonstrated the feasibility of the concept up to lithium volume fractions of 50 percent. In particular the feasibility of a heat rejection system based on a binary Li/NaK pumped loop was demonstrated during transient operating conditions representative of both the cool-down (Li freezing) and the warmup (Li melting) phases of typical alkali metal heat rejection pumped loops. The thawing process during the warmup condition had to be controlled

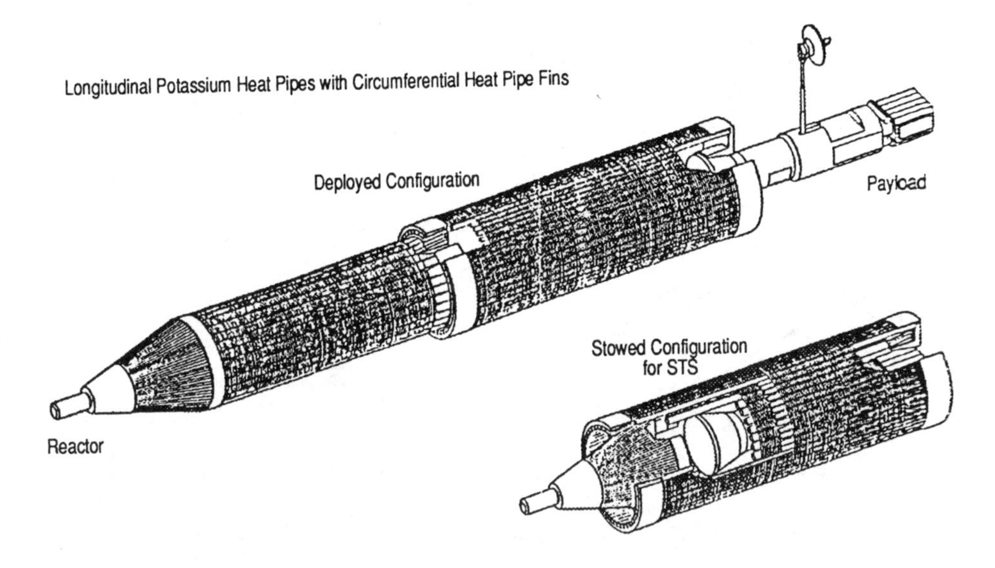

FIGURE 15. Multi-Megawatt Telescoping Cylinder Heat Pipe Radiator Concept..

FIGURE 16. Folding Panel Heat Pipe Radiator Concept.

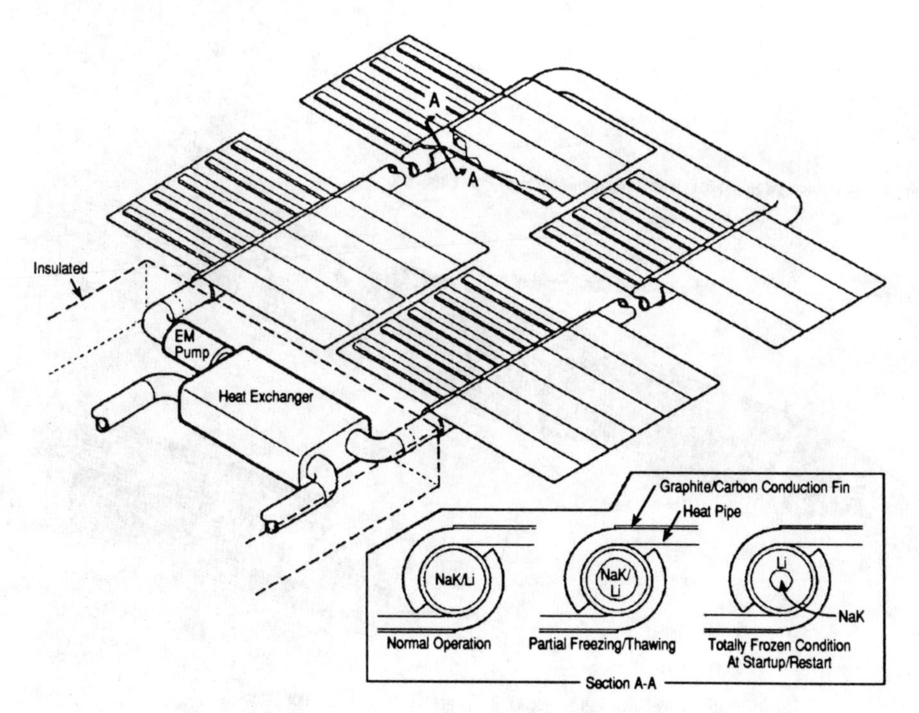

FIGURE 17. Pumped Li/NaK Binary Loop Radiator Concept.

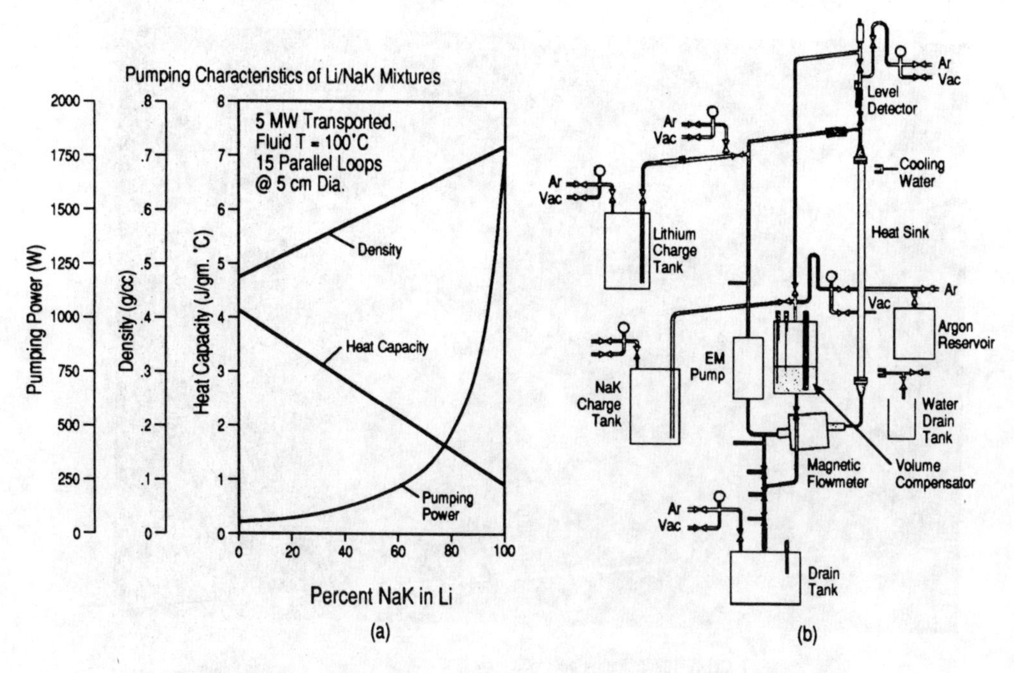

FIGURE 18. Binary Li/NaK Radiator Development (a) Design Characteristics (b) Test Loop.

very closely to avoid plugging of the flow loop due to molten Li re-freezing downstream at certain operating conditions. Current efforts focus on widening the operating envelope by a variety of techniques. One of these involves the use of fine mesh screens which act as semi-permeable membranes to NaK under certain operating conditions.

A two dimensional computer analysis of the cool-down (freeze) and the warmup (thaw) processes has also been completed, including color graphics output. Although the flat flow channel cross sectional geometry assumed in the analysis deviated from the cylindrical flow channel used in the experimental loop, this computer code nevertheless permitted visualization of basic phenomena taking place within the binary loop during the warmup and cool-down periods. A video tape was produced illustrating several cases with and without plugging of the flow channel due to lithium freezing over the entire channel cross section. Time and funding permitting, freeze-thaw behavior at higher than 50 volume percent Li and with two parallel channels will also be briefly explored (Koester et al. 1994).

High Conductivity Fin Development

Progress was also made by SPI in working with innovative subcontractors, namely Applied Sciences Inc. (ASI) and Science Applications International Co. (SAIC), who have demonstrated considerable success in the development and fabrication of high thermal conductivity composite materials for space radiator fin applications. In particular, ASI has produced a composite by chemical vapor deposition (CVD) and densification of closely packed (up to 60 volume percent) vapor grown carbon fibers (VGCF) which was shown to have high thermal conductivity, near 560 W/m K, at room temperature, and a density of 1.65 g/cc. Similarly SAIC has successfully fabricated composites using short (near 0.01 m long) VGCF fibers with a density of 1.6 g/CC and a demonstrated conductivity of 470 W/m K at near 600 K operating temperature. A comparison of these properties with those of copper (density of 8.9 g/cc; thermal conductivity of 380 W/m K) reveals the significantly higher conductivity-to-density ratio of these composites.

Use of composite materials which exhibit specific thermal conductivity values at these levels for heat pipe fin applications permits increasing the fin length at constant fin efficiency, and it thereby has the potential of reducing radiator specific mass by over 60 percent for radiators that are radiative heat transfer surface limited. Technology for joining the high conductivity fins to the heat pipes by advanced brazing or welding techniques and processes that will lead to even higher composite thermal conductivity values have also been recently demonstrated (Denham et al. 1994).

Contract NAS 3-25209 with RI

A sketch of the "Petal-Cone" radiator concept being developed at RI (Rovang et al. 1988) is shown in Figure 19. Since each of the "petals", or radiator panels, are composed of a large number (384) of variable length C-C heat pipes mounted transverse to the panel axis, a major objective of this effort is the development of these integrally woven graphite carbon tubes with an internal metallic barrier that is compatible with the intended potassium working fluid.

Carbon-carbon (C-C) heat pipe tube sections with integrally woven fins were fabricated under phase III. Highlights of the fabrication process are illustrated in figure 20. Because of its low cost, commercial availability, and ease of weaving a T-300 fiber was selected for this demonstration of C-C heat pipe preform fabrication. This polyacrylonitrile (PAN) fiber was judged to represent a tradeoff between high elastic modulus, and consequently ease of handling and weaving, at medium thermal conductivity (80 W/m K), as contrasted to some very high conductivity fibers which, however, may be brittle and difficult to weave.

Several fiber architectures were investigated before settling on the angle interlock, integrally woven concept. In this design, the axial fiber bundles, referred to as warp weavers, are woven in an angle interlock pattern, repeatedly traversing from the ID to the OD surface of the tube. An unfilled "Novolack"/resole prepreg resin was selected for prepregging the woven preforms followed by a low pressure impregnation and carbonization process for densification of the composite.

Considerable progress was made in the development of internal metallic coatings to ensure compatibility of the heat pipe surface with the potassium working fluid. A coating consisting of a 2 to 3 micron rhenium sublayer with a 70 - 80 micron niobium overlayer emerged as the final recommended coating design. Due to funding and time limitations, however, this final coating design and the recommended method to achieve it by a novel chemical vapor deposition (CVD) process utilizing a moving heat source could not be fully implemented during phase III. Other coating approaches which were tried achieved incremental improvements over each previous coating attempt, but a constant thickness coating over the full length of the tube without any flaws or imperfections could not be achieved.

Due to the problems encountered with achieving a flawless metallic coating on the internal C-C tube surface, with the inception of phase IV a technical direction was issued to RI. This directive required that the safe containment of the heat pipe working fluid be accomplished by means of a thin walled metallic liner, rather than the metallic

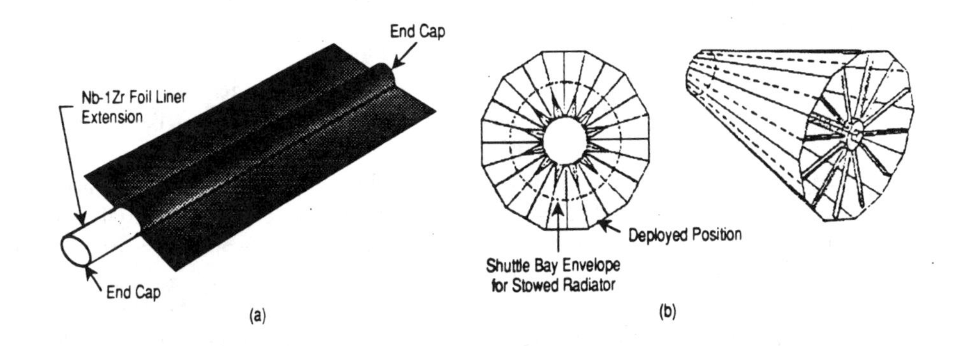

FIGURE 19. Cone- Petal Heat Pipe Radiator Concept
(a) Heat Pipe (b) Cone Radiator (12 Panels)

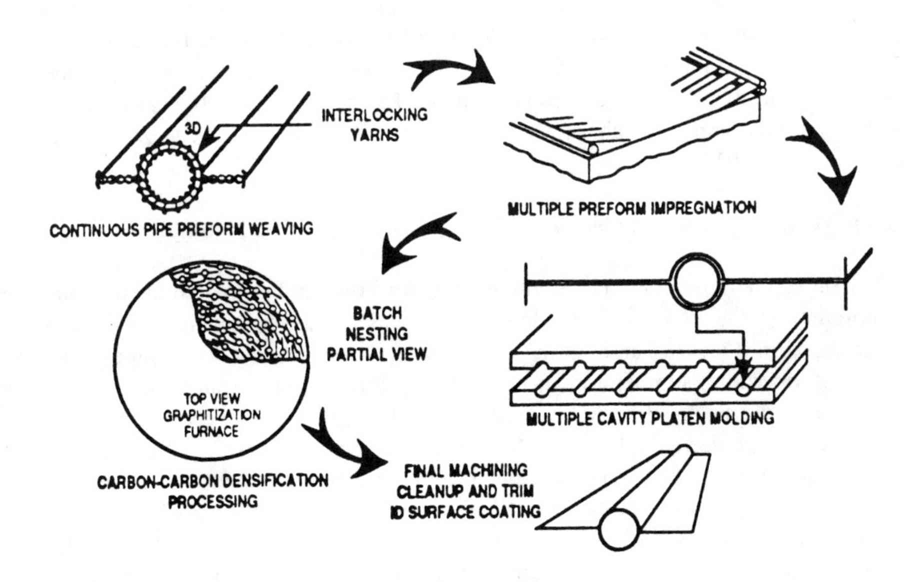

FIGURE 20. Integral Finned Graphite-Carbon Heat Pipe Fabrication Process.

coating that had been under development during phase III. Another advantage of the metal liner approach was that the heat pipe evaporator could be formed by simple extension of the liner beyond the C-C shell.

Concurrently with this task a high temperature braze or other joining process was developed, in order to insure good mechanical and thermal contact between the thin metallic liner and the C-C internal tube surface. Bonding of the entire liner surface to a finned C-C tube of 0.3 m length has also been achieved by use of ternary braze alloys, such as Silver ABA or "Cusil" ABA. This was necessary to prevent partial separation and collapse of the liner at conditions where the external atmospheric pressure exceeds the internal pressure of the working fluid.

Concerning the integral fin weaving process using T-300 fibers, significant improvements were made in the weaving process which lead to the elimination of a troublesome internal cusp formed at the fin-tube interface. This was accomplished by changing the weave architecture, so that the outer, rather than the inner plies were used to form the fins.

In addition to fabricating and testing a complete 2.5 cm OD heat pipe with a niobium/zirconium alloy liner using the T-300 composite shell, the fabrication of a higher conductivity composite (P95WG) finned heat pipe shell was also completed. With a measured conductivity of over 300 W/m K for this composite, fin length could be doubled from 2.5 cm to 5 cm, resulting in a reduction of specific mass (for one sided heat rejection) to 2.9 kg/m^2, as compared to 4.2 kg/m^2 for the T-300 composite. A recent update summarizing the fabrication and testing of the RI C-C heat pipe is given by Rovang et al. (1994).

Light Weight Advanced Ceramic Fiber (ACF) Heat Pipe Radiators

The objective of this joint NASA/LeRC and Air Force program with Pacific Northwest Laboratory (DOE/PNL) was to demonstrate the feasibility of light weight flexible ceramic fabric (such as aluminum borosilicate) metal foil lined heat pipes for a wide range of operating temperatures and working fluids. Specifically the NASA LeRC objectives were to develop this concept for application to Stirling space radiators with operating temperatures below 500 K, using water as the working fluid. The specific mass goals for these heat pipes were < 3 kg/m^2 at a surface emissivity of at least 0.80. Several heat pipes were built using titanium and copper foil material for containment of the water working fluid. A heat pipe with an eight mil (0.2 mm) titanium liner, designed to operate at temperatures up to 475 K, was demonstrated at the 8th Symposium on Space Nuclear Power Systems (SSNPS) in Albuquerque, in early January 1991 (Antoniak et al. 1991).

An innovative "Uniskan Roller Extrusion" process has been developed at PNL and used to draw 30 mil (0.75 mm) wall tubing to a 2 mil (0.05 mm) foil liner in one pass. Moreover, this process eliminated the need for joining the thin foil section to a heavier tube section for the heat pipe evaporator, which needs to be in tight mechanical and thermal contact with the heat rejection system transport duct. The heavier end sections are also used for attachment of the end caps. The liner fabrication technique was also applied to the RI heat pipe fabrication discussed above, and it is expected to have broad application beyond the scope of this program. The water heat pipes fabricated for LeRC by PNL have been subjected to a test program to evaluate performance and reliability at demanding operating conditions, including operating pressures up to 25 bar. Tests were conducted with and without wicks with the heat pipes in various gravity tilt orientations from vertical to horizontal. In addition a number of wick designs were tested for capillary pumping capability, both in ground test and under low G conditions produced during KC-135 aircraft testing. The work was reported in a paper presented at the 27th National Heat Transfer Conference (Antoniak et al. 1991). Future work in this area needs to focus on perfecting the heat pipe fabrication procedure, using very thin (1 to 2 mil (0.025 to 0.05 mm)) foil liners, internally texturized by exposure to high pressures. Because of the high operating pressures, hyper-velocity and ballistic velocity simulated micro-meteoroid impact tests will need to be conducted with thin walled pressurized tubes enclosed in woven fabric, in order to ascertain if secondary fragments from a penetrated heat pipe will result in failures of neighboring heat pipes. Another major challenge will be to design and fabricate a heat pipe with high conductivity, light weight fins, as a first step toward the fabrication of light weight radiator panels.

Supporting Project Elements

Space limitations prevent a detailed discussion of the remaining project elements referred to in Figure 13. However, a brief paragraph highlighting these activities is warranted. As mentioned previously, the system integration studies performed in house, serve to guide the overall thermal management work by providing the proper framework for it. As shown by Juhasz (1991), for example, a liquid sheet radiator (LSR) with lighter specific mass than a heat pipe radiator, will not necessarily benefit all power conversion systems equally. As discussed in the reference, the LSR (or the LDR) concept is not suited to the heat rejection temperature profile of a CBC power system. However, it does work well with a Stirling power system which rejects heat at a near constant temperature. As a further example of how radiator-power system integration studies are used to ascertain radiator induced power system performance degradation, the reader is referred to Figure 21. The curves shown here illustrate the reduction in power output and efficiency for both Brayton and Stirling power systems, resulting from a reduction in cycle temperature ratio due to a loss of radiator area, caused, for example, by micrometeoroid damage. It is reassuring to note, that even with a loss of 50 % of

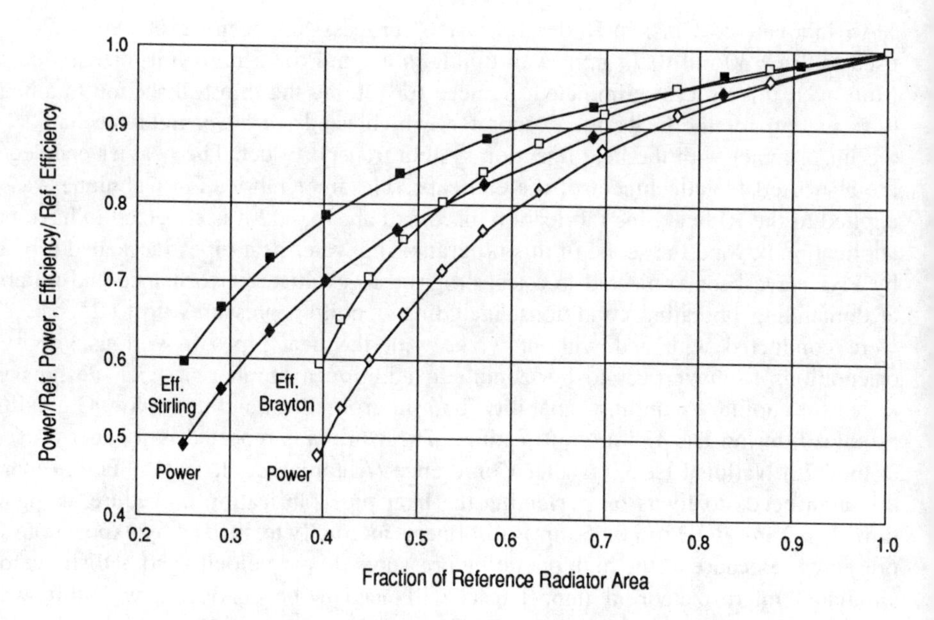

FIGURE 21. Sample Results from Radiator-Power System Integration Study

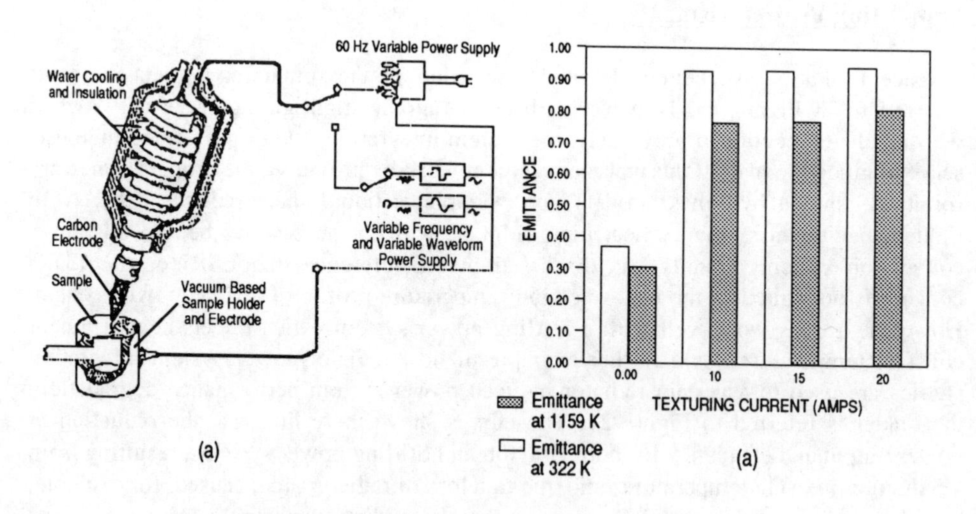

FIGURE 22. Surface Emittance Enhancement by Arc Texturing.
(a) Texturing Equipment Setup
(b) Emittance of Graphite/Copper Samples

radiator area, a Stirling is still capable of producing over 75 percent of its design power, while a CBC produces over 65 percent of its rated output.

A typical example of radiator surface morphology alteration by arc texturing for emissivity enhancement purposes is shown in Figure 22. With the arc texturing apparatus shown in Figure 22 (a), the adjoining bar chart, Figure 22 (b), shows the surface emittance results achieved with various arc current values for graphite-copper samples produced under the in-house materials program. Additional details on the emittance enhancement process and measurement techniques are discussed by Rutledge et al. (1991). A detailed treatment of emittance enhancement of C-C composite surfaces by atomic oxygen (AO) beam texturing is also given by Rutledge et al. (1989), and a brief overview of the same topic is included in a CSTI status report (Winter et al. 1991).

Heat pipe performance modelling, both under steady state and transient operating conditions is being conducted in-house and under university grants respectively, with University of California, Los Angeles (UCLA), University of New Mexico (UNM), and Wright State University (WSU). The objective of this work is to develop a capability to analytically predict transient operation of heat pipes, particularly during the startup and cool-down phases. An especially important feature of this work is the development of an analysis code capable of modeling startup with the working fluid initially in the frozen state for a variety of working fluids, including water and liquid metals. Working versions of the codes have been developed, and validation of predicted performance by laboratory testing of heat pipes is under way at LeRC, Los Alamos National Laboratory (LANL), and Wright Laboratories (WL). Efforts have also been initiated to compile an experimental database by a systematic literature search and close communication with other researchers in the field.

CONCLUDING REMARKS

The LeRC CSTI Thermal Management Program was designed to combine a number of project oriented elements in order to accomplish the overall objective of reducing radiator specific mass by at least a factor of two at a subsystem reliability of 0.99 over a ten year life. Although the main focus was on support of the SP-100 program by advances in heat rejection technology, the concepts and hardware developed under this program are expected to benefit space power systems in general, ranging from solar photovoltaic or solar dynamic systems with a power level of a few kilowatts to future multi-megawatt power systems with nuclear heat sources for planetary surface and nuclear (electric) propulsion applications. In spite of the termination of the SP-100 program, its major subsystem technology advances, especially in the thermal management area, are judged to be ready for on orbit demonstration before the end of the decade, so that NASA's and the nation's long term goals in space exploration and utilization may be realized some time during the next century. In the mean time, terrestrial and small spacecraft applications of these technologies will also be pursued.

Acknowledgments

The authors gratefully acknowledge the support provided by the TAMU Center for the Commercialization of Space Power, the U. S. Naval Research Laboratory, Washington, D.C., and the NASA Lewis Research Center, Cleveland, Ohio. The dedicated work and excellent cooperation of ARC contract participants, SPI, RI, PNL, SAIC, and ASI, contributed greatly to the success of the ARC program.

References

Alario, J. (1984) "Monogroove Heat Pipe Radiator Shuttle Flight Experiment: Design, Analysis and Testing", SAE Paper 840959, 1984.

Antoniak, Z. I., Webb, B. J., Bates, J. M., Cooper, M. F., and Pauley, K. A. (1991) "Testing of Advanced Ceramic Fabric Wick Structures and their Use in a Water Heat Pipe" 27th NHTC, Minneapolis, Minn., July 28-31, 1991.

Apley, W. J. and Babb A. L. (1984) "Rotating Solid Radiative Coolant System for Space Nuclear Reactors", Technical Report PNL-SA-15433, Pacific Northwest Laboratory, Richland, Washington, 1984.

Begg, L. I. and Engdahl, E. H. (1989) "Advanced Radiator Concepts - Phase II Final Report for Spacecraft Radiators Rejecting Heat at 875 K and 600 K" NASA Contractor Report 182172, May 1989.

Bennett, G. and Cull, R. C. (1991) "Enabling the Space Exploration Initiative: NASA's Exploration Technology Program in Space Power" AIAA/NASA/OAI Conference on SEI Technologies, Cleveland, Ohio, Sept 4-6, 1991.

Brandhorst, H. W.; Juhasz, A. J.; and Jones, B. I. (1986) "Alternative Power Generation Concepts in Space", NASA TM-88876 (1986).

Carelli, M. D., Gibson, G., Flaherty, R., Wright, R.F., Markley, R. A., and Schmidt, J. E. (1986) "A Novel Heat Rejection System for Space Applications: The Curie Point Radiator", *Proc. 21st Intersociety Energy Conversion Engineering Conference*, Paper

No. 789592, 1986.

Chittenden, D., Grossman, G., Rossel, E., Van Etten, P., and Williams, G. (1988) "High Power Inflatable Radiator for Thermal Rejection from Space Power Systems", *Proc. 23rd Intersociety Energy Conversion E Engineering Conference*, Vol. 1, pp. 353-358, 1988.

Chow, L. C., Mahefkey, E. T., and Yokajty, J. E. (1985) "Low Temperature Expandable Megawatt Pulse Power Radiator", Paper No. 85-1078, 1985.

Chow, L. C., and Mahefkey, E. T. (1985) "Fluid Recirculation, Deployment and Retraction of an Expandable Pulse Power Radiator", 1986.

Chubb, D. L.; and White, K. A. (1987) "Liquid Sheet Radiator", NASA TM-89841, NASA Lewis Research Center, 1987. (Also, AIAA Paper 87-1525, 1987).

Chubb, D. L.; Calfo, F. D.; and McMaster, M. F. (1993) "Current Status Of Liquid Sheet Radiator Research", NASA TM-105764.

Denham, H. B.; Koester, K. J.; Clarke W.; and Juhasz A. J. (1994) "NASA Advanced Radiator C-C Fin Development" 11th Symposium on Space Nuclear Power and Propulsion, Albuquerque, New Mexico, Jan 10-14, 1994.

Edelstein, F. (1984) "Thermal Bus Subsystem for Large Space Platforms", Technical Report PR-3147-13, Grumman Aerospace Corporation, Bethpage, N.Y, 1984.

Elliott, D. G. (1984) "External Flow Radiators for Reduced Space Powerplant Temperature", *Proc. 1st Symposium on Space Nuclear Power Systems*, University of New Mexico, Albuquerque, New Mexico, 1984.

Ellis, W. (1989) "The Space Station Active Thermal Control Technical Challenge", AIAA Paper No. 89-0073, 1989.

Gustafson, E.; and Carlson, A. W. (1987) "Solar Dynamic Heat Rejection Technology", Technical Report NASA CR-179618, Grumman Aerospace Corporation, Bethpage, N.Y, 1987.

Harwell, W. (1983) "The Heat Pipe Thermal Cannister: An Instrument Thermal Control System", SAE Paper 831124, 1983.

Juhasz, A. J.; El-Genk, M. S.; and Harper, W. B. Jr. (1993) "Closed Brayton Cycle

Power System with a High Temperature Pellet Bed Reactor Heat Source for NEP Applications", in Proc. 10th Symposium on Space Nuclear Power and Propulsion Systems, DOE CONF-930103, M.S. El-Genk and M. D. Hoover, eds., American Institute of Physics, New York, AIP Conference Proc. No. 271, 2: 1055-1064.

Juhasz, A. J. and Jones, B. I. (1986) "Analysis of Closed Cycle Megawatt Class Space Power Systems with Nuclear Reactor Heat Sources" Proceedings of the Third Symposium on Space Nuclear Power Systems, Albuquerque, NM, Jan. 13-16, 1986.

Juhasz, A. J. and Chubb, D. L. "Design Considerations for Space Radiators Based on the Liquid Sheet (LSR) Concept", NASA TM 106158; Presented at the 26th IECEC, Boston, Massachusetts, August 4-9, 1991.

Koenig, D. R. (1985) "Rotating Film Radiators for Space Applications", *Proc. 20th Intersociety Energy Conversion Engineering Conference*, Vol. 1, pp. 439-445, 1985.

Koester, J. K. and Juhasz, A. J. (1991) "The Telescoping Boom Radiator Concept for Multimegawatt Space Power Systems" AIAA/NASA/OAI Conference on Advanced SEI Technologies, Cleveland, Ohio, Sept 4-6, 1991.

Koester, J. K.; and Juhasz, A. J. (1994) "NASA Advanced Radiator Technology Development" 11th Symposium on Space Nuclear Power and Propulsion, Albuquerque, New Mexico, Jan 9-13, 1994.

Kossan, R., Brown, R., and Ungar, E. (1990) "Space Station Heat Pipe Advanced Radiator Element (SHARE) Flight Test Results and Analysis", AIAA Paper No. 90-0059, 1990.

Leach, J. W., and Cox, R. L. (1978) "Flexible Deployable Retractable Space Radiators", *Progress in Astronautics and Aeronautics*, Vol. 60, 1978.

Mahefkey, T. (1982) "Military Spacecraft Thermal Management, The Evolving Requirements and Challenges", Paper No. 82-0827, 1982.

Mattick, A. T. and Hertzberg, A. (1981) "Liquid Droplet Radiators for Heat Rejection in Space", *Journal of Energy*, Vol. 5, No. 6, pp. 387-393, 1981.

Mertesdorf, S. J.; Pohner, J. A.; Herold, L. M.; and Busby, M. S. (1987) "High Power Spacecraft Thermal Management" Technical Report AFWAL-TR-86-2121, TRW Inc., Redondo Beach, California, 1987.

Oren, J. A. (1982) "Flexible Radiator System", Vought Report No. 2-19200/3R-1195 B, 1982.

Peterson, G.P. (1987) " Thermal Control Systems for Spacecraft Instrumentation", *AIAA Journal of Spacecraft and Rockets*, Vol. 24, No. 1, pp. 7-13, 1987.

Ponnappan, R.; Beam, J. E.; and Mahefkey, E. T. (1984) "Conceptual Design of a 1 m Long 'Roll Out Fin' Type Expandable Space Radiator", Paper No. 86-1323, 1984.

Prenger, F. C., and Sullivan, J. A. (1982) "Conceptual Designs for 100-Megawatt Space Radiators", *Symp. on Advanced Compact Reactors Systems*, National Academy of Sciences, Washington, D. C., November 15-17, 1982.

Rankin, J. G., Ungar, E. K. and Glenn, S. D. (1989) "Development and Integration of the SHARE Payload Bay Flight Experiment", *AIChE Symposium Series*, Vol. 85, No. 265, 1989.

Rovang, R. D. (1988) "Advanced Radiator Concepts for SP-100 Space Power Systems", NASA Contractor Report 182174, October 1988.

Rovang, R. D.; Moriarty, M. P.; Ampaya, J. P.; Dirling, R. B.; and Hoelzl, R. (1990) "Advanced Radiator Concepts for SP-100 Space Power Systems - Phase III Final Report", NASA Contractor Report 187170, August 1990.

Rovang, R. D.; Hunt, M. E.; and Juhasz A. J. (1994) "Testing of a Liquid Metal Carbon-Carbon Heat Pipe" 11th Symposium on Space Nuclear Power and Propulsion, Albuquerque, New Mexico, Jan 10-14, 1994.

Rutledge, S. K.; Hotes, D. L.; and Paulsen, P. L.; (1989) "The Effects of Atomic Oxygen on the Thermal Emittance of High Temperature Radiator Surfaces", NASA TM 103224, 1989.

Rutledge, S. K.; Forkapa, M.; and Cooper, J. (1991) "Thermal Emittance Enhancement of Graphite - Copper Composites for High Temperature Space Based Radiators", AIAA/NASA/OAI Conference on SEI Technologies, Cleveland, Ohio, Sept 4-6, 1991.

Siegel, R. and Howell, J. R. (1980) *Thermal Radiation Heat Transfer*, Hemisphere Publishing, New York, NY, pp. 258-261, 1980.

Song, S. J.; and Louis, J. F. (1988) "Rotating Film Radiator for Heat Rejection in Space", *Proc. 23rd Intersociety Energy Conversion Engineering Conference*, Vol. 1, pp. 385-390, 1988.

Swako, P. M. (1983) "Flexible Thermal Protection Materials", NASA Ames Research Center, NASA Conference Publication #2315, NASA, 1983.

Teagan, W. P. and Fitzgerald, K. (1984) "Preliminary Evaluation of A Liquid Belt Radiator for Space Applications", Technical Report NASA CR-174807, Arthur D. Little, Cambridge, Massachusetts, 1984.

Weatherston, R. C. and Smith, W. E. (1960) "A Method for Heat Rejection for Space Power Plants", *ASR Journal*, Vol. 20, pp. 268-269, 1960.

Weatherston, R. C. and Smith, W. E. (1960) "A Method for Heat Rejection for Space Power Plants", Cornell Aeronautical Laboratory Report, No. DK-1369-A-1, 1960.

Webb, B. J. and Antoniak, Z. I. (1988) "Rotating Bubble Membrane Radiator for Space Applications", *Proc. 21st Intersociety Energy Conversion Engineering Conference*, Paper No. 869426, 1988.

White, K. A. III. (1988) "Moving Belt Radiator Technology Issues", Proc. 23rd Intersociety Energy Conversion Engineering Conference, Vol. 1, Denver, CO, pp. 365-371, 1988.

Winter, J. M. (1991) "The NASA CSTI High Capacity Power Program" AIAA/NASA /OAI Conference on Advanced SEI Technologies, Cleveland, Ohio, Sept 4-6, 1991.

ADVANCED STATIC ENERGY CONVERSION FOR SPACE NUCLEAR POWER SYSTEMS

C. P. Bankston
Jet Propulsion Laboratory
California Institute of Technology
4800 Oak Grove Dr.
Pasadena, CA. 91109
(818) 354-6793

ABSTRACT

Several advanced static thermal-to-electric conversion technologies have been investigated as potential alternatives to conventional thermoelectrics and thermionic converters. The goal has been to identify static converters with efficiencies that are competitive with dynamic systems. Thermophotovoltaic, AMTEC, HYTEC, thermo-acoustic and liquid metal MHD are the principal technologies studied. Of these, AMTEC has received the most attention and is now the most mature. Several key fundamental issues have been resolved and engineering issues are now being addressed. Thermophotovoltaics has the advantage of extensive photovoltaic systems heritage, however the requirements to maintain a highly efficient optical cavity for long periods and to provide an efficient means of cooling the cells to near room temperature are challenges that must be addressed. Experiments have begun to characterize the HYTEC converter and basic technical issues are now being identified for further study. Thermoacoustics has received less attention and system concept designs are not yet developed. Liquid metal MHD experiments have not yet demonstrated performance consistent with theoretical projections and substantial obstacles to high efficiency performance remain.

INTRODUCTION

Static thermal-to-electric energy converters have historically been the mainstay for space nuclear-electric power systems. Radioisotope power systems have extensively utilized thermoelectric converters based upon the Seebeck effect to provide reliable, long lived power sources for a wide range of space exploration missions. Static devices have also been the converters of choice for space reactor-based systems, including those under development today that would employ either thermionic or thermoelectric converters. Both thermoelectric and thermionic conversion systems offer the durability of a static system without moving mechanical parts and the reliability afforded by the redundant connection of many devices in series-parallel networks having no single point failure mechanisms.

However, both thermoelectric and thermionic systems operate at thermal-to-electric conversion efficiencies that are typically below 10%. This can result in significant mass penalties in both nuclear heat source and heat rejection systems when compared to system concepts that would utilize dynamic conversion systems. The cost and scarcity of radioisotope fuel exacerbates the impact of low conversion efficiency in radioisotope-based systems. In the case of thermionics, very high temperatures are typically required, giving rise to stringent materials requirements.

While high efficiency dynamic conversion systems are viable options for a variety of space nuclear power systems, work has also been carried out on other conversion concepts. The goal has been to achieve both the reliability attributes of a static converter and the high efficiencies of dynamic conversion systems, at heat source temperatures in the 1000K to 1300K range.

This paper will review the status of the most prominent static conversion concepts that have been studied in the laboratory and in system designs during the past ten years. Only closed loop concepts and those requiring no moving mechanical components throughout the entire cycle are considered. Three of those discussed (AMTEC, HYTEC and Thermophotovoltaics) provide the possibility of a high level of redundancy via the connection of many cells in series-parallel networks. Otherwise, efforts to improve the performance of thermoelectric (especially silicon-germanium) and thermionic converters have also been carried out. These results are reported elsewhere in this volume by Vining and Fleurial (1993), and Dahlberg et al. (1993).

THERMOPHOTOVOLTAICS

Thermophotovoltaic(TPV) energy conversion is based on the response of a photovoltaic cell to infrared photons emitted from a high temperature source. System concepts have considered both concentrated solar energy and nuclear-based heat sources. A schematic of a TPV converter configuration is shown in Figure 1. Efficient thermal-to-electric energy conversion requires that the source heat flux energy spectrum be matched to the response of the photovoltaic cell and also requires that the system design effectively contains photons within the cavity between emitter and cell.

The status of thermophotovoltaic conversion was reviewed in the proceedings of the First Symposium on Space Nuclear Power Systems (Ewell and Mondt, 1985). Since then significant progress has been made in the development of new photovoltaic cells, some having the potential for use in thermophotovoltaic applications. The result of these developments has been to expand the temperature range of possible thermophotovoltaic application.

Previously, studies were limited to silicon solar cells having a high (indirect) band gap requiring heat source emitters of more than 2000K (Horne et al. 1980). More recently, lower (direct and indirect) bandgap photovoltaic cells such as GaSb, GaInAs, InAs, AlInAs, InAsP and Ge have been considered for static thermal-to-electric conversion in the temperature regime around 1000K (Wolf 1986). The degree to which these materials have been characterized and the number of cells fabricated varies widely. Only GaSb has been both characterized and cell fabrication processes well developed.

The band gap of GaSb, developed for use in a GaAs/GaSb tandem solar cell, is 0.73eV. It is in the optimal band gap range calculated by Woolf (1986) for cells operating between 300K and 400K and using a 1473K radiant emitter. Present GaSb cell efficiencies are near 35% to wavelengths of 1.5 to 1.6 micron. Based upon these data, a GaSb-based radioisotope TPV converter concept design was proposed (Day et al. 1990 and Morgan et al. 1993). The concept is similar to the General Purpose Heat Source (GPHS) - Radioisotope Thermoelectric Generator (RTG), but would replace conventional SiGe converters with TPV cells. The design features the GPHS at a temperature of 1473K, radiating to cells at 350K giving a system efficiency of 12-14% at a specific power near 10W/kg. Most recent concept designs require active cooling of the cells using heat pipes.

These predictions are based on current cell technology, assuming that the cells can be backed and surrounded by a highly reflective (>95%) surface. Higher system efficiencies could be possible through the implementation of a narrow band pass filter between emitter and cell that would transmit infrared radiation tuned to the band gap of the TPV cell (see Fig. 1) and reflect all other wavelengths back to the emitter. Research on such filters for TPV application is still in the exploratory phase, however system efficiencies near 20% could be enabled by the development of such materials.

While the GaSb cell technology is relatively well characterized, no self-contained modules have been built or tested to verify system performance models. Such a demonstration is necessary in order to completely ascertain the feasibility of the GPHS-TPV concept. Questions involve the ability to maintain an efficient optical cavity over the lifetime of the mission, and a system design which would efficiently cool the converter elements to 350K. Radiation data on GaSb cells predicts an 8-10% power loss in 10 years using the GPHS. Otherwise, if the model projections are accurate, significant mass and fuel savings would be possible in comparison with the efficiency (7%) and specific power (6 W/kg) of the GPHS-RTG.

No recent studies of TPV conversion for space <u>reactor</u> power systems have been reported. Such a system could also utilize the results of research now underway on selective emitter materials that are tuned to the band gap of specific solar cells. Such materials, including rare earth oxides, are being investigated for terrestrial TPV applications and would have an impact similar to the filter approach discussed above. Again, the goal is to identify an emitter-converter combination that would enable system efficiencies near 20%, or more.

ALKALI METAL THERMOELECTRIC CONVERSION - AMTEC

The alkali metal thermoelectric converter, AMTEC, is a thermally regenerative electrochemical cell based on the sodium (or possibly potassium) ion conductive properties of beta"-alumina solid electrolyte (BASE) (Weber 1974). The operating cycle of the AMTEC is illustrated in Figure 2a. A closed vessel is divided into a high-temperature/pressure region in contact with a heat source and a low-temperature/pressure region in contact with a heat sink. These regions are separated by a barrier of BASE which has an ionic conductivity much larger than its electronic conductivity. The high-temperature/pressure region contains liquid sodium a T_2, and the low-pressure region contains mostly sodium vapor and a small amount of liquid sodium at T_1. Electrical leads make contact with a (positive) porous electrode which covers the low pressure surface of the BASE and with the high temperature liquid sodium (negative electrode). When the circuit is closed, sodium ions are conducted through the BASE due to the difference in vapor pressures (or chemical activity) across the BASE, while electrons flow to the porous electrode surface through the load producing electrical work. Sodium is recirculated by an electromagnetic pump or capillary pumped loop.

AMTEC systems have been considered in several space and terrestrial power system applications with nuclear, solar and combustion heat sources. System efficiencies have typically been projected to be near 20%. Bankston et al. (1984) reviewed the status of AMTEC technology at the First Symposium on Space Nuclear Power Systems, and many papers have appeared in these and other Symposia volumes. During this period, significant progress has been made on some key technical issues. Of particular importance has been progress in long life electrode development, and demonstration of high efficiency performance in laboratory cells.

Two families of electrodes have been identified as being capable of operating at high power density with very long lifetime potential (ten years or more). Rhodium-tungsten electrodes (Williams et al. 1990 and Ryan et al. 1993) and titanium-nitride electrodes (Hunt et al. 1990) have both demonstrated near 0.5W/cm^2 power density. These are in the range of areal power densities needed to achieve thermal-to-electric conversion efficiencies near 20% in most system concepts studied that operate at 1100-1300K. In addition, modeling of grain growth of these materials, utilizing experimental data for several thousand hours at high temperature, shows that these electrodes should function for ten years or more.

A wide range of single and multi-cell AMTEC devices have been operated. A 36-cell system has been operated at 550W (Hunt et al. 1990), however, most test results are with single cell devices having output powers of less than 25W. Numerous cells have now been operated at high temperature for 1000 hours or more, with almost 1900 hours being achieved under load at 1050K and 13% efficiency in a cell at JPL (Underwood et al. 1993). This is the highest efficiency ever achieved in a self-contained AMTEC cell; it produced 14W. While most cells have been operated with electromagnetic pumps, several small cells have been operated using wick return pumping for sodium recirculation (Hunt et al. 1993). These have usually been small cells (1-5W), however, the wick return demonstration has significant implications for space (zero gravity) operation. In fact, a wick return cell has been operated in an inverted position pumping against gravity.

Since AMTEC parasitic losses generally scale with area, it should be possible to build cells over a wide range of sizes without significantly affecting efficiency. As a result, AMTEC space nuclear power system concept studies have considered the range from 10's of watts in radioisotope systems to a megawatt class, reactor based system (Sievers et al. 1992). All utilize multiple (smaller) unit cells connected in series-parallel networks to achieve voltage and reliability requirements. Most recent studies have been concerned with radioisotope systems using the General Purpose Heat Source. Efficiencies near 20% and specific powers near 20W/kg are projected for most such concepts studied to date. These include both cells integrated with and radiatively coupled to the GPHS for small ($\leq$1 kW) systems, and remotely heated cells via an intermediate heat transfer loop for larger (multi-kilowatt) systems. Figure 2b shows a unit cell design for space applications.

In summary, AMTEC technology has progressed to the point where most fundamental issues have been resolved. However, other issues relating to system design and performance must now be addressed based on specific mission requirements. For example, optimum cell designs and multi-cell module designs must be developed, fabricated, and tested. In this way, system performance and lifetime projections can be verified or modified. Also, a space flight demonstration may be required to verify liquid sodium flow and management techniques and to evaluate the stability of the BASE under launch conditions.

HYDROGEN THERMAL-TO-ELECTRIC CONVERSION - HYTEC

The relatively recent hydrogen thermal-to-electric converter, HYTEC, is another in the family of thermally regenerative electrochemical systems that has the potential to produce power in practical devices (Roy 1987). In this case it utilizes a lithium or lithium-sodium liquid circulating loop to circulate the working fluid, hydrogen. Power is produced via a metal electrode/molten salt/metal electrode electrochemical cell. A schematic of the cycle is shown in Figure 3a. The electrochemical cell diagram is in Figure 3b. Key points in the cycle are the electrochemical cell and a decomposition retort. A voltage is produced in the electrical cell due to hydrogen activity difference across the electrode/electrolyte/electrode system. Power is produced when hydride ions are conducted through the molten salt electrolyte and then react to form a metal hydride mixture with the circulating liquid metal(s). Heat is input at the decomposition retort where hydrogen is separated from the hydride mixture to return to the electrochemical cell in the gas phase for subsequent reaction in the electrochemical cell. Recirculation of the working fluid is achieved by an electromagnetic pump.

The advantages offered by a HYTEC system are similar to those for AMTEC. They include the potential for high efficiency (near 20%), relatively lightweight converter mass, no moving mechanical parts, and modularity (since parasitic losses are dependent on electrode area). Again, multiple cell systems are preferred to meet voltage and reliability requirements. An advantage is the fact that the power producing step takes place at relatively moderate heat rejection temperatures (approximately 850-900K for some reactor applications).

Recently, progress has been made in selecting a promising HYTEC working fluid which has desirable properties in terms of the equilibrium hydrogen vapor pressure of the working fluid as a function of temperature (Roy et al. 1993). Also, progress has been made in evaluating metal membranes which have fairly favorable hydrogen permeability, as well as some degree of stability under cell operating conditions. To date, demonstration of anticipated high device power densities in the system configurations proposed has not been achieved (0.014 W/cm^2 has been achieved and accurately predicted, compared with 0.02 - 0.04 W/cm^2 employed in system analyses) in the laboratory and a quantitative theoretical model is not yet formulated. Thus, a better understanding of the potential losses occurring at the power producing step is needed, including information on hydride ion mobility in the electrolyte, the kinetics of charge transfer processes at the electrode/electrolyte interface, and hydrogen permeability through the selected electrodes under HYTEC conditions. Experiments that provide this information will allow the ultimate potential of HYTEC to be assessed.

System studies typically place HYTEC in a reactor-based power system (Salamah et al. 1992). Coupled with an SP-100 type reactor system efficiencies between 17% and 23% have been predicted with a specific mass as a low as 12.5kg/kw. If the, as yet unknown, losses observed in the laboratory in the power producing step can be identified and reduced, then HYTEC could become a viable, high efficiency static converter option for future space nuclear power systems.

THERMOACOUSTIC POWER CONVERSION

Thermoacoustic energy converters convert thermal energy to acoustic energy; a transducer then converts the acoustic energy to electricity (Wheatley et al. 1983 and Swift 1988). A schematic of an acoustic engine is shown in Figure 4 and "a dozen or so working engines have been built in several laboratories" (Ward and Merrigan 1992). The device utilizes an acoustic resonator in which a standing wave is established. A stack of short plates are positioned at one end of the resonator, with heat added at one end of the length of the plates, and heat removed at the other end, the ends being held at the cycle hot and cold temperatures, respectively. The position and length of the plates is designed to place the hot end of the plates at a pressure maximum and the cold end at a pressure minimum in the standing acoustic wave. The effect is to add energy to the gas between hot and cold end, resulting in a traveling acoustic wave which transfers energy to the transducer.

The plate stack is the most important component in the thermoacoustic converter as its design has the greatest affect on converter performance. The spacing between plates is determined by the thermal penetration depth of the fluid, which in turn is dependent on operating temperature. Actual devices have utilized spiraled sheet material, honeycomb and blocks with parallel square channels. The power output required from the converter determines the diameter of the resonator at the plate stack end.

Working fluids for the thermoacoustic converter may be either gas or liquid metal with engines having been built with air, helium, argon, helium/argon, or helium/xenon mixtures, and liquid sodium. For space power systems the high heat rejection temperatures required to minimize radiator area would likely dictate the use of liquid metals. Piezoelectric transducers used for gas systems would depolarize at high temperature, whereas a liquid metal system could utilize an MHD transducer.

Performance characteristics for 1kW engine designs using helium and liquid sodium as working fluids were given by Ward and Merrigan (1992). They project overall thermal-to-electric conversion efficiencies in the 15%-21% range for these engines at T_{hot}=1100K and T_{cold}=330K. No studies have been reported that integrate thermoacoustic engines with a radioisotope or reactor-based heat source. Thus, this technology is at a very early state of development with little data being available to fully ascertain the technical issues involved in reducing thermoacoustic engines to practice in space systems. For example, the optimum design for integration with a heat source and lifetime limiting processes in a system must be determined. Also, vibration may be an issue for some applications, and thus, its effects and possible means to minimize vibration require study. Component and converter experimental data must be combined with a system design to begin to answer these questions and permit evaluation of thermoacoustic engines as an option for space nuclear power systems.

LIQUID METAL MAGNETOHYDRODYNAMICS

Blumeneau et al. (1988), and Fabris (1992) reviewed the status of liquid metal magnetohydrodynamics at the 1987 and 1989 Symposia, respectively. Work on this technology was carried out from the 1960's into the 1980's, but little experimental work has been carried out in recent years. However, the cycle does offer the benefits of a static device and has the potential for operating efficiencies of 10-15% over a range of power levels. The simplest form of an LMMHD cycle is shown in Figure 5a. Here, a single component liquid metal is heated via reactor (or possibly radioisotope) heat source to a partial boiling state, whereupon it enters the two-phase MHD generator (Figure 5b).

Work is extracted as the two-phase mixture expands isentropically through the divergent duct of the MHD generator. In the expansion, the pressure drops, additional liquid vaporizes, and the temperature falls. The two-phase mixture is subsequently cooled in the rejection heat exchanger and pumped, via electromagnetic pump, back to the heat source. This cycle is called by Blumeneau et al. (1988) a "wet-vapor" cycle and could incorporate multiple MHD stages for higher efficiencies.

There are different variations on the LMMHD cycle, including some that add a second liquid metal fluid, of high vapor pressure, upstream of the MHD generator that vaporizes to aid in accelerating the liquid flow. Variations in cycle concepts involve whether or not to place a liquid/vapor separator before the MHD generator. The separator is a source for losses, if used; while the MHD generator may not operate efficiently if operated in a two phase flow regime. Most recent concepts have focused on operating the MHD generator in the two-phase mode. However, the losses encountered remain so substantial that experimental efficiencies reported to date are usually substantially less than 10%. These losses usually involve inefficient energy tranfer between vapor and liquid phases in the two phase flow or drag losses at various points in the cycle.

System concepts for LMMHD have typically addressed reactor-based systems from tens of kilowatts to hundreds of megawatts, with single MHD-stage system efficiencies possibly reaching 10%. Efficiencies above 10% would probably require two- or three- stages of MHD generators. For example, a 100-MW_e, three-stage, cesium wet-vapor cycle system was projected to have a system efficiency of 11.6%. On the other hand, the radiator area required for such a system was competitive with most other concepts at the time since the heat rejection temperature was 1123K (T_h=1673K).

The highest projected efficiencies are predicated on achieving high (approx 90%) vapor volume fraction in the MHD generator. Research work has recently been directed toward this issue, focused on the possibility of adding surfactants to the liquid metal to promote foaming (Fabris et al. 1990). Projections show that a high efficiency (70+%), high void fraction (92%) MHD generator should be feasible (Fabris et al. 1992). However, until these calculations are verified by basic experiments, the relatively small increases in efficiency now offered by LMMHD do not justify a development program for space power applications. Also, further concept work is needed at lower temperatures and in small size ranges to evaluate systems that might be applicable to radioisotope heat sources.

SUMMARY

The combination of efficiency (13%) and lifetime demonstrations (several thousand hours) with AMTEC cells establishes it as the most advanced of the technologies reviewed here. AMTEC technology work is continuing to demonstrate key system lifetime and performance goals, to be followed by a space experiment. Thermophotovoltaic conversion is based on a well developed solar cell technology and needed cavity tests are presently being planned to answer key performance issues before further development would be initiated. HYTEC is in the same family of conversion technologies as AMTEC, but is based on a different working fluid and is still in its early stages of basic development work. Thermoacoustic generators have not been studied extensively for space applications and an experimental plan based upon space requirements is needed before a development program is warranted. Finally, liquid metal MHD requires a basic advance in two phase flow technology to warrant significant development attention. Thus, AMTEC is a definite candidate for space power applications; thermophotovoltaics could become viable if upcoming experiments are successful; HYTEC experiments must provide a complete understanding of that cycle to establish its viability; and thermoacoustics and liquid metal MHD are still far from establishing feasibility for space applications.

Acknowledgement

This work was carried out by the Jet Propulsion Laboratory, California Institute of Technology under contract with the National Aeronautics and Space Administration.

References

Bankston, C.P., T. Cole, S. K. Khanna, and A. P. Thakoor (1985) "Alkali Metal Thermoelectric Conversion (AMTEC) for Space Nuclear Power Systems," in Space Nuclear Power Systems 1984, M. S. El-Genk and M. D. Hoover, eds. Orbit Book Co., Malabar, Fl, pp. 393-402.

Blumenau, L., H. Branover, A. El-Boher, E. Spero, S. Sukoriansky, E. Greenspan, V. A. Walker, J. R. Bilton, and N. J. Hoffman (1988) "Liquid Metal MHD Power Conversion for Space Electric Systems," in Space Nuclear Power Systems 1987, M. S. El-Genk and M. D. Hoover, eds. Orbit Book Co., Malabar, Fl, pp. 267-279.

Dahlberg, R. C., L. L. Begg, J. N. Smith, G. O. Fitzpatrick, D. T. Allen, G. L. Hatch, J. B. McVey, D. P. Lieb and G. Miskolczy (1993) "Reviews of Thermionic Technology: 1983 to 1992," This Volume.

Day, A. C., W. E. Horne, and M. D. Morgan (1990) "Application of the GaSb Solar Cell in Isotope-Heated Power Systems," in Proc. 21st IEEE Photovoltaics Specialist Conf., IEEE, New York, pp. 1320-1325.

Ewell, R. and J. Mondt (1985) "Static Conversion Systems," in Space Nuclear Power Systems 1984, M. S. El-Genk and M. D. Hoover, eds. Orbit Book Co., Malabar, Fl, pp. 385-391.

Fabris, G., E. Kwack, K. Harstad, and L. H. Back (1990) "Two-Phase Flow Bubbly Mixing for Liquid Metal Magnetohydrodynamic Energy Conversion," Proc. 25th Intersociety Energy Conversion Engineering Conference, Am. Inst. Chem. Engineers, New York, Vol. 2, pp. 486-493.

Fabris, G. (1992) "Review of Two-Phase Flow Liquid Metal MHD and Turbine Energy Conversion Concepts for Space Applications" in Space Nuclear Power Systems 1989, M. S. El-Genk and M. D. Hoover, eds. Orbit Book Co., Malabar, Fl. pp. 407-415.

Fabris, G. and L. Back (1992) "Prediction of Performance of Two-Phase Flow Nozzle and Liquid Metal Magnetohydrodynamic (LMMHD) Generator for No Slip Condition," in Proc. 27th Intersociety Energy Conversion Engineering Conference, Soc. Automotive Eng., Warrendale, PA, Vol. 3, pp. 391-399.

Horne, W. E., A. C. Day, R. B. Greegor, and L. D. Milliman (1980) "Solar Thermophotovoltaic Space Power Systems," in Proc. 15th Intersociety Energy Conversion Engineering Conference, AIAA, New York, pp. 377-382.

Horne, W. E. (1988) "Experimental Evaluation of a Solar Thermophotovoltaic Energy Conversion Module," in Proc. 23rd Intersociety Energy Conversion Engineering Conference, Vol. 3, pp. 37-42.

Hunt, T. K, R. F. Novak, J. R. McBride, D. J. Schmatz, W. B. Copple, J. T. Brockway, N. Arnon, and G. A. Grab (1990) "Test Results on a Kilowatt-Scale Sodium Heat Engine," in Proc. 25th Intersociety Energy Conversion Engineering Conference, Am. Inst. Chem. Engineers, New York, Vol. 2, pp. 420-425.

Hunt, T. K., R. K. Sievers, D. A. Butkiewicz, J. E. Pantolin and J. F. Ivanenok (1993) "Small Capillary Pumped AMTEC Systems," in Proc. Tenth Symposium on Space Nuclear Power and Propulsion, Conf. Proc. 271, American Institute of Physics, New York, pp. 891-896.

Morgan, M. D., W. E. Horne, and P. R. Brothers (1993) "Radioisotope Thermophotovoltaic Power System Utilizing the GaSb IR Photovoltaic Cell," in Proc. Tenth Symposium on Space Nuclear Power and Propulsion, Conf. Proc. 271, American Institute of Physics, New York, pp. 313 -318.

Roy, P. (1987) "Hydrogen Thermo-Electrochemical Converter, U.S. Patent No. 4,692,390, 8 September 1987.

Roy, P., S. A. Salamah, J. Maldonado, and Regina S. Narkiewicz (1993) "HYTEC -- A Thermally Regenerative Fuel Cell," in Proc. Tenth Symposium on Space Nuclear Power and Propulsion, Conf. Proc. 271, American Institute of Physics, New York, pp. 913-921.

Ryan, M. A., R. M. Williams, M. L. Underwood, B. Jeffries-Nakamura and D. O'Connor (1993) "Electrode, Current Collector,and Electrolyte Studies for AMTEC Cells," in Proc. Tenth Symposium on Space Nuclear Power and Propulsion, Conf. Proc. 271, American Institute of Physics, New York, pp. 905-912.

Salamah, S. A., D. N. Rodgers, D. G. Hoover, and P. Roy (1991) "SP-100 Reactor/ HYTEC - A High Efficiency Static Conversion Power System," in Proc. 26th Intersociety Energy Conversion Engineering Conference, Am. Nuclear Soc., La Grange Park, IL, Vol. 1, pp. 471-475.

Sievers, R. K., R. M. Williams, M. L. Underwood, B. Jeffries-Nakamura, and C. P. Bankston (1992) "AMTEC System Performance Studies Using the Detailed Electrode Kinetic Transport Model," in Space Nuclear Power Systems 1989, M. S. El-Genk and M. D. Hoover, eds. Orbit Book Co., Malabar, Fl, pp. 373-378.

Swift, G. W. (1988) "Thermoacoustic Engines," J. Acous. Soc. Am., Vol. 84, pp. 1145-1178.

Underwood, M. L., R. M. Williams, M. A. Ryan, B. Jeffries-Nakamura, and D. O'Connor (1993) "Recent Advances in AMTEC Recirculating Test Cell Performance," in Proc. Tenth Symposium on Space Nuclear Power and Propulsion, Conf. Proc. 271, American Institute of Physics, New York, pp. 885-890.

Vining, C. B. and J. P. Fleurial (1993) "Silicon Germanium: An Overview of Recent Developments," This volume.

Ward, W. C. and M. A. Merrigan (1992) "Thermoacoustic Power Conversion for Space Power Applications," in Proc. Ninth Symposium on Space Nuclear Power Systems, Conf. Proc. 246, American Institute of Physics, New York, pp. 1338-1343.

Weber, N. (1974), "A Thermoelectric Device Based on Beta-Alumina Solid Electrolyte," Energy Conversion, Vol. 14, pp. 1-8.

Wheatley, J. C., T. Hofler, G. W. Swift and A. Migliori (1983) "An Intrinsically Irreversible Thermoacoustic Heat Engine," J. Acous. Soc. Am., Vol. 74, p. 153.

Williams, R. M., B. Jeffries-Nakamura, M. L. Underwood, D. O'Connor, M. A. Ryan, S. Kikkert and C. P. Bankston (1990) "Lifetime Studies of High Power Rhodium/Tungsten and Molybdenum Electrodes for Application to AMTEC," in Proc. 25th Intersociety Energy Conversion Engineering Conference, Am. Inst. Chem. Engineers, New York, Vol. 2, pp. 413-419.

Woolf, L. D. (1986) "Optimum Efficiency of Single and Multiple Bandgap Cells in Thermophotovoltaic Energy Conversion," Solar Cells, Vol. 19, No. 1, p. 19.

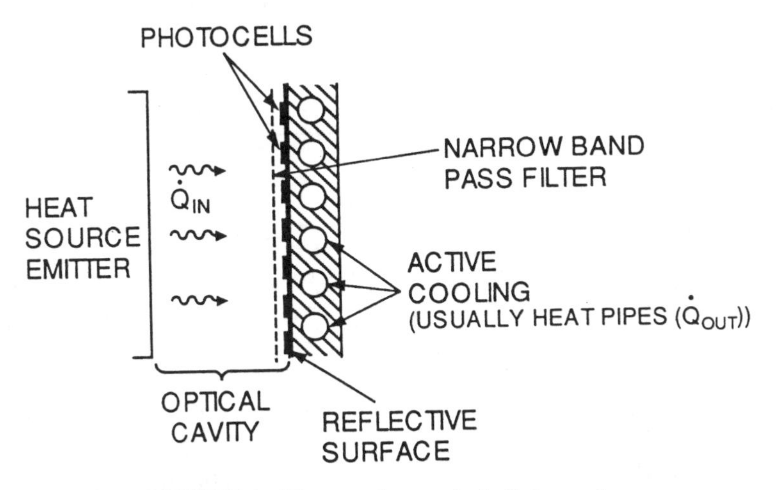

FIGURE 1. Thermophotovoltaic Schematic.

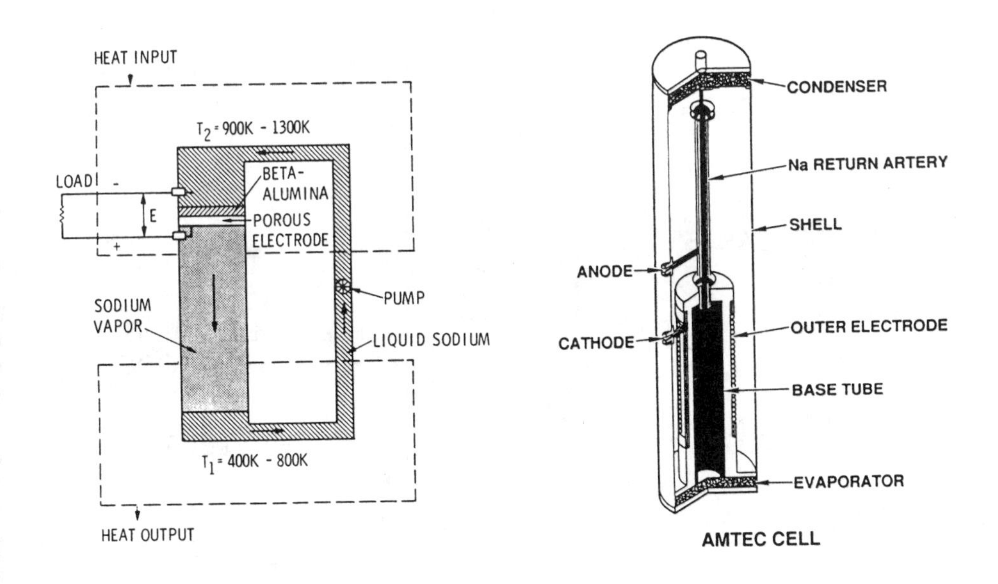

FIGURE 2a. AMTEC Schematic.

FIGURE 2b. AMTEC Space Cell Concept Design.

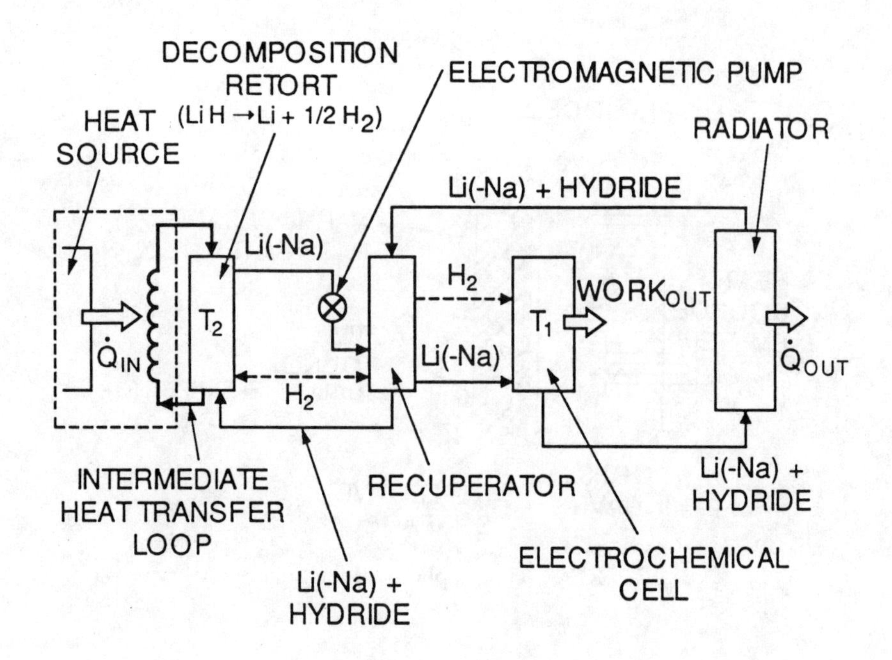

FIGURE 3a. HYTEC Cycle.

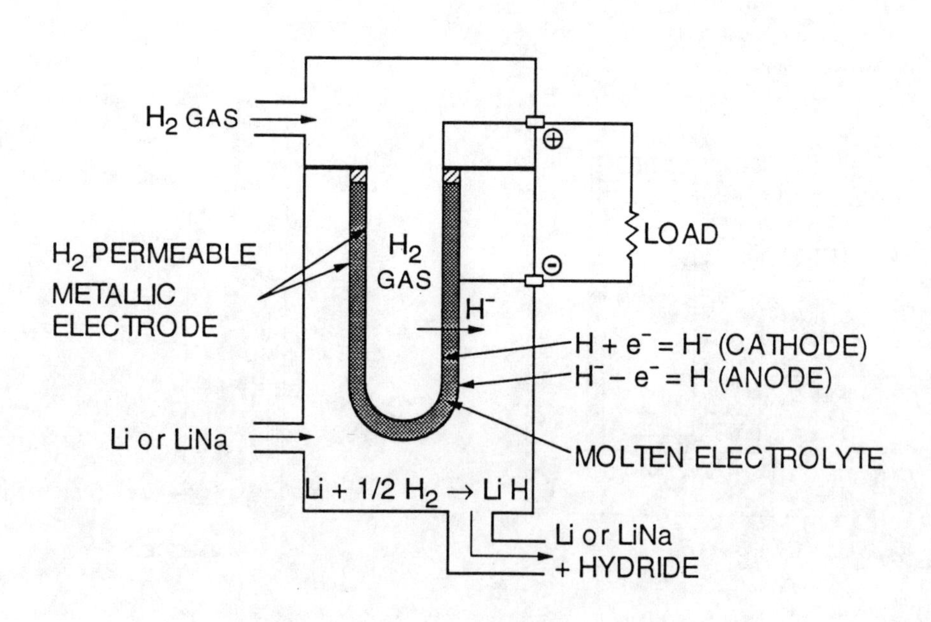

FIGURE 3b. HYTEC Electrochemical Cell.

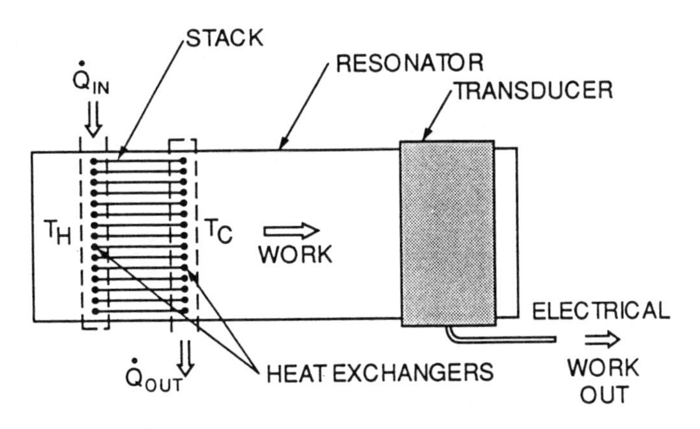

FIGURE 4. Thermoacoustic Converter Schematic.

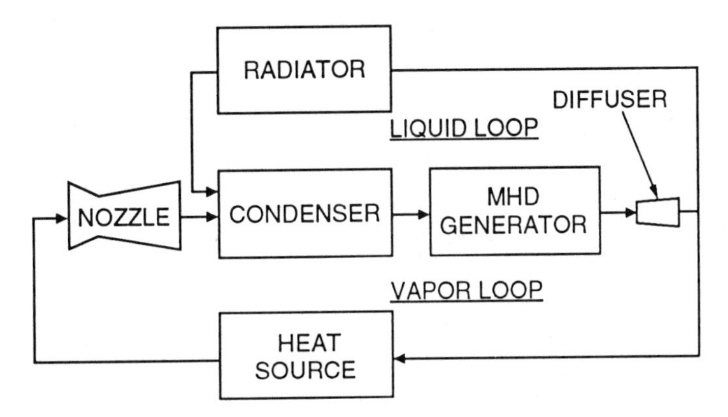

FIGURE 5a. LMMHD Condensing Cycle.

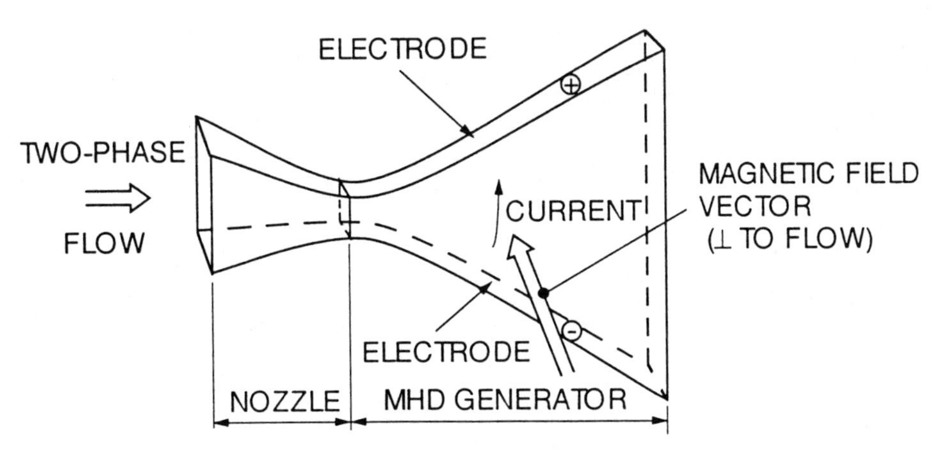

FIGURE 5b. MHD Generator.

LIST OF PEER REVIEWERS

(We gratefully acknowledge the following individual technical reviewers, along with those whose names we do not have, for their contributions to the peer review of the manuscripts published in this book.)

Wayne R. Amos
EG&G Mound Applied Technologies

Zenen I. Antoniak
Pacific Northwest Laboratory

J. Sam Armijo
Martin Marietta Corporation

H. Sterling Bailey
Martin Marietta Astro Space

C. Perry Bankston
Jet Propulsion Laboratory

Robert A. Bari
Brookhaven National Laboratory

Gary L. Bennett
NASA Headquarters

Samit K. Bhattacharyya
Argonne National Laboratory

David L. Black
Consultant

Harvey S. Bloomfield
NASA Lewis Research Center

Richard J. Bohl
Los Alamos National Laboratory

W. Jerry Bowman
U. S. Air Force

Wayne M. Brittain
Teledyne Energy Systems

R. William Buckman, Jr.
Refractory Metals Technology

David Buden
Idaho National Engineering Laboratory

Richard L. Coats
Sandia National Laboratories

Elden H. Cross
Rockwell International/Rocketdyne Division

Paul R. Davis
Linfield College

Mark J. Dibben
U. S. Air Force/Phillips Laboratory

James R. Distefano
Oak Ridge National Laboratory

Dean Dobranich
Sandia National Laboratories

Dale S. Dutt
Westinghouse Hanford Company

Mohamed S. El-Genk
University of New Mexico

Frank J. Halfen
Martin Marietta Corporation

William B. Harper
AlliedSignal Fluid Systems

Robert Hartman
Martin Marietta Corporation

Richard B. Harty
Rockwell International/Rocketdyne Division

Edwin A Harvego
Idaho National Engineering Laboratory

George Haverly
Martin Marietta Astro Space

Thomas Hirons
Los Alamos National Laboratory

Richard Hobbins
EG&G Idaho, Inc.

Steven D. Howe
Los Alamos National Laboratory

Maribeth E. Hunt
Rockwell International/Rocketdyne Division

Thomas K. Hunt
Advanced Modular Power Systems

DeWayne L. Husser
Babcock & Wilcox Company

Allan T. Josloff
Martin Marietta Corporation

Vahe Keshishian
Rockwell International/Rocketdyne Division

Ehsan Khan
U. S. Department of Energy

Thomas R. Lamp
Wright Laboratory

Robert G. Lange
U. S. Department of Energy

Thomas E. Mahefkey
Wright Laboratory

Donald N. Matteo
Martin Marietta Corporation

Michael A. Merrigan
Los Alamos National Laboratory

John D. Metzger
Grumman Aerospace Corporation

Joseph C. Mills
Rockwell International/Rocketdyne Division

Edward A. Mock
AlliedSignal Fluid Systems

Michael P. Moriarty
Rockwell International/Rocketdyne Division

Ronald E. Murata
Martin Marietta Corporation

John C. Newcomb
Rockwell International/Rocketdyne Division

Alexander G. Parlos
Texas A&M University

Keith A. Pauley
Pacific Northwest Laboratory

G. P. "Bud" Peterson
Texas A&M University

Jerry Peterson
Martin Marietta Astro Space

Philip R. Pluta
Martin Marietta Astro Space

James R. Powell
Brookhaven National Laboratory

Richard D. Rovang
Rockwell International/Rocketdyne Division

Prodyot Roy
Martin Marietta Corporation

Lyle L. Rutger
U. S. Department of Energy

Samir A. Salamah
Martin Marietta Corporation

Kurt Schoenburg
Los Alamos National Laboratory

Robert P. Santandrea
Babcock & Wilcox Company

Michael Schuller
Phillips Laboratory

Gene E. Schwarze
NASA Lewis Research Center

James R. Stone
NASA Lewis Research Center

Leonard Tower
Sverdrup Technologies, Inc.

Mark L. Underwood
Jet Propulsion Laboratory

John E. Van Hoomissen
Martin Marietta Corporation

Carl E. Walter
Lawrence Livermore National Laboratory

David M. Woodall
University of Idaho

Stevan A. Wright
Sandia National Laboratories

AUTHOR INDEX

(Bold chapter numbers indicate senior authorship)

Alger, Donald L., 305
Allen, Daniel T., 121

Baars, Ralph E., 179
Bankston, C. Perry, **443**
Begg, Lester L., 121
Bennett, Gary L., **221**
Blair, H. Thomas, 179
Bloomfield, Harvey S., 305
Brophy, John R., 381
Brown, Neil W., 269
Buden, David, **21**
Butt, Darryl P., 179

Cairelli, James E., 305
Connell, Leonard W., 269

Dahlberg, Richard C., **121**
Dudenhoefer, James E., **305**

Finger, Harold B., 221
Fitzpatrick, Gary O., 121
Fleurial, Jean-Pierre, 87

Geng, Steven M., 305
Gilland, James H., **381**

Hatch, Laurie, 121

Juhasz, Albert J., **407**

Klein, Milton, 221

Lange, Robert G., **1**
Lieb, David P., 121

Marshall, Albert C., 269
Mason, Richard E., 179
Mastal, Edward F., 1
Matthews, R. Bruce, **179**
McCulloch, William H., 269
McVey, John B., 121
Merrigan, Michael A., **167**
Miller, Thomas J., 221
Mims, James E., 269
Miskolczy, Gabor, 121
Myers, Roger M., 381

Overholt, David M., 305

Peterson, G. P., 407
Potter, Andrew, 269

Rasor, Ned, 121
Rauch, Jeffrey S., 305
Robbins, William H., 221

Schreiber, Jeffrey G., 305
Sholtis, Joseph A., Jr., **269**
Smith, Joe N., Jr., 121
Sovey, James S., 381
Stark, Walter A., 179
Storms, Edmund K., 179

Tew, Roy C., 305
Thieme, Lanny G., 305

Vining, Cronin B., **87**

Wallace, Terry C., 179
Winchester, Robert O., 269
Wong, Wayne A., 305

KEY WORD INDEX

(Numbers indicate chapters covering these subjects)

Advanced propulsion 381
Alkali metals 167
AMTEC 443

Brayton cycle 305

Carbon-carbon 167
Carbon-carbon heat pipes 407
Converter 121

Diode 121

Electric propulsion 381
Energy conversion 87

Heat pipes 167
Heat pipe materials 167
Heat pipe wicks 167
Heat transfer 167
HYTEC 443

Isotope heat source 1, 305

Kiwi 221

Liquid metal MHD 443

Multi-megawatt 21

NERVA 221
Nuclear reactor 305
Nuclear electric propulsion 381
Nuclear thermal propulsion 221

Peewee 221
Phoebus 221
Plasma applications 381
Plutonium-238 1
Power conversion 121
Power generation 87

RTG 1, 87
Radioisotopes 1
Rankine cycle 305
Risk assessment 269

SP-100 21, 87
Safety analyses 269
Safety review and approval 269
Safety testing 269
Silicon-Germanium 87
Space nuclear mission safety 269
Space nuclear power 21, 167
Space nuclear systems safety 269
Space power 121, 305
Space power system heat rejection 407
Space radiators 407
Space reactor fuels 179
Spacecraft propulsion 381
Static energy conversion 1
Stirling cycle 305

Thermionics 21, 87, 121
Thermoacoustics 443
Thermoelectric 1
Thermophotovoltaics 443
Topaz 21
Transient modeling 167

Uranium carbide 179
Uranium nitride 179